Discrete Mathematics

Jones and Bartlett Books in Mathematics

Advanced Calculus, Revised Edition
Lynn H. Loomis and Shlomo Sternberg

Calculus with Analytic Geometry, Fourth Edition
Murray H. Protter and Philip E. Protter

College Geometry
Howard Eves

Differential Equations: Theory and Applications
Ray Redheffer

Discrete Mathematics
James L. Hein

Discrete Structures, Logic, and Computability
James L. Hein

Fundamentals of Modern Elementary Geometry
Howard Eves

Introduction to Differential Equations
Ray Redheffer

Introduction to Fractals and Chaos
Richard M. Crownover

Introduction to Numerical Analysis
John Gregory and Don Redmond

Lebesgue Integration on Euclidean Space
Frank Jones

Logic, Sets, and Recursion
Robert L. Causey

Mathematical Models in the Social and Biological Sciences
Edward Beltrami

The Poincaré Half-Plane: A Gateway to Modern Geometry
Saul Stahl

The Theory of Numbers: A Text and Source Book of Problems
Andrew Adler and John E. Coury

Wavelets and Their Applications
Mary Beth Ruskai, et al.

The Way of Analysis
Robert S. Strichartz

Discrete Mathematics

James L. Hein

Department of Computer Science
Portland State University

Jones and Bartlett Publishers
Sudbury, Massachusetts

Boston   London   Singapore

Editorial, Sales, and Customer Service Offices
Jones and Bartlett Publishers
40 Tall Pine Drive
Sudbury, MA 01776
800-832-0034
508-443-5000

Jones and Bartlett Publishers International
7 Melrose Terrace
London W6 7RL
England

Library of Congress Cataloging-in-Publication Data

Hein, James L.
    Discrete mathematics / James L. Hein
      p.   cm.
    Includes bibliographical references and index.
    ISBN 0-86720-496-6
    1. Computer science--Mathematics.   I. Title.
QA76.9.M35H43   1995
004'.01'511--dc20   95-45893
                                                    CIP

Printed in the United States of America
99 98 97 96 95   10 9 8 7 6 5 4 3 2 1

To My Mother

Ruth Holzer Hein

She taught all eight grades in a one-room country
school house in Minnesota. One cold winter morn-
ing after a snowstorm, she was riding a two-horse
sleigh to school when one of the horses fell dead in
its tracks. So she ran to the nearest farm and called
home. Her father and brother harnessed a fresh
horse and met her back at the sleigh, where they
harnessed the new team and drove her to school so
that she could open the door and start a fire for her
waiting students.

I hope that this book opens up the door and starts a
fire for waiting students too.

Contents

Preface

*The last thing one discovers in writing a book
is what to put first.*
—Blaise Pascal (1623–1662)

This book is designed for an introductory course in discrete mathematics that serves a variety of majors, including mathematics, computer science, and engineering. The book is the outgrowth of a course at Portland State University that has evolved over 15 years from a course for upper-division students into a course for sophomores. The book can be read by anyone with a good background in high school mathematics. Therefore it could also be used at the freshman level or at the advanced high school level.

This book differs in several ways from current books about discrete mathematics. It presents an elementary and unified introduction to a collection of topics that, up to now, have not been available in a single source. A major feature of the book is the unification of the material so that it doesn't fragment into a vast collection of seemingly unrelated ideas. This is accomplished by organization and focus. The book is organized more along the lines of technique than on a subject-by-subject basis. The focus throughout the book is on the computation and construction of objects. Therefore many traditional topics are dispersed throughout the text to places where they fit naturally with the techniques under discussion. For example, to read about properties of—and techniques for processing—natural numbers, lists, strings, graphs, or trees, it's necessary to look in the index or scan the table of contents to find the several places where they are found.

The logic coverage is much more extensive than in current books at this level because of its fundamental importance in problem solving, programming, automatic reasoning systems, and logic programming languages. Logic is also dispersed throughout the text. For example, we introduce informal proof techniques in the first section of Chapter 1. Then we use informal logic without much comment until Chapter 4, where inductive proof techniques are presented. After the informal use of logic is well in hand, we move to the formal aspects of logic in Chapters 6 and 7, where equivalence proofs and inference-based proofs are introduced. Formal logic is applied to proving correctness properties of programs in Chapter 8. We introduce automatic reasoning and logic programming in Chapter 9.

The coverage of algebraic structures in Chapter 10 differs from that in other texts. After giving an elementary introduction to algebras, we then introduce algebras and algebraic techniques that apply directly to computer science. In addition to the important ideas of Boolean algebra, we also introduce abstract data types as algebras, computational algebras, and some other algebraic ideas that are used in computational problems.

Some Notes

- The writing style is intended to be informal. For example, the definitions and many of the examples are integrated into the prose.
- The book contains over 700 exercises. Answers are provided for about half of the exercises.
- Each chapter begins with a chapter guide, which gives a brief outline of the topics to be covered in each section.
- Each chapter ends with a chapter summary, which gives a brief description of the main ideas covered in the chapter.
- Everyday mathematics and logic may be the ultimate computer languages. The book supports this idea by introducing—in a natural way as part of the discourse—some aspects of the functional and logical styles of declarative programming.
- Some people like to learn a new game by first learning all the rules. Others like to learn a few basics first and then start to play the game, referring to the rules when questions arise. Still others like to jump right in and play, asking questions and learning the rules as they go. *The main point:* People learn in different ways. The book tries to accommodate different learning styles. For example, general (abstract) ideas are useful because they apply to many things. Yet we all crave examples. The ideas in the book are made concrete by first giving informal definitions and examples. Then we consider general properties and abstractions.
- Algorithms in the text are presented in a variety of ways. Some are simply a few sentences of explanation. Others are presented in a more traditional notation. For example, we use assignment statements like $x := t$ and control statements like **if** A **then** B **else** C **fi**, **while** A **do** B **od**, and **for** $i := 1$ **to** 10 **do** C **od**. We avoid the use of **begin-end** or {-} pairs by using indentation. We'll also present some algorithms as logic programs after logic programming has been introduced.

- The word "proof" makes some people feel uncomfortable. It shouldn't, but it does. Maybe words like "show" or "verify" make you cringe. Most of the time we'll discuss things informally and incorporate proofs as part of the prose. At times we'll start a proof with the word "Proof," and we'll end it with QED. QED is short for the Latin phrase *quod erat demonstrandum*, which means "which was to be proved." We'll formalize the idea of proof in the logic chapters.

- A laboratory component for a course is a natural way to motivate and study the topics of the book. The course at Portland State University has evolved into a laboratory course. Labs have worked quite well when the lab experiments are short and specific. Interactive labs give students instant feedback in trying to solve a problem. A few handouts and a few sample problems usually suffice to get students started in an interactive programming environment. Some samples of the current lab experiments can be found in the paper by Hein [1993].

- I wish to apologize in advance for any errors found in the book. They were not intentional, and I would appreciate hearing about them. As always, we should read the printed word with some skepticism.

Using the Book

You should feel free to jump into the book at whatever topic suits your fancy and then refer back to unfamiliar definitions. There are some natural dependencies that occur among the topics in the book. For example, all parts of the book depend on the material contained in Chapter 1 and Section 2.1. Here is a listing of the topics along with the associated dependencies:

Inductive Definitions: They are used throughout the text after being introduced in Section 3.1.

Recursively Defined Functions: They are discussed in Section 3.3 and depend somewhat on inductive definitions in Section 3.1.

Logic: Informal proof techniques are introduced in Section 1.1. The technique of proof by induction is covered in Section 4.4, which can be read independently with only a few references back to unfamiliar definitions. Chapter 6 and Chapter 7 should be read in order. Then Chapters 8 and 9 can be read in either order.

Analysis: Chapter 5 introduces some tools that are necessary for analyzing algorithms. It uses proof by induction, which is discussed in Section 4.4.

Algebra: Chapter 10 uses recursively defined functions as discussed in Section 3.3, and it uses proof by induction as discussed in Section 4.4.

Course Suggestions

The topics in the book can be presented in a variety of ways, depending on the length of the course, the emphasis, and student background. Most of the manuscript has been used for several years in a sophomore course at Portland State University that consists of two 10-week terms. Parts of the text have also been used at Portland Community College. Here are a few suggestions for courses of various lengths and emphases.

Emphasis on discrete mathematics, with some logic

10-week course: 1, 2.1–2.3, 3.1, 3.3, 4, 5.1, 5.2, 6.1, 6.2, 7.1, 7.2, 10.1, 10.2.
15-week course: 1, 2.1–2.3, 3, 4, 5.1–5.3, 6, 7.1, 7.2, 8.1, 8.2, 10.1, 10.2, and
                  one of 2.4, 5.4, 10.3, 10.4, or 10.5.

Emphasis on discrete mathematics and logic

10-week course: 1, 2.1, 2.2, 3.1, 3.3, 4, 5.1, 5.2, 6, 7.1, 7.2, 10.1, 10.2.
15-week course: 1, 2.1–2.3, 3, 4, 5.1–5.3, 6, 7.1, 7.2, 8, 9.1, 10.1, 10.2.
20-week course: 1–9, 10.1–10.3.

Acknowledgments

Many people helped me create this book. I received numerous suggestions and criticisms from the students and teaching assistants who used drafts of the manuscript. Five of these people—Janet Vorvick, Roger Shipman, Yasushi Kambayashi, Mike Eberdt, and Tom Hanrahan—deserve special mention for their help. Several reviewers of the manuscript gave very good suggestions that I incorporated into the text. In particular I would like to thank David Mix Barrington, Larry Christensen, Norman Neff, Karen Lemone, Michael Barnett, and James Crook. I also wish to thank Carl Hesler at Jones and Bartlett for his entrepreneurial spirit. Finally, I wish to thank my family—Janice, Gretchen, and Andrew—for their constant support.

J.L.H.
Portland, Oregon

1

Elementary Notions and Notations

'Excellent!' I cried. 'Elementary,' said he.
—Watson in *The Crooked Man*
by Arthur Conan Doyle (1859–1930)

To communicate, we sometimes need to agree on the meaning of certain terms. If the same idea is mentioned several times in a discussion, we often replace it with some shorthand notation. The choice of notation can help us avoid wordiness and ambiguity, and it can help us achieve conciseness and clarity in our written and oral expression.

Many problems of computer science, as well as other areas of thought, deal with reasoning about things and representing things. Since much of our communication involves reasoning about things, we'll begin the chapter with a short discussion about the notions of informal proof. Next we'll introduce the basic notions and notations for sets. In the last section we'll discuss the notions and notations for some other fundamental structures that are used to represent things. We'll include informal descriptions of lists, strings, relations, graphs, and trees. The treatment here is introductory in nature, and we'll expand on these ideas in later chapters as the need arises.

Chapter Guide

Section 1.1 introduces some informal proof techniques that are used throughout the book. We'll practice each technique with a proof about numbers.

Section 1.2 introduces the basic ideas of sets. We'll see how to compare them and how to combine them, and we'll introduce some elementary ways to count them. We'll also introduce bags, which are sets that might contain

1

redundant elements, and we'll have a little discussion on why we should stick with uncomplicated sets.

Section 1.3 introduces the following basic structures that are used to represent things: tuples, products of sets, lists, strings, relations, graphs, and trees. Unlike sets and bags, these structures have some kind of order to them. We'll also look at some simple counting techniques.

1.1   A Proof Primer

For our purposes an *informal proof* is a demonstration that some statement is true. We normally communicate an informal proof in an Englishlike language that mixes everyday English with symbols that appear in the statement to be proved. In the next few paragraphs we'll discuss some basic techniques for doing informal proofs. These techniques will come in handy in trying to understand someone's proof or in trying to construct a proof of your own, so keep them in your mental tool kit.

We'll start off with a short refresher on logical statements followed by a short discussion about numbers. This will give us something to talk about when we look at examples of informal proof techniques.

Logical Statements

For this primer we'll consider only statements that are either true or false. Let's look at some familiar ways to construct statements. If S represents some statement, then the *negation* of S is the statement "not S," whose truth value is opposite that of S. We can represent this relationship with a *truth table* in which each row gives a value for S and the corresponding value for not S:

S	not S
true	false
false	true

We often paraphrase the negation of a statement to make it more understandable. For example, to negate the statement "x is odd," we normally write "x is not odd" or "it is not the case that x is odd" rather than "not x is odd."

The *conjunction* of A and B is the statement "A and B," which is true when both A and B are true. The *disjunction* of A and B is the statement "A or B," which is true if either or both of A and B are true. The truth tables for conjunction and disjunction are given in Table 1.1.

A	B	A and B	A or B
true	true	true	true
true	false	false	true
false	true	false	true
false	false	false	false

Table 1.1

Sometimes we paraphrase conjunctions and disjunctions. For example, instead of "x is odd and y is odd," we write "x and y are odd." Instead of "x is odd or y is odd," we might write "either x or y is odd." But we can't do much paraphrasing with a statement like "x is odd and y is even."

We can combine negation with either conjunction or disjunction to obtain alternative ways to write the same thing. For example, the two statements "not (A and B)" and "(not A) or (not B)" have the same truth tables. So they can be used interchangeably. We can state the rule as

The negation of a conjunction is a disjunction of negations.

For example, the statement "it is not the case that x and y are odd" can be written as "either x is not odd or y is not odd."

Similarly, the two statements "not (A or B)" and "(not A) and (not B)" have the same truth tables. We can state this rule as

The negation of a disjunction is a conjunction of negations.

For example, the statement "it is not the case that x or y is odd" can be written as "x is not odd and y is not odd."

Many statements are written in the general form "If A then B," where A and B are also logical statements. Such a statement is called a *conditional statement* in which A is the *hypothesis* and B is the *conclusion*. We can read "If A then B" in several other ways: "A is a sufficient condition for B," or "B is a necessary condition for A," or simply "A implies B." The truth table for the conditional is contained in Table 1.2.

Let's make a few comments about this table. Notice that the conditional is false only when the hypothesis is true and the conclusion is false. It's true in the other three cases. The conditional truth table gives some people fits because they interpret the statement "If A then B" to mean "B can be proved from A," which assumes that A and B are related in some way. But we've all

A	B	if A then B
true	true	true
true	false	false
false	true	false
false	false	false

Table 1.2

heard statements like "If the moon is made of green cheese then $1 = 2$." We nod our heads and agree that the statement is true, even though there is no relationship between the hypothesis and conclusion. Similarly, we shake our heads and don't agree with a statement like "If $1 = 1$ then the moon is made of green cheese."

When the hypothesis of a conditional is false, we say that the conditional is *vacuously* true. For example, the statement "If $1 = 2$ then $39 = 12$" is vacuously true because the hypothesis is false. If the conclusion is true, we say that the conditional is *trivially* true. For example, the statement "If $1 = 2$ then $2 + 2 = 4$" is trivially true because the conclusion is true. We leave it to the reader to convince at least one person that the conditional truth table is defined properly.

The *converse* of "If A then B" is "If B then A." The converse does not always have the same truth value. For example, we know that the following statement about numbers is true:

If x and y are odd then $x + y$ is even.

The converse of this statement is

If $x + y$ is even then x and y are odd.

This converse is false because $x + y$ could be even with both x and y even.

The *contrapositive* of "If A then B" is "If not B then not A." These two statements have the same truth table. So they can always be used interchangeably. For example, the following statement and its contrapositive are both true:

If x and y are odd then $x + y$ is even.

If $x + y$ is not even then not both x and y are odd.

Something to Talk About

When we discuss proof techniques, we'll be giving sample proofs about numbers. The numbers that we'll be discussing are called *integers*, and we can list them as follows:

$$..., -3, -2, -1, 0, 1, 2, 3, ...\ .$$

For two integers m and n we say that m *divides* n if $m \neq 0$ and n can be written in the form $n = m \cdot k$ for some integer k. We also say that m is a *divisor* of n or n is *divisible* by m. For example, the number 9 has six divisors: $-9, -3, -1, 1, 3, 9$. If m divides n, we write

$$m \mid n.$$

If m does not divide n, we write $m \nmid n$. For example, $3 \mid 12$ and $6 \mid 12$, but $5 \nmid 12$ and $8 \nmid 12$. Let's record two elementary facts about divisibility:

Divisibility Properties   (1.1)

    a) If $d \mid m$ and $m \mid n$, then $d \mid n$.

    b) If $d \mid m$ and $d \mid n$, then $d \mid (a \cdot m + b \cdot n)$ for any integers a and b.

For example, we know that $3 \mid 12$ and $12 \mid 144$, so we can use (1.1a) to conclude that $3 \mid 144$. We know that $7 \mid 14$ and $7 \mid 21$, so we can use (1.1b) to conclude that $7 \mid 91$ because $91 = 2 \cdot 14 + 3 \cdot 21$.

Any positive integer n has at least two positive divisors, 1 and n. For example, 9 has three positive divisors: 1, 3, and 9. The number 7 has exactly two positive divisors: 1 and 7. A positive integer p is said to be *prime* if $p > 1$ and its only positive divisors are 1 and p. For example, 9 is not prime because it has a positive divisor other than 1 and 9. The first eight prime numbers are

$$2, 3, 5, 7, 11, 13, 17, 19.$$

Many of us are aware of prime numbers, and we know that any integer greater than 1 either is a prime number or can be written as a product of prime numbers. For example, $315 = 3 \cdot 3 \cdot 5 \cdot 7$. Another interesting fact is that there is no largest prime number. In other words, there are infinitely many prime numbers. Let's record these two fundamental facts.

Prime Number Properties   (1.2)

    a) Every integer greater than 1 is a product of primes.

    b) There are infinitely many prime numbers.

Proof Techniques

Now that we have something to talk about, let's look at some proof techniques.

Proof by Example

When can an example prove something? Examples can be used to prove statements that claim the existence of an object. Suppose someone says, "There is a prime number between 80 and 88." This statement can be proved by observing that the number 83 is prime.

An example can also be used to disprove (show false) statements that assert that some property is true in many cases. Suppose someone says, "Every prime number is odd." We can disprove the statement by finding a prime number that is not odd. Since the number 2 is prime and even, it can't be the case that every prime number is odd. An example that disproves a statement is often called a *counterexample*.

Proof by Exhaustive Checking

Suppose we want to prove the statement "The sum of any two of the numbers 1, 3, and 5 is an even number." We can prove the statement by checking that each possible sum is an even number. In other words, we notice that each of the following sums represents an even number:

$$1 + 1, 1 + 3, 1 + 5, 3 + 3, 3 + 5, 5 + 5.$$

Of course, exhaustive checking can't be done if there are an infinity of things to check. But even in the finite case, exhaustive checking may not be feasible. For example, the statement "The sum of any two of the odd numbers ranging from 1 to 23195 is even" can be proved by exhaustive checking. But not many people are willing to do it.

Proof Using Variables

When proving a general statement like "The sum of any two odd integers is even," we can't list all possible sums of two odd integers to check to see whether they are even. So we must introduce variables and formulas to represent arbitrary odd integers.

For example, we can represent two arbitrary odd integers as expressions of the form $2k + 1$ and $2m + 1$, where k and m are arbitrary integers. Now we can use elementary algebra to compute the sum

$$(2k + 1) + (2m + 1) = 2(k + m + 1).$$

Since the expression on the right-hand side contains 2 as a factor, it represents an even integer. Thus we've proven that the sum of two odd integers is an even integer.

Direct Proofs

A *direct proof* of the conditional statement "If A then B" starts with the assumption that A is true. We then try to find another statement, say C, that is true whenever A is true. This means that the statement "If A then C" is true. If C happens to be B, then we're done. Otherwise, we try to find a statement D whose truth follows from that of C. This means that "If C then D" is true. From the truth of the two statements "If A then C" and "If C then D," we conclude that "If A then D" is true. If D happens to be B, then we're done. Otherwise, we continue the process until we eventually reach the goal B. When we're done, we have a direct chain of statements reaching from A to B.

It's often useful to work from both ends to find a direct proof of "If A then B." For example, we might find a sufficient condition C for B to be true. This gives us the true statement "If C then B." Now all we need is a proof of the statement "If A then C." We may be able to work forward from A or backward from C to finish constructing the proof.

Let's give an example of a direct proof of statement (1.1b):

If $d \mid m$ and $d \mid n$, then $d \mid (a{\cdot}m + b{\cdot}n)$ for any integers a and b.

Proof: Assume that $d \mid m$ and $d \mid n$. Then there are integers x and y such that $m = d{\cdot}x$ and $n = d{\cdot}y$. Now we can write the expression $a{\cdot}m + b{\cdot}n$ as follows:

$$a{\cdot}m + b{\cdot}n = a{\cdot}d{\cdot}x + b{\cdot}d{\cdot}y = d(a{\cdot}x + b{\cdot}y).$$

This says that $d \mid (a{\cdot}m + b{\cdot}n)$.   QED.

Indirect Proofs

A proof is *indirect* if it contains an assumption that some statement is false. For example, since a conditional and its contrapositive have the same truth table, we can prove "If A then B" by proving its contrapositive statement "If not B then not A." In this case we assume that B is false and then try to show that A is false. This is called *proving the contrapositive*. For example, suppose we want to prove the following statement about integers:

If x^2 is even, then x is even.

Proof: We'll prove the contrapositive statement: If x is odd, then x^2 is odd. So assume that x is odd. Then x can be written in the form $x = 2k + 1$ for some integer k. Squaring x, we obtain

$$x^2 = (2k + 1)(2k + 1) = 4k^2 + 4k + 1 = 2(2k^2 + 2k) + 1.$$

The expression on the right side of the equation is an odd number. Therefore x^2 is odd.   QED.

The second indirect method is called *proof by contradiction*. A *contradiction* is a false statement. A proof by contradiction starts with the assumption that the entire statement to be proved is false. Then we argue until we reach a contradiction. Such an argument is often called a *refutation*. A contradiction often occurs in two parts. One part assumes or proves that some statement S is true. The second part proves that S is false. Since S can't be both true and false, a contradiction occurs.

For example, let's give a proof (due to Euclid) of (1.2b), which we'll restate for convenience:

There are infinitely many prime numbers.

Proof: Assume that the statement is false. Then there is a largest prime number p. Let m be the product of all the prime numbers: $m = 2 \cdot 3 \cdot 5 \cdots p$. Now consider the number $m + 1$. Since $m + 1$ is bigger than 1, we can use (1.2a) to conclude that $m + 1$ is itself a product of prime numbers. So $m + 1$ is divisible by a prime number q. But q is one of the prime numbers in the product making up m. Therefore q divides m. Since q divides m and q divides $m + 1$, we can use (1.1b) to conclude that q divides 1, which is impossible. We've reached a contradiction. Therefore we conclude that there are infinitely many prime numbers.   QED.

We now have three different ways to prove the conditional "If A then B." We can use the direct method, or we can use either of the two indirect methods. Among the indirect methods, contradiction is often easier than proving the contrapositive because it allows us to assume more. We start by assuming that the whole statement "If A then B" is false. But this means that we can assume that A is true and B is false (from the truth table for conditional). Then we can wander wherever the proof takes us to find a contradiction.

If and Only If Proofs

The statement "A if and only if B" is shorthand for the two statements "If A then B" and "If B then A." The abbreviation "A iff B" is often used for "A if and only if B." Instead of "A iff B," some people write "A is a necessary and sufficient condition for B" or "B is a necessary and sufficient condition for A." Two separate proofs are required for an iff statement, one for each conditional statement.

For example, let's prove the following statement about integers:

x is even if and only if x^2 is even.

To prove this iff statement, we must prove the following two statements:

(1) If x is even then x^2 is even.

(2) If x^2 is even then x is even.

Proof of (1): If x is even, then we can write x in the form $x = 2k$ for some integer k. Squaring x, we obtain $x^2 = (2k)^2 = 4k^2 = 2(2k^2)$. Thus x^2 is even. Proof of (2): We proved this statement in a preceding paragraph, so we won't repeat it here.   QED.

On Constructive Existence

If a statement asserts that some object exists, then we can try to prove the statement in either of two ways. One way is to use proof by contradiction, in which we assume that the object does not exist and then come up with some kind of contradiction. The second way is to use proof by example, in which we construct an instance of the object. In either case we know that the object exists, but the second way also gives us an instance of the object. Computer science leans toward the construction of objects by algorithms. So the constructive approach is usually preferred, although it's not always possible.

Important Note

Always try to write out your proofs. Use complete sentences. If your proof seems to consist only of a bunch of equations or expressions, you still need to describe how they contribute to the proof. Try to write your proofs the same way you would write a letter to a friend who wants to understand what you have written.

Exercises

1. See whether you can convince yourself, or a friend, that the conditional truth table is correct by making up English sentences of the form "If A then B."

2. Verify that the truth tables for each of the following pairs of statements are identical.
   a. "not $(A$ and $B)$" and "(not A) or (not B)."
   b. "not $(A$ or $B)$" and "(not A) and (not B)."
   c. "if A then B" and "if (not B) then (not A)."

3. Prove or disprove each of the following statements by giving an example or by exhaustive checking.

   a. There is a prime number between 45 and 54.

   b. The product of any two of the four numbers 2, 3, 4, and 5 is even.

   c. Every odd integer greater than 1 is prime.

   d. Every integer greater than 1 is either prime or the sum of two primes.

4. Prove the following statement about divisibility of integers (1.1a): If $d \mid m$ and $m \mid n$ then $d \mid n$.

5. Prove each of the following statements about the integers.

   a. The sum of two even integers is even.

   b. The sum of an even integer and an odd integer is odd.

   c. If x and y are odd then $x - y$ is even.

6. Write down the converse of the following statement about integers:

$$\text{If } x \text{ and } y \text{ are odd then } x - y \text{ is even.}$$

   Is the statement that you wrote down true or false? Prove your answer.

7. Prove that numbers having the form $3n + 4$, where n is any integer, are closed under multiplication. In other words, if two numbers of this form are multiplied, then the result is a number of the same form.

8. Prove that numbers having the form $3n + 2$, where n is any integer, are not closed under multiplication. In other words, if two numbers of this form are multiplied, then the result need not be of the same form.

9. Prove the following statement about the integers: x is odd if and only if x^2 is odd.

1.2 Sets

Informally, a *set* is a collection of things called its *elements, members,* or *objects*. Sometimes the word *collection* is used in place of *set* to clarify a sentence. For example, "a collection of sets" seems clearer than "a set of sets." We say that a set contains its elements, or that the elements belong to the set, or that the elements are in the set. If S is a set and x is an element in S, then we write

$$x \in S.$$

If x is not an element of S, then we write $x \notin S$. If $x \in S$ and $y \in S$, we often denote this fact by the shorthand notation $x, y \in S$.

A set is defined by describing its elements in some way. For example, a *gaggle* is a set whose members are geese, a *pride* is a set whose members are lions, and a *pod* is a set whose members are whales. One way to define a set is to explicitly name its elements. A set defined in this way is denoted by listing its elements, separated by commas, and surrounding the listing with braces. For example, the set S consisting of the letters x, y, and z is denoted by

$$S = \{x, y, z\}.$$

Sets can have other sets as elements. For example, the set $A = \{x, \{x, y\}\}$ has two elements. One element is x, and the other element is $\{x, y\}$. So we can write $x \in A$ and $\{x, y\} \in A$.

We often use the three-dot ellipsis, ..., to informally denote a sequence of elements that we do not wish to write down. For example, the set

$$\{1, 2, 3, 4, 5, 6, 7, 8, 9, 10, 11, 12\}$$

can be denoted in several different ways with ellipses, two of which are

$$\{1, 2, ..., 12\} \quad \text{and} \quad \{1, 2, 3, ..., 11, 12\}.$$

The set with no elements is called the *empty set*—some people refer to it as the *null set*. The empty set is denoted by { } or more often by the symbol

$$\varnothing.$$

A set with one element is called a *singleton*. For example, $\{a\}$ and $\{b\}$ are singletons.

Two sets A and B are *equal* if each element of A is an element of B and conversely each element of B is an element of A. We denote the fact that A and B are equal sets by writing

$$A = B.$$

We can use equality to demonstrate two important characteristics of sets:

> *There is no particular order or arrangement of the elements.*
> *There are no redundant elements.*

For example, the set whose elements are are g, h, and u can be represented in many ways, four of which are

$$\{u, g, h\} = \{h, u, g\} = \{h, u, g, h\} = \{u, g, h, u, g\}.$$

In other words, there are many ways to represent the same set.

If the sets A and B are not equal, we write

$$A \neq B.$$

For example, $\{a, b, c\} \neq \{a, b\}$ because c is an element of only one of the sets. We also have $\{a\} \neq \varnothing$ because the empty set doesn't have any elements.

Suppose we start counting the elements of a set S, one element per second of time with a stop watch. If $S = \varnothing$, then we don't need to start, because there are no elements to count. But if $S \neq \varnothing$, we agree to start the counting after we have started the timer. If a point in time is reached when all the elements of S have been counted, then we stop the timer, or in some cases we might need to have one of our descendants stop the timer. In this case we say that S is a *finite* set. If the counting never stops, then S is an *infinite* set. All the examples that we have discussed to this point are finite sets. We will discuss counting finite and infinite sets in other parts of the book as the need arises.

Familiar infinite sets are sometimes denoted by listing a few of the elements followed by an ellipsis. We reserve some letters to denote specific sets that we'll refer to throughout the book. For example, the set of natural numbers will be denoted by $\mathbb{N}^{\dagger}$ and the set of integers by $\mathbb{Z}$. So we can write

$$\mathbb{N} = \{0, 1, 2, 3, ...\} \quad \text{and} \quad \mathbb{Z} = \{..., -3, -2, -1, 0, 1, 2, 3, ...\}.$$

Many sets are hard to describe by a listing of elements. Examples that come to mind are the rational numbers, which we denote by $\mathbb{Q}$, and the real numbers, which we denote by $\mathbb{R}$. On the other hand, any set can be defined by stating a property that the elements of the set must satisfy. If P is some property, then there is a set S whose elements have property P, and we denote S by

$$S = \{x \mid x \text{ has property } P\},$$

which we read as "S is the set of all x such that x has property P." For example, we can write Gaggle = $\{x \mid x$ is a goose$\}$ and say that Gaggle is the set of

---

† Some people consider the natural numbers to be the set $\{1, 2, 3, ...\}$. If you are one of these people, then think of $\mathbb{N}$ as the Nonnegative integers.

all x such that x is a goose. The set Odd = {..., –5, –3, –1, 1, 3, 5, ...} of odd integers can be defined by

$$\text{Odd} = \{x \mid x = 2k + 1 \text{ and } k \in \mathbb{Z}\}.$$

Similarly, the set {1, 2, ..., 12} can be defined by writing

$$\{x \mid x \in \mathbb{N} \text{ and } 1 \le x \le 12\}.$$

Let's try to represent the rational numbers as a set. Recall that a rational number can be represented by a fraction $\frac{p}{q}$, where $p, q \in \mathbb{Z}$ and $q \ne 0$. So we might represent the rational numbers as the following set of fractions:

$$\left\{ \frac{p}{q} \mid p, q \in \mathbb{Z} \text{ and } q \ne 0 \right\}.$$

Notice, for example, that the fractions $\frac{1}{2}, \frac{-1}{-2}, \frac{2}{4}, \frac{-2}{-4}, \frac{3}{6}, \ldots$ are all distinct elements in the set, but they represent the same rational number. Suppose we want our set to contain exactly one fraction for each rational number. One way to do this is to restrict the fractions to be those that are in lowest terms. Recall that a fraction $\frac{p}{q}$ is in lowest terms when p and q have no common factors other than 1. For example, the fractions $\frac{1}{2}$ and $\frac{8}{3}$ are in lowest terms, but $\frac{2}{4}$, $\frac{3}{6}$, and $\frac{34}{40}$ are not. Thus we can represent the rational numbers as the following set of fractions with this added property:

$$\left\{ \frac{p}{q} \mid p, q \in \mathbb{Z} \text{ and } q \ne 0 \text{ and } \frac{p}{q} \text{ is in lowest terms} \right\}.$$

If A and B are sets and every element of A is also an element of B, then we say that A is a *subset* of B and write

$$A \subset B.$$

For example, we have {a, b} $\subset$ {a, b, c}, {0, 1, 2} $\subset \mathbb{N}$, and $\mathbb{N} \subset \mathbb{Z}$. It follows from the definition that every set A is a subset of itself. Thus we have $A \subset A$. It also follows from the definition that the empty set is a subset of any set A. So we have $\varnothing \subset A$. Can you see why? We'll leave this as an exercise.

If $A \subset B$ and there is some element in B that does not occur in A, then A is called a *proper* subset of B. For example, {a, b} is a proper subset of {a, b, c}. We also conclude that $\mathbb{N}$ is a proper subset of $\mathbb{Z}$, $\mathbb{Z}$ is a proper subset of $\mathbb{Q}$, and $\mathbb{Q}$ is a proper subset of $\mathbb{R}$.

If A is not a subset of B, we sometimes write

$$A \not\subset B.$$

For example, $\{a, b\} \not\subset \{a, c\}$ and $\{-1, -2\} \not\subset \mathbb{N}$. Remember that the idea of subset is different from the idea of membership. For example, if $A = \{a, b, c\}$, then $\{a\} \subset A$ and $a \in A$. But $\{a\} \notin A$ and $a \not\subset A$. For another example, let $A = \{a, \{b\}\}$. Then $a \in A$, $\{b\} \in A$, $\{a\} \subset A$, and $\{\{b\}\} \subset A$. But $b \notin A$ and $\{b\} \not\subset A$.

The collection of all subsets of a set S is called the *power set* of S, which we denote by power(S). For example, if $S = \{a, b, c\}$, then the power set of S can be written as follows:

$$\text{power}(S) = \{\varnothing, \{a\}, \{b\}, \{c\}, \{a, b\}, \{a, c\}, \{b, c\}, S\}.$$

An interesting programming problem is to construct the power set of a finite set. We'll discuss this problem later, once we've developed some tools to help build an easy solution.

In dealing with sets, it's often useful to draw a picture in order to visualize the situation. A *Venn diagram*—named after the logician John Venn (1834–1923)—consists of one or more closed curves in which the interior of each curve represents a set. For example, the Venn diagram in Figure 1.1 represents the fact that A is a proper subset of B and x is an element of B that does not occur in A.

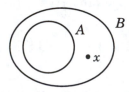

Figure 1.1   Venn diagram of proper subset $A \subset B$.

We can use the subset definition to give a precise definition of set equality: Two sets are equal if they are subsets of each other. In more concise form we can write

$$A = B \quad \text{means} \quad A \subset B \text{ and } B \subset A. \tag{1.3}$$

Now let's look at three simple but useful proof techniques for comparing two sets. The first technique is used to show that $A \subset B$. We start by letting x stand for an arbitrary element of A. Then we give an argument to show that x is also an element of B. We'll record the technique as follows:

To prove that $A \subset B$, let $x \in A$ and show that $x \in B$.   (1.4)

The second technique is used to show $A \not\subset B$:

To prove that $A \not\subset B$, find an $x \in A$ such that $x \notin B$.   (1.5)

This third technique is used to show that $A = B$:

To prove that $A = B$, show that $A \subset B$ and $B \subset A$.   (1.6)

EXAMPLE 1. We'll show that $A \subset B$, where A and B are defined as follows:

$$A = \{x \mid x \text{ is a prime number and } 42 \leq x \leq 51\}$$
$$B = \{x \mid x = 4k + 3 \text{ and } k \in \mathbb{N}\}.$$

We start the proof by letting $x \in A$. Then either $x = 43$ or $x = 47$. We can write $43 = 4(10) + 3$ and $47 = 4(11) + 3$. So in either case, x has the form of an element of B. Thus $x \in B$. Therefore $A \subset B$.   ♦

EXAMPLE 2. We'll show that $A \not\subset B$ and $B \not\subset A$, where A and B are defined by

$$A = \{3k + 1 \mid k \in \mathbb{N}\} \quad \text{and} \quad B = \{4k + 1 \mid k \in \mathbb{N}\}.$$

By listing a few elements from each set we can write A and B as follows:

$$A = \{1, 4, 7, \dots\} \quad \text{and} \quad B = \{1, 5, 9, \dots\}.$$

Now it's easy to prove that $A \not\subset B$ because $4 \in A$ and $4 \notin B$. We can also prove that $B \not\subset A$ by observing that $5 \in B$ and $5 \notin A$.   ♦

EXAMPLE 3. We'll show that $A = B$, where A and B are defined as follows:

$$A = \{x \mid x \text{ is prime and } 12 \leq x \leq 18\},$$
$$B = \{x \mid x = 4k + 1 \text{ and } k \in \{3, 4\}\}.$$

First we'll show that $A \subset B$. Let $x \in A$. Then either $x = 13$ or $x = 17$. We can write $13 = 4(3) + 1$ and $17 = 4(4) + 1$. It follows that $x \in B$. Therefore $A \subset B$. Next we'll show that $B \subset A$. Let $x \in B$. Then $x = 4(3) + 1$ or $x = 4(4) + 1$. In either case, x is a prime number between 12 and 18. Thus $B \subset A$. So $A = B$.   ♦

Operations on Sets

If A and B are sets, then the *union* of A and B is the set of all elements that either are in A or in B or in both A and B. The union of A and B is denoted by $A \cup B$. So we can write

$$A \cup B = \{x \mid x \in A \text{ or } x \in B\}. \tag{1.7}$$

The use of the word "or" in the definition is taken to mean "either or both." For example, if $A = \{a, b, c\}$ and $B = \{c, d\}$, then $A \cup B = \{a, b, c, d\}$. The union of two sets A and B is represented by the shaded regions of the Venn diagram in Figure 1.2.

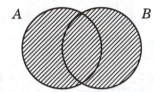

Figure 1.2   Venn diagram of $A \cup B$

The following properties give the basic facts about the union operation.

Properties of Union $\hspace{6cm}$ (1.8)

a) $A \cup \varnothing = A$.

b) $A \cup B = B \cup A$ $\hspace{3cm}$ ($\cup$ is commutative).

c) $A \cup (B \cup C) = (A \cup B) \cup C$ $\hspace{1cm}$ ($\cup$ is associative).

d) $A \cup A = A$.

e) $A \subset B$ if and only if $A \cup B = B$.

Proof: We'll include a proof for part (e) and leave the other parts as exercises. The "if and only if" means that we actually have two things to prove. First we'll prove "if $A \subset B$ then $A \cup B = B$." Assume that $A \subset B$. With this assumption we must show that $A \cup B = B$. Let $x \in A \cup B$. Then $x \in A$ or $x \in B$. Since we have assumed that $A \subset B$, it follows that $x \in B$. Thus $A \cup B \subset B$. Since we always have $B \subset A \cup B$, it follows from (1.3) that $A \cup B = B$.

Now we'll prove "if $A \cup B = B$ then $A \subset B$." Assume that $A \cup B = B$. If $x \in A$, then certainly $x \in A \cup B$. Since $A \cup B = B$, it follows that $x \in B$. Therefore $A \subset B$. We have proven both parts of the "if and only if" statement.   QED.

The union operation can be defined for an arbitrary collection of sets in a natural way. We know that if A and B are sets, then the union of A and B is the set denoted by $A \cup B$. If we have a finite collection of sets $\{A_1, ..., A_n\}$, then the union of the n sets in this collection is well defined because $\cup$ is associative. We denote the union by writing

$$A_1 \cup ... \cup A_n.$$

Other notations for this union are either

$$\bigcup_{i=1}^{n} A_i \quad \text{or} \quad \bigcup_{1 \le i \le n} A_i \ .$$

Now suppose we have an infinite collection of sets $\{A_i \mid i \in \mathbb{N}\}$. Does it make sense to form the union of infinitely many sets? Yes it does. We define the union of the given collection as the set

$$\{x \mid x \in A_i \text{for some } i \in \mathbb{N}\},$$

which we denote by

$$A_0 \cup A_1 \cup ... \cup A_i \cup ... \ .$$

There are several ways to denote this union, three of which are

$$\bigcup_{i \ge 0} A_i = \bigcup_{i \in \mathbb{N}} A_i = \bigcup_{i=0}^{\infty} A_i \ .$$

EXAMPLE 4. Let W be the set of all words in the English language that are contained in your favorite dictionary. Then W can be represented as an infinite union of sets. For each $i > 0$, let B_i denote the set of all words with i letters. Then

$$W = \bigcup_{i=1}^{\infty} B_i.$$

If your dictionary does not have any words with more than 25 letters, then most of the sets B_i are empty, $B_{26} = \varnothing$, $B_{27} = \varnothing$, and so on. In this case we could write W as the finite union

$$W = \bigcup_{i=1}^{25} B_i. \quad \blacklozenge$$

We can make a general definition for the union of any nonempty collection of sets. If C is a nonempty collection of sets, then the *union of the collection* C is the set of all elements that occur in each set of C:

$$\bigcup_{A \in C} A = \{x \mid x \in A \text{ for some } A \in C\}. \tag{1.9}$$

For example, if $C = \{\{a\}, \{b, c\}, \{d, e, f\}, \{g, h, i, j\}\}$, then

$$\bigcup_{A \in C} A = \{a, b, c, d, e, f, g, h, i, j\}.$$

If I is a set of indices and A_i is a set for each $i \in I$, then we can write the union of the sets in the collection of sets $\{A_i \mid i \in I\}$ as

$$\bigcup_{i \in I} A_i.$$

For example, if for each $i \in \text{Odd} = \{1, 3, 5, ...\}$ we let $A_i = \{-2i, 2i\}$, then we have

$$\bigcup_{i \in Odd} A_i = \{..., -10, -6, -2, 2, 6, 10, ...\}.$$

If A and B are sets, then the *intersection* of A and B is the set of all elements that are in both A and B and is denoted by $A \cap B$. We can write

$$A \cap B = \{x \mid x \in A \text{ and } x \in B\}. \tag{1.10}$$

For example, if $A = \{a, b, c\}$ and $B = \{c, d\}$, then $A \cap B = \{c\}$. If $A \cap B = \varnothing$, then A and B are said to be *disjoint*. The nonempty intersection of two sets A and B is represented by the shaded region of the Venn diagram in Figure 1.3.

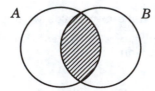

Figure 1.3   Venn diagram of $A \cap B$.

The basic facts about intersection are given next. The proofs are left as an exercise.

Properties of Intersection (1.11)

a) $A \cap \varnothing = \varnothing$.

b) $A \cap B = B \cap A$ ($\cap$ is commutative).

c) $A \cap (B \cap C) = (A \cap B) \cap C$ ($\cap$ is associative).

d) $A \cap A = A$.

e) $A \subset B$ if and only if $A \cap B = A$.

Intersection can be defined for any nonempty collection of sets. If C is a nonempty collection of sets, then the *intersection of the collection C* is the set of all elements that occur in every set in C, which we write as

$$\bigcap_{A \in C} A = \{x \mid x \in A \text{ for every } A \in C\}.$$ (1.12)

If I is a set of indices for a collection of sets $C = \{A_i \mid i \in I\}$, then we can write the intersection of the sets in the collection C as

$$\bigcap_{i \in I} A_i.$$

If more is known about the index set then, as with union, there are other possibilities. For example, if $C = \{A_i \mid i \in \mathbb{N}\}$, then the intersection of the sets in C can be written as

$$\bigcap_{i=0}^{\infty} A_i = A_0 \cap \ldots \cap A_i \cap \ldots .$$

For example, suppose for each $i \in \mathbb{N}$ we let $A_i = \{0, 1, \ldots, i\}$. Then we have

$$\bigcap_{i=0}^{\infty} A_i = \{0\}.$$

Venn diagrams are often quite useful in trying to visualize the sets constructed with different operations. For example, the set $A \cap (B \cup C)$ is represented by the shaded regions of the Venn diagram in Figure 1.4.

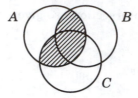

Figure 1.4   Venn diagram of $A \cap (B \cup C)$.

Two useful properties that combine the operations of union and intersection are given next.

Distributive Properties   (1.13)

a) $A \cap (B \cup C) = (A \cap B) \cup (A \cap C)$ ($\cap$ distributes over $\cup$).

b) $A \cup (B \cap C) = (A \cup B) \cap (A \cup C)$ ($\cup$ distributes over $\cap$).

Proof: We'll prove part (a) and leave part (b) as an exercise. If $x \in A \cap (B \cup C)$, then $x \in A$ and $x \in B \cup C$. Therefore $x \in A$, and either $x \in B$ or $x \in C$. This says that either ($x \in A$ and $x \in B$) or ($x \in A$ and $x \in C$), which is the same as saying $x \in (A \cap B) \cup (A \cap C)$. Therefore $A \cap (B \cup C) \subset (A \cap B) \cup (A \cap C)$. To show that $(A \cap B) \cup (A \cap C) \subset A \cap (B \cup C)$, just reverse the above argument. QED.

If A and B are sets, then the *difference* between A and B (also called the *relative complement* of B in A) is the set of elements in A that are not in B, and it is denoted by $A - B$. That is,

$$A - B = \{x \mid x \in A \text{ and } x \notin B\}. \tag{1.14}$$

For example, if $A = \{a, b, c\}$ and $B = \{c, d\}$, then $A - B = \{a, b\}$. We can picture the difference $A - B$ of two general sets A and B by the shaded region of the Venn diagram in Figure 1.5.

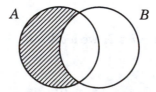

Figure 1.5   Venn diagram of $A - B$.

Venn diagrams can be used to discover or verify relationships between sets that use difference. For example, it is easy to see that $A \cap B = A - (A - B)$. Can you discover some other relationships?

A natural extension of the difference $A - B$ is the *symmetric difference* of sets A and B, which is the union of $A - B$ with $B - A$ and is denoted by $A \oplus B$. The set $A \oplus B$ is represented by the shaded regions of the Venn diagram in Figure 1.6.

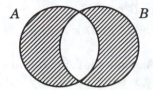

Figure 1.6   Venn diagram of $A \oplus B$.

We can define the symmetric difference by using the "exclusive" form of "or" as follows:

$$A \oplus B = \{x \mid x \in A \text{ or } x \in B \text{ but not both}\}. \tag{1.15}$$

As usual, there are many relationships to discover. For example, it's easy to see that

$$A \oplus B = (A \cup B) - (A \cap B).$$

Can you verify that $(A \oplus B) \oplus C = A \oplus (B \oplus C)$? For example, try to draw two Venn diagrams, one for each side of the equation.

If the discussion always refers to sets that are subsets of a particular set U, then U is called the *universe of discourse*, and the difference $U - A$ is called the *complement* of A, which we denote by A'. The Venn diagram in Figure 1.7 pictures the universe U as a rectangle, with two subsets A and B, where the shaded region represents the complement $(A \cup B)'$.

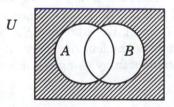

Figure 1.7   Venn diagram of $(A \cup B)'$.

The following properties give the basic facts about complement.

Properties of Complement   (1.16)

a) $(A')' = A$.

b) $\varnothing' = U$ and $U' = \varnothing$.

c) $A \cap A' = \varnothing$ and $A \cup A' = U$.

d) $A \subset B$ if and only if $B' \subset A'$.

e) $(A \cup B)' = A' \cap B'$ and $(A \cap B)' = A' \cup B'$ (De Morgan's laws).

Proof: We'll prove part (d). Assume that $A \subset B$ and show that $B' \subset A'$. If $x \in U - B$, then $x \notin B$. Therefore x can't be in any subset of B. So $x \in U - A$. Therefore $B' \subset A'$. Now assume that $B' \subset A'$ and show that $A \subset B$. If $x \in A$ and it also happens that $x \notin B$, then the assumption $B' \subset A'$ gives the contradiction that $x \notin A$. So if $x \in A$, then it must follow that $x \in B$. Thus $B' \subset A'$ implies that $A \subset B$.   QED.

There are, as usual, many relationships between sets that use the complement, and they are easy to discover by using Venn diagrams. For example, convince yourself that the following *absorption laws* hold:

$$A \cap (A' \cup B) = A \cap B \quad \text{and} \quad A \cup (A' \cap B) = A \cup B.$$

Counting Finite Sets

Let's apply some of our knowledge about sets to counting finite sets. The size of a set S is called its *cardinality*, which we'll denote by

$$|S|.$$

For example, if $S = \{a, b, c\}$ then $|S| = |\{a, b, c\}| = 3$. We can say "the cardinality of S is 3," or "3 is the cardinal number of S," or simply "S has three elements."

Suppose we want to count the union of two sets. For example, suppose we have $A = \{1, 2, 3, 4, 5\}$ and $B = \{2, 4, 6, 8\}$. Since $A \cup B = \{1, 2, 3, 4, 5, 6, 8\}$, it follows that $|A \cup B| = 7$. Similarly, since $A \cap B = \{2, 4\}$, it follows that $|A \cap B| = 2$. If we know any three of the four numbers $|A|$, $|B|$, $|A \cap B|$, and $|A \cup B|$, then we can find the fourth by using the following counting rule for finite sets:

Union Rule

$$|A \cup B| = |A| + |B| - |A \cap B|.$$ (1.17)

It's easy to discover this rule by drawing a Venn diagram.

The union rule extends to three or more sets. For example, the following calculation gives the union rule for three finite sets:

$$|A \cup B \cup C| \tag{1.18}$$
$$= |A \cup (B \cup C)|$$
$$= |A| + |B \cup C| - |A \cap (B \cup C)|$$
$$= |A| + |B| + |C| - |B \cap C| - |A \cap (B \cup C)|$$
$$= |A| + |B| + |C| - |B \cap C| - |(A \cap B) \cup (A \cap C)|$$
$$= |A| + |B| + |C| - |B \cap C| - |A \cap B| - |A \cap C| + |A \cap B \cap C|.$$

The popular name for the union rule and its extensions to three or more sets is the *principle of inclusion and exclusion*. The name is appropriate because the rule says to add (include) the count of each individual set. Then subtract (exclude) the count of all intersecting pairs of sets. Next, include the count of all intersections of three sets. Then exclude the count of all intersections of four sets, and so on.

EXAMPLE 5 (*A Building Project*). Suppose A, B, and C are sets of tools needed by three workers on a job. For convenience let's call the workers A, B, and C. Suppose further that the workers share some of the tools (for example, on a housing project, all three workers might share a single table saw). Suppose that A uses 8 tools, B uses 10 tools, and C uses 5 tools. Suppose further that A and B share 3 tools, A and C share 2 tools, and B and C share 2 tools. Finally, suppose that A, B, and C share the use of 2 tools. How many distinct tools are necessary to do the job? Thus we want to find the value

$$|A \cup B \cup C|.$$

We can apply the inclusion exclusion principle (1.18) to compute the answer:

$$|A \cup B \cup C|$$
$$= |A| + |B| + |C| - |A \cap B| - |A \cap C| - |B \cap C| + |A \cap B \cap C|$$
$$= 8 + 10 + 5 - 3 - 2 - 2 + 2$$
$$= 18 \text{ tools.} \quad \blacklozenge$$

EXAMPLE 6 (*Surveys*). Suppose we survey 200 students to see whether they are taking courses in computer science, mathematics, or physics. The results show that 90 students take computer science, 110 take mathematics, and 60 take physics. Further, 20 students take computer science and mathematics, 20 take computer science and physics, and 30 take mathematics and physics. We are interested in those students that take courses in all three areas. Do we have enough information? Let C, M, and P stand for the sets of students that take computer science, mathematics, and physics. So we are interested in the number

$$|C \cap M \cap P|.$$

From the information given we know that $200 \geq |C \cup M \cup P|$. The reason for the inequality is that some students may not be taking any courses in the three areas. The statistics also give us the following information:

$$|C| = 90$$
$$|M| = 110$$
$$|P| = 60$$
$$|C \cap M| = 20$$
$$|C \cap P| = 20$$
$$|M \cap P| = 30.$$

Applying the inclusion exclusion principle (1.18), we have

$$
\begin{aligned}
200 \geq \ & |C| + |M| + |P| - |C \cap M| - |C \cap P| - |M \cap P| \\
& + |C \cap M \cap P| \\
= \ & 90 + 110 + 60 - 20 - 20 - 30 + |C \cap M \cap P| \\
= \ & 190 + |C \cap M \cap P|.
\end{aligned}
$$

Therefore $|C \cap M \cap P| \leq 10$. So we can say that at most 10 of the students polled take courses in all three areas. We would get an exact answer if we knew that each student in the survey took at least one course from one of the three areas. ◆

Suppose we want to count the difference of two sets. For example, let $A = \{1, 3, 5, 7, 9\}$ and $B = \{2, 3, 4, 8, 10\}$. Then $A - B = \{1, 5, 7, 9\}$, so $|A - B| = 4$. If we know any two of the numbers $|A|$, $|A - B|$, and $|A \cap B|$, then we can find the third by using the following counting rule for finite sets:

Difference Rule

$$|A - B| = |A| - |A \cap B|. \tag{1.19}$$

It's easy to discover this rule by drawing a Venn diagram. Two special cases of (1.19) are also intuitive and can be stated as follows:

$$\text{If } B \subset A, \text{ then } |A - B| = |A| - |B|. \tag{1.20}$$

$$\text{If } A \cap B = \emptyset, \text{ then } |A - B| = |A|. \tag{1.21}$$

EXAMPLE 7 (*Tool Boxes*). Continuing with the data from Example 5, suppose we want to know how many personal tools each worker needs (tools not shared with other workers). For example, Worker A needs a tool box of size

$$|A - (B \cup C)|.$$

We can compute this value using both the difference rule (1.19) and the union rule (1.17) as follows:

$$
\begin{aligned}
|A - (B \cup C)| &= |A| - |A \cap (B \cup C)| \\
&= |A| - |(A \cap B) \cup (A \cap C)| \\
&= |A| - (|(A \cap B)| + |(A \cap C)| - |A \cap B \cap C|) \\
&= 8 - (3 + 2 - 2) \\
&= 5 \text{ personal tools for } A. \quad \blacklozenge
\end{aligned}
$$

Bags (Multisets)

A *bag* (or *multiset*) is a collection of objects that may contain a finite number of redundant occurrences of elements. The two important characteristics of a bag are:

There is no particular order or arrangement of the elements.

There may be a finite number of redundant occurrences of elements.

To differentiate bags from sets, we'll use brackets to enclose the elements. For example, $[h, u, g, h]$ is a bag with four elements. Two bags A and B are *equal* if the number of occurrences of each element in A or B is the same in either bag. If A and B are equal bags, we write $A = B$. For example, $[h, u, g, h] = [h, h, g, u]$, but $[h, u, g, h] \neq [h, u, g]$.

We can also define the subbag notion. Define A to be a *subbag* of B, and write $A \subset B$, if the number of occurrences of each element x in A is less than or equal to the number of occurrences of x in B. For example, $[a, b] \subset [a, b, a]$, but $[a, b, a] \not\subset [a, b]$. It follows from the definition of subbag that two bags A and B are equal if and only if A is a subbag of B and B is a subbag of A.

If A and B are bags, we define the *sum* of A and B, denoted by $A + B$, as follows: If x occurs m times in A and n times in B, then x occurs $m + n$ times in $A + B$. For example,

$$[2, 2, 3] + [2, 3, 3, 4] = [2, 2, 2, 3, 3, 3, 4].$$

We can define union and intersection for bags also (we will use the same symbols as for sets). Let A and B be bags, and let m and n be the number of times x occurs in A and B, respectively. Put the larger of m and n occurrences of x in $A \cup B$. Put the smaller of m and n occurrences of x in $A \cap B$. For example, we have

$$[2, 2, 3] \cup [2, 3, 3, 4] = [2, 2, 3, 3, 4]$$

and

$$[2, 2, 3] \cap [2, 3, 3, 4] = [2, 3] .$$

EXAMPLE 8. Let $p(x)$ denote the bag of prime numbers that occur in the prime factorization of the natural number x. For example, we have

$$p(54) = [2, 3, 3, 3] \text{ and } p(12) = [2, 2, 3].$$

Let's compute the union and intersection of these two bags. The union gives $p(54) \cup p(12) = [2, 2, 3, 3, 3] = p(108)$, and 108 is the least common multiple of 54 and 12 (i.e., the smallest positive integer that they both divide). Similarly, we get $p(54) \cap p(12) = [2, 3] = p(6)$, and 6 is the greatest common divisor of 54 and 12 (i.e., the largest positive integer that divides them both). Can we discover anything here? It appears that $p(x) \cup p(y)$ and $p(x) \cap p(y)$ compute the least common multiple and the greatest common divisor of x and y. Can you convince yourself?   ◆

Sets Should Not Be Too Complicated

Set theory was created by the mathematician Georg Cantor (1845–1918) during the period 1874 to 1895. Later some contradictions were found in the theory. Everything works fine as long as we don't allow sets to be too complicated. Basically, we never allow a set to be defined by a test that checks whether a set is a member of itself. If we allowed such a thing, then we could not decide some questions of set membership. For example, suppose we define the set T as follows:

$$T = \{A \mid A \text{ is a set and } A \notin A\}.$$

In other words, T is the set of all sets that are not members of themselves. Now ask the question "Is $T \in T$?" If so, then the condition for membership in T must hold. But this says that $T \notin T$. On the other hand, if we assume that $T \notin T$, then we must conclude that $T \in T$. In either case we get a contradiction. This example is known as *Russell's paradox*—after the philosopher and mathematician Bertrand Russell (1872–1970).

This kind of paradox led to a more careful study of the foundations of set theory. For example, Whitehead and Russell [1910] developed a theory of sets based on a hierarchy of levels that they called *types*. The lowest type contains individual elements. Any other type contains only sets whose elements are from the next lower type in the hierarchy. We can list the hierarchy of types as $T_0, T_1, ..., T_k, ...$, where T_0 is the lowest type containing individual elements and in general T_{k+1} is the type consisting of sets whose elements are from T_k. So any set in this theory belongs to exactly one type T_k for some $k \geq 1$.

As a consequence of the definition, we can say that $A \notin A$ for all sets A in the theory. To see this, suppose A is a set of type T_{k+1}. This means that the elements of A are of type T_k. If we assume that $A \in A$, we would have to conclude that A is also a set of type T_k. This says that A belongs to the two types T_k and T_{k+1}, contrary to the fact that A must belong to exactly one type.

Let's examine why Russell's paradox can't happen in this new theory of sets. Since $A \notin A$ for all sets A in the theory, the original definition of T can be simplified to $T = \{A \mid A \text{ is a set}\}$. This says that T contains all sets. But T itself isn't even a set in the theory because it contains sets of different types. In order for T to be a set in the theory, each A in T must belong to the same type. For example, we could pick some type T_k and define $T = \{A \mid A \text{ has type } T_k\}$. This says that T is a set of type T_{k+1}. Now since T is a set in the theory, we know that $T \notin T$. But this fact doesn't lead us to any kind of contradictory statement.

Exercises

1. Give a definition of the form $\{x \mid x \text{ has property } P\}$ for each of the following sets.

    a. A pride of lions.

    b. $\{1, 2, 3, 4, 5, 6, 7\}$.

    c. The set D of dates in the month of January.

    d. The set of even integers EVEN $= \{..., -4, -2, 0, 2, 4, ...\}$.

    e. $\{1, 3, 5, 7, 9, 11, 13, 15\}$.

2. Let $A = \{a, \varnothing\}$. Answer true or false for each of the following statements.

    a. $a \in A$.   b. $\{a\} \in A$.   c. $a \subset A$.   d. $\{a\} \subset A$.

    e. $\varnothing \subset A$.   f. $\varnothing \in A$.   g. $\{\varnothing\} \subset A$.   h. $\{\varnothing\} \in A$.

3. Write down a simpler description of the set $\{a, 4, x, a, 3, b, 4, a, c, b, d\}$.

4. Show that $\varnothing \subset A$ for every set A.

5. Find two finite sets A and B such that $A \in B$ and $A \subset B$.

6. Write down the power set for each of the following sets.
   a. $\{x, y, z, w\}$.   b. $\{a, \{a, b\}\}$.   c. $\varnothing$.   d. $\{\varnothing\}$.   e. $\{\{a\}, \varnothing\}$.

7. For each collection of sets, find the smallest set A such that the collection is a subset of power(A).
   a. $\{\{a\}, \{b, c\}\}$.   b. $\{\{a\}, \{\varnothing\}\}$.   c. $\{\{a\}, \{\{a\}\}\}$.   d. $\{\{a\}, \{\{b\}\}, \{a, b\}\}$.

8. Suppose A and B are sets defined as follows:
$$A = \{x \mid x = 4k + 1 \text{ and } k \in \mathbb{N}\},$$
$$B = \{x \mid x = 3k + 5 \text{ and } k \in \mathbb{N}\}.$$
   a. List the first ten elements of $A \cup B$.
   b. List the first four elements of $A \cap B$.

9. Prove each of the following facts about the union operation (1.8). Use subset arguments that are written in complete sentences.
   a. $A \cup \varnothing = A$.
   b. $A \cup B = B \cup A$.
   c. $A \cup A = A$.
   d. $A \cup (B \cup C) = (A \cup B) \cup C$.

10. Prove each of the following facts about the intersection operation (1.11). Use subset arguments that are written in complete sentences.
    a. $A \cap \varnothing = \varnothing$.
    b. $A \cap B = B \cap A$.
    c. $A \cap (B \cap C) = (A \cap B) \cap C$
    d. $A \cap A = A$.
    e. $A \subset B$ if and only if $A \cap B = A$.

11. Use subset arguments to show that power($A \cap B$) = power(A) $\cap$ power(B).

12. Is power($A \cup B$) = power(A) $\cup$ power(B)?

13. Prove the distribution result: $A \cup (B \cap C) = (A \cup B) \cap (A \cup C)$.

14. Prove each of the following statements twice. The first proof should use subset arguments. The second proof should use an already known result. These equations are examples of *absorption laws*.

a. $A \cap (B \cup A) = A$.

b. $A \cup (B \cap A) = A$.

15. Show that $(A \cap B) \cup C = A \cap (B \cup C)$ if and only if $C \subset A$.

16. Give a proof or a counterexample for each of the following statements.

a. $A \cap (B \cup A) = A \cap B$?

b. $A - (B \cap A) = A - B$?

c. $A \cap (B \cup C) = (A \cup B) \cap (A \cup C)$?

d. $A \oplus A = A$?

17. For each integer i, define A_i as follows:

$$\text{If } i \text{ is even then } A_i = \{x \mid x \in \mathbb{Z} \text{ and } x < -i \text{ or } i < x\}.$$

$$\text{If } i \text{ is odd then } A_i = \{x \mid x \in \mathbb{Z} \text{ and } -i < x < i\}.$$

a. Describe each of the sets A_0, A_1, A_2, A_3, A_{-2}, and A_{-3}.

b. Find the union of the collection $\{A_i \mid i \in \{1, 3, 5, 7, 9\}\}$.

c. Find the union of the collection $\{A_i \mid i \text{ is even}\}$.

d. Find the union of the collection $\{A_i \mid i \text{ is odd}\}$.

e. Find the union of the collection $\{A_i \mid i \in \mathbb{N}\}$.

f. Find the intersection of the collection $\{A_i \mid i \in \{1, 3, 5, 7, 9\}\}$.

g. Find the intersection of the collection $\{A_i \mid i \text{ is even}\}$.

h. Find the intersection of the collection $\{A_i \mid i \text{ is odd}\}$.

i. Find the intersection of the collection $\{A_i \mid i \in \mathbb{N}\}$.

18. For each natural number n, let A_n be defined by

$$A_n = \{x \mid x \in \mathbb{N} \text{ and } x \text{ divides } n \text{ with no remainder}\}.$$

a. Describe each of the sets A_0, A_1, A_2, A_3, A_4, A_5, A_6, A_7, and A_{100}.

b. Find the union of the collection $\{A_n \mid n \in \{1, 2, 3, 4, 5, 6, 7\}\}$.

c. Find the intersection of the collection $\{A_n \mid n \in \{1, 2, 3, 4, 5, 6, 7\}\}$.

d. Find the union of the collection $\{A_n \mid n \in \mathbb{N}\}$.

e. Find the intersection of the collection $\{A_n \mid n \in \mathbb{N}\}$.

19. For each of the following expressions, use a Venn diagram representing a universe U and two subsets A and B. Shade the part of the diagram that corresponds to the given set.

a. A'.

b. B'.

c. $(A \cup B)'$.

d. $A' \cap B'$.

e. $A' \cup B'$.

f. $(A \cap B)'$.

20. Each Venn diagram in Figure 1.8 represents a set whose regions are indicated by the letter x. Write each of the three sets in terms of set operations. Try to simplify your answers to as few symbols as you can.

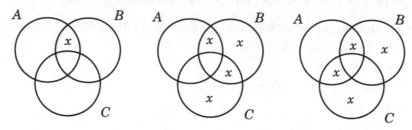

Figure 1.8

21. Prove each of the following properties of the complement (1.16).
    a. $(A')' = A$.   b. $\varnothing' = U$ and $U' = \varnothing$.
    c. $A \cap A' = \varnothing$ and $A \cup A' = U$.
    d. $(A \cup B)' = A' \cap B'$ and $(A \cap B)' = A' \cup B'$.

22. Discover an inclusion exclusion formula for the number of elements in the union of four sets A, B, C, and D.

23. Given three sets A, B, and C. Suppose the union of the three sets has cardinality 280. Suppose also that $|A| = 100$, $|B| = 200$, and $|C| = 150$. And suppose we also know $|A \cap B| = 50$, $|A \cap C| = 80$, and $|B \cap C| = 90$. Find the cardinality of the intersection of the three given sets.

24. Suppose A, B, and C represent three bus routes through a suburb of your favorite city. Let A, B, and C also be sets whose elements are the bus stops for the corresponding bus route. Suppose A has 25 stops, B has 30 stops, and C has 40 stops. Suppose further that A and B share (have in common) 6 stops, A and C share 5 stops, and B and C share 4 stops. Lastly, suppose that A, B, and C share 2 stops. Answer each of the following questions.

    a. How many distinct stops are on the three bus routes?
    b. How many stops for A are not stops for B?
    c. How many stops for A are not stops for both B and C?
    d. How many stops for A are not stops for any other bus?

25. Suppose a highway survey crew noticed the following information about 500 vehicles: In 100 vehicles the driver was smoking, in 200 vehicles the driver was talking to a passenger, and in 300 vehicles the driver was tuning the radio. Further, in 50 vehicles the driver was smoking and talking, in 40 vehicles the driver was smoking and tuning the radio, and

in 30 vehicles the driver was talking and tuning the radio. What can you say about the number of drivers who were smoking, talking, and tuning the radio?

26. Suppose the following people went to a summer camp: 27 boys, 15 city children, 27 men, 21 noncity boys, 42 people from the city, 18 city males, and 21 noncity females. How many people went to summer camp?

27. Find the union and intersection of each of the following pairs of bags.

    a. $[x, y]$ & $[x, y, z]$.   b. $[x, y, x]$ & $[y, x, y, x]$.

    c. $[a, a, a, b]$ & $[a, a, b, b, c]$.   d. $[1, 2, 2, 3, 3, 4, 4]$ & $[2, 3, 3, 4, 5]$.

    e. $[x, x, [a, a], [a, a]]$ & $[a, a, x, x]$.

    f. $[a, a, [b, b], [a, [b]]]$ & $[a, a, [b], [b]]$.

28. Find a bag B that solves the following two simultaneous bag equations.
$$B \cup [2, 2, 3, 4] = [2, 2, 3, 3, 4, 4, 5]$$
$$B \cap [2, 2, 3, 4, 5] = [2, 3, 4, 5].$$

29. How would you define the difference operation for bags? Try to make your definition agree with the difference operation for sets whenever the bags are like sets (without repeated elements).

30. Find a description of a set A satisfying the equation $A = \{a, A, b\}$. Notice in this case that $A \in A$.

31. Let C denote a collection of sets.

    a. Does the union of the sets in C make sense when C is empty?

    b. Does the intersection of the sets in C make sense when C is empty?

1.3 Other Structures

Now that we have a working knowledge of sets, let's introduce some other fundamental structures for representing things.

Tuples

When we write down a sentence, it always has a sequential nature. For example, in the previous sentence the word "When" is the first word, the word "we" is the second word, and so on. We often need to work with structures that have a first element, a second element, and so on. Informally, a *tuple* is a collection of things, called its *elements,* where there is a first element, a second element, and so on. The elements of a tuple are also called *members, objects,* or *components.* We'll denote a tuple by writing down its elements, separated by commas, and surrounding everything with the two symbols "(" and

">." For example, the tuple $\langle 12, R, 9 \rangle$ has three elements. The first element is 12, the second element is the letter R, and third element is 9. The beginning sentence of this paragraph can be represented by the following tuple:

$$\langle \text{When, we, write, down, ..., sequential, nature} \rangle.$$

If a tuple has n elements, we say that its *length* is n, and we call it an *n-tuple*. So the tuple $\langle 8, k, \text{hello} \rangle$ is a 3-tuple, and $\langle x_1, ..., x_8 \rangle$ is an 8-tuple. The 0-tuple is denoted by $\langle \ \rangle$, and we call it the *empty* tuple. A 2-tuple is often called an *ordered pair*, and a 3-tuple might be called an *ordered triple*. Other words used in place of the word *tuple* are *vector* and *sequence*, possibly modified by the word *ordered*. Some notations for a tuple use the two parentheses "(" and ")" in place of "$\langle$" and "$\rangle$."

Two n-tuples $\langle x_1, ..., x_n \rangle$ and $\langle y_1, ..., y_n \rangle$ are said to be *equal* if $x_i = y_i$ for $1 \le i \le n$, and we denote this by $\langle x_1, ..., x_n \rangle = \langle y_1, ..., y_n \rangle$. Thus the ordered pairs $\langle 3, 7 \rangle$ and $\langle 7, 3 \rangle$ are not equal, and we write $\langle 3, 7 \rangle \neq \langle 7, 3 \rangle$. Since tuples convey the idea of order, they are different from sets and bags. Here are some examples:

Sets: $\{h, u, g, h\} = \{h, u, g\} = \{u, g, h\}$.

Bags: $[h, u, g, h] = [h, h, g, u] \neq [h, u, g] = [u, g, h]$.

Tuples: $\langle h, u, g, h \rangle \neq \langle h, h, g, u \rangle$ and $\langle h, u, g \rangle \neq \langle u, g, h \rangle$.

The two important characteristics of a tuple are

There is an order or arrangement of the elements.

There may be redundant occurrences of elements.

Even though the meanings of tuple and set are different, we'll show in the following example that tuples can be defined in terms of sets. Feel free to skip the example if it's confusing.

EXAMPLE 1 (*Tuples Are Sets*). Can we define the idea of order in terms of sets? For example, can we find some way to define ordered pairs as special sets so that two ordered pairs $\langle a, b \rangle$ and $\langle c, d \rangle$ are equal if and only if $a = c$ and $b = d$? Let's try it. We start with the 0-tuple $\langle \ \rangle$ and define it to be the empty set $\varnothing$. If x is an object, then the 1-tuple $\langle x \rangle$ is defined as the singleton set $\{x\}$. Now the interesting part begins because we have to capture the idea of order with 2-tuples. An ordered pair of objects $\langle x, y \rangle$ can be defined as the following set:

$$\langle x, y \rangle = \{\{x\}, \{x, y\}\}. \tag{1.22}$$

Notice with this definition that $\langle a, b \rangle = \{\{a\}, \{a, b\}\}$ and $\langle b, a \rangle = \{\{b\}, \{b, a\}\}$. Using equality of sets, it's easy to see that two ordered pairs $\langle a, b \rangle$ and $\langle c, d \rangle$ are equal if and only if $a = c$ and $b = d$. Can we find a way to represent a 3-tuple $\langle x, y, z \rangle$ as a set? Let's proceed by letting S be the set representing the ordered pair $\langle x, y \rangle$. Then define $\langle x, y, z \rangle$ as the set

$$\langle x, y, z \rangle = \{\{S\}, \{S, z\}\}. \tag{1.23}$$

If we replace S by its value $\{\{x\}, \{x, y\}\}$, we obtain the terrible-looking result

$$\begin{aligned} \langle x, y, z \rangle &= \{\{S\}, \{S, z\}\} \\ &= \{\{\{\{x\}, \{x, y\}\}\}, \{\{\{x\}, \{x, y\}\}, z\}\}. \end{aligned}$$

We can use equality of sets to show that two 3-tuples $\langle a, b, c \rangle$ and $\langle d, e, f \rangle$ are equal if and only if $a = d$, $b = e$, and $c = f$. We'll leave the details as an exercise. We could continue in the manner we have been going and define n-tuples as sets for any natural number n. Although defining a tuple as a set is not at all intuitive, it does illustrate how sets can be used as a foundation from which to build objects and ideas. It also shows why good notation is so important for communicating ideas.   ◆

Products of Sets

We often need to represent information in the form of tuples, in which the elements in each tuple come from known sets. If A and B are sets, then the *product* of A and B is the set of all 2-tuples with first components from A and second components from B. The product is denoted by $A \times B$. So we can write the definition as follows:

$$A \times B = \{\langle a, b \rangle \mid a \in A \text{ and } b \in B\}.$$

For example, if $A = \{x, y\}$ and $B = \{0, 1\}$, then we have

$$A \times B = \{\langle x, 0 \rangle, \langle x, 1 \rangle, \langle y, 0 \rangle, \langle y, 1 \rangle\}.$$

Suppose we let $A = \varnothing$ and $B = \{0, 1\}$ and then ask the question "What is $A \times B$?" If we apply the definition of product, we must conclude that there aren't any 2-tuples with first elements from the empty set. Therefore $A \times B = \varnothing$. So it's easy to generalize and say that $A \times B$ is nonempty if and only if both A and B are nonempty sets. The product is sometimes called the *cross product* or the *Cartesian product*—after the mathematician René Descartes (1596–1650), who introduced the idea of graphing ordered pairs. The product of two sets is easily extended to any number of sets $A_1, ..., A_n$ by writing

$$A_1 \times \cdots \times A_n = \{\langle x_1, ..., x_n \rangle \mid x_i \in A_i\}.$$

If all the sets A_i in a product are the same set A, then we use the abbreviated notation $A^n = A \times \cdots \times A$. With this notation we have the following definitions for the sets A^1 and A^0:

$$A^1 = \{\langle a \rangle \mid a \in A\} \quad \text{and} \quad A^0 = \{\langle \, \rangle\}.$$

So we must conclude that $A^1 \neq A$ and $A^0 \neq \emptyset$.

EXAMPLE 2. Let $A = \{a, b, c\}$. Then we have the following products:

$A^0 = \{\langle \, \rangle\}$,

$A^1 = \{\langle a \rangle, \langle b \rangle, \langle c \rangle\}$,

$A^2 = \{\langle a, a \rangle, \langle a, b \rangle, \langle a, c \rangle, \langle b, a \rangle, \langle b, b \rangle, \langle b, c \rangle, \langle c, a \rangle, \langle c, b \rangle, \langle c, c \rangle\}$,

A^3 is bigger yet, with 27 3-tuples.   ♦

When working with tuples, we need the ability to randomly access any component. The components of an n-tuple can be indexed in several different ways depending on the problem at hand. For example, if $t \in A \times B \times C$, then we might represent t in any of the following ways:

$$\langle t_1, t_2, t_3 \rangle,$$
$$\langle t(1), t(2), t(3) \rangle,$$
$$\langle t[1], t[2], t[3] \rangle,$$
$$\langle t(A), t(B), t(C) \rangle,$$
$$\langle A(t), B(t), C(t) \rangle.$$

Let's look at an example that shows how products and tuples are related to some familiar objects of programming.

EXAMPLE 3 (*Arrays, Matrices, and Records*). In computer science, a 1-dimensional array of size n with elements in the set A can be represented by an n-tuple in the product A^n. So we can think of the product A^n as the set of all 1-dimensional arrays of size n over A. If $x = \langle x_1, ..., x_n \rangle$, then the component x_i is usually denoted—in programming languages—by $x[i]$.

A 2-dimensional array—also called a *matrix*—can be thought of as a table of objects that are indexed by rows and columns. If x is a matrix with m rows and n columns, we say that x is an m by n matrix. For example, if x is a 3 by 4 matrix, then x can be represented by the following diagram:

$$x = \begin{bmatrix} x_{11} & x_{12} & x_{13} & x_{14} \\ x_{21} & x_{22} & x_{23} & x_{24} \\ x_{31} & x_{32} & x_{33} & x_{34} \end{bmatrix}.$$

We can also represent x as a 3-tuple whose components are 4-tuples as follows:

$$x = \langle\langle x_{11}, x_{12}, x_{13}, x_{14}\rangle, \langle x_{21}, x_{22}, x_{23}, x_{24}\rangle, \langle x_{31}, x_{32}, x_{33}, x_{34}\rangle\rangle.$$

In programming, the component x_{ij} is usually denoted by $x[i, j]$. We can think of the product $(A^4)^3$ as the set of all 2-dimensional arrays over A with 3 rows and 4 columns. Of course, this idea extends to higher dimensions. For example, the product $((A^5)^7)^4$ represents the set of all 3-dimensional arrays over A consisting of 4-tuples whose components are 7-tuples whose components are 5-tuples of elements of A.

For another example we can think of the product $A \times B$ as the set of all records, or structures, with two fields A and B. For a record $r = \langle a, b\rangle \in A \times B$ the components a and b are normally denoted by $r.A$ and $r.B$.   ◆

There are at least three nice things about tuples: They are easy to understand; they are basic building blocks for the representation of information; and they are easily implemented by a computer. In the remainder of the section we'll introduce several important structures that are used to represent information, and we'll see how these structures can be represented as tuples.

Lists

A *list* is a sequence of zero or more elements that can be redundant and that is ordered. In other words, a list is just like a tuple. In fact, we'll use tuple notation to represent lists. The *empty list* is $\langle \ \rangle$, and the number of elements in a list is called its *length*.

So what's the difference between tuples and lists? The difference is in what parts can be randomly accessed. In the case of tuples we can randomly access any component. In the case of lists we can randomly access only two things: the first component of a list, which is called its *head*, and the list made up of everything except the first component, which is called its *tail*. For example, given the list $\langle x, y, z\rangle$, its head is x, and its tail is the list $\langle y, z\rangle$.

An important property of lists is the ability to easily construct a new list from an element and another list. For example, given the element x and the list $\langle y, z \rangle$, we can easily construct the new list $\langle x, y, z \rangle$. This construction process can be done efficiently and dynamically during the execution of a program. In Chapter 3 we'll discuss the access and construction of lists more thoroughly.

We often need to work with lists whose elements are from a single set A. A *list over the set A* is a finite sequence of elements from A. We'll denote the collection of all lists over A by Lists[A]. For example, if $A = \{a, b, c\}$, then three of the lists in Lists[A] are $\langle\,\rangle$, $\langle a, a, b \rangle$, and $\langle b, c, a, b, c \rangle$. If we forget that lists and tuples are accessed in different ways, then we can demonstrate an interesting relationship between the two ideas by writing the set Lists[A] as the union of the products $A^0, A^1, A^2, \ldots$. In other words, we have the following equation:

$$\text{Lists}[A] = A^0 \cup A^1 \cup \ldots \cup A^n \cup \ldots . \tag{1.24}$$

Suppose we want to allow lists to be elements of other lists. A *generalized list* over A is either a regular list over A or a list whose elements either are from A or are generalized lists over A. We'll denote the collection of all generalized lists over A by GenLists[A]. We can certainly say that Lists[A] $\subset$ GenLists[A].

EXAMPLE 4. If $A = \{a, b, c\}$, then the following lists are some examples of generalized lists over A:

$$\langle\,\rangle,$$
$$\langle a, b, b, a \rangle,$$
$$\langle a, \langle b \rangle \rangle,$$
$$\langle \langle a, \langle\,\rangle \rangle, b, \langle a, a \rangle \rangle. \quad \blacklozenge$$

If A is a finite set, then it's possible to give a systematic description of the general lists over A. We'll base our listing on the number of symbols used to express a list. For each list we'll count the number of occurrences of elements from A together with the number of occurrences of the symbols "$\langle$" and "$\rangle$." We won't count commas. We'll start by writing down the list $\langle\,\rangle$, which has two symbols. Next we'll write down all the lists with three symbols. These lists must be of the form $\langle a \rangle$ for each $a \in A$. What about the lists with four symbols? If $A = \{a, b\}$, then there are five lists over A with four symbols as follows:

$$\langle \langle\,\rangle \rangle, \quad \langle a, a \rangle, \quad \langle a, b \rangle, \quad \langle b, a \rangle, \quad \text{and} \quad \langle b, b \rangle.$$

We'll leave it as an exercise to write down the general lists consisting of five and six symbols.

The word *stream* (or infinite list, or infinite sequence) is often used in computer science to describe an infinite list of objects. We'll use the tuple notation to represent streams. For example, a program to compute the decimal expansion of pi would compute the following stream of integers:

$$\langle 3, 1, 4, 1, 5, 9, 2, 6, 5, 3, 5, 8, 9, 7, 9, 3, ... \rangle.$$

Streams are useful in programming as inputs and outputs to computations. They normally have the same access and construction properties as lists. In other words, we can randomly access the first element and the stream consisting of everything except the first element. Similarly, if we are given an element and a stream, then we can construct a new stream.

Strings

A *string* is a finite sequence of zero or more elements that are placed next to each other in juxtaposition. The individual elements that make up a string are taken from a finite set called an *alphabet*. For example, the set {a, b, c} is an alphabet for the string

$$aacabb.$$

The string with no elements is called the *empty string*, and we denote it by the Greek letter lambda

$$\Lambda.$$

The number of elements that occur in a string is called the *length* of the string. For example, over the alphabet {a, b, c}, the string *aacabb* has length 6. We sometimes denote the length of a string s by

$$|s|.$$

Strings are used in the world of written communication to represent information: computer programs; written text in all the languages of the world; and formal notation for logic, mathematics, and the sciences.

There is a strong association between strings and lists because both are defined as finite sequences of elements. This association is important in computer science because computer programs must be able to recognize certain kinds of strings. This means that a string must be decomposed into its individual elements, which can then be represented by a list. For example, the string *aacabb* can be represented by the list $\langle a, a, c, a, b, b \rangle$. Similarly, the empty string Λ corresponds to the empty list $\langle \, \rangle$.

If A is an alphabet, then the set of all strings over A is denoted by A^*. In other words, A^* is the set of all possible strings made up from the elements of A. For example, if $A = \{a\}$, then we have

$$A^* = \{\Lambda, a, aa, aaa, \ldots\}.$$

For a natural number n and a string w we often let w^n denote the string of n w's. For example, we have

$$w^0 = \Lambda, \quad w^1 = w, \quad w^2 = ww, \quad \text{and} \quad w^3 = www.$$

The exponent notation allows us to represent some sets of strings in a nice concise manner. For example, if $A = \{a\}$, then we can write

$$A^* = \{a^n \mid n \in \mathbb{N}\}.$$

We should note that if the empty string occurs as part of another string, then it does not contribute anything new to the string. In other words, for any string w, we have

$$w\Lambda = \Lambda w = w.$$

EXAMPLE 5. If $A = \{a, b\}$, then A^* can be described by writing down a few strings of small length followed by an ellipsis. For example, in the following description of A^* we've explicitly listed the strings of length 0, 1, 2, and 3.

$$A^* = \{\Lambda, a, b, aa, ab, ba, bb, aaa, aab, aba, abb, baa, bab, bba, bbb, \ldots\}.$$

Some subsets of A^* can be represented concisely by using exponents. For example, here are three subsets of A^*:

$$\{ab^n \mid n \in \mathbb{N}\} = \{a, ab, abb, abbb, \ldots\}$$

$$\{a^n b^n \mid n \in \mathbb{N}\} = \{\Lambda, ab, aabb, aaabbb, \ldots\}$$

$$\{(ab)^n \mid n \in \mathbb{N}\} = \{\Lambda, ab, abab, ababab, \ldots\}. \quad \blacklozenge$$

EXAMPLE 6 (*Numerals*). A *numeral* is a written number. In terms of strings, we can say that a numeral is a nonempty string of symbols that represents a number. Most of us are familiar with the following three numeral systems. The *Roman numerals* represent the nonnegative integers by using the alphabet

$$\{I, V, X, L, C, D, M\}.$$

The *decimal numerals* represent the natural numbers by using the alphabet

$$\{0, 1, 2, 3, 4, 5, 6, 7, 8, 9\}.$$

The *binary numerals* represent the natural numbers by using the alphabet

$$\{0, 1\}.$$

For example, the Roman numeral MDCLXVI, the decimal numeral 1666, and the binary numeral 11010000010 all represent the same number.   ♦

EXAMPLE 7 (*Real Numbers*). Sometimes it makes sense to represent things with infinite strings. For example, it's natural to represent the number pi as an infinite string of decimal digits containing a decimal point:

$$\pi = 3 . 1 4 1 5 9 2 6 5 3 5 8 9 7 9 3 \ldots .$$

Any integer or rational number can also be represented this way. For example, $7 = 7.00\ldots$ and $\frac{1}{3} = 0.333\ldots$. Notice that we can have different representations of the same number with this scheme. For example, $0.333\ldots$ and $00.333\ldots$ both represent the same number $\frac{1}{3}$. For a more interesting example, recall (or notice for the first time) that the number 1 can be represented by $1.000\ldots$ and by $0.999\ldots$. Similarly, the strings $45.89000\ldots$ and $45.88999\ldots$ both represent the same number. To see this, just use the school method of changing a repeating decimal into a fraction. For example, if we let $x = 45.88999\ldots$, then $1000x = 45889.999\ldots$ and $100x = 4588.999\ldots$. Subtracting, we obtain $900x = 41301$. So $x = \frac{41301}{900}$, which in lowest terms is $\frac{4589}{100}$. This is exactly the fraction represented by $45.89000\ldots$.

So if we want to represent the real numbers as a set in which each infinite string represents a distinct number, then we have to exclude some representations. For example, we might require at least one digit to the left of the decimal point but no other leading 0's. In this case, $0.597\ldots$ is OK, and $00.587\ldots$ is not. We might also require that infinite sequences of 9's at the end of a string are not allowed. So 1.0 is OK, and $0.999\ldots$ is not.   ♦

Relations

Ideas such as kinship, connection, and association of objects are key to the concept of a relation. Informally, a *relation* is a set of tuples in which the elements of each tuple are related in some way.

For example, suppose we let "Family" be the relation whose tuples have the form

$$\langle \text{father, mother, child1, child2, ... } \rangle.$$

If we want to indicate that James and Janice are the parents of Gretchen and Andrew, we would write

$$\langle \text{James, Janice, Gretchen, Andrew} \rangle \in \text{Family}.$$

As another example, let "Less" be the relation defined as follows:

$$\text{Less} = \{\langle 1, 2\rangle, \langle 1, 3\rangle, \langle 2, 3\rangle\}.$$

Notice that each tuple $\langle x, y \rangle \in$ Less satisfies the property "x is less than y."

For another example, suppose we let P be the set of all Pythagorean triples. In other words, P is the set of 3-tuples $\langle x, y, z \rangle$ of real numbers satisfying the property $x^2 + y^2 = z^2$. For example, we have $\langle 3, 4, 5 \rangle \in P$ and $\langle 6, 8, 10 \rangle \in P$.

If R is a relation, then the statement $\langle x_1, ..., x_n \rangle \in R$ is also denoted by writing the prefix form $R(x_1, ..., x_n)$. For example, we could write

$$\text{Family(James, Janice, Gretchen, Andrew)},\quad \text{Less}(1, 3),\quad \text{and}\quad P(3, 4, 5).$$

It's normal practice to explicitly state where the tuples in a relation come from. For example, we have the following subsets:

$$\text{Family} \subset \text{Lists[People]}$$
$$\text{Less} \subset \mathbb{N} \times \mathbb{N}$$
$$P \subset \mathbb{R} \times \mathbb{R} \times \mathbb{R}.$$

The formal definition of a relation—which we'll give shortly—states that a relation is a subset of some product set. But how can Family be considered a subset of a product set? With a little thought it seems that there are just three parts to a family, where the third part can be considered a tuple of children. For example, the 3-tuple

$$\langle \text{James, Janice, } \langle \text{Gretchen, Andrew} \rangle \rangle$$

represents a family with this new representation. Using this definition for Family, we can see that

$$\text{Family} \subset \text{People} \times \text{People} \times \text{Lists[People]}.$$

If we represented the children in as a set, then we could represent Family as

$$\text{Family} \subset \text{People} \times \text{People} \times \text{power(People)}.$$

A relation R may also be defined by giving a property that each tuple satisfies. When this is the case, a more descriptive notation can be given for R. For example, suppose we want to represent a collection of parts in an inventory. Then we might use the names of the properties (*attributes*) of each part as the indices for a tuple. For example, we'll assume that each part has five attributes:

$$\text{PartNumber, Price, Cost, DateBought, NumberOnHand.}$$

Then a part can be represented by a 5-tuple. For example, some part might have the representation

$$\langle 1131, \$49.95, \$29.95, 3/12/88, 25 \rangle.$$

If we let "Parts" be the relation whose elements are these 5-tuples, then

$$\text{Parts} \subset \text{PartNumber} \times \text{Price} \times \text{Cost} \times \text{DateBought} \times \text{NumberOnHand.}$$

If $t \in$ Parts, then we can refer to the element of t with attribute A by $t(A)$. For example, if

$$t = \langle 1131, \$49.95, \$29.95, 3/12/88, 25 \rangle,$$

then we have

$$t(\text{PartNumber}) = 1131$$
$$t(\text{Price}) = \$49.95$$
$$t(\text{Cost}) = \$29.95$$
$$t(\text{DateBought}) = 3/12/88$$
$$t(\text{NumberOnHand}) = 25.$$

We could also refer to the element of t with attribute A by $A(t)$. With the above example we write

$$\text{PartNumber}(t) = 1131$$

$$\text{Price}(t) = \$49.95$$

$$\text{Cost}(t) = \$29.95$$

$$\text{DateBought}(t) = 3/12/88$$

$$\text{NumberOnHand}(t) = 25.$$

We often use the term *n-ary relation* when referring to a relation R whose tuples are elements of a product set $A_1 \times \cdots \times A_n$. So an n-ary relation R over the product set $A_1 \times \cdots \times A_n$ is just a subset of $A_1 \times \cdots \times A_n$. Thus there are lots of n-ary relations over $A_1 \times \cdots \times A_n$, ranging from the smallest subset $\varnothing$, called the *empty relation*, to the largest subset $A_1 \times \cdots \times A_n$ itself, called the *universal relation*. The terms *unary*, *binary*, and *ternary* are often used instead of 1-ary, 2-ary, and 3-ary. If R is a subset of the product A^n, then R is called an *n-ary relation on A*.

If R is a binary relation and $\langle x, y \rangle \in R$, we often denote this fact by writing the following *infix expression*—the name appears between its two arguments:

$$x \, R \, y.$$

For example, we write $1 < 2$ instead of $\langle 1, 2 \rangle \in <$ and $<(1, 2)$. For another example the divides relation, $|$, is a binary relation on $\mathbb{Z}$, and we write $3 \,|\, 24$ rather than $\langle 3, 24 \rangle \in |$ or $|(3, 24)$. Suppose we define the "isParentOf" relation on People by letting isParentOf(a, b) mean "a is a parent of b." If Lloyd is a parent of Jim, then we can write "Lloyd isParentOf Jim."

An important example of a binary relation is the *equality relation* on a set. For example, if $A = \{a, b, c\}$, then the equality relation on A, which of course is denoted by the symbol $=$, is the set $\{\langle a, a \rangle, \langle b, b \rangle, \langle c, c \rangle\}$. In this case we normally write $a = a$ instead of $=(a, a)$. Unary relations are of less interest because there are no elements to related. However, a unary relation is similar to a test for membership in a set. To see this, suppose R is a unary relation over the set A. Then R is just a set of 1-tuples of elements from A. So we can write

$$R = \{\langle x \rangle \mid x \in A \text{ and } x \text{ satisfies some property}\}.$$

If we replace each tuple $\langle x \rangle$ in R with x, then we obtain a subset B of A, where $B = \{x \mid \langle x \rangle \in R\}$. So the statements $R(x)$ and $x \in B$ are equivalent, meaning that they are either both true or both false.

Graphs

Informally, a *graph* is a set of objects in which some of the objects are connected to each other in some way. The objects are called *vertices* or *nodes*, and the connections are called *edges*. For example, the United States can be represented by a graph where the vertices are states and the edges are the common borders between adjacent states. In this case, Hawaii and Alaska would be vertices without any edges connected to them. We say that two vertices are *adjacent* if there is an edge connecting them.

We can picture a graph in several ways. For example, Figure 1.9 shows two ways to represent the graph with vertices 1, 2, and 3 and edges connecting 1 to 2 and 1 to 3.

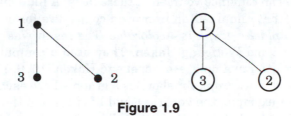

Figure 1.9

Let's do another example. Figure 1.10 represents a graph of those states in the United States and those provinces in Canada that touch the Pacific Ocean or that touch states and provinces that touch the Pacific Ocean.

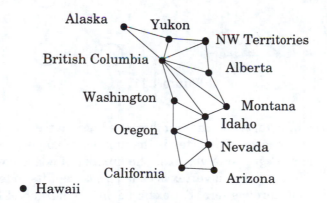

Figure 1.10

An interesting problem dealing with maps is to try to color a map with the fewest number of colors subject to the restriction that any two adjacent areas must have distinct colors. From a graph point of view, this means that any two distinct adjacent vertices must have different colors. Before reading any further, try to color the graph in Figure 1.10 with the fewest colors. It's usually easier to represent the colors by numbers like 1, 2, ...

A graph is *n-colorable* if there is an assignment of *n* colors to its vertices such that any two distinct adjacent vertices have distinct colors. The *chromatic number* of a graph is the smallest *n* for which it is *n*-colorable. For example, the chromatic number of the graph in Figure 1.10 is 3. A graph whose edges are the connections between all pairs of distinct vertices is called a *complete graph*. It's easy to see that the chromatic number of a complete graph with *n* vertices is *n*.

A graph is *planar* if it can be drawn on a plane such that no edges intersect. For example, the graph in Figure 1.10 is planar. A complete graph with four vertices is planar, but a complete graph with five vertices is not planar. See whether you can convince yourself of these facts. A fundamental result on graph coloring—that remained an unproven conjecture for over 100 years—states that *every planar graph is 4-colorable*. The result was proven in 1976 by Kenneth Appel and Wolfgang Haken. They used a computer to test over 1900 special cases. For example, see Appel and Haken [1976, 1977].

A *directed graph* (*digraph* for short) is a graph where each edge points in one direction. For example, the vertices could be cities and the edges could be the one-way air routes between them. For digraphs we use arrows to denote the edges. For example, Figure 1.11 shows two ways to represent the digraph with three vertices *a*, *b*, and *c* and edges from *a* to *b*, *c* to *a*, and *c* to *b*.

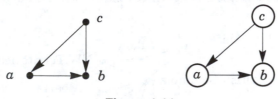

Figure 1.11

The *degree* of a vertex is the number of edges that it touches. For directed graphs the *indegree* of a vertex is the number of edges pointing at the vertex, while the *outdegree* of a vertex is the number of edges pointing away from the vertex. In a digraph a vertex is called a *source* if its indegree is zero and a *sink* if its outdegree is zero. For example, in the digraph of Figure 1.11, *c* is a source and *b* is a sink.

If a graph has more than one edge between some pair of vertices, the graph is called a *multigraph*, or a *directed multigraph* in case the edges point in the same direction. For example, there are usually two or more road routes between most cities. So a graph representing road routes between a set of cites is most likely a multigraph.

From a computational point of view, we need to represent graphs as data. This is easy to do because we can define a graph in terms of tuples, sets, and bags. For example, we can define a graph G as an ordered pair $\langle V, E \rangle$, where V is a set of vertices and E is a set or bag of edges. If G is a digraph, then the edges in E can be represented by ordered pairs, where $\langle a, b \rangle$ represents the edge with an arrow from a to b. In this case the set E of edges is a subset of $V \times V$. In other words, E is a binary relation on V. For example, the digraph in Figure 1.11 has vertex set $\{a, b, c\}$ and edge set

$$\{\langle a, b \rangle, \langle c, b \rangle, \langle c, a \rangle\}.$$

If G is a directed multigraph, then we can represent the edges as a bag (or multiset) of ordered pairs. For example, the bag $[\langle a, b \rangle, \langle a, b \rangle, \langle b, a \rangle]$ represents three edges: two from a to b and one from b to a.

If a graph is not directed, we have more ways to represent the edges. We could still represent an edge as an ordered pair $\langle a, b \rangle$ and agree that it represents an undirected line between a and b. But we can also represent an edge between vertices a and b by a set $\{a, b\}$. For example, the graph in Figure 1.9 has vertex set $\{1, 2, 3\}$ and edge set $\{\{1, 2\}, \{1, 3\}\}$.

We often encounter graphs that have information attached to each edge. For example, a good road map places distances along the roads between major intersections. A graph is called *weighted* if each edge is assigned a number, called a *weight*. We can represent an edge $\langle a, b \rangle$ that has weight w by the 3-tuple $\langle a, b, w \rangle$. In some cases we might want to represent an unweighted graph as a weighted graph. For example, if we have a multigraph in which we wish to distinguish between multiple edges that occur between two vertices, then we can assign a different weight to each edge, thereby creating a weighted multigraph.

We can observe from our discussion of graphs that any binary relation R on a set A can be thought of as a digraph $G = \langle A, R \rangle$ with vertices A and edges R. For example, let $A = \{1, 2, 3\}$ and

$$R = \{\langle 1, 2 \rangle, \langle 1, 3 \rangle, \langle 2, 3 \rangle, \langle 3, 3 \rangle\}.$$

Figure 1.12 shows the digraph corresponding to this binary relation.

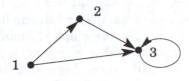

Figure 1.12

Representing a binary relation as a graph is often quite useful in trying to establish properties of the relation. We'll see this when we discuss properties of binary relations in Chapter 4.

Sometimes it's useful to talk about portions of a graph. A *subgraph* of a graph $\langle V, E \rangle$ is any graph of the form $\langle V', E' \rangle$ where $V' \subset V$ and $E' \subset E$. For example, the four graphs in Figure 1.13 are subgraphs of the graph in Figure 1.12.

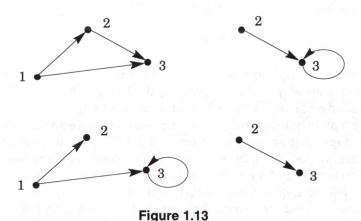

Figure 1.13

Walks, Trails, and Paths

Some graph problems involve "walking" from one vertex to another by moving along a sequence of edges, where each edge shares a vertex with the next edge in the sequence. In formal terms, a *walk* from x_0 to x_n is a sequence of edges that we denote by a sequence of vertices $x_0, x_1, ..., x_n$ such that there is an edge from x_{i-1} to x_i for $1 \leq i \leq n$. A walk allows the possibility that some edge or some vertex occurs more than once. For example, in the graph of Figure 1.14, the sequence b, c, d, b, a is a walk that visits b twice and the sequence a, b, c, b, d is a walk that visits b twice and also uses the edge between b and c twice.

A *trail* is a walk in which no edge appears more than once. So vertices can be revisited in a trail, but edges cannot. For example, in the graph of Figure 1.14, the sequence b, c, d, b, a is a trail that visits b twice but the sequence a, b, c, b, d is not a trail.

A *path* is a walk in which no edge appears more than once and no vertex appears more than once, except possibly when the beginning vertex is also the ending vertex. For example, in the graph of Figure 1.14, the sequence a, b, c, d is a path, but the sequence a, b, c, d, b is not a path because the vertex b occurs twice. A *cycle* is a path whose beginning and ending vertices are equal. The sequence a, b, c, a is a cycle, but the sequence a, b, a is not a cycle because the edge from a to b occurs twice.

The *length* of a walk (also trail and path) is the number of edges on it. So the length of the walk $x_0, ..., x_n$ is n. For example, in the graph of Figure 1.14, the path a, b, d has length two and the cycle a, b, c, a has length three.

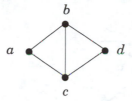

Figure 1.14

Let's emphasize here that the definitions for walk, trail, path, and cycle apply to both graphs and directed graphs. But now we come to an idea that needs a separate definition for each type of graph. A graph is *connected* if there is a path between any two vertices. A directed graph is *connected* if, when direction is ignored, the resulting undirected graph is connected. A graph with no cycles is called *acyclic*. A directed graph that is acyclic is sometimes referred to as a "DAG" to mean a directed acyclic graph.

Let's look at two famous graph problems that you may have seen. For the first problem you are to trace the first diagram in Figure 1.15 without taking your pencil off the paper and without retracing any line.

Figure 1.15

After some fiddling, it's easy to see that it can be done by starting at one of the bottom two corners and finishing at the other bottom corner. The second diagram in Figure 1.15 emphasizes the graphical nature of the problem. From this point of view we can say that there is a walk that travels each edge exactly once. In other words, there is a trail that includes every edge. Since the trail visits some vertices more than once, our walk is not a path.

The second famous problem is named after the seven bridges of Königsberg that, in the early 1800s, connected two islands in the river Pregel to the rest of the town. The problem is to find a trail through the town by walking across each of the seven bridges exactly once. In Figure 1.16 we've pictured the two islands and seven bridges of Königsberg together with a multigraph representing the situation. The vertices of the multigraph represent the four land areas and the edges represent the seven bridges.

Figure 1.16

The mathematician Leonhard Euler (1707–1783) proved that there aren't any such trails by finding a general condition for such trails to exist. In his honor, any trail that contains every edge of a graph is called an *Euler trail*. For example, the graph for the tracing problem has an Euler trail, but the seven bridges problem does not. We can go one step further and define an *Euler circuit* to be any trail that begins and ends at the same vertex and contains every edge of the graph. There are no Euler circuits in the graphs of Figure 1.15 and Figure 1.16. We'll discuss conditions for the existence of Euler trails and Euler circuits in the exercises.

Graph Traversals

A *graph traversal* starts at some vertex v and visits all vertices x that can be reached from v by traveling along some path from v to x. If a vertex has already been visited, it is not visited again. Two popular traversal algorithms are called *breadth-first* and *depth-first*.

To describe *breadth-first traversal*, we'll let visit(v, k) denote the procedure that visits every vertex x not yet visited for which there is a length k path from v to x. If the graph has n vertices, a breadth-first traversal starting at v can be described as follows:

$$\textbf{for } k := 0 \textbf{ to } n - 1 \textbf{ do } \text{visit}(v, k) \textbf{ od}.$$

Since we haven't specified how visit(v, k) does it's job, there are usually several different traversals from any given starting vertex. For example, let's consider the graph in Figure 1.17.

If we start at vertex a, then there are four possible breadth-first traversals, which we've represented by the following strings:

$$a\,b\,c\,d\,e\,f\,g \qquad a\,b\,c\,d\,f\,e\,g \qquad a\,c\,b\,d\,e\,f\,g \qquad a\,c\,b\,d\,f\,e\,g.$$

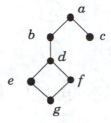

Figure 1.17

Of course, we can start a breadth-first traversal at any vertex. For example, one of several breadth-first traversals that start with d is represented by the string

$$d\,b\,e\,f\,a\,g\,c.$$

We can describe *depth-first traversal* with a recursive procedure–one that calls itself. Let DF(v) denote the depth-first procedure that traverses the graph starting at vertex v. Then DF(v) can be defined as follows:

DF(v):   **if** v has not been visited **then**
   visit v;
   for each edge from v to x **do** DF(x) **od**
fi

Since we haven't specified how each edge from v to x is picked in the **for** loop, there are usually several different traversals from any given starting vertex. For example, starting from vertex a in the graph of Figure 1.17, there are four possible depth-first traversals, which are represented by the following strings:

$$a\,b\,d\,e\,g\,f\,c \qquad a\,b\,d\,f\,g\,e\,c \qquad a\,c\,b\,d\,e\,g\,f \qquad a\,c\,b\,d\,f\,g\,e.$$

Trees

From an informal point of view, a *tree* is a structure that looks like a real tree. For example, a family tree and an organizational chart for a business are both trees. In computer science, as well as many other disciplines, trees are usually drawn upside down, as in Figure 1.18. From a formal point of view we can say that a *tree* is a graph that is connected and has no cycles.

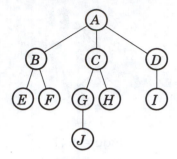

Figure 1.18

Trees have their own terminology. The elements of a tree are called *nodes*, and the lines between nodes are called *branches*. The node at the top is called the *root*. The nodes that hang immediately below a given node are its *children*, and the node immediately above a given node is its *parent*. If a node is childless, then it is a *leaf*. The *height* or *depth* of a tree is the length of the longest path from the root to the leaves. So the height of the tree in Figure 1.18 is 3.

If x is a node in a tree T, then x together with all its descendants forms a tree S with x as its root. S is called a *subtree* of T. If y is the parent of x, then S is sometimes called a subtree of y. For example, Figure 1.19 shows a tree that is a subtree of the tree in Figure 1.18. This tree is also called a subtree of node A.

Figure 1.19

If we don't care about the ordering of the children of a tree, then the tree is called an *unordered tree*. A tree is *ordered* if there is a unique ordering of the children of each node. Algebraic expressions have ordered tree representations. For example, the expression $x - y$ can be represented by a tree whose root is the minus sign and with two subtrees, one for x on the left and one for y on the right. For example, the expression $3 - (4 + 8)$ can be represented by the ordered tree in Figure 1.20.

The tree must be ordered because the subtraction operation is not commutative. For example, $3 - (4 + 8) \neq (4 + 8) - 3$. The expression $(4 + 8) - 3$ is represented by the ordered tree in Figure 1.21.

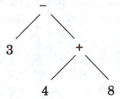

Figure 1.20

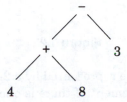

Figure 1.21

How can we represent a tree as a data object? The key to any representation is that we should be able to recover the tree from its representation. One method is to let the tree be a tuple whose first element is the root and whose next elements are the subtrees of the root, in order from left to right. For example, the tree with a single node r is represented by $\langle r \rangle$, and the tuple representation of the tree for the algebraic expression $a - b$ is

$$\langle -, \langle a \rangle, \langle b \rangle \rangle.$$

The tuple representation of the tree for the expression $(4 + 8) - 3$ is

$$\langle -, \langle +, \langle 4 \rangle, \langle 8 \rangle \rangle, \langle 3 \rangle \rangle.$$

For a more complicated example, let's consider the tree represented by the following tuple:

$$T = \langle r, \langle b, \langle c \rangle, \langle d \rangle \rangle, \langle x, \langle y, \langle z \rangle \rangle, \langle w \rangle \rangle, \langle e, \langle u \rangle \rangle \rangle.$$

Notice that T has root r, which has the following three subtrees:

$$\langle b, \langle c \rangle, \langle d \rangle \rangle$$
$$\langle x, \langle y, \langle z \rangle \rangle, \langle w \rangle \rangle$$
$$\langle e, \langle u \rangle \rangle.$$

Similarly, the subtree $\langle b, \langle c \rangle, \langle d \rangle \rangle$ has root b, which has two children c and d. We can continue in this way to recover the picture of T in Figure 1.22.

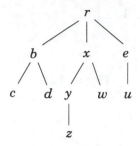

Figure 1.22

An unordered tree can be represented by a 2-tuple whose first element is the root and whose second element, if there is one, is a nonempty set containing the subtrees of the root. For example, the tuple $\langle r, \{\langle s \rangle, \langle q \rangle\} \rangle$ represents the unordered tree whose root is r and whose children are s and q. This tree can be represented by either of the trees in Figure 1.23.

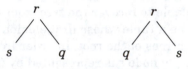

Figure 1.23

A *binary tree* is an ordered tree that may be empty or else has the property that each node has two subtrees, called the *left* and *right* subtrees of the node, which are binary trees. We can represent the empty binary tree by the empty tuple $\langle \, \rangle$. Since each node has two subtrees, we represent non-empty binary trees as 3-tuples of the form

$$\langle L, x, R \rangle, \tag{1.25}$$

where x is the root, L is the left subtree, and R is the right subtree. For example, the tree with one node x is represented by the tuple $\langle \langle \, \rangle, x, \langle \, \rangle \rangle$.

When we draw a picture of a binary tree, it is common practice to omit the empty subtrees. For example, the binary tree $\langle \langle \langle \, \rangle, a, \langle \, \rangle \rangle, b, \langle \, \rangle \rangle$ is usually, but not always, pictured as the simpler tree in Figure 1.24.

Using pictures and tuples together, we can say that the 3-tuple

   represents the tree

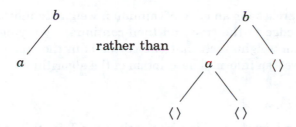

Figure 1.24

Binary trees can be used to represent sets whose elements have some ordering. Such a tree is called a *binary search tree* and has the property that for each node of the tree, each element in its left subtree precedes the node element and each element in its right subtree succeeds the node element. For example, the binary search tree in Figure 1.25 holds the three-letter abbreviations for six of the months, where we are using the dictionary ordering of the words.

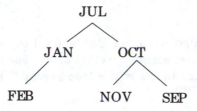

Figure 1.25

A *spanning tree* for a connected graph is a subgraph that is a tree and contains all the vertices of the graph. For example, Figure 1.26 shows a graph followed by two of its spanning trees. This example shows that a graph can have many spanning trees. A *minimal spanning tree* for a connected weighted graph is a spanning tree such that the sum of the edge weights is minimum among all spanning trees.

Figure 1.26

A famous algorithm, due to Prim [1957], constructs a minimal spanning tree for any undirected connected weighted graph. Starting with any vertex, the algorithm searches for an edge of minimum weight connected to the vertex. It adds the edge to the tree and then continues by trying to find new edges of minimum weight such that one vertex is in the tree and the other vertex is not. Here's an informal description of the algorithm:

Prim's Algorithm   (1.26)

Construct a minimal spanning tree with edges T for the undirected, connected, weighted graph with vertices V.

1. Initialize $T := \varnothing$.

2. Pick any vertex x and set $W := \{x\}$.

3. **while** $W \neq V$ **do**

      Find an edge $\{a, b\}$ of minimum weight such that $a \in W$

         and $b \in V - W$;

      $T := T \cup \{\{a, b\}\}$;

      $W := W \cup \{b\}$

   od

Of course, Prim's algorithm can also be used to find a spanning tree for an unweighted graph. Just assign a weight of 1 to each edge of the graph. Or modify the first statement in the **while** loop to read "Find an edge $\{a, b\}$ such that $a \in W$ and $b \in V - W$."

Counting Tuples

Suppose we want to find the cardinality of the product $A \times B$, where $A = \{a, b, c\}$ and $B = \{0, 1, 2, 3\}$. The sets A and B are small enough so that we can write down all 12 tuples. The exercise might also help us notice that each element of A can be paired with any one of the four elements in B. Since there are three elements in A, it follows that there are $3 \cdot 4 = 12$ ordered pairs in $A \times B$. This is an example of a general counting technique called the product rule, which we'll state as follows for any two finite sets A and B:

Product Rule

$$|A \times B| \;=\; |A|\,|B|. \qquad\qquad (1.27)$$

It's easy to see that (1.27) generalizes to a product of three or more finite sets. For example, the sets $A \times B \times C$ and $A \times (B \times C)$ are not actually equal because an arbitrary element in $A \times \mathrm{B} \times C$ is a 3-tuple $\langle a, b, c \rangle$, while an arbitrary element in $A \times (B \times C)$ is a 2-tuple $\langle a, \langle b, c \rangle \rangle$. Still the two sets have the same cardinality. Can you convince yourself of this fact? Now proceed as follows:

$$
\begin{aligned}
|A \times B \times C| &= |A \times (B \times C)| \\
&= |A| \, |B \times C| \\
&= |A| \, |B| \, |C|.
\end{aligned}
$$

The extension of (1.27) to any number of sets allows us to obtain other useful formulas for counting tuples of things. For example, for any finite set A and any natural number n we have the following product rule:

$$|A^n| = |A|^n. \tag{1.28}$$

We can use product rules to count strings of things as well as tuples of things because a string can be represented as a tuple. For example, to count the number of strings of length 5 over the alphabet $A = \{a, b, c\}$, we notice that any string of length 5 can be considered as a 5-tuple. For example, the string $abcbc$ can be represented by the tuple $\langle a, b, c, b, c \rangle$. So the number of strings of length 5 over A equals the number of 5-tuples over A, which by product rule (1.28) is

$$|A^5| = |A|^5 = 3^5 = 243.$$

Let's look at a few ways to count strings that have special properties. For example, suppose we want to count the number of strings of length n over an alphabet A that contain at least one occurrence of the letter $x \in A$. An easy way to proceed is to count the number of strings that do not contain any occurrence of x and subtract this number from the total number of strings of length n. The first number is easy to count because it is the total number of strings of length n over the alphabet $A - \{x\}$. Using (1.28) and (1.20), we can write this number as $|(A - \{x\})^n| = |A - \{x\}|^n = (|A| - 1)^n$. Therefore the number of strings of length n over A containing at least one occurrence of a particular letter from A is given by the formula

$$|A|^n - (|A| - 1)^n. \tag{1.29}$$

Let's extend this formula to count the number of strings of length n over A that contain at least one occurrence of an element from a subset B of A. Using (1.28) and (1.20), it's easy to see that the number of strings of length n that contain no occurrences of elements from B is

$$|(A - B)^n| = |A - B|^n = (|A| - |B|)^n.$$

Therefore the number of strings of length n over A that contain at least one occurrence of an element from a subset B of A is given by

$$|A|^n - (|A| - |B|)^n. \qquad (1.30)$$

In the next example we'll add a little spice by putting some more restrictions on the strings we want to count.

EXAMPLE 8. Suppose we need to count the strings of length 6 over the alphabet $A = \{a, b, c, d\}$ that begin with either a or c and contain at least one occurrence of b. In this case it's easy to count the total number of strings of length 6 that begin with a or c. It's the cardinality of the set $\{a, c\} \times A^5$, which is $2 \cdot 4^5$. It's also easy to count the number of these strings that do not contain any occurrences of b. It's the cardinality of the set $\{a, c\} \times \{a, c, d\}^5$, which is $2 \cdot 3^5$. Subtracting this latter number from the former gives us the desired value:

$$2 \cdot 4^5 - 2 \cdot 3^5 = 1,562.$$

Let's modify the problem and compute the number of strings of length 6 over A that start with a or c and contain at least one occurrence of either b or d. The total number of strings of length 6 that start with a or c is still $2 \cdot 4^5$. The number of these strings that do not contain b and d is the cardinality of the product $\{a, c\}^6$, which is 2^6. Thus the desired value is

$$2 \cdot 4^5 - 2^6 = 1,984.$$

Let's modify the problem again and compute the number of strings of length 6 over A that start with a or c and contain at least one occurrence of b and at least one occurrence of d. The total number of strings of length 6 that start with a or c is still $2 \cdot 4^5$. There are $2 \cdot 3^5$ of these strings that don't contain b and $2 \cdot 3^5$ that don't contain d. Since each of these latter two counts contains the count of the strings that don't contain b and d, which is 2^6, we need to add one of these counts back to obtain the desired value:

$$2 \cdot 4^5 - 2 \cdot 3^5 - 2 \cdot 3^5 + 2^6 = 1,140. \quad \blacklozenge$$

We'll leave a few more counting problems as exercises. We've barely scratched the surface when it comes to techniques to count things, and we'll discuss more counting techniques as the need arises.

Exercises

1. Write down all possible 3-tuples over the set $\{x, y\}$.

2. Let $A = \{a, b, c\}$ and $B = \{a, b\}$. Compute each of the following sets.
    a. $A \times B$.
    b. $B \times A$.
    c. A^0.
    d. A^1.
    e. A^2.
    f. $A^2 \cap (A \times B)$.

3. Prove each of the following statements about the interaction of the set operations and the product.
    a. $(A \cup B) \times C = (A \times C) \cup (B \times C)$.
    b. $(A - B) \times C = (A \times C) - (B \times C)$.
    c. Find and prove a similar equality using the intersection operation.

4. Use the set definition of tuples (1.22) and (1.23) to verify each of the following statements.
    a. $\langle 3, 7 \rangle \neq \langle 7, 3 \rangle$.
    b. $\langle x, y \rangle = \langle u, v \rangle$ if and only if $x = u$ and $y = v$.
    c. $\langle x_1, x_2, x_3 \rangle = \langle y_1, y_2, y_3 \rangle$ if and only if $x_i = y_i$ for each $i = 1, 2, 3$.

5. For each of the following proposed definitions of a tuple in terms of sets, find an example to show that the definition does not distinguish between distinct tuples.
    a. $\langle x, y \rangle = \{x, \{y\}\}$.
    b. $\langle x, y, z \rangle = \{\{x\}, \{x, y\}, \{x, y, z\}\}$.

6. Write down all possible strings of length 2 over the set $A = \{a, b, c\}$.

7. For each of the following strings, find an appropriate alphabet of elements for the string.
    a. $x+3$.
    b. $x+y = 3y+4$.
    c. 12310.
    d. 297.34001.
    e. This is a sentence ending with a period..

8. Represent each relation as a set by listing each individual tuple.
    a. $R = \{\langle x, y, z \rangle \mid x = y + z \text{ where } x, y, z \in \{1, 2, 3\}\}$.
    b. Let $\langle x, y \rangle \in S$ if and only if $x \leq y$ and $x, y \in \{1, 2, 3\}$.
    c. Let $\langle x, y \rangle \in U$ if and only if $x \in \{a, b\}$ and $y \in \{1, 2\}$.

9. Using only the set $\mathbb{N}$ in your answer, write down an appropriate product set that contains the relation Parts, where the five attributes of each tuple are: PartNumber, Price, Cost, DateBought, and NumberOnHand.

10. Draw a picture of a graph that represents those states of the United States and those provinces of Canada that touch the Atlantic Ocean or touch states or provinces that do.

11. Find planar graphs with the smallest possible number of vertices that have chromatic numbers of 1, 2, 3, and 4.

12. What is the chromatic number of the graph representing the map of the United States? Explain your answer.

13. Draw a picture of the directed graph that corresponds to each of the following binary relations.
    a. $\{\langle a, a \rangle, \langle b, b \rangle, \langle c, c \rangle\}$.
    b. $\{\langle a, b \rangle, \langle b, b \rangle, \langle b, c \rangle, \langle c, a \rangle\}$.
    c. The relation $\leq$ on the set $\{1, 2, 3\}$.

14. Given the following graph:

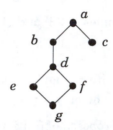

    a. Write down all breadth-first traversals that start at vertex f.
    b. Write down all depth-first traversals that start at vertex f.

15. Given the following graph:

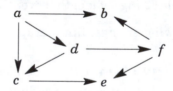

    a. Write down a breadth-first traversal that starts at vertex a.
    b. Write down a depth-first traversal that starts at vertex a.

16. Given the algebraic expression $a \times (b + c) - (d / e)$. Draw a picture of the tree representation of this expression. Then convert the tree into a tuple representation of the expression.

17. Draw a picture of the ordered tree that is represented by the list

$$\langle a, \langle b, \langle c \rangle, \langle d, \langle e \rangle \rangle \rangle, \langle r, \langle s \rangle, \langle t \rangle \rangle, \langle x \rangle \rangle.$$

18. Draw a picture of a binary search tree containing the three-letter abbreviations for the 12 months of the year in dictionary order. Make sure that your tree has the least possible depth.

19. Find two distinct minimal spanning trees for the following weighted graph:

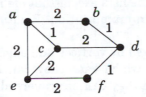

20. For each of the following cases, find the number of strings over the alphabet $\{a, b, c, d, e\}$ that satisfy the given conditions.

    a. Length 4, begins with a or b, contains at least one c.

    b. Length 5, begins with a, ends with b, contains at least one c or d.

    c. Length 6, begins with d, ends with b or d, contains no c's.

    d. Length 6, contains at least one a and at least one b.

21. Let $A = \{a, b\}$. Write down all the lists in GenLists[A] that have five symbols, where the symbols that we count are either a or b or $\langle$ or $\rangle$. Can you do the same for general lists over A that have six symbols? There are quite a few of them.

22. Try to find a necessary and sufficient condition for an undirected graph to have an Euler trail. *Hint:* Look at the degrees of the vertices.

23. Try to find a necessary and sufficient condition for an undirected graph to have an Euler circuit. *Hint:* Look at the degrees of the vertices.

Chapter Summary

We normally prove things informally, and we use a variety of proof techniques: proof by example, proof by exhaustive checking, proof using variables, direct proofs, indirect proofs (e.g., proving the contrapositive and proof by contradiction), and iff proofs.

Sets are characterized by lack of order and no redundant elements. There are easy techniques for comparing sets by subset and by equality. Sets can be combined by the operations of union, intersection, difference, and complement. Venn diagrams are useful for representing these operations. Two useful rules for counting sets are the union rule—also called the inclusion-exclusion principle—and the difference rule.

Bags—also called multisets—are characterized by lack of order, and they may contain a finite number of redundant elements.

Tuples are characterized by order, and they may contain a finite number of redundant elements. A useful rule for counting tuples is the product rule.

Many useful structures are related to tuples. Products of sets are collections of tuples. Lists are similar to tuples except that lists can be accessed only by head and tail. Strings are like lists except that elements from an alphabet are placed next to each other in juxtaposition. Relations are sets of tuples that are related in some way.

Graphs are characterized by vertices and edges, where the edges may be undirected or directed, in which case they can be represented as tuples. Graphs can be colored, and they can be traversed. Trees are special graphs that look like real trees. Prim's algorithm constructs a minimal spanning tree for an undirected, connected, weighted graph.

2

Facts About Functions

*All my discoveries were simply improvements
in notation.*
—Gottfried Wilhelm von Leibniz (1646–1716)[†]

Functions can often make life simpler. In this chapter we'll start with the
basic notions and notations for functions. Then we'll introduce some functions
that are especially important for solving problems in computer science. Since
programs can be functions and functions can be programs, we'll pay special
attention to the basic techniques for combining functions and constructing
new functions from simpler ones. We'll also discuss those properties of functions that will help us compare the cardinality of two sets.

Chapter Guide

Section 2.1 introduces the basic ideas of functions—what they are, and how to
represent them. We'll give many examples, including several functions
that are especially useful to computer scientists.

Section 2.2 discusses techniques for constructing functions from existing
functions by composition and tupling. We also introduce the idea of
higher-order functions—those that can take functions as arguments.

Section 2.3 introduces three important properties of functions—injective,
surjective, and bijective. We'll see how these properties are used when
we discuss the pigeonhole principle and hash functions.

[†] Leibniz introduced the word "function" around 1692. He is responsible for such diverse
ideas as binary arithmetic, symbolic logic, combinatorics, and calculus. Around 1694 he built
a calculating machine that could add and multiply.

61

Section 2.4 gives a brief introduction to techniques for comparing infinite sets. We'll discuss the ideas of countable and uncountable sets. We'll introduce the diagonalization technique, and we'll discuss whether we can compute everything.

2.1   Definitions and Examples

Suppose A and B are sets and for each element in A we associate exactly one element in B. Such an association is called a *function* from A to B. The key point is that each element of A is associated with *exactly one* element of B. In other words, if $x \in A$ is associated with $y \in B$, then x is not associated with any other element of B.

Functions are normally denoted by letters like f, g, and h or other descriptive names or symbols. If f is a function from A to B and f associates $x \in A$ with $y \in B$, then we write $f(x) = y$ or $y = f(x)$. The notation $f(x)$ is read, "f of x," or "f at x," or "f applied to x." When $f(x) = y$, we often say, "f maps x to y." Some other words for "function" are *mapping*, *transformation*, and *operator*.

Functions can be described in many ways. Sometimes a formula will do the job. For example, if we want the function f to map every natural number x to its square, then we can define f by the formula $f(x) = x^2$ for all $x \in \mathbb{N}$. Other times, we'll have to write down all possible associations. For example, we can define a function g from $A = \{a, b, c\}$ to $B = \{1, 2, 3\}$ as follows:

$$g(a) = 1, \quad g(b) = 1, \quad \text{and} \quad g(c) = 2.$$

We can also describe a function by a drawing a figure. For example, Figure 2.1 shows three ways to represent the function g. The top figure uses Venn diagrams together with a digraph. The lower left figure is a digraph. The lower right figure is the familiar Cartesian graph, in which each ordered pair $\langle x, g(x) \rangle$ is plotted as a point. Figure 2.2 shows two associations that are not functions. Be sure to explain why these associations do not represent functions from A to B.

Terminology

To talk about functions, we need to introduce some more terminology. The symbol

$$A \to B$$

denotes the set of all functions from A to B. If f is a function from A to B, we write

$$f : A \to B.$$

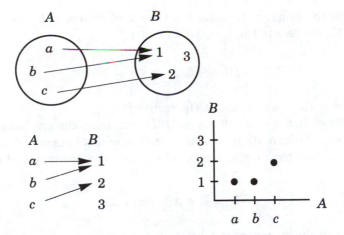

Figure 2.1

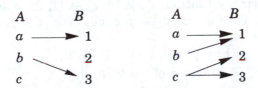

Figure 2.2

We also say that f has *type* $A \to B$. The set A is called the *domain* of f, and B is the *codomain* of f. If $f(x) = y$, then x is called an *argument* of f, and y is called a *value* of f.

If the domain of a function f is a product of n sets, then we say that f has *arity n*, or f has n *arguments*. For example, if f's domain is $A \times B \times C$, then we say that f has arity 3, or that f has 3 arguments. If this is the case and if $\langle a, b, c \rangle \in A \times B \times C$, then the expression $f(a, b, c)$ denotes the value of f at the three arguments a, b, and c.

A function f with two arguments is called a *binary* function. Binary functions give us the option of writing $f(x, y)$ in the popular *infix* form $x f y$. For example, $4 + 5$ is usually preferable to $+(4, 5)$ for representing values of the function $+ : \mathbb{R} \times \mathbb{R} \to \mathbb{R}$.

The *range* of f is the set of elements in B that are associated with some element of A. In other words, the range of f is the set of all values that f can take in B. We denote the range of f by range(f):

$$\text{range}(f) = \{f(a) \mid a \in A\}.$$

If $C \subset A$, then the *image* of C under f is the set of values in B associated with elements of C. We denote the image of C under f by $f(C)$:

$$f(C) = \{f(x) \mid x \in C\}.$$

We always have the special case $f(A) = \text{range}(f)$.

Now let's go in the other direction. If $D \subset B$, then the *pre-image* of D under f is the set of elements in A that associate with elements of D. The pre-image is also called the *inverse image*. We denote the pre-image of D under f by $f^{-1}(D)$:

$$f^{-1}(D) = \{a \in A \mid f(a) \in D\}.$$

We always have the special case $f^{-1}(B) = A$.

EXAMPLE 1. Consider the function $f : \{a, b, c\} \to \{1, 2, 3\}$ defined by $f(a) = f(b) = 1$ and $f(c) = 2$. Then we can make the following statements about f:

> f has type $\{a, b, c\} \to \{1, 2, 3\}$.
> The domain of f is $\{a, b, c\}$.
> The codomain of f is $\{1, 2, 3\}$.
> The range of f is $\{1, 2\}$.

Some sample images are

> $f(\{a\}) = \{1\}$,
> $f(\{a, b\}) = \{1\}$,
> $f(A) = f(\{a, b, c\}) = \{1, 2\} = \text{range}(f)$.

Some sample pre-images are

> $f^{-1}(\{1, 2\}) = \{a, b, c\}$,
> $f^{-1}(\{1, 3\}) = \{a, b\}$,
> $f^{-1}(\{3\}) = \varnothing$,
> $f^{-1}(B) = f^{-1}(\{1, 2, 3\}) = \{a, b, c\} = A$.   ◆

If f and g are both functions of type $A \to B$, then f and g are said to be *equal* if $f(x) = g(x)$ for all $x \in A$. If f and g are equal, we write

$$f = g.$$

For example, suppose f and g are functions of type $\mathbb{N} \to \mathbb{N}$ and they are defined by the formulas $f(x) = x + x$ and $g(x) = 2x$. It's easy to see that $f = g$.

Functions can often be defined by cases. For example, the absolute value function "abs" has type $\mathbb{R} \to \mathbb{R}$, and it can be defined by the following rule:

$$\text{abs}(x) = \begin{cases} x & \text{if } x \geq 0 \\ -x & \text{if } x < 0. \end{cases}$$

A definition by cases can also be written in terms of the *if-then-else* rule. For example, we can write the preceding definition in the following form:

$$\text{abs}(x) = \text{if } x \geq 0 \text{ then } x \text{ else } -x.$$

The if-then-else rule can be used more than once if there are several cases to define. For example, suppose we want to classify the roots of a quadratic equation having the following form:

$$ax^2 + bx + c = 0.$$

We can define the function "classifyRoots" to give the appropriate statements as follows:

$$\text{classifyRoots}(a, b, c) = \quad \text{if } b^2 - 4ac > 0 \text{ then}$$
$$\text{"The roots are real and distinct."}$$
$$\text{else if } b^2 - 4ac < 0 \text{ then}$$
$$\text{"The roots are complex conjugates."}$$
$$\text{else}$$
$$\text{"The roots are real and repeated."}$$

Let's look a few more examples to help us get used to using the terminology of functions.

EXAMPLE 2.

a) Let P denote the set of all people on earth or who have lived on earth. If $g(x)$ is defined to be the child of x, then g is not a function. Why? Because some person has two children, and some person has no children. However, we can define a function $g : P \to \text{Power}(P)$ by letting $g(x)$ be the set of all children of x. Then g is a function. Can you see why?

b) Sine, cosine, and tangent are functions. Sine and cosine both have type $\mathbb{R}$ $\to \mathbb{R}$. Recall that the tangent is the sine divided by the cosine. Therefore tangent(x) is not defined whenever cosine(x) = 0. Thus the domain of tangent function is the difference set

$$\mathbb{R} - \{..., -\tfrac{3\pi}{2}, -\tfrac{\pi}{2}, \tfrac{\pi}{2}, \tfrac{3\pi}{2}, ...\}.$$

c) Any relation gives rise to a function that returns one of the words "true" or "false." For example, if R is a relation over $A \times B \times C$, then we can define the function "testR" as follows:

$$\text{testR}(a, b, c) = \text{if } \langle a, b, c \rangle \in R \text{ then true else false.}$$

This function has type $A \times B \times C \to \{\text{true, false}\}$.

d) Any computer program that has input and output can be thought of as a function of type $I \to O$, where I is the set of valid inputs to the program and O is a set containing all possible outputs.   ◆

EXAMPLE 3 (*Tuples Are Functions*). Any sequence of objects can be thought of as a function. For example, the tuple $\langle 22, 14, 55, 1, 700, 67 \rangle$ can be considered a listing of the values of a function

$$f : \{0, 1, 2, 3, 4, 5\} \to \mathbb{N}.$$

That is, we defined f by the equality

$$\langle f(0), f(1), f(2), f(3), f(4), f(5) \rangle = \langle 22, 14, 55, 1, 700, 67 \rangle.$$

So the the tuple $\langle 22, 14, 55, 1, 700, 67 \rangle$ is just a listing of the values of f.

An infinite sequence can also be considered a function. For example, suppose we have the following sequence of things from a set S:

$$\langle b_0, b_1, ..., b_n, ... \rangle.$$

The elements b_n can be considered values of the function $b : \mathbb{N} \to S$ defined by $b(n) = b_n$.   ◆

EXAMPLE 4 (*Functions and Binary Relations*). Any function can be defined as a special kind of binary relation. A function $f : A \to B$ is a binary relation from A to B such that no two ordered pairs have the same first element. We can also describe this uniqueness condition as: If $\langle a, b \rangle, \langle a, c \rangle \in f$, then $b = c$.

The functional notation $f(a) = b$ is normally preferred over the relational notations $f(a, b)$ and $\langle a, b \rangle \in f$. ♦

Some Useful Functions

Let's look at some functions that are especially useful in computer science. These functions are used for tasks such as analyzing properties of data, analyzing properties of programs, and constructing programs.

The Floor and Ceiling Functions

Let's discuss two important functions that "integerize" real numbers by going down or up to the nearest integer. The *floor* function has type $\mathbb{R} \to \mathbb{Z}$ and is defined by setting floor(x) to the closest integer less than or equal to x. For example, floor(8) = 8, floor(8.9) = 8, and floor(−3.5) = −4. A useful shorthand notation for floor(x) is

$$\lfloor x \rfloor.$$

The *ceiling* function also has type $\mathbb{R} \to \mathbb{Z}$ and is defined by setting ceiling(x) to the closest integer greater than or equal to x. For example, ceiling(8) = 8, ceiling(8.9) = 9, and ceiling(−3.5) = −3. The shorthand notation for ceiling(x) is

$$\lceil x \rceil.$$

Table 2.1 gives a few sample values for floor and ceiling:

x	−2.0	−1.7	−1.3	−1.0	−0.7	−0.3	0.0	0.3	0.7	1.0	1.3	1.7	2.0
$\lfloor x \rfloor$	−2	−2	−2	−1	−1	−1	0	0	0	1	1	1	2
$\lceil x \rceil$	−2	−1	−1	−1	0	0	0	1	1	1	2	2	2

Table 2.1

Can you find some simple relationships between floor and ceiling? For example, is $\lfloor x \rfloor = \lceil x - 1 \rceil$?

Greatest Common Divisor

The *greatest common divisor* of two integers, not both zero, is the largest number that divides them both. For example, the common divisors of 12 and 18 are ±1, ±2, ±3, ±6. So the greatest common divisor of 12 and 18 is 6. We denote the greatest common divisor of a and b by

$$(a, b).$$

This is the traditional notation for the greatest common divisor function, and we'll stick with it. For example, we can write $(12, 18) = 6$, $(-44, -12) = 4$, and $(5, 0) = 5$. If $a \neq 0$, we can observe that $(a, 0) = |a|$. If $(a, b) = 1$, we say a and b are *relatively prime*. For example, 9 and 4 are relatively prime. Let's list a few important properties of the greatest common divisor function.

Greatest Common Divisor Properties   (2.1)

a) $(a, b) = (b, a) = (a, -b)$.

b) $(a, b) = (b, a - bq)$ for any integer q.

c) If $g = (a, b)$, then there are integers m and n such that

$$g = m{\cdot}a + n{\cdot}b.$$

d) If $d \mid a{\cdot}b$ and $(d, a) = 1$, then $d \mid b$.

Property (2.1a) confirms that the ordering of the arguments doesn't matter and that negative numbers have positive greatest common divisors. For example, $(-4, -6) = (-4, 6) = (6, -4) = (6, 4) = 2$. We'll see shortly how property (2.1b) can help us compute greatest common divisors. Property (2.1c) says that we can write (a, b) in terms of a and b. For example, $(126, 45) = 9$, and we can write 9 in terms of 126 and 45 as follows:

$$9 = (-1){\cdot}126 + 3{\cdot}45.$$

Property (2.1d) is a divisibility property that we'll be using later.

Now let's get down to brass tacks and describe an algorithm to compute the greatest common divisor. Most of us recall from elementary school that we can divide an integer a by a nonzero integer b to obtain two other integers, a quotient q and a remainder r, which satisfy an equation like the following:

$$a = bq + r.$$

For example, if $a = -16$ and $b = 3$, then we can write many equations, each with different values for q and r. For example, the following four equations all have the form $a = bq + r$:

$$-16 = 3\,(-4) + (-4)$$
$$-16 = 3\,(-5) + (-1)$$
$$-16 = 3\,(-6) + 2$$
$$-16 = 3\,(-7) + 5.$$

In mathematics and computer science the third equation is by far the most useful. In fact it's a result of a theorem called the *division algorithm*, which we'll state for the record:

Division Algorithm   (2.2)

If a and b are integers and $b \neq 0$, then there are unique integers
q and r such that $a = bq + r$, where $0 \leq r < |b|$.

The division algorithm together with property (2.1b) gives us the seeds
of an algorithm to compute the greatest common divisor function. Suppose a
and b are integers and $b \neq 0$. The division algorithm gives us the equation $a =
bq + r$, where $0 \leq r < |b|$. Solving the equation for r gives $r = a - bq$. This fits
the form of (2.1b). So we have the nice equation

$$(a, b) = (b, r).$$

The important point about this equation is that the numbers in (b, r) are get-
ting closer to zero. Let's see how we can use this equation to compute the
greatest common divisor. For example, to compute (315, 54), we apply the division algorithm to obtain the equation $315 = 54 \cdot 5 + 45$. Thus we know that

$$(315, 54) = (54, 45).$$

Now apply the division algorithm again to obtain $54 = 45 \cdot 1 + 9$. So we have

$$(315, 54) = (54, 45) = (45, 9).$$

Continuing, we have $45 = 9 \cdot 5 + 0$, which extends our computation to

$$(315, 54) = (54, 45) = (45, 9) = (9, 0) = 9.$$

The algorithm that we have been demonstrating is called *Euclid's algorithm*. Since greatest common divisors are always positive, we'll describe the
algorithm to calculate (a, b) for the case in which a and b are natural numbers that are not both zero.

Euclid's Algorithm   (2.3)

Input two natural numbers a and b, not both zero.

while $b > 0$ **do**
    Use the division algorithm to compute q and r such that
        $a = bq + r$, where $0 \leq r < b$;
    $a := b$;
    $b := r$
od;
Output a.

We can use Euclid's algorithm to show how property (2.1c) is satisfied. The idea is to keep track of the equations $a = bq + r$ from each execution of the loop. Then work backwards through the equations to solve for (a, b) in terms of a and b. For example, in our calculation of (315, 54) we obtained the three equations

$$315 = 54 \cdot 5 + 45$$
$$54 = 45 \cdot 1 + 9$$
$$45 = 9 \cdot 5 + 0.$$

Starting with the second equation, we can solve for 9. Then use the first equation to replace 45. The result is an expression for $9 = (315, 54)$ written in terms of 315 and 54 as $9 = (-1) \cdot 315 + 6 \cdot 54$.

The Mod Function

Let's look at the division algorithm again. It states that for any integer a and any nonzero integer b there are two unique integers q and r such that

$$a = bq + r \quad \text{where} \quad 0 \leq r < |b|.$$

Solving the equation for q, we obtain

$$q = \frac{a - r}{b}.$$

If $b > 0$, we can write the inequality as $0 \leq r < b$. From this we obtain the following inequalities:

$$-b < -r \leq 0$$
$$a - b < a - r \leq a$$
$$\frac{a - b}{b} < \frac{a - r}{b} \leq \frac{a}{b}$$
$$\frac{a}{b} - 1 < q \leq \frac{a}{b}.$$

Since q is an integer, the last inequality implies that q can be written as the floor expression

$$q = \left\lfloor \frac{a}{b} \right\rfloor.$$

Since $r = a - b \cdot q$, we have the following representation:

$$\text{If } b > 0, \text{ then } r = a - b \left\lfloor \frac{a}{b} \right\rfloor.$$

This representation of r forms the basis for the definition of the *mod* function. If a and b are integers and $b \neq 0$, then $a \bmod b$ is defined by

$$a \bmod b = a - b \left\lfloor \frac{a}{b} \right\rfloor.$$

We should note that $a \bmod b$ is the remainder upon division of a by b, where the remainder has the same sign as b. For example,

$$5 \bmod 4 = 1$$
$$-5 \bmod 4 = 3$$
$$5 \bmod -4 = -3$$
$$-5 \bmod -4 = -1.$$

So the mod function has type $\mathbb{Z} \times (\mathbb{Z} - \{0\}) \to \mathbb{Z}$. Table 2.2 shows a few values of the mod function. The entry in row a and column b is $a \bmod b$:

a\\b	1	2	3	4	5	6	7	8	9
1	0	1	1	1	1	1	1	1	1
2	0	0	2	2	2	2	2	2	2
3	0	1	0	3	3	3	3	3	3
4	0	0	1	0	4	4	4	4	4
5	0	1	2	1	0	5	5	5	5
6	0	0	0	2	1	0	6	6	6
7	0	1	1	3	2	1	0	7	7
8	0	0	2	0	3	2	1	0	8
9	0	1	0	1	4	3	2	1	0

Table 2.2

Can you see any patterns here? For example, look at the second column, whose entries alternate between 0 and 1. We can conclude that

$$a \bmod 2 = \text{if } a \text{ is even then } 0 \text{ else } 1.$$

So we have a test for oddness or evenness. See whether you can find some other properties of mod. What about the rows? Anything there?

If we agree to fix n as a positive integer constant, then we can define a function $f : \mathbb{Z} \to \mathbb{N}$ by $f(x) = x \bmod n$. The range of f is $\{0, 1, ..., n - 1\}$, which is the set of possible remainders obtained upon division of x by n. We sometimes let $\mathbb{N}_n$ denote the set

$$\mathbb{N}_n = \{0, 1, 2, ..., n - 1\}.$$

For example, $\mathbb{N}_0 = \varnothing$, $\mathbb{N}_1 = \{0\}$, and $\mathbb{N}_2 = \{0, 1\}$. Notice that the numbers in the nth column of Table 2.2 make up the set $\mathbb{N}_n$.

EXAMPLE 5 (*Converting Decimal to Binary*). How can we convert a decimal number to binary? For example, the decimal number 53 has the binary representation 110101. The rightmost bit (binary digit) in this representation of 53 is 1 because 53 is an odd number. In general, we can find the rightmost bit (binary digit) of the binary representation of a natural decimal number x by evaluating the expression $x \bmod 2$. In our example, $53 \bmod 2 = 1$, which is the rightmost bit.

So we can apply the division algorithm, dividing 53 by 2, to obtain the rightmost bit as the remainder. We have the following equation:

$$53 = 2 \cdot 26 + 1.$$

Now do the same thing for the quotient 26 and the succeeding quotients as follows:

$$
\begin{aligned}
53 &= 2 \cdot 26 + 1 \\
26 &= 2 \cdot 13 + 0 \\
13 &= 2 \cdot 6 + 1 \\
6 &= 2 \cdot 3 + 0 \\
3 &= 2 \cdot 1 + 1 \\
1 &= 2 \cdot 0 + 1 \\
0 & \quad \text{(done)}.
\end{aligned}
$$

We can read off the remainders in the above equations from bottom to top to obtain the binary representation of 53: 110101. The important point to notice is that any natural number x can be written in the form

$$x = 2 \left\lfloor \frac{x}{2} \right\rfloor + x \bmod 2.$$

Thus an algorithm to convert x to binary can be implemented with the floor and mod functions. ◆

The Characteristic Function

Another useful function is the *characteristic* function on a subset B of a set S. It is denoted by $\chi_B : S \to \{0, 1\}$ and is defined as follows:

$$\chi_B(x) = \text{if } x \in B \text{ then } 1 \text{ else } 0.$$

So the characteristic function is just a test for set membership. Table 2.3 shows the first few values of the characteristic functions on the following five subsets of $\mathbb{N}$:

Odd	$=$	the set of odd natural numbers.
Even	$=$	the set of even natural numbers.
Prime	$=$	the set of prime numbers.
$4k + 1$	$=$	$\{4k + 1 \mid k \in \mathbb{N}\}$.
$P + P$	$=$	the set of natural numbers that can be written as a sum of two primes.

$B \setminus x$	0	1	2	3	4	5	6	7	8	9	10	11	12	13	14	15	16	17
Odd	0	1	0	1	0	1	0	1	0	1	0	1	0	1	0	1	0	1
Even	1	0	1	0	1	0	1	0	1	0	1	0	1	0	1	0	1	0
Prime	0	0	1	1	0	1	0	1	0	0	0	1	0	1	0	0	0	1
$4k + 1$	0	1	0	0	0	1	0	0	0	1	0	0	0	1	0	0	0	1
$P + P$	0	0	0	0	1	1	1	1	1	1	1	0	1	1	1	1	1	0

Table 2.3

The characteristic function for the union of two sets can be written in terms of the characteristic functions for the two sets. For example, it's easy to verify the following equality, where B and C are subsets of some universe:

$$\chi_{B \cup C}(x) = \chi_B(x) + \chi_C(x) - \chi_B(x)\,\chi_C(x).$$

What about characteristic functions for the intersection or the difference of two sets? Can you find general formulas for $\chi_{B \cap C}(x)$ and $\chi_{B-C}(x)$? We'll leave these problems as exercises. Once we have these formulas, we can use them with Table 2.3 to find tests for membership in sets such as the even primes, the odd primes, the primes of the form $4k + 1$, the numbers that are either primes or of the form $4k + 1$, the primes that are not of the form $4k + 1$, the numbers of the form $4k + 1$ that are the sum of two primes, and so on.

The Log Function

The *log* function—which is shorthand for logarithm—measures the size of exponents. If b is a positive real number, then $\log_b x = y$ means $b^y = x$, and we say, "log base b of x is y." Notice that x must be a positive real number. The base 2 log function $\log_2$ occurs frequently in computer science because many algorithms and data representations use a binary decision (two choices). For example, suppose we have a binary search tree with 16 nodes having the structure shown in Figure 2.3.

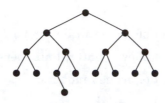

Figure 2.3

Then the depth of the tree is 4. So a maximum of 5 comparisons are needed to find any element in the tree. Notice in this case that $16 = 2^4$, so we can write the depth in terms of the number of nodes: $4 = \log_2 16$. Table 2.4 gives a few choice values for the $\log_2$ function.

x	1	2	4	8	16	32	64	128	256	512	1024
$\log_2 x$	0	1	2	3	4	5	6	7	8	9	10

Table 2.4

Of course, the $\log_2$ function takes on real values also. For example, if $8 < x < 16$, then

$$3 < \log_2 x < 4.$$

For any positive real number b the function $\log_b$ is an increasing function with the positive real numbers as its domain and the real numbers as its range. No matter what the value of b, the graph of $\log_b$ has the general form shown in Figure 2.4.

The log function has many properties. For example, it's easy to see that

$$\log_b 1 = 0 \quad \text{and} \quad \log_b b = 1.$$

The following list contains some of the most useful properties of the log function. We'll leave the proofs as exercises in applying the definition of log.

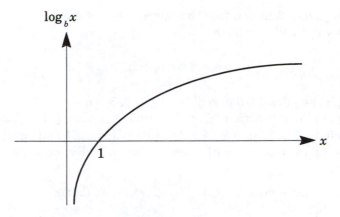

Figure 2.4

$$\log_b (b^x) = x$$

$$\log_b (x\,y) = \log_b x + \log_b y$$

$$\log_b(x^y) = y \log_b x$$

$$\log_b (x/y) = \log_b x - \log_b y$$

$$\log_a x = (\log_a b) (\log_b x) \qquad \text{(change of base)}.$$

These properties are useful in the evaluation of log expressions. For example, suppose we need to evaluate the expression $\log_2(2^7 3^4)$. Make sure you can justify each step in the following evaluation:

$$\log_2(2^7 3^4) = \log_2(2^7) + \log_2(3^4) = 7 \log_2(2) + 4 \log_2(3) = 7 + 4 \log_2(3).$$

At this point we're stuck for an exact answer. But we can make an estimate. We know that $1 = \log_2(2) < \log_2(3) < \log_2(4) = 2$. Therefore $1 < \log_2(3) < 2$. Thus we have the following estimate of the answer:

$$11 < \log_2(2^7 3^4) < 15.$$

Sequence, Distribute, and Pairs

The next three functions can be quite useful as tools to construct more complicated functions. Most functional programming languages come equipped with these functions or something similar.

The *sequence* function "seq" has type $\mathbb{N} \rightarrow$ Lists[$\mathbb{N}$] and is defined as follows for any natural number n:

$$\text{seq}(n) = \langle 0, 1, ..., n \rangle.$$

For example, $\text{seq}(0) = \langle 0 \rangle$ and $\text{seq}(4) = \langle 0, 1, 2, 3, 4 \rangle$.

The *distribute* function "dist" has type $A \times$ Lists[B] $\rightarrow$ Lists[$A \times B$]. It takes an element x from A and a list y from Lists[B] and returns the list of pairs made up by pairing x with each element of y. For example,

$$\text{dist}(x, \langle r, s, t \rangle) = \langle \langle x, r \rangle, \langle x, s \rangle, \langle x, t \rangle \rangle.$$

The *pairs* function takes two lists of equal length and returns the list of pairs of corresponding elements. For example,

$$\text{pairs}(\langle a, b, c \rangle, \langle d, e, f \rangle) = \langle \langle a, d \rangle, \langle b, e \rangle, \langle c, f \rangle \rangle.$$

Since the domain of pairs is a proper subset of Lists[A] $\times$ Lists[B], it is not a function of type Lists[A] $\times$ Lists[B] $\rightarrow$ Lists[$A \times B$]. However, we'll see in the next paragraph that pairs is a "partial" function of this type.

Partial Functions

A *partial function* from A to B is like a function except that it might not be defined for some elements of A. In other words, some elements of A might not be associated with any element of B. But we still have the requirement that if $x \in A$ is associated with $y \in B$, then x can't be associated with any other element of B. For example, we know that division by zero is not allowed. Therefore $\div$ is a partial function of type $\mathbb{R} \times \mathbb{R} \rightarrow \mathbb{R}$ because $\div$ is not defined for all pairs of the form $\langle x, 0 \rangle$.

When discussing partial functions, to avoid confusion we use the term *total function* to mean a function that is defined on all its domain. Any partial function can be transformed into a total function. One simple technique is to shrink the domain to the set of elements for which the partial function is defined. For example, $\div$ is a total function of type $\mathbb{R} \times (\mathbb{R} - \{0\}) \rightarrow \mathbb{R}$.

A second technique keeps the domain the same but increases the size of the codomain. For example, suppose $f : A \rightarrow B$ is a partial function. Pick some symbol that is not in B, say $\# \notin B$, and assign $f(x) = \#$ whenever $f(x)$ is not defined. Then we can think of f as the total function of type $A \rightarrow B \cup \{\#\}$. In programming, the analogy would be to pick an error message to indicate that an incorrect input string has been received.

Exercises

1. Write down all possible functions of type $\{a, b, c\} \to \{1, 2\}$.

2. Suppose we have a function $f : \mathbb{N} \to \mathbb{N}$ defined by $f(x) = 2x + 1$. Describe each of the following sets, where E and O denote the even and odd natural numbers, respectively.

   a. range(f).   b. $f(E)$.   c. $f(O)$.   d. $f^{-1}(E)$.   e. $f^{-1}(O)$.

3. Evaluate each of the following expressions.

   a. $\lfloor -4.1 \rfloor$.   b. $\lceil -4.1 \rceil$.   c. $\lfloor 4.1 \rfloor$.   d. $\lceil 4.1 \rceil$.

4. Evaluate each of the following expressions.

   a. $(-12, 15)$.   b. $(98, 35)$.   c. $(872, 45)$.

5. Calculate $(296, 872)$. Then write the answer in the form $m \cdot 296 + n \cdot 872$.

6. Evaluate each of the following expressions.

   a. 15 mod 12.   b. −15 mod 12.

   c. 15 mod (−12).   d. −15 mod (−12).

7. Let $f : \mathbb{N}_6 \to \mathbb{N}_6$ be defined by $f(x) = 2x \bmod 6$. Find the image and the pre- image for each of the following sets.

   a. $\{0, 2, 4\}$.   b. $\{1, 3\}$.   c. $\{0, 5\}$.

8. For a real number x, let trunc(x) denote the truncation of x, which is the integer obtained from x by deleting the part of x to the right of the decimal point.

   a. Write the floor function in terms of trunc.

   b. Write the ceiling function in terms of trunc.

9. Suppose we define the function f as follows: $f(x, y) = x - y\ \text{trunc}(x/y)$. How does f compare to the mod function?

10. Does it make sense to extend the definition of the mod function to real numbers? What would be the range of the function $f : \mathbb{R} \to \mathbb{R}$ defined by $f(x) = x \bmod 2.5$?

11. Evaluate each of the following expressions.

    a. $\log_5 625$.   b. $\log_2 8192$.

    c. $\log_3 (1/27)$.   d. 93467 mod 3.

    e. dist(4, seq(3)).   f. dist(2, pairs(seq(3), seq(3))).

12. Suppose B and C are subsets of some set. Find expressions for the characteristic functions $\chi_{B \cap C}$ and χ_{B-C} in terms of the two characteristic functions χ_B and χ_C.

13. Given a function $f : A \to A$. An element $a \in A$ is called a *fixed point* of f if $f(a) = a$. Find the set of fixed points for each of the following functions.

   a. $f : A \to A$ where $f(x) = x$.
   b. $f : \mathbb{N} \to \mathbb{N}$ where $f(x) = x + 1$.
   c. $f : \mathbb{N}_6 \to \mathbb{N}_6$ where $f(x) = 2x \bmod 6$.
   d. $f : \mathbb{N}_6 \to \mathbb{N}_6$ where $f(x) = 3x \bmod 6$.

14. Prove each of the following statements about floor and ceiling.

   a. $\lfloor x + 1 \rfloor = \lfloor x \rfloor + 1$.
   b. $\lceil x - 1 \rceil = \lceil x \rceil - 1$.
   c. $\lceil x \rceil = \lfloor x \rfloor$ if and only if $x \in \mathbb{Z}$.
   d. $\lceil x \rceil = \lfloor x \rfloor + 1$ if and only if $x \notin \mathbb{Z}$.

15. Use the definition of the logarithm function to prove each of the following facts.

   a. $\log_b 1 = 0$.
   b. $\log_b b = 1$.
   c. $\log_b (b^x) = x$.
   d. $\log_b (x\,y) = \log_b x + \log_b y$.
   e. $\log_b(x^y) = y \log_b x$.
   f. $\log_b (\frac{x}{y}) = \log_b x - \log_b y$.
   g. $\log_a x = (\log_a b)\,(\log_b x)$ (change of base).

16. Prove each of the following facts about greatest common divisors.

   a. $(a, b) = (b, a) = (a, -b)$.
   b. $(a, b) = (b, a - bq)$ for any integer q.
   c. If $d \mid a \cdot b$ and $(d, a) = 1$, then $d \mid b$. *Hint:* Use (2.1c).

17. Given the result of the division algorithm $a = bq + r$, where $0 \le r < |b|$, prove the following statement:

$$\text{If } b < 0 \text{ then } r = a - b \cdot \left\lceil \frac{a}{b} \right\rceil.$$

18. Let $f : A \to B$ be a function, and let E and F be subsets of A. Prove each of the following facts about images.

   a. $f(E \cup F) = f(E) \cup f(F)$.
   b. $f(E \cap F) \subset f(E) \cap f(F)$.
   c. $E \subset f^{-1}(f(E))$.
   d. Find examples to show that parts (b) and (c) can be proper subsets.

19. Let $f : A \to B$ be a function, and let G and H be subsets of B. Prove each of the following facts about pre-image.

    a. $f^{-1}(G \cup H) = f^{-1}(G) \cup f^{-1}(H)$.

    b. $f^{-1}(G \cap H) = f^{-1}(G) \cap f^{-1}(H)$.

    c. $f(f^{-1}(G)) \subset G$.

    d. Find an example to show that part (c) can be a proper subset.

2.2 Constructing Functions

We can often construct a new function by combining other simpler functions. In this section we'll discuss two methods of combining functions: composition and tupling. These methods will help us understand the behavior of existing functions, and they will give us the powerful tools necessary to create new functions. We'll also discuss the important idea of higher-order functions—those that can have other functions as arguments or values. Many programming systems and languages rely on the ideas of this section.

Composition and Tupling

The composition of functions is a natural process that we often use without even thinking. For example, the expression floor($\log_2(5)$) involves the composition of the two functions $\log_2$ and floor. To evaluate the expression, we first apply $\log_2$ to its argument 5, obtaining a value somewhere between 2 and 3. Then we apply the floor function to this number, obtaining the value 2.

To give a formal definition of composition, we start with two functions in which the domain of one is the codomain of the other:

$$f : A \to B \quad \text{and} \quad g : B \to C.$$

The *composition* of f and g is the function $g \circ f : A \to C$ defined by

$$(g \circ f)(x) = g(f(x)).$$

In other words, first apply f to x and then apply g to the resulting value. Composition also makes sense in the more general setting in which $f : A \to B$ and $g : D \to C$, and $B \subset D$. Can you see why?

Composition of functions is associative. In other words, if $f, g,$ and h are functions, then

$$(f \circ g) \circ h = f \circ (g \circ h).$$

This is easy to establish by noticing that the two expressions $((f \circ g) \circ h)(x)$ and $(f \circ (g \circ h))(x)$ are equal:

$$((f \circ g) \circ h)(x) = (f \circ g)(h(x)) = f(g(h(x))).$$
$$(f \circ (g \circ h))(x) = f(g \circ h)(x) = f(g(h(x))).$$

So we can feel free to write the composition of three or more functions without the use of parentheses. For example, the composition $f \circ g \circ h$ has exactly one meaning.

Notice that composition is usually not a commutative operation. For example, if f and g are defined by $f(x) = x + 1$ and $g(x) = x^2$, then

$$(g \circ f)(x) = g(f(x)) = g(x + 1) = (x + 1)^2$$

and

$$(f \circ g)(x) = f(g(x)) = f(x^2) = x^2 + 1.$$

The *identity* function "id" always returns its argument. For a particular set A we sometimes write "id_A" to denote the fact that $id_A(a) = a$ for all $a \in A$. If $f : A \to B$, then we certainly have the following equation:

$$f \circ id_A = f = id_B \circ f.$$

Another way to combine functions is to create a *tuple* of functions. For example, given the pair of functions $f : A \to B$ and $g : A \to C$, we can define the function $\langle f, g \rangle$ by

$$\langle f, g \rangle(x) = \langle f(x), g(x) \rangle.$$

The function $\langle f, g \rangle$ has type $A \to B \times C$. The definition for a tuple of two functions can be extended easily to an *n-tuple of functions*, $\langle f_1, f_2, ..., f_n \rangle$.

We can also compose tuples of functions with other functions. Suppose we are given the following three functions:

$$f : A \to B, \quad g : A \to C, \quad \text{and} \quad h : B \times C \to D.$$

We can form the composition $h \circ \langle f, g \rangle : A \to D$, where for $x \in A$ we have

$$(h \circ \langle f, g \rangle)(x) = h(\langle f, g \rangle(x)) = h(f(x), g(x)).$$

People often use composition and tupling of functions without thinking about the concepts. This usually happens when binary operations are used in infix form. For example, suppose we have the following definition:

$$h(x) = f(x) + g(x).$$

If we agree to write + as a prefix operator, then h can be written explicitly as the composition of the two functions + and $\langle f, g \rangle$. We proceed as follows:

$$
\begin{aligned}
h(x) &= f(x) + g(x) \\
&= +(f(x), g(x)) \\
&= +(\langle f, g \rangle(x)) \\
&= (+ \circ \langle f, g \rangle)(x).
\end{aligned}
$$

This clearly demonstrates that h is a composition of + and $\langle f, g \rangle$ because we have the equation

$$h(x) = (+ \circ \langle f, g \rangle)(x).$$

We can also "cancel" x from both sides of this equation to obtain a definition of h as a composition of functions without using variables:

$$h = + \circ \langle f, g \rangle.$$

This little example demonstrates a simple and important idea: We can often construct a function by first writing down an informal definition and then proceeding by stages to transform the definition into a formal one that suits our needs. For example, we might start with an informal definition of some function f such as

$$f(x) = \text{expression involving } x.$$

Now we try to transform the right side of the equality into an expression that has the degree of formality that we need. For example, we might try to reach a composition of known functions applied to x as follows:

$$
\begin{aligned}
f(x) &= \text{expression involving } x \\
&= \text{another expression involving } x \\
&= \ldots \\
&= g(h(x)) \\
&= (g \circ h)(x).
\end{aligned}
$$

From a programming point of view, our goal would be find an expression that fits the syntax of the language. In this example our goal was to find the expression $(g \circ h)(x)$, so we can make the definition

$$f(x) = (g \circ h)(x).$$

This example demonstrates a construction technique that we'll call *cancellation*. We can cancel the argument x from both sides of the equation $f(x) = (g \circ h)(x)$ to obtain the definition

$$f = g \circ h.$$

Is cancellation good for anything? Sure. Some programming systems allow users to define new functions in terms of existing ones without having to use variables. For example, the UNIX operating system allows users to define the composition $g \circ h$ by writing the command $h \mid g$. Some functional programming languages also allow function definitions without variables.

Let's do some examples to demonstrate how composition can be useful in solving problems. In each of the following examples we'll construct a function to solve a given problem. We'll represent each function in two ways, one with variables and one without variables obtained by cancellation.

EXAMPLE 1. Suppose we want to find the minimum depth of a binary tree in terms of the numbers of nodes. Table 2.5 lists a few sample cases in which the trees are as compact as possible, which means that they have the least depth for the number of nodes. Let n denote the number of nodes. Notice that when $4 \leq n < 8$, the depth is 2. Similarly, the depth is 3 whenever $8 \leq n < 16$. At the same time we know that $\log_2(4) = 2$, $\log_2(8) = 3$, and for $4 \leq n < 8$ we have $2 \leq \log_2(n) < 3$. So $\log_2(n)$ almost works as the depth function. The problem is that the depth must be exactly 2 whenever $4 \leq n < 8$. Can we make this happen? Sure—just apply the floor function to $\log_2(n)$ to get floor($\log_2(n)$) = 2 if $4 \leq n < 8$. This idea extends to the other intervals that make up $\mathbb{N}$. For example, if $8 \leq n < 16$, then floor($\log_2(n)$) = 3. So it makes sense to define our minimum depth function as the composition of the floor function and the $\log_2$ function:

$$\text{minDepth}(n) = \text{floor}(\log_2(n)).$$

We can also write

$$\text{minDepth}(n) = \text{floor}(\log_2(n)) = (\text{floor} \circ \log_2)(n).$$

So we can cancel the variable n to obtain the following form:

$$\text{minDepth} = \text{floor} \circ \log_2. \quad \blacklozenge$$

binary tree	nodes	depth
	1	0
	2	1
	3	1
	4	2
	7	2
	15	3

Table 2.5

EXAMPLE 2. Suppose we want to construct the function f defined informally as follows:

$$f(n) = \langle\langle 0, 0\rangle, \langle 1, 1\rangle, ..., \langle n, n\rangle\rangle \text{ for any } n \in \mathbb{N}.$$

To discover a solution, we'll start with this informal definition and transform it into a composition of known functions:

$$
\begin{aligned}
f(n) &= \langle\langle 0, 0\rangle, \langle 1, 1\rangle, ..., \langle n, n\rangle\rangle \\
&= \text{pairs}(\langle 0, 1, ..., n\rangle, \langle 0, 1, ..., n\rangle) \\
&= \text{pairs}(\text{seq}(n), \text{seq}(n)) \\
&= \text{pairs}(\langle \text{seq}, \text{seq}\rangle(n)) \\
&= (\text{pairs} \circ \langle \text{seq}, \text{seq}\rangle)(n).
\end{aligned}
$$

From these equations we can define f in several ways. Perhaps the most familiar definition comes from the third equation, which contains the argument n as follows:

$$f(n) = \text{pairs}(\text{seq}(n), \text{seq}(n)).$$

The last equation allows us to cancel the variable n to obtain a definition that is free of variables:

$$f = \text{pairs} \circ \langle \text{seq}, \text{seq} \rangle.$$

Notice that f has type $\mathbb{N} \to \text{Lists}[\mathbb{N} \times \mathbb{N}]$. ♦

EXAMPLE 3. Suppose we want to construct the function g defined informally as follows:

$$g(k) = \langle \langle k, 0 \rangle, \langle k, 1 \rangle, ..., \langle k, k \rangle \rangle \quad \text{for any } k \in \mathbb{N}.$$

We'll start with this informal definition and transform it into a composition of known functions:

$$
\begin{aligned}
g(k) \;&=\; \langle \langle k, 0 \rangle, \langle k, 1 \rangle, ..., \langle k, k \rangle \rangle \\
&=\; \text{dist}(k, \langle 0, 1, ..., k \rangle) \\
&=\; \text{dist}(k, \text{seq}(k)) \\
&=\; \text{dist}(\text{id}(k), \text{seq}(k)) \\
&=\; \text{dist}(\langle \text{id}, \text{seq} \rangle(k)) \\
&=\; (\text{dist} \circ \langle \text{id}, \text{seq} \rangle)(k).
\end{aligned}
$$

The third equation gives us the more familiar definition that contains the argument k:

$$g(k) = \text{dist}(k, \text{seq}(k)).$$

The last equation allows us to cancel the variable k to obtain the following definition:

$$g = \text{dist} \circ \langle \text{id}, \text{seq} \rangle.$$

Can you figure out the type of g? ♦

EXAMPLE 4 (*The Selector Functions*). A *selector function* is a function that selects one component from a tuple. If $n \in \mathbb{N}$, we'll let boldface **n** represent the selector function that picks out the nth component of a tuple of length n or more. For example, we have

$$\mathbf{2}(x, y, z) = y \quad \text{and} \quad \mathbf{3}(x, y, z, w) = z.$$

Selector functions can sometimes help us write a function as a composition of simpler functions. For example, suppose we define the function "max," having type $\mathbb{R} \times \mathbb{R} \to \mathbb{R}$, as follows:

$$\max(x, y) \;=\; \text{if } x < y \text{ then } y \text{ else } x.$$

Now let's use max to define the function "max3," which returns the maximum of three real numbers, as follows:

$$\max3(x, y, z) = \max(\max(x, y), z).$$

Now let's use selector functions to help us write max3 as a composition of other functions without using variables.

$$
\begin{aligned}
\max3(x, y, z) \;&=\; \max(\max(x, y), z) \\
&=\; \max(\max(\mathbf{1}(x, y, z), \mathbf{2}(x, y, z)), \mathbf{3}(x, y, z)) \\
&=\; \max(\max(\langle \mathbf{1}, \mathbf{2}\rangle(x, y, z)), \mathbf{3}(x, y, z)) \\
&=\; \max((\max \circ \langle \mathbf{1}, \mathbf{2}\rangle)(x, y, z), \mathbf{3}(x, y, z)) \\
&=\; \max(\langle \max \circ \langle \mathbf{1}, \mathbf{2}\rangle, \mathbf{3}\rangle(x, y, z)) \\
&=\; (\max \circ \langle \max \circ \langle \mathbf{1}, \mathbf{2}\rangle, \mathbf{3}\rangle)(x, y, z).
\end{aligned}
$$

We can cancel (x, y, z) to obtain the following definition of max3 as a composition of known functions:

$$\max3 = \max \circ \langle \max \circ \langle \mathbf{1}, \mathbf{2}\rangle, \mathbf{3}\rangle. \quad \blacklozenge$$

Higher-Order Functions

A function is called a *higher-order* function if its arguments and values are allowed to be functions. This is an important property that most good programming languages possess. The composition and tupling operations are examples of functions that take other functions as arguments and return functions as results. For example, suppose we have two functions $f : A \to B$ and $g : B \to C$. Then we can write $g \circ f$ in prefix form as

$$\circ (f, g).$$

Viewed in this way, composition is a higher-order function with the following type expression:

$$(A \to B) \times (B \to C) \to (A \to C).$$

We can do the same thing with a tuple of functions. For example, suppose we have the tuple of functions $\langle f, g, h \rangle$, where $f : A \to B$, $g : A \to C$ and $h : A \to D$. If we define $T(f, g, h) = \langle f, g, h \rangle$, then T is a higher-order function and its type is given by the following expression:

$$(A \to B) \times (A \to C) \times (A \to D) \to (A \to B \times C \times D).$$

The higher-order functions that we discuss next are very useful in constructing programs. They occur in one form or another in most functional programming languages.

The Map Function (Apply to All)

The *map* function takes one argument, a function $f : A \to B$, and returns as a result the function $\text{map}(f) : \text{Lists}[A] \to \text{Lists}[B]$, where $\text{map}(f)$ applies f to each element in its argument list. For example, let $f : \{a, b, c\} \to \{1, 2, 3\}$ be defined by $f(a) = f(b) = 1$ and $f(c) = 2$. Then $\text{map}(f)$ applied to the list $\langle a, b, c, a \rangle$ can be calculated as follows:

$$\text{map}(f)(\langle a, b, c, a \rangle) = \langle f(a), f(b), f(c), f(a) \rangle = \langle 1, 1, 2, 1 \rangle.$$

The type of the map function is $(A \to B) \to (\text{Lists}[A] \to \text{Lists}[B])$. The map function is sometimes called the "applyToAll" function.

For a specific example we'll consider the + function and apply $\text{map}(+)$ to a list of pairs of integers:

$$\text{map}(+)(\langle \langle 1, 2 \rangle, \langle 3, 4 \rangle, \langle 5, 6 \rangle \rangle) = \langle +\langle 1, 2 \rangle, +\langle 3, 4 \rangle, +\langle 5, 6 \rangle \rangle = \langle 3, 7, 11 \rangle.$$

The map function is quite useful. For example, suppose we define the function squares : $\mathbb{N} \to \text{Lists}[\mathbb{N}]$ by $\text{squares}(n) = \langle 0, 1, 4, ..., n^2 \rangle$. We can construct squares as follows:

$$\text{squares} = \text{map}(*) \circ \text{pairs} \circ \langle \text{seq}, \text{seq} \rangle.$$

Let's calculate the value of the expression squares(3).

$$
\begin{aligned}
\text{squares}(3) \;&=\; (\text{map}(*) \circ \text{pairs} \circ \langle \text{seq}, \text{seq} \rangle)(3) \\
&=\; \text{map}(*)(\text{pairs}(\langle \text{seq}, \text{seq} \rangle(3))) \\
&=\; \text{map}(*)(\text{pairs}(\text{seq}(3), \text{seq}(3))) \\
&=\; \text{map}(*)(\text{pairs}(\langle 0, 1, 2, 3 \rangle, \langle 0, 1, 2, 3 \rangle)) \\
&=\; \text{map}(*)(\langle \langle 0, 0 \rangle, \langle 1, 1 \rangle, \langle 2, 2 \rangle, \langle 3, 3 \rangle \rangle) \\
&=\; \langle *(0, 0), *(1, 1), *(2, 2), *(3, 3) \rangle \\
&=\; \langle 0, 1, 4, 9 \rangle.
\end{aligned}
$$

The Apply Function

The *apply* function takes two arguments, a function $f : A \rightarrow B$ and an element $x \in A$, and returns the result of applying f to x. In other words, we have

$$\text{apply}(f, x) = f(x).$$

On the surface, this doesn't appear too useful. But it does allow a function to be used as a parameter to another function. Consider the following simple example. The function $g(f, x) = f(x) + x$ can be constructed as follows:

$$
\begin{aligned}
g(f, x) &= f(x) + x \\
&= +(f(x), x) \\
&= +(\text{apply}(f, x), x) \\
&= +(\langle \text{apply}, \mathbf{2} \rangle (f, x)) \\
&= (+ \circ \langle \text{apply}, \mathbf{2} \rangle) (f, x).
\end{aligned}
$$

So we can cancel the arguments (f, x) and define $g = + \circ \langle \text{apply}, \mathbf{2} \rangle$.

For another example we'll use apply to construct an alternative version of the map function, which we'll call "altMap." AltMap takes two arguments, a function f and a list L, and returns the list of elements $f(x)$ for each $x \in L$. For example, we can write

$$\text{altMap}(f, \langle a, b, c \rangle) = \langle f(a), f(b), f(c) \rangle.$$

To get the same result with the map function, we would write

$$\text{map}(f)(\langle a, b, c \rangle) = \langle f(a), f(b), f(c) \rangle.$$

AltMap has type $(A \rightarrow B) \times \text{Lists}[A] \rightarrow \text{Lists}[B]$. We can use the apply, map, and selector functions to define altMap as follows:

$$
\begin{aligned}
\text{altMap}(f, \langle a, b, c \rangle) &= \langle f(a), f(b), f(c) \rangle \\
&= \text{map}(f)(\langle a, b, c \rangle) \\
&= \text{apply}(\text{map}(f), \langle a, b, c \rangle) \\
&= \text{apply}((\text{map} \circ \mathbf{1})(f, \langle a, b, c \rangle), \mathbf{2}(f, \langle a, b, c \rangle)) \\
&= \text{apply}(\langle \text{map} \circ \mathbf{1}, \mathbf{2} \rangle (f, \langle a, b, c \rangle)) \\
&= (\text{apply} \circ \langle \text{map} \circ \mathbf{1}, \mathbf{2} \rangle)(f, \langle a, b, c \rangle).
\end{aligned}
$$

Therefore we can cancel the arguments $(f, \langle a, b, c \rangle)$ to obtain the definition

$$\text{altMap} = \text{apply} \circ \langle \text{map} \circ \mathbf{1}, \mathbf{2} \rangle.$$

EXAMPLE 5 (*Graphing*). We'll look at a graphing example to illustrate the use of composition of known functions to build new functions. Suppose we have a function f defined on the closed interval $[a, b]$ and we have a sequence of points $\langle x_0, ..., x_n \rangle$ that form a regular partition of $[a, b]$. We want to find the following set of $n + 1$ points:

$$\langle x_0, f(x_0) \rangle, ..., \langle x_n, f(x_n) \rangle.$$

The partition is defined by $x_i = a + d \cdot i$ for $0 \leq i \leq n$, where $d = (b - a)/n$. So the sequence is a function of a, d, and n. If we can somehow create the two sequences $\langle x_0, ..., x_n \rangle$ and $\langle f(x_0), ..., f(x_n) \rangle$, then the desired set of points can be obtained by applying the "pairs" function to these two sequences. Let "makeSeq" be the function that returns the sequence $\langle x_0, ..., x_n \rangle$. We will start by trying to define makeSeq in terms of functions that are already at hand. First we write down the desired value of the expression, makeSeq(a, d, n) and then try to gradually transform the value into an expression involving known functions and the arguments a, d, and n.

$$
\begin{aligned}
\text{makeSeq}(a, d, n) & \\
&= \langle x_0, x_1, ..., x_n \rangle \\
&= \langle a, a + d, a + 2d, ..., a + nd \rangle \\
&= \text{map}(+)(\langle \langle a, 0 \rangle, \langle a, d \rangle, \langle a, 2d \rangle, ..., \langle a, nd \rangle \rangle) \\
&= \text{map}(+)(\text{dist}(a, \langle 0, d, 2d, ..., nd \rangle)) \\
&= \text{map}(+)(\text{dist}(a, \text{map}(*)(\langle \langle d, 0 \rangle, \langle d, 1 \rangle, \langle d, 2 \rangle, ..., \langle d, n \rangle \rangle))) \\
&= \text{map}(+)(\text{dist}(a, \text{map}(*)(\text{dist}(d, \langle 0, 1, 2, ..., n \rangle)))) \\
&= \text{map}(+)(\text{dist}(a, \text{map}(*)(\text{dist}(d, \text{seq}(n))))).
\end{aligned}
$$

The last expression contains only known functions and the arguments a, d, and n. So we have a definition for makeSeq. We'll leave it as an exercise to write the expression for makeSeq as a composition of known functions without using variables. Now it's an easy matter to build the second sequence. Just notice that

$$
\begin{aligned}
\langle f(x_1), ..., f(x_n) \rangle &= \text{map}(f)(\langle x_0, x_1, ..., x_n \rangle) \\
&= \text{map}(f)(\text{makeSeq}(a, d, n)).
\end{aligned}
$$

Now let "makeGraph" be the name of the function that returns the desired set of points. Then makeGraph can be written as follows:

$$\text{makeGraph}(f, a, d, n) = \langle\langle x_0, f(x_0)\rangle, ..., \langle x_n, f(x_n)\rangle\rangle$$
$$= \text{pairs}(\text{makeSeq}(a, d, n), \text{map}(f)(\text{makeSeq}(a, d, n)))$$

This gives us a definition of makeGraph in terms of known functions and the variables $f, a, d,$ and n. We'll leave it as an exercise to write makeGraph as a composition of functions without using variables.   ◆

The Insert Function

It's important from a computational point of view to be able to easily extend a function such as + to handle more that two arguments. For example, if we want to add up the numbers 8, 2, 9, 4, then we can associate pairs of numbers as in the following expression:

$$8 + (2 + (9 + 4))).$$

The insert function is designed to handle this situation.

The *insert function* extends a binary function to take two or more arguments. For example, if f is a binary function, then $\text{insert}(f)(x, y) = f(x, y)$ and for $n \geq 2$ we have

$$\text{insert}(f)(y_1, ..., y_n) = f(y_1, f(y_2, ..., f(y_{n-1}, y_n)\cdots)).$$

For example, if we write + as a prefix operation, then we have

$$
\begin{aligned}
\text{insert}(+)(8, 2, 9, 4) &= +(8, +(2, +(9, 4))) \\
&= +(8, +(2, 13)) \\
&= +(8, 15) \\
&= 23.
\end{aligned}
$$

EXAMPLE 6 (*A Sum of Functional Values*). Suppose we want to construct the sum $f(x_1) + \cdots + f(x_n)$ for an arbitrary function $f : \mathbb{N} \to \mathbb{N}$ and an arbitrary list $\langle x_1, ..., x_n \rangle$. We'll let sumFun be the name of the desired function. Starting with the desired output expression, we'll transform it into a composition of simpler functions, all of which are known to us. We'll continue the process until we reach a point where we read off a definition for sumFun as a composition of functions without using variables. It goes something like this:

$$\text{sumFun}(f, \langle x_1, ..., x_n \rangle)$$

$$= f(x_1) + \cdots + f(x_n)$$

$$= \text{insert}(+)(f(x_1), ..., f(x_n))$$

$$= \text{insert}(+)(\text{map}(f)(\langle x_1, ..., x_n \rangle))$$

$$= \text{insert}(+)(\text{apply}(\text{map}(f), \langle x_1, ..., x_n \rangle))$$

$$= \text{insert}(+)(\text{apply}(\langle \text{map} \circ \mathbf{1}, \mathbf{2}\rangle(f, \langle x_1, ..., x_n \rangle)))$$

$$= (\text{insert}(+) \circ \text{apply} \circ \langle \text{map} \circ \mathbf{1}, \mathbf{2}\rangle)(f, \langle x_1, ..., x_n \rangle)))$$

So we can cancel and make the following definition:

$$\text{sumFun} = \text{insert}(+) \circ \text{apply} \circ \langle \text{map} \circ \mathbf{1}, \mathbf{2}\rangle. \quad \blacklozenge$$

From the programming point of view there are many other interesting ways to combine functions. But they will take us too far afield. The primary purpose now is to get a feel for what a function is and to grasp the idea of building a function from other functions.

Exercises

1. a. Write down the 16 values of floor($\log_2(x)$) for x in {1, 2, 3, ..., 16}.
    b. Write down the 16 values of ceiling($\log_2(x)$) for x in {1, 2, 3, ..., 16}.

2. a. Describe the set of natural numbers x such that floor($\log_2(x)$) = 7.
    b. Describe the set of natural numbers x such that ceiling($\log_2(x)$) = 7.

3. Prove each of the following statements.
    a. floor(ceiling(x)) = ceiling(x).
    b. ceiling(floor(x)) = floor(x).
    c. floor($\log_2(x)$) = floor($\log_2$(floor(x))) for $x \geq 1$.

4. Find a formula for the number of binary digits in the binary representation of a nonzero natural number x. For example, the number 15 requires four binary digits 1111.

5. Suppose that $f : \mathbb{N} \to \text{GenLists}[\mathbb{N}]$ is defined by $f(m) = \langle \langle 0, 1, ..., m \rangle, m \rangle$. Find a definition of f as a combination of known functions.

6. Find a definition for each function in terms of known functions.
    a. The function $f : \text{Lists}[\mathbb{R}] \to \text{Lists}[\mathbb{R}]$ that takes a list of real numbers and returns the list of absolute values of the given list.
    b. The function cubes : $\mathbb{N} \to \text{Lists}[\mathbb{N}]$ that takes a number n and returns the list of cubes $\langle 0, 1, 8, 27, ..., n^3 \rangle$.

7. Write down each step in the evaluation of max3(4, 9, 7) using the definition max3 = max ∘ ⟨max ∘ ⟨**1, 2**⟩, **3**⟩.

8. Use the apply function to define each of the following functions in terms of known functions.

   a. $h(f, g, x) = f(x) + g(x)$.

   b. $h(f, g, x, y) = f(x) + g(y)$.

   c. $h(f, g, \langle x, y \rangle) = f(x) + g(y)$.

9. Find a definition for each of the following functions as a composition of functions without using variables.

   a. makeSeq(a, d, n) = map(+)(dist(a, map(∗)(dist(d, seq(n))))).

   b. makeGraph(f, a, d, n)
   $$= \text{pairs(makeSeq}(a, d, n), \text{map}(f)(\text{makeSeq}(a, d, n)))$$

10. Let h be the function that takes as arguments a triple consisting of two functions and a list. It returns a list of pairs of functional values as indicated:

   $$h(f, g, \langle x_1, ..., x_n \rangle) = \langle \langle f(x_1), g(x_1) \rangle, ..., \langle f(x_n), g(x_n) \rangle \rangle.$$

   Write down a definition of h as a composition of known functions.

11. Combine known functions to give a definition for the function sumCubes : $\mathbb{N} \to \mathbb{N}$ defined by sumCubes(n) = $1^3 + 2^3 + \cdots + n^3$.

12. Given that + has the type $\mathbb{N} \times \mathbb{N} \to \mathbb{N}$, find the type of insert(+). Find an expression for the type of the insert function.

2.3 Properties of Functions

We're going to look at some properties of functions that can help us compare the cardinality of sets. Sometimes it's not so easy to count the elements of a set. For example, suppose we want to count the elements in the following set:

$$A = \{2, 5, 8, 11, 14, 17, ..., 44, 47\}.$$

One way to count A is to observe that adjacent numbers in A differ by 3. This allows us to write down a second representation of A as follows:

$$A = \{2 + 3k \mid 0 \leq k \leq 15\}.$$

Now it's easy to count A by counting the range of k, which is the same as counting the set $\{0, 1, ..., 15\}$. From a functional point of view, we've defined a function $f : \mathbb{N}_{16} \to A$ by $f(k) = 2 + 3k$. Is there some property of f that allows us to conclude that $\mathbb{N}_{16}$ and A have the same number of elements? Yes, and that's what we're about to discuss.

Injective and Surjective

A function $f : A \to B$ is called *injective* (also *one-to-one*, or an *embedding*) if no two elements in A map to the same element in B. Formally, f is injective if for all $x, y \in A$, whenever $x \neq y$, then $f(x) \neq f(y)$. Another way to say this is that for all $x, y \in A$, if $f(x) = f(y)$, then $x = y$. An injective function is called an *injection*. Figure 2.5 illustrates an injection from A to B.

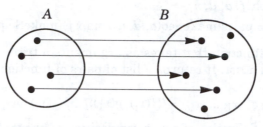

Figure 2.5

For example, let $A = \{6k + 4 \mid k \in \mathbb{N}\}$ and $B = \{3k + 4 \mid k \in \mathbb{N}\}$. Define the function $f : A \to B$ by $f(x) = x + 3$. Then f is injective, since $x \neq y$ implies that $x + 3 \neq y + 3$. Notice that if A and B are finite sets and $f : A \to B$ is injective, then $|A| \leq |B|$.

EXAMPLE 1 (*Real Functions*). A glance at the graph of a function $f : \mathbb{R} \to \mathbb{R}$ is an easy way to decide properties of f. For example, Figure 2.6 represents a function f that is not injective because there are numbers a and b such that $f(a) = f(b)$ with $a \neq b$. If a function's graph is always increasing or always decreasing, then the function is injective. Consider the following examples.

a) A familiar example of an injective function $f : \mathbb{R} \to \mathbb{R}$ is defined by $f(x) = ax + b$ ($a \neq 0$). In other words, f is just the equation of a line that is not horizontal in an x-y coordinate system where $y = f(x)$.

b) The function $f : \mathbb{R} \to \mathbb{R}$ defined by $f(x) = ax^2 + bx + c$ ($a \neq 0$) is not injective. Recall that the graph of f is a parabola opening up or down, depending on the sign of a.

c) The function $f : \mathbb{R} \to \mathbb{R}$ defined by $f(x) = x^3$ is injective. But if we define $f(x) = x^3 + 2x^2$, then f is not injective.   ◆

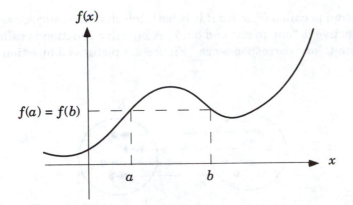

Figure 2.6

A function $f : A \rightarrow B$ is called *surjective* (also *onto*) if each element $b \in B$ can be written as $b = f(x)$ for some element x in A. Another way to say this is that f is surjective if range$(f) = B$. A surjective function is called a *surjection*. Figure 2.7 pictures a surjection from A to B.

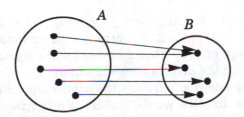

Figure 2.7

For example, the floor function from $\mathbb{R}$ to $\mathbb{Z}$ is surjective. For another example, the function $f : \mathbb{Q} \rightarrow \mathbb{Z}$ defined by $f(\frac{m}{n}) = m$ is surjective. Notice that if A and B are finite and $f : A \rightarrow B$ is surjective, then $|A| \geq |B|$.

EXAMPLE 2. Let $f : A \times A \rightarrow A$ be the function that projects every pair $\langle x, y \rangle$ to its first coordinate x. In other words, $f(x, y) = x$. Clearly, f is surjective. Is it injective?   ◆

EXAMPLE 3. Let $g : A \rightarrow A \times A$ be the function defined by $g(a) = \langle a, a \rangle$. Clearly, g is injective. Is it surjective?   ◆

A function is called *bijective* if it is both injective and surjective. Another term for bijective is "one to one and onto." A bijective function is called a *bijection* or a "one-to-one correspondence." Figure 2.8 pictures a bijection from A to B:

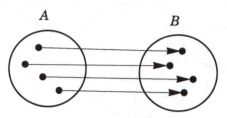

Figure 2.8

Whenever we have a bijection $f : A \to B$, then we always have an *inverse* function $g : B \to A$ defined by $g(b) = a$ if $f(a) = b$. The inverse is also a bijection and we always have the two equations $g \circ f = \mathrm{id}_A$ and $f \circ g = \mathrm{id}_B$. For example, the function $f :$ Odd $\to$ Even defined by $f(x) = x - 1$ is a bijection, and its inverse is the function

$$g : \text{Even} \to \text{Odd defined by } g(x) = x + 1.$$

In this case we have $g \circ f = \mathrm{id}_{\text{Odd}}$ and $f \circ g = \mathrm{id}_{\text{Even}}$. For example, we have $g(f(3)) = g(3 - 1) = (3 - 1) + 1 = 3$. A function with an inverse is often called *invertible*. The inverse of f is often denoted by the symbol f^{-1}, which we should not confuse with the inverse image notation.

We say that two sets A and B have the *same cardinality* if there is a bijection between them. In other words, if there is a function $f : A \to B$ that is bijective, then A and B have the same cardinality. In this case we write

$$|A| = |B|.$$

The term *equipotent* is often used to indicate that two sets have the same cardinality. Since bijections might occur between infinite sets, the idea of equipotence applies not only to finite sets, but also to infinite sets.

For example, let Odd and Even denote the sets of odd and even natural numbers. Then $|\text{Odd}| = |\text{Even}|$ because the function $f :$ Odd $\to$ Even defined by $f(x) = x - 1$ is a bijection. Similarly, $|\text{Even}| = |\mathbb{N}|$ because the function $g :$ Even $\to \mathbb{N}$ defined by $g(y) = \frac{y}{2}$ is a bijection. We'll see in Section 2.4 that not all infinite sets have the same cardinality.

A function may or may not be injective, surjective, or bijective. For example, there are nine functions of type $\{a, b\} \to \{1, 2, 3\}$. Six are injective, and none are surjective. Figure 2.9 shows diagrams of the nine functions.

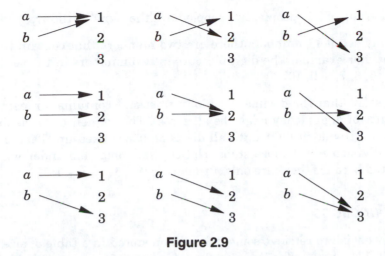

Figure 2.9

The composition $g \circ f$ will always inherit the property of injectivity if both f and g are injective. A similar result holds for the property of surjectivity. The results can be stated as follows:

If f and g are injective, then $g \circ f$ is injective. (2.4)

If f and g are surjective, then $g \circ f$ is surjective. (2.5)

Proof: We'll prove (2.4) and leave (2.5) as an exercise. Let f and g be injective, and assume that $g \circ f(x) = g \circ f(y)$ for some $y \in A$. Since g is injective, it follows that $f(x) = f(y)$, and it follows that $x = y$ because f is injective. Therefore $g \circ f$ is injective.   QED.

The Pigeonhole Principle

We're going to describe a useful rule that we often use without thinking. For example, suppose 20 people receive 21 pieces of mail. It makes sense to conclude that one person received at least two letters. This is an example of the *pigeonhole principle*. Suppose we think of office mail boxes as pigeonholes. If m letters are delivered to n boxes and $m > n$, then one box will get more than one letter. We can describe the pigeonhole principle in more formal terms as follows: If A and B are finite sets with $|A| > |B|$, then every function from A to B maps at least two elements of A to a single element of B. This is the same as saying that no function from A to B is an injection.

This simple idea is used often in many different settings. We'll be using it at several places in the book. Here are a few statements that can be justified by the pigeonhole principle:

In a group of 367 people, two people have the same birthday.

In any set of 11 numbers there are two numbers that contain the same digit. For example, the digit 0 occurs in two numbers in the set {0, 1, 2, 3, 4, 5, 6, 7, 8, 9, 10}.

We know that the decimal form of a fraction $\frac{m}{n}$ contains a repeating sequence of digits (they might be all zeros). The sequence can be found as follows: Divide m by n until all digits of m are used up. Then continue the division $n + 1$ more steps. Notice that some remainder will be repeated because there are only n remainders 0, 1, ..., $n - 1$.

Hash Functions

Suppose we wish to retrieve some information stored in a table of size n with indexes 0, 1, ..., $n - 1$. The items in the table can be very general things. For example, the items might be strings of letters, or they might be large records with many fields of information. To look up a table item we need a *key* to the information we desire. The key is normally an actual piece of the information that we need. For example, if the table contains records of information for the 12 months of the year, the keys might be the three-letter abbreviations for the 12 months. To look up the record for January, we would present the key JAN to a lookup program. The program uses the key to find the table entry for the January record of information. Then the information would be displayed for our use.

An easy way to look up the January record is to search the table until the key JAN is found. This might be OK for a small table with 12 entries. But it may be impossibly slow for large tables with thousands of entries. Here is the general problem that we want to solve:

Given a key, find the table entry containing the key without searching.

This may seem impossible at first glance. But let's consider a way to use a function to map each key directly to its table location.

A *hash function* is a function that maps a set of keys to a finite set of table indexes. For example, let S be the set of all three-letter strings. If X is any letter of the alphabet, let ord(X) denote the integer value of the ASCII code for X. A hash function $h : S \to \{0, 1, ..., n - 1\}$ might be defined as follows, where XYZ represents a string of three letters:

$$h(XYZ) = \text{ord}(X) \bmod n.$$

In other words, h picks off the first letter, ords it, and then mods the result to get a table index. Most programming languages have efficient implementations of the ord and mod functions.

If a hash function is injective, then every key maps to its table index, and no searching is involved. Often this is not possible. When two keys map to the same table index, the result is called a *collision*. So if a hash function is not injective, it has collisions. Our example hash function has several collisions if we agree to let S be the three-letter abbreviations for the 12 months, in capital letters. For example, the keys JAN, JUN, JUL all begin with the letter J. So they all map to the same table value.

When collisions occur, we store one of the records in the common table location and find another table location for the other records. There are many ways to find the location for a key that has collided with another key. One technique is called *linear probing*. With this technique the program searches the remaining locations in a "linear" manner. For example, if location i is the collision index, then the following sequence of table locations is searched:

$$(i + 1) \bmod n, (i + 2) \bmod n, ..., (i + n) \bmod n.$$

In creating the table in the first place, these locations would be searched to find the first open table entry. Then the key would be placed in that location. The above probe sequence is not the best for certain kinds of keys. Often it's better to probe with a "gap" between table locations in order to "scatter" or "hash" the information to different parts of the table. The idea is to keep the number of searches to a minimum. Let g be a gap, where $1 \leq g < n$. Then the following sequence of table locations is searched in case a collision occurs at location i:

$$(i + g) \bmod n, (i + 2g) \bmod n, ..., (i + ng) \bmod n.$$

Some problems can occur if we're not careful with our choice of g. For example, suppose $n = 12$ and $g = 3$. Then the probe sequence can skip some table entries. For example, if $i = 0$, the above sequence becomes

$$3, 6, 9, 0, 3, 6, 9, 0, 3, 6, 9, 0.$$

So we would miss table entries 1, 2, 4, 5, 7, 8, 10, and 11. Let's try another value for g. Suppose we try $g = 5$. Then we obtain the following probe sequence starting at $i = 0$:

$$5, 10, 3, 8, 1, 6, 11, 4, 9, 2, 7, 0.$$

In this case we cover the entire set $\{0, 1, ..., 11\}$. In other words, we've defined a bijection $f : \mathbb{N}_{12} \to \mathbb{N}_{12}$ by $f(x) = 5x \bmod 12$. Can we always find a probe sequence that hits all the elements of $\{0, 1, ..., n - 1\}$? Happily, the answer is yes. Just pick g and n so that they are relatively prime, $(g, n) = 1$. Exercise 8 is devoted to this topic. For example, if we pick n to be a prime number, then $(g, n) = 1$ for any g in the interval $1 \le g < n$.

There are many ways to define hash functions and to resolve collisions. The paper by Cichelli [1980] examines some bijective hash functions.

Exercises

1. For each of the following properties, construct a function with that property. Choose your domains and codomains from the three sets

    $$A = \{a, b, c\}, \quad B = \{x, y, z\}, \quad C = \{1, 2\}.$$

    a. Injective but not surjective.
    b. Surjective but not injective.
    c. Bijective.

2. For each of the following types, compile some statistics: the number of functions of that type; the number that are injective; the number that are surjective; the number that are bijective; the number that are neither injective, surjective, nor bijective.

    a. $\{a, b, c\} \to \{1, 2\}$.
    b. $\{a, b\} \to \{1, 2, 3\}$.
    c. $\{a, b, c\} \to \{1, 2, 3\}$.

3. The fatherOf function from People to People is neither injective nor surjective. Why?

4. For each of the following functions, state which one of the four properties holds: injective, surjective, bijective, and none (meaning not injective, not surjective, and not bijective). For each bijective function, find its inverse.

    a. $f : \mathbb{N} \to \mathbb{N}$, where $f(x) = x + 1$.
    b. $f : \mathbb{R} \to \mathbb{Z}$, where $f(x) = \text{floor}(x)$.
    c. $f : \mathbb{N} \to \mathbb{N}$, where $f(x) = x \bmod 10$.
    d. $f : \mathbb{N} \to \mathbb{N}$, where $f(x) = $ if x is odd then $x - 1$ else $x + 1$.
    e. $f : A \to \text{Power}(A)$, where A is any set and f is defined by $f(x) = \{x\}$.
    f. $f : \text{Lists}[A] \to \text{Power}(A)$, where A is a finite set and f is defined by
        $$f(\langle x_1, ..., x_n \rangle) = \{x_1, ..., x_n\}.$$
    g. $f : \text{Lists}[A] \to \text{Bags}(A)$, where A is a finite set and f is defined by
        $$f(\langle x_1, ..., x_n \rangle) = [x_1, ..., x_n].$$

h. $f : \text{Bags}(A) \rightarrow \text{Power}(A)$, where A is a finite set and f is defined by
$$f([x_1, ..., x_n]) = \{x_1, ..., x_n\}.$$

i. The sequence function $\text{seq} : \mathbb{N} \rightarrow \text{Lists}[\mathbb{N}]$.

j. The distribute function $\text{dist} : A \times \text{Lists}[B] \rightarrow \text{Lists}[A \times B]$.

k. $f : \mathbb{Z} \rightarrow \mathbb{N}$ defined by $f(x) = |x + 1|$.

l. $f : \mathbb{N}_6 \rightarrow \mathbb{N}_6$ where $f(x) = 2x \bmod 6$.

m. $f : \mathbb{N}_5 \rightarrow \mathbb{N}_5$ where $f(x) = 2x \bmod 5$.

n. $f : \mathbb{N}_6 \rightarrow \mathbb{N}_6$ where $f(x) = 5x \bmod 6$.

5. Show that each function is a bijection from the positive real numbers to the open interval $(0, 1)$.

   a. $f(x) = \dfrac{1}{x + 1}$.   b. $g(x) = \dfrac{x}{x + 1}$.

6. Show that each function is a bijection from the open interval $(0, 1)$ to the positive real numbers.

   a. $f(x) = \dfrac{x}{1 - x}$.   b. $g(x) = \dfrac{1 - x}{x}$.

7. Let $S = \{\text{one, two, three, four, five, six, seven, eight, nine}\}$. Suppose we want to create a hash table containing the strings of S, where the table is indexed from 0 to 8 and the hash function $h : S \rightarrow \{0, 1, ..., 8\}$ is defined as follows:

x	one	two	three	four	five	six	seven	eight	nine
$h(x)$	2	4	6	8	8	6	4	2	0

   Starting with an empty table, place the elements of S in the table using the following input sequence: one, two, three, four, five, six, seven, eight, nine. Resolve collisions by linear probing. Write down the table obtained for each of the following linear probing gaps.

   a. Linear probing with gap = 1.

   b. Linear probing with gap = 4.

   c. Linear probing with gap = 3.

8. Let k and m be relatively prime positive integers, $(k, m) = 1$. Define the function $f : \mathbb{N}_k \rightarrow \mathbb{N}_k$ by $f(x) = (mx) \bmod k$. Prove that f is a bijection. *Hint:* Since $\mathbb{N}_k$ is finite, just show that f is an injection.

9. Let $f : A \rightarrow B$ and $g : B \rightarrow C$. Prove that if f and g are surjective, then the composition $g \circ f$ is surjective.

10. Suppose that the two functions f and g can be formed into a composition $g \circ f$.

a. What can you tell about f or g if you know that $g \circ f$ is surjective?

b. What can you tell about f or g if you know that $g \circ f$ is injective?

11. Given the functions $g : A \to B$ and $h : A \to C$, let $f = \langle g, h \rangle$. Show that each of the following statements holds.

a. If f is surjective, then g and h are surjective. Find an example to show that the converse is false.

b. If g or h is injective, then f is injective. Find an example to show that the converse is false.

2.4 Counting Infinite Sets

We want to show that some infinite sets have more elements than others. As a consequence, we'll see that many things cannot be computed. To do this, we need to give a meaning to the word "more." For example, does $\mathbb{N}$ have more elements than the set A^*, where $A = \{a, b\}$? To answer this question, we need to define "more" in terms of mappings between sets.

For two sets A and B, if there is an injection $f : A \to B$, then we say, "The cardinality of A is less than or equal to the cardinality of B," and we write

$$|A| \leq |B|.$$

For example, let Even denote the set of even natural numbers. Then the function $f : \text{Even} \to \mathbb{N}$ defined by $f(x) = x$ is an injection from the even natural numbers to $\mathbb{N}$. So we can write $|\text{Even}| \leq |\mathbb{N}|$.

Now we can give a precise meaning to the idea of "more." If A and B are sets and $|A| \leq |B|$ and $|A| \neq |B|$, then we say, "The cardinality of A is less than the cardinality of B," and we write

$$|A| < |B|.$$

In other words, $|A| < |B|$ means that there is an injection from A to B but <u>no</u> bijection between them.

Why can't we simply say that $|A| < |B|$ means that there is an injection from A to B that is not a bijection? To see why not, let Odd be the set of odd natural numbers, and let $g : \mathbb{N} \to \text{Odd}$ be defined by $g(x) = 4x + 1$. Then g is an injection, but g is not a bijection because $3 \in \text{Odd}$ and $g(x) \neq 3$ for any $x \in \mathbb{N}$. We surely don't want to conclude from this that $|\mathbb{N}| < |\text{Odd}|$. For example, the function $f : \mathbb{N} \to \text{Odd}$ defined by $f(x) = 2x + 1$ is a bijection. So $|\mathbb{N}| = |\text{Odd}|$. Thus we will stick with our definition of "more."

It's easy to find infinite sets having different cardinalities. Georg Cantor showed that any set A has less cardinality than its power set, which we can write as follows:

$$|A| < |\text{power}(A)|. \tag{2.6}$$

We know that this is true for finite sets. But it's also true for infinite sets. The easy part of the proof can be seen by observing that each element $x \in A$ corresponds to the singleton set $\{x\} \in \text{power}(A)$. This correspondence is an injection. Therefore $|A| \leq |\text{power}(A)|$. The hard part, which we won't describe, was Cantor's proof that no bijections exist between the two sets.

We can apply (2.6) to any set. For example, we have $|\mathbb{N}| < |\text{power}(\mathbb{N})|$. In other words, there are "more" subsets of $\mathbb{N}$ than there are elements of $\mathbb{N}$, even though both sets are infinite.

Countable and Uncountable

We want to describe those infinite sets that can be counted even if it takes forever to count them. A set C is *countable* if it's finite or if $|C| = |\mathbb{N}|$. In the case $|C| = |\mathbb{N}|$ we sometimes say that C is *countably infinite*. So $\mathbb{N}$ is the fundamental example of a countably infinite set. We could also define C to be countable if $|C| \leq |\mathbb{N}|$. Thus we can show that a set C is countable by finding an injection from C to $\mathbb{N}$ or by finding a surjection from $\mathbb{N}$ to C. Can you see why? If a set is not countable, it is called *uncountable*. For example, (2.6) tells us that $|\mathbb{N}| < |\text{power}(\mathbb{N})|$. Since $\mathbb{N}$ is countable, it follows that $\text{power}(\mathbb{N})$ is uncountable.

A simple way to show that an infinite set is countable is by listing the elements in some way. The listing may include repetitions, since all we need to do is exhibit a surjection from $\mathbb{N}$ to the set. For example, the positive rational numbers can be listed by first writing the number $\frac{1}{1}$. Then we write the two numbers whose numerator and denominator sum to 3. Next, write down the three numbers whose numerator and denominator sum to 4, and so on. We can write down the first few numbers in the list as follows:

$$\frac{1}{1}$$
$$\frac{1}{2} \quad \frac{2}{1}$$
$$\frac{1}{3} \quad \frac{2}{2} \quad \frac{3}{1}$$
$$\frac{1}{4} \quad \frac{2}{3} \quad \frac{3}{2} \quad \frac{4}{1}$$
$$\vdots$$

Notice that repetitions occur in the listing. For example,$\frac{1}{1} = \frac{2}{2} = \cdots$. But all the positive rationals are listed. Therefore the positive rationals are countable.

The following result can often be used to show that a set is countable if it can be represented as a countable union of countable sets.

Counting Unions of Countable Sets   (2.7)

If A is the union of a countable collection of sets, where each
set in the collection is countable, then A is countable.

Proof: We'll start by listing the sets in the countable collection of sets as follows: $A_0, A_1, ..., A_n, ...$. Since each set A_i is countable, we can list its elements as follows: $a_{i0}, a_{i1}, a_{i2}, ...$. This allows us to list the elements in the union of all the sets A_i as follows:

$$a_{00}, a_{01}, a_{02}, ...$$

$$a_{10}, a_{11}, a_{12}, ...$$

$$\vdots$$

$$a_{i0}, a_{i1}, a_{i2}, ...$$

$$\vdots$$

We can count these elements by starting with a_{00}. Then we count the diagonal elements a_{01} and a_{10}. Continue in this manner to count each northeast to southwest diagonal. For example, the next diagonal consists of the three elements a_{02}, a_{11}, a_{20}. In this way we have established a bijection between $\mathbb{N}$ and the listed elements. Thus they are countable.   QED.

An important consequence of (2.7) is the following fact about the countability of the set of all strings over a finite alphabet:

If A is a finite alphabet, then A^* is countably infinite.   (2.8)

Proof: For each $n \in \mathbb{N}$, let A_n be the set of strings over A having length n. It follows that A^* is the union of the sets $A_0, A_1, ..., A_n, ...$. Since each set A_n is finite, we can apply (2.7) to conclude that A^* is countable.   QED.

As a result of (2.8) and (2.6), it follows that

If A is a finite alphabet, then power(A^*) is uncountable.   (2.9)

As another application of (2.8) we can answer the question: What is the cardinality of the set of all programs written in your favorite programming language? The answer is countably infinite. One way to see this is to consider each program as a finite string of symbols over a fixed finite alphabet A. For example, A might consist of all characters that can be typed from a keyboard. Now we can proceed as in the proof of (2.8). For each natural number n, let P_n denote the set of all programs that are strings of length n over A. For example, the program {print(4)} is in P_{10} because it's a string of length 10. So the set of all programs is the union of the sets $P_0, P_1, ..., P_n, ...$. Since each P_n is finite, we can use (2.7) to give the following result:

<blockquote>The set of all programs for a programming language   (2.10)
is countably infinite.</blockquote>

We know that infinite sets can have different cardinalities. But some infinite sets have the same cardinality. Let's do an example.

EXAMPLE 1. We'll show that the closed unit interval of real numbers [0, 1] has the same cardinality as power($\mathbb{N}$). To do this, we'll define a bijection between the two sets. Notice that any number in [0, 1] can be written in the binary form

$$0.d_0 d_1 ... d_i ...,$$

where each $d_i = 0$ or 1. This binary form represents the real number

$$\frac{1}{2} d_0 + \frac{1}{4} d_1 + \cdots + \frac{1}{2^{i+1}} d_i + \cdots.$$

For example, the binary form 0.101000... represents the number

$$\frac{1}{2}1 + \frac{1}{4}0 + \frac{1}{8}1 = \frac{5}{8}.$$

Now define the function h : power($\mathbb{N}$) $\rightarrow$ [0, 1] by $h(S) = 0.d_0 d_1 ... d_i ...,$ where $d_i = 1$ if and only if $i \in S$. For example, the empty set corresponds to the binary form 0.000..., which is the number 0. The set of odd natural numbers corresponds to the binary form 0.010101.... $\mathbb{N}$ corresponds to the binary form 0.111..., which represents decimal number 1, just as the decimal form 0.999... represents the decimal number 1. The function h is a bijection (convince yourself). Therefore $|\text{power}(\mathbb{N})| = |[0, 1]|$.   ♦

We know by (2.6) that power($\mathbb{N}$) is uncountable. So as a result of Example 1 we can say that [0, 1] is uncountable. It's not always easy to construct a bijection to show that two sets have the same cardinality. The next result—which was conjectured by Cantor, proved by Bernstein, and independently proved by Schröder—gives us a technique to show that two sets have the same cardinality without explicitly exhibiting a bijection between them:

$$\text{If } |A| \leq |B| \text{ and } |B| \leq |A|, \text{ then } |A| = |B|. \tag{2.11}$$

In other words, to show that a bijection exists between sets A and B, it suffices to find two injections, one from A to B and one from B to A. Let's do some examples.

EXAMPLE 2. We'll use (2.11) to show that the two intervals $(0, 1)$ and $[0, 1]$ have the same cardinality. Since $(0, 1)$ is a subset of $[0, 1]$, the identity mapping from $(0, 1)$ to $[0, 1]$ is an injection. Now we need an injection from $[0, 1]$ to $(0, 1)$. Define $f : [0, 1] \rightarrow (0, 1)$ by $f(x) = \frac{x}{2} + \frac{1}{4}$. It follows that f is an injection. So we can apply (2.11) to conclude that $|[0, 1]| = |(0, 1)|$.   ◆

EXAMPLE 3. We'll show that $|\mathbb{R}| = |(0, 1)|$. Since $(0, 1) \subset \mathbb{R}$, it follows that $|(0, 1)| \leq |\mathbb{R}|$. On the other hand, we can define an injection from $\mathbb{R}$ to $(0, 1)$ as follows. The mapping $f(x) = 2^x$ is an injection from $\mathbb{R}$ to the set of positive real numbers, and the mapping $g(x) = 1/(x + 1)$ is an injection (actually, it's a bijection) from the positive real numbers to $(0, 1)$. Therefore the composition $g \circ f$ is an injection from $\mathbb{R}$ to $(0, 1)$. Thus $|\mathbb{R}| \leq |(0, 1)|$. So $|\mathbb{R}| = |(0, 1)|$ by (2.11).   ◆

Now we can say that the following sets are uncountable and have the same cardinality:

$$|\text{power}(\mathbb{N})| = |[0, 1]| = |(0, 1)| = |\mathbb{R}|.$$

Let's show that not everything is computable. We'll do this by considering the computation of real numbers. The problem can be described as follows:

The Computable Number Problem (2.12)

Compute a real number to any given number of decimal places.

Can any real number be computed? The answer is no. The reason is that there are "only" a countably infinite number of computer programs (2.10).

Therefore there are only a countable number of computable numbers in $\mathbb{R}$ because each computable number needs a program to compute it. Since $\mathbb{R}$ is uncountable, most real numbers cannot be computed. The rational numbers can be computed, and there are also many irrational numbers that can be computed. Pi is the most famous example of a computable irrational number. In fact, there are countably infinitely many computable irrational numbers.

One way to show that a set is uncountable is find another uncountable set and exhibit a bijection between the two as our examples have shown. Another way to proceed that works in some cases is to argue by way of contradiction. It's a classic method that does not need another set for comparison. In the next example we'll use this method to show that the open interval (0, 1) is uncountable.

EXAMPLE 4 (*The Open Unit Interval Is Uncountable*). We'll show that the set of real numbers in the open unit interval (0, 1) is uncountable. We start with the assumption that (0, 1) is countable. With this assumption we can list the elements of (0, 1) as the following sequence:

$$r_0, r_1, r_2, ..., r_n, ... \ .$$

Each number in the sequence will be represented in its decimal form. (For example, $\frac{1}{2} = 0.50000000...$.) Thus we can write down the following listing of the numbers in (0, 1):

$$
\begin{aligned}
r_0 &= 0 . d_{00}\, d_{01}\, d_{02} \cdots \\
r_1 &= 0 . d_{10}\, d_{11}\, d_{12} \cdots \\
r_2 &= 0 . d_{20}\, d_{21}\, d_{22} \cdots \\
&\ \ \vdots \\
r_n &= 0 . d_{n0}\, d_{n1} \quad \cdots \quad d_{nn} \cdots \\
&\ \ \vdots
\end{aligned}
$$

Now we'll construct a new element of (0, 1). To start off, pick your favorite two distinct digits between 1 and 9. For example, we'll use 4 and 5. Now construct a new number, say s, that uses only 4's and 5's in its representation as follows:

$$s = 0 .s_0\, s_1\, s_2 \cdots \ ,$$

where each digit is defined in terms of the "diagonal" digits in the listing as follows:

$$s_i = \text{if } d_{ii} = 4 \text{ then } 5 \text{ else } 4.$$

Certainly, s represents a number in the open interval (0, 1). But s does not occur in the listing because the ith decimal place of s differs from the ith decimal place of r_i for each i. So the listing does not exhaust all the numbers in (0,
1). Since we obtained this contradiction by assuming that (0, 1) is countable,
it follows that (0, 1) is uncountable.   ◆

The preceding example uses a technique known as *diagonalization*. It
was first used by Cantor. We'll state the main result and introduce the technique in the proof.

Diagonalization   (2.13)

Suppose we have a countable listing of objects in which each object is
represented as a stream (i.e., infinite list) over an alphabet A with at
least two symbols. Then the listing is not exhaustive. In other words,
there is some object that is represented as a stream over A but is NOT in
the original listing.

We'll prove (2.13) for an alphabet with two symbols.

Proof: Suppose we have a countable listing of objects in which each object can
be represented as a stream over $A = \{x, y\}$. Then we can represent the listing
as an infinite matrix M, where each row of M represents the stream for an object in the listing. For example, in Figure 2.10, row 0 represents the stream
$\langle x, y, y, x, ... \rangle$, which represents the first object in the listing. Row 1 represents
the second object, and so on. We've emphasized the main diagonal entries be-
cause they will be used to define a stream that is not in the listing.

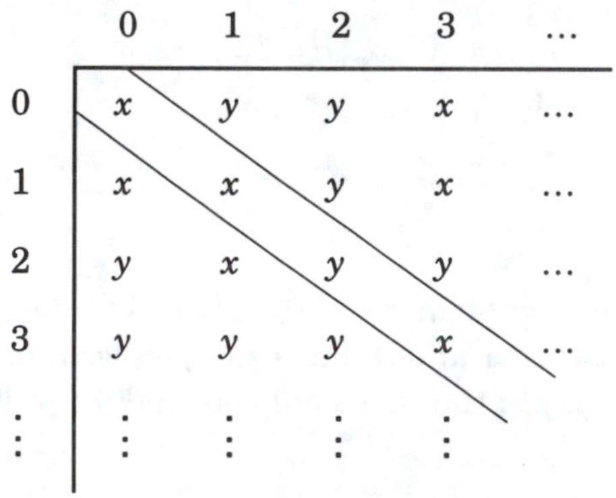

Figure 2.10

We'll define a new stream $\langle d_0, d_1, ... \rangle$ by using the diagonal entries of M as follows:

$$d_i = \text{if } M_{ii} = x \text{ then } y \text{ else } x.$$

For the example matrix we have the stream $\langle y, y, x, y, ... \rangle$. The new stream differs from the ith row of M at the ith position. So it can't occur as a row of M. Therefore we've defined a stream over A that represents an object that is NOT in the listing.   QED.

Let's see how (2.13) could have been applied to the preceding example. We could have represented each real number in the open interval (0, 1) as the stream of its decimal digits $\langle d_{n0}, d_{n1}, ..., d_{nn}, ... \rangle$. Therefore (2.13) applies to tell us that any countable listing of these numbers excludes one of them. So we can't have a countable listing of (0, 1). Therefore (0, 1) is uncountable.

We can use (2.13) in two different ways. We can use it to assert that certain listings are not exhaustive. We can also use the diagonalization technique presented in the proof of (2.13) as a general purpose method to construct a new element not in the listing.

Let's find some more sets that have different infinite cardinalities. Since $|\mathbb{R}| = |(0, 1)|$ and $|(0, 1)| = |\text{power}(\mathbb{N})|$, we can apply (2.6) two times to obtain the following inequality:

$$|\mathbb{N}| < |\mathbb{R}| < |\text{power}(\text{power}(\mathbb{N}))|.$$

Is there any "well-known" set S such that $|S| = |\text{power}(\text{power}(\mathbb{N}))|$? Since the real numbers are hard enough to imagine, how can we comprehend all the elements in $\text{power}(\text{power}(\mathbb{N}))$? If we keep applying (2.6), we can obtain an infinite set of sets of higher and higher cardinality:

$$|\mathbb{N}| < |\text{power}(\mathbb{N})| < |\text{power}(\text{power}(\mathbb{N}))| < |\text{power}(\text{power}(\text{power}(\mathbb{N})))| < \cdots$$

Luckily, in computer science we will seldom, if ever, have occasion to worry about sets having higher cardinality than the set of real numbers.

We'll close the discussion with a question: Is there a set S whose cardinality is between that of $\mathbb{N}$ and that of (0, 1)? In other words, does there exist a set S such that $|\mathbb{N}| < |S| < |(0, 1)|$? The answer is that no one knows. Interestingly, it has been shown that people who assume that the answer is yes won't run into any contradictions by using the assumption in their reasoning. Similarly it has been shown that people who assume that the answer is no won't run into any contradictions by using the assumption in their arguments! The assumption that the answer is no is called the *continuum hypothesis*. If we accept the continuum hypothesis, then we can use it as part of a proof technique. For example, suppose that for some set S we can show that $|\mathbb{N}| \leq |S| < |(0, 1)|$. Then we can conclude that $|\mathbb{N}| = |S|$ by the continuum hypothesis.

Exercises

1. Show that $\mathbb{N} \times \mathbb{N}$ is countable. *Hint:* Find a representation for $\mathbb{N} \times \mathbb{N}$ as a countable union of countable sets, and then apply the result of (2.7).

2. Let A be a countably infinite alphabet $A = \{a_0, a_1, a_2, \ldots\}$. Let A^* denote the set of all finite strings over A. For each $n \in \mathbb{N}$, let A_n denote the set of all strings in A^* having length n.

   a. Show that A_n is countable for $n \in \mathbb{N}$. *Hint:* Use (2.7).

   b. Show that A^* is countable. *Hint:* Use (2.7) and part (a).

3. Let a and b be real numbers with $a < b$. Show that $|(a, b)| = |(0, 1)|$ by finding a bijection between $(0, 1)$ and (a, b).

4. Let a and b be real numbers with $a < b$. Show that the intervals (a, b) and $[a, b]$ have the same cardinality.

5. Show that $|\mathbb{R}| = |(0, 1)|$ by finding a bijection between the two sets.

6. Let $F(\mathbb{N})$ denote the set of all finite subsets of $\mathbb{N}$. We'll show that $F(\mathbb{N})$ is countable. Let B be the set of binary strings with no leading zeros, except for the number 0. B is countable because each string corresponds to a natural number. For example, $0 = 0$, $1 = 1$, $10 = 2$, $11 = 3$, and so on. We'll show that $F(\mathbb{N})$ is countable by constructing a bijection from B to $F(\mathbb{N})$. Define $g : B \to F(\mathbb{N})$ by $g(b_k \ldots b_1 b_0) = \{i \mid b_i = 1\}$. For example, $g(0) = \varnothing$ and $g(1) = \{0\}$. A few more values of g are

$$g(10) = \{1\}$$
$$g(11) = \{0, 1\}$$
$$g(100) = \{2\}$$
$$g(101) = \{0, 2\}.$$

Show that g is a bijection.

7. In Exercise 6 we saw that the set $F(\mathbb{N})$ of finite subsets of $\mathbb{N}$ is countable. Use this result to argue that $F(A^*)$ is also a countable set, where A is an alphabet that may be finite or infinite. *Hint:* Note that A^* is countable in either case by (2.8) or Exercise 2b.

8. Let A be a finite alphabet. Use diagonalization (2.13) to show that power(A^*) is uncountable.

9. Use diagonalization (2.13) to prove that the set of functions of type $\mathbb{N} \to \mathbb{N}$ is uncountable.

10. Prove Cantor's result about the cardinality of sets: If A is any set, then $|A| < |\text{power}(A)|$. *Hint:* Assume that there is some surjection from A to power(A), and try to arrive at a contradiction.

Chapter Summary

Functions allow us to associate different sets of objects. They are character-ized by associating each domain element with a unique codomain element. For any function $f : A \rightarrow B$, subsets of the domain A have images in the codomain B, and subsets of B have pre-images—also called inverse images—in A. The image of A is the range of f. Partial functions need not be defined for all domain elements. Two ways to combine functions are composition and tupling.

Some functions that are particularly useful in computer science are floor, ceiling, greatest common divisor (with the associated division algorithm), mod, characteristic, and log. Three functions that deal with programming with lists are sequence, distribute, and pairs.

Higher-order functions can take functions as arguments or return a function as a result. Some useful higher-order functions for programming are map (apply to all), apply, and insert.

Three important properties of functions that allow us to compare sets are injective, surjective, and bijective. These properties are useful in describ-ing the pigeonhole principle and in discussing hash functions. These proper-ties are also useful in comparing the cardinality of sets. Any set has smaller cardinality than its power set, even when the set is infinite. Countable unions of countable sets are still countable. The diagonalization technique can be used to show that a countable listing is not exhaustive. It can also be used to show that some sets are uncountable, like the real numbers. So we can't com-pute all the real numbers.

3

Construction
Techniques

When we build, let us think that we
build forever.
—John Ruskin (1819–1900)

To construct an object, we need some kind of description. If we're lucky, the description might include a technique to construct the object. Construction techniques are what this chapter is all about. We'll begin by introducing the technique of inductive definition for sets of objects. Then we'll introduce some techniques for constructing languages—sets of strings. Lastly, we'll discuss techniques for describing recursively defined functions.

Although technique is the focus of the chapter, the only way to learn a technique is to use it on a wide variety of problems. Each technique is presented in the framework of objects that occur in computer science. So as we go along, we'll extend our knowledge of these objects.

Chapter Guide

Section 3.1 introduces the inductive definition technique. We'll apply the technique by defining various sets of natural numbers, lists, strings, binary trees, and products of sets.

Section 3.2 introduces two construction techniques for languages. We'll see how to combine languages by concatenating strings. Then we'll concentrate on constructing languages by recursive grammar rules.

Section 3.3 introduces the technique of recursive definition for functions and procedures. We'll apply the technique to functions and procedures that

111

process natural numbers, lists, strings, and binary trees. We'll solve the redundant element problem and the power set problem, and we'll construct some stream functions.

3.1 Inductively Defined Sets

When we write down an informal statement such as $A = \{3, 5, 7, 9, ...\}$, most of us will agree that we mean the set $A = \{2k + 3 \mid k \in \mathbb{N}\}$. Another way to describe A is to observe that $3 \in A$, that whenever $x \in A$, then $x + 2 \in A$, and that the only way an element gets in A is by the previous two steps. This description has three ingredients, which we'll state informally as follows:

1. There is a starting element (3 in this case).
2. There is a construction operation to build new elements from existing elements (addition by 2 in this case).
3. There is a statement that no other elements are in the set.

This process is an example of an *inductive definition* of a set. The set of objects defined is called an *inductive set*. An inductive set consists of objects that are constructed, in some way, from objects that are already in the set. So nothing can be constructed unless there is at least one object in the set to start the process. Inductive sets are important in computer science because the objects can be used to represent information, and the construction rules can often be programmed. We give the following formal definition:

An *inductive definition* of a set S consists of three steps: (3.1)

Basis: List some specific elements of S (at least one element must be listed).

Induction: Give one or more rules to construct new elements of S from existing elements of S.

Closure: State that S consists exactly of the elements obtained by the basis and induction steps. This step is usually assumed rather than stated explicitly.

The closure step is a very important part of the definition. Without it, there could be lots of sets satisfying the first two steps of an inductive definition. For example, the two sets $\mathbb{N}$ and $\{3, 5, 7, ...\}$ both contain the number 3, and if x is in either set, then so is $x + 2$. It's the closure statement that tells us that the only set defined by the basis and induction steps is $\{3, 5, 7, ...\}$. So the closure statement tells us that we're defining exactly one set, namely, the smallest set satisfying the basis and induction steps. We'll always omit the specific mention of closure in our inductive definitions.

The *constructors* of an inductive set are the basis elements and the rules for constructing new elements. For example, the inductive set {3, 5, 7, 9, ...} has two constructors, the number 3 and the operation of adding 2 to a number.

For the rest of this section we'll use the technique of inductive definition to construct sets of objects that are often used in computer science.

Natural Numbers

The set of natural numbers $\mathbb{N}$ = {0, 1, 2, ...} is an inductive set. Its basis element is 0, and we can construct a new element from an existing one by adding the number 1. So we can write an inductive definition for $\mathbb{N}$ in the following way.

> *Basis:* $0 \in \mathbb{N}$.
>
> *Induction:*   If $n \in \mathbb{N}$, then $n + 1 \in \mathbb{N}$.

The constructors of $\mathbb{N}$ are the integer 0 and the operation that adds 1 to an element of $\mathbb{N}$. The operation of adding 1 to n is called the *successor* function, which we write as follows:

$$\text{succ}(n) = n + 1.$$

Using the successor function, we can rewrite the induction step in the above definition of $\mathbb{N}$ in the alternative form

$$\text{If } n \in \mathbb{N}, \text{ then succ}(n) \in \mathbb{N}.$$

So we can say that $\mathbb{N}$ is an inductive set with two constructors, 0 and succ.

EXAMPLE 1. We'll give an inductive definition of A = {1, 3, 7, 15, 31, ...}. Of course, the basis case should place 1 in A. If $x \in A$, then we can construct another element of A with the expression $2x + 1$. So the constructors of A are the number 1 and the operation of multiplying by 2 and adding 1. An inductive definition of A can be written as follows:

> *Basis:* $1 \in A$.
>
> Induction: If $x \in A$, then $2x + 1 \in A$.   ◆

EXAMPLE 2. Is the following set inductive?

$$A = \{2, 3, 4, 7, 8, 11, 15, 16, ...\}.$$

It might be easier if we think of A as the union of the two sets $B = \{2, 4, 8, 16, ...\}$ and $C = \{3, 7, 11, 15, ...\}$. Both these sets are inductive. The constructors of B are the number 2 and the operation of multiplying by 2. The constructors of C are the number 3 and the operation of adding by 4. We can combine these definitions to give an inductive definition of A.

Basis: $2, 3 \in A$.

Induction: If $x \in A$ and x is odd, then $x + 4 \in A$ else $2x \in A$.

This example shows that there can be more than one basis element, and the induction step can include tests.   ♦

EXAMPLE 3 (*Communicating with a Robot*). Suppose we want to communicate the idea of the natural numbers to a robot that knows about functions, has a loose notion of sets, and can follow an inductive definition. Symbols like 0, 1, ..., and + make no sense to the robot. How can we convey the idea of $\mathbb{N}$? We'll tell the robot that N is the name of the set we want to construct.

Suppose we start by telling the robot to put the symbol 0 in N. For the induction case we need to tell the robot about the successor function. We tell the robot that $s : N \to N$ is a function, and whenever an element $x \in N$, then put the element $s(x) \in N$. After a pause, the robot communicates with us and says, "$N = \{0\}$ because I'm letting s be the function defined by $s(0) = 0$." Since we don't want $s(0) = 0$, we have to tell the robot that $s(0) \neq 0$. Then the robot says, "$N = \{0, s(0)\}$ because $s(s(0)) = 0$." So we tell the robot that $s(s(0)) \neq 0$. Since this could go on forever, let's tell the robot that $s(x)$ does not equal any previously defined element.

Do we have it? Yes. The robot responds with "$N = \{0, s(0), s(s(0)), s(s(s(0))), ...\}$." So we can give the robot the following definition:

Basis: $0 \in N$.

Induction: If $x \in N$, then put $s(x) \in N$, where $s(x) \neq 0$ and $s(x)$ is not equal to any previously defined element of N.

This definition of the natural numbers—along with a closure statement—is due to the mathematician and logician Giuseppe Peano (1858–1932).   ♦

EXAMPLE 4 (*Communicating with Another Robot*). Suppose we want to define the natural numbers for a robot that knows about sets and can follow an inductive definition. How can we convey the idea of $\mathbb{N}$ to the robot? Since we can use only the notation of sets, let's use $\varnothing$ to stand for the number 0. What about the number 1? Can we somehow convey the idea of 1 using the empty set? Let's let $\{\varnothing\}$ stand for 1. What about 2? We can't use $\{\varnothing, \varnothing\}$, because $\{\varnothing, \varnothing\} = \{\varnothing\}$. Let's let $\{\varnothing, \{\varnothing\}\}$ stand for 2 because it has two distinct elements. Notice the little pattern we have going: If s is the set standing for a number, then $s \cup \{s\}$ stands for the successor of the number, as follows:

$$
\begin{array}{lll}
0 & \text{is represented by} & \varnothing, \\
1 & \text{is represented by} & \varnothing \cup \{\varnothing\} = \{\varnothing\}, \\
2 & \text{is represented by} & \{\varnothing\} \cup \{\{\varnothing\}\} = \{\varnothing, \{\varnothing\}\}.
\end{array}
$$

The construction $s \cup \{s\}$ always creates a new set containing one more element than s. Starting with $\varnothing$ as the basis element, we have an inductive definition. Letting Nat be the set that we are defining for the robot, we have the following inductive definition:

Basis:　　$\varnothing \in$ Nat.

Induction:　If $s \in$ Nat, then $s \cup \{s\} \in$ Nat.

For example, since 2 is represented by the set $\{\varnothing, \{\varnothing\}\}$, the number 3 is represented by the set

$$\{\varnothing, \{\varnothing\}\} \cup \{\{\varnothing, \{\varnothing\}\}\} = \{\varnothing, \{\varnothing\}, \{\varnothing, \{\varnothing\}\}\}.$$

This is not fun. After a while we might try to introduce some of our own notation to the robot. For example, we might introduce the decimal numerals as follows:

$$
\begin{array}{rcl}
0 & = & \varnothing, \\
1 & = & 0 \cup \{0\}, \\
2 & = & 1 \cup \{1\}, \\
& \vdots &
\end{array}
$$

With this correspondence we have another way to think about the natural numbers, as follows:

$$
\begin{array}{l}
1 = 0 \cup \{0\} = \varnothing \cup \{0\} = \{0\}, \\
2 = 1 \cup \{1\} = \{0\} \cup \{1\} = \{0, 1\}, \\
\quad \vdots
\end{array}
$$

Thus each number is the set of numbers that precede it. We might also introduce the robot to the idea of order by writing $\text{succ}(x) = x \cup \{x\}$. Assume that the robot is familiar with functions. Then we can teach the robot statements like "$\text{succ}(0) = 1$" and "$\text{succ}(1) = 2$."   ♦

Lists

Let's try to find an inductive definition for the set of lists with elements from a set A. In Chapter 1 we denoted the set of all lists over A by Lists[A], and we'll continue to do so. We also mentioned that from a computational point of view the only parts of a nonempty list that can be accessed randomly are its *head* and its *tail*. Head and tail are sometimes called *destructors*, since they are used to destroy a list (take it apart). For example, the list $\langle x, y, z \rangle$ has x as its head and $\langle y, z \rangle$ as its tail, which we write as

$$\text{head}(\langle x, y, z \rangle) = x \quad \text{and} \quad \text{tail}(\langle x, y, z \rangle) = \langle y, z \rangle.$$

But we need to construct lists. The idea is to take an element h and a list t and construct a new list whose head is h and whose tail is t. We'll denote this newly constructed list by the expression

$$\text{cons}(h, t).$$

So *cons* is a constructor of lists. For example, we have

$$\text{cons}(x, \langle y, z \rangle) = \langle x, y, z \rangle$$
$$\text{cons}(x, \langle \, \rangle) = \langle x \rangle.$$

The operations cons, head, and tail work nicely together. For example, we can write

$$\langle x, y, z \rangle = \text{cons}(x, \langle y, z \rangle) = \text{cons}(\text{head}(\langle x, y, z \rangle), \text{tail}(\langle x, y, z \rangle)).$$

So if l is any nonempty list, then we have the equation

$$l = \text{cons}(\text{head}(l), \text{tail}(l)).$$

Now we have the proper tools, so let's get down to business and write an inductive definition for Lists[A]. The empty list, $\langle \, \rangle$, is certainly a basis element of Lists[A]. Using $\langle \, \rangle$ and cons as constructors, we can write the inductive definition of Lists[A] for any set A as follows:

Basis: $\langle \, \rangle \in$ Lists[*A*].   (3.2)

Induction: If $x \in A$ and $t \in$ Lists[*A*], then cons(*x*, *t*) $\in$ Lists[*A*].

EXAMPLE 5. Let $A = \{a, b\}$. We'll use (3.2) to see how some lists become members of Lists[*A*]. The basis case puts $\langle \, \rangle \in$ Lists[*A*]. Since $a \in A$ and $\langle \, \rangle \in$ Lists[*A*], the induction step gives $\langle a \rangle = $ cons(*a*, $\langle \, \rangle$) $\in$ Lists[*A*]. In the same way we get $\langle b \rangle \in$ Lists[*A*]. Now since $a \in A$ and $\langle a \rangle \in$ Lists[*A*], the induction step puts $\langle a, a \rangle \in$ Lists[*A*]. In the same way we get $\langle b, a \rangle$, $\langle a, b \rangle$, and $\langle b, b \rangle$ as elements of Lists[*A*].   ◆

A popular infix notation for cons is the double colon symbol

$$:: \, .$$

Thus the infix form of cons(*x*, *t*) is $x :: t$. For example, the list $\langle a, b, c \rangle$ can be constructed using cons as follows:

$$
\begin{aligned}
\text{cons}(a, \text{cons}(b, \text{cons}(c, \langle \, \rangle))) &= \text{cons}(a, \text{cons}(b, \langle c \rangle)) \\
&= \text{cons}(a, \langle b, c \rangle) \\
&= \langle a, b, c \rangle.
\end{aligned}
$$

Using the infix form, we construct $\langle a, b, c \rangle$ as follows:

$$a :: (b :: (c :: \langle \, \rangle)) = a :: (b :: \langle c \rangle) = a :: \langle b, c \rangle = \langle a, b, c \rangle.$$

The infix form of cons allows us to omit parentheses by agreeing that $::$ is right associative. In other words, $a :: b :: t = a :: (b :: t)$. Thus we can represent the list $\langle a, b, c \rangle$ by writing

$$a :: b :: c :: \langle \, \rangle \quad \text{instead of} \quad a :: (b :: (c :: \langle \, \rangle)).$$

Many programming problems involve processing data represented by lists. The operations cons, head, and tail provide basic tools for writing programs to create and manipulate lists. Thus they are necessary for programmers. Now let's look at a few examples.

EXAMPLE 6. Suppose we need to define the set S of all nonempty lists over the set $\{0, 1\}$, where the elements in each list alternate between 0 and 1. We can get an idea about S by listing a few elements:

$$S = \{\langle 0 \rangle, \langle 1 \rangle, \langle 1, 0 \rangle, \langle 0, 1 \rangle, \langle 0, 1, 0 \rangle, \langle 1, 0, 1 \rangle, ...\}.$$

Let's try $\langle 0 \rangle$ and $\langle 1 \rangle$ as basis elements of S. Then we can construct a new list from a list $x \in S$ by testing whether head(x) is 0 or 1. If head(x) = 0, then we place 1 at the left of x. Otherwise, we place 0 at the left of x. So an inductive definition for S can be written as follows:

Basis: $\langle 0 \rangle, \langle 1 \rangle \in S$.

Induction: If $x \in S$ and head(x) = 0, then cons($1, x$) $\in S$
else cons($0, x$) $\in S$.

The infix form of this induction statement reads

If $x \in S$ and head(x) = 0, then $1 :: x \in S$ else $0 :: x \in S$.   ◆

EXAMPLE 7. Suppose we need to define the set S of all lists over $\{a, b\}$ that begin with the single letter a followed by zero or more occurrences of b. We can describe S informally by writing a few of its elements:

$$S = \{\langle a \rangle, \langle a, b \rangle, \langle a, b, b \rangle, \langle a, b, b, b \rangle, \ldots\}.$$

It seems appropriate to make $\langle a \rangle$ the basis element of S. Then we can construct a new list from any list $x \in S$ by attaching the letter b on the right end of x. But cons places new elements at the left end of a list. We can overcome the problem by using the tail operation together with cons as follows:

If $x \in S$, then cons(a, cons(b, tail(x))) $\in S$.

In infix form the statement reads as follows:

If $x \in S$, then $a :: b :: \text{tail}(x) \in S$.

For example, if $x = \langle a \rangle$, then we construct the list

$$a :: b :: \text{tail}(\langle a \rangle) = a :: b :: \langle \, \rangle = a :: \langle b \rangle = \langle a, b \rangle.$$

So we have the following inductive definition of S:

Basis: $\langle a \rangle \in S$.

Induction: If $x \in S$, then $a :: b :: \text{tail}(x) \in S$.   ◆

EXAMPLE 8 (*Generalized Lists*). Recall that a generalized list over a set A can contain lists as components. For example, if $A = \{a, b\}$, then we can find lists like the following in GenLists[A]:

$$\langle\langle b, a\rangle, b\rangle, \quad \langle\langle\langle\,\rangle, a\rangle, \langle b\rangle\rangle, \quad \text{and} \quad \langle a, b, a\rangle.$$

Can we write down the elements of GenLists[A] in some systematic way, to see whether we can define the set inductively? Yes. If we start with lists having a small number of symbols, including the tuple markers $\langle$ and $\rangle$, then for each $n \geq 2$ we can write down the lists made up of n symbols (not including commas). Table 3.1 shows these listings for the first few values of n.

2	3	4	5	6
$\langle\,\rangle$	$\langle a\rangle$	$\langle\langle\,\rangle\rangle$	$\langle\langle a\rangle\rangle$	$\langle\langle\langle\,\rangle\rangle\rangle$
	$\langle b\rangle$	$\langle a, b\rangle$	$\langle\langle b\rangle\rangle$	$\langle\langle\,\rangle, \langle\,\rangle\rangle$
		$\langle b, a\rangle$	$\langle\langle\,\rangle, a\rangle$	$\langle\langle a, b\rangle\rangle$
		$\langle a, a\rangle$	$\langle\langle\,\rangle, b\rangle$	$\langle\langle a, a\rangle\rangle$
		$\langle b, b\rangle$	$\langle a, \langle\,\rangle\rangle$	$\langle a, b, \langle\,\rangle\rangle$
			$\langle b, \langle\,\rangle\rangle$	$\langle a, \langle\,\rangle, b\rangle$
			$\langle a, b, a\rangle$	$\langle a, b, a, b\rangle$
			$\langle a, a, a\rangle$	$\langle b, b, a, a\rangle$
			$\vdots$	$\vdots$

Table 3.1

If we allow the cons operation to take a list as its first argument, then we can build $\langle\langle\,\rangle\rangle$ from $\langle\,\rangle$ and $\langle\,\rangle$ by writing cons($\langle\,\rangle$, $\langle\,\rangle$). Similarly, we can build $\langle\langle a\rangle\rangle$ from $\langle a\rangle$ and $\langle\,\rangle$ by writing cons($\langle a\rangle$, $\langle\,\rangle$). We'll construct a few more sample lists as follows:

$$\text{cons}(a, \langle b\rangle) = \langle a, b\rangle$$
$$\text{cons}(\langle a\rangle, \langle b\rangle) = \langle\langle a\rangle, b\rangle$$
$$\text{cons}(a, \langle\langle b\rangle\rangle) = \langle a, \langle b\rangle\rangle$$
$$\text{cons}(\langle a\rangle, \langle\langle b\rangle\rangle) = \langle\langle a\rangle, \langle b\rangle\rangle.$$

This more general cons operation is all we need. If A is any set, then an inductive definition for GenLists[A] can be written as follows:

Basis: $\langle\,\rangle \in$ GenLists[*A*].   (3.3)

Induction: If $x \in A \cup$ GenLists[*A*] and $t \in$ GenLists[*A*],
              then cons(x, t) $\in$ GenLists[*A*].

The general versions of the head and tail functions work the same way as for Lists[*A*]. For example, we have

$$\text{head}(\langle\langle a, b\rangle, c\rangle) = \langle a, b\rangle$$
$$\text{tail}(\langle\langle a, b\rangle, c\rangle) = \langle c\rangle. \quad \blacklozenge$$

Strings

Suppose we want to give an inductive definition for a set of strings. To do so, we need to have some way to construct strings. The situation is similar to that of lists, in which the constructors are the empty list and cons (::). For strings the constructors are the empty string Λ together with the operation of *appending* a letter to the left end of a string in juxtaposition. We'll denote the append operation by the dot symbol. For example, to append the letter a to the string s, we'll use the following notation:

$$a \cdot s.$$

For example, if $s = aba$, then the evaluation of the expression $a \cdot s$ is given by

$$a \cdot s = a \cdot aba = aaba.$$

When a letter is appended to the empty string, the result is the letter. In other words, for any letter a we have

$$a \cdot \Lambda = a\Lambda = a.$$

To get along without parentheses, we'll agree that appending is right associative. For example, $a \cdot b \cdot \Lambda$ means $a \cdot (b \cdot \Lambda)$.

Now we have the tools to give inductive definitions for some sets of strings. For example, if A is an alphabet, then an inductive definition of A^* can be written as follows:

Basis: $\Lambda \in A^*$.   (3.4)

Induction: If $a \in A$ and $s \in A^*$, then $a \cdot s \in A^*$.

For example, if $A = \{a, b\}$, then the string *bab* can be constructed by the following expression:

$$
\begin{aligned}
b \cdot a \cdot b \cdot \Lambda &= b \cdot a \cdot b\Lambda \\
&= b \cdot a \cdot b \\
&= b \cdot ab \\
&= bab.
\end{aligned}
$$

As we did with lists, we'll use the same two words *head* and *tail* to pick the appropriate parts of a nonempty string. For example, we have

$$\text{head}(abc) = a \quad \text{and} \quad \text{tail}(abc) = bc.$$

EXAMPLE 9. Suppose that $A = \{0, 1\}$ and we want to define the set of strings L such that each string in L contains exactly one occurrence of 0 on the right. For example, L should contain strings like

$$0, 10, 110, 1110, \ldots$$

Can we define L inductively? Sure. Let the digit 0 be the basis element of L. If s is an element of L, then we can construct a new element of L by appending the digit 1 to s. Thus the inductive definition of L can be written as follows:

Basis: $0 \in L$.

Induction: If $s \in L$, then $1 \cdot s \in L$.   ♦

EXAMPLE 10 (*Appending on the Right*). Let $A = \{0, 1\}$, and suppose that S is the set of strings over A with the following property: No string contains a leading zero except 0 itself. Then S should contain strings like

$$0, 1, 10, 11, 100, 101, 110, 111, \ldots$$

If we let 0 and 1 be basis elements of S, then we can append 1 to each of these strings to obtain the strings 10 and 11. Similarly, by appending 1 to each of these latter strings we obtain the strings 110 and 111. But how can we construct the two strings 100 and 101?

What if we could append an element on the right end of a string? Let's assume that we have such an operation, which we'll denote by the same dot that we are using for the regular append operation. In other words, if s is a string and a is a letter, then

$$s \cdot a$$

denotes the string obtained by juxtaposing the letter a to the right of s. For example, if $s = abc$, then

$$s \cdot a = abc \cdot a = abca.$$

Now let's continue with the problem. If s is any string in S, then as long as $s \neq 0$, we can construct new strings by appending 0 or 1 to the right of s. For example, $100 = 10 \cdot 0$, and $101 = 10 \cdot 1$. The inductive definition of S can be written as follows:

Basis: $0, 1 \in S$.

Induction: If $s \in S$ and $s \neq 0$, then $s \cdot 0, s \cdot 1 \in S$.

The first few strings constructed by this definition are listed as follows:

$$0, 1, 10, 11, 100, 101, 110, 111, 1000, 1001, \ldots$$

EXAMPLE 11. Let $A = \{a, b\}$, and let S be the following set of strings over A:

$$S = \{a, b, ab, ba, aab, bba, aaab, bbba, \ldots\}.$$

Suppose we start with a and b as basis elements. From a we can construct the element ba, from ba the element bba, and so on. Similarly, from b we construct ab, then aab, and so on. Another way to see this is to think of S as the union of two simpler sets $\{a, ba, bba, \ldots\}$, and $\{b, ab, aab, \ldots\}$. To describe the construction, we can use the head function. Given a string $s \in S$, if $s = a$, then construct ba, while if $s = b$, then construct ab. Otherwise, if $\text{head}(s) = a$, then construct $a \cdot s$, and if $\text{head}(s) = b$, then construct $b \cdot s$. The definition of S can be written as follows:

Basis: $a, b \in S$.

Induction: Let $s \in S$. Construct a new element of S as follows:
          If $s = a$, then $b \cdot a \in S$.
          If $s = b$, then $a \cdot b \in S$.
          If $s \neq a$ and $\text{head}(s) = a$, then $a \cdot s \in S$.
          If $s \neq b$ and $\text{head}(s) = b$, then $b \cdot s \in S$.

Can you find another way to define S?   ◆

Binary Trees

Let's look at binary trees. In Chapter 1 we represented binary trees by tuples, where the empty binary tree is denoted by the empty tuple and a non-empty binary tree is denoted by a 3-tuple $\langle L, x, R \rangle$, in which x is the root, L is the left subtree, and R is the right subtree. Thus a new binary tree can be constructed from binary trees that already exist. This gives us the ingredients for an inductive definition of the set of all binary trees.

We'll let tree(L, x, R) denote the binary tree with root x, left subtree L, and right subtree R. If we still want to represent binary trees as tuples, then of course we can write

$$\text{tree}(L, x, R) = \langle L, x, R \rangle.$$

Now suppose A is any set. Let BinTrees[A] be the set of all binary trees whose nodes come from A. We can write down an inductive definition of BinTrees[A] using the two constructors $\langle \rangle$ and tree:

Basis:	$\langle \rangle \in$ BinTrees[A].	(3.5)

Induction: If $x \in A$ and $L, R \in$ BinTrees[A], then tree(L, x, R) $\in$ BinTrees[A].

We also have destructor operations for binary trees. Suppose we let "left," "root," and "right" denote the operations that return the left subtree, the root, and the right subtree, respectively, of a nonempty tree. For example, if $t = $ tree(L, x, R), then left(t) = L, root(t) = x, and right(t) = R.

EXAMPLE 12 *(Twins)*. Let $A = \{0, 1\}$. Suppose we need to work with the set Twins of all binary trees T over A that have the following property: The left and right subtrees of each node in T are identical in structure and node content. For example, Twins contains the empty tree and any single-node tree. Twins also contains the two trees shown in Figure 3.1.

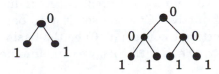

Figure 3.1

We can give an inductive definition of Twins by simply making sure that each new tree has the same left and right subtrees. Here's the definition:

Basis: ⟨ ⟩ ∈ Twins.

Induction: If $x \in A$ and $T \in$ Twins, then tree(T, x, T) ∈ Twins.   ◆

EXAMPLE 13 (*Opposites*). Let $A = \{0, 1\}$, and suppose that Opps is the set of all non-empty binary trees T over A with the following property: The left and right subtrees of each node of T have identical structures, but the 0's and 1's are interchanged. For example, the single node trees are in Opps, as well as the two trees shown in Figure 3.2.

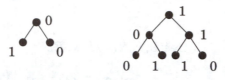

Figure 3.2

Since our set does not include the empty tree, the two singleton trees with nodes 1 and 0 should be the basis trees in Opps. The inductive definition of Opps can be given as follows:

Basis: Put both tree(⟨ ⟩, 0, ⟨ ⟩) and tree(⟨ ⟩, 1, ⟨ ⟩) in Opps.

Induction: If $x \in A$ and $T \in$ Opps, then
                 if root(T) = 0, then
                     tree(T, x, tree(right(T), 1, left(T))) ∈ Opps.
                 else
                     tree(T, x, tree(right(T), 0, left(T))) ∈ Opps.

Does this definition work? Try out some examples. See whether the definition builds the four possible three-node trees.   ◆

Product Sets

Let's see whether we can define some product sets inductively. For example, we know that $\mathbb{N}$ is inductive. Can the set $\mathbb{N} \times \mathbb{N}$ be inductively defined? Suppose we start by letting the tuple ⟨0, 0⟩ be the basis element. For the induction case, if a pair $\langle x, y \rangle \in \mathbb{N} \times \mathbb{N}$, then we can build new pairs

$$\langle x + 1, y + 1 \rangle, \quad \langle x, y + 1 \rangle \quad \text{and} \quad \langle x + 1, y \rangle.$$

The graph in Figure 3.3 shows an arbitrary point $\langle x, y \rangle$ together with the three new points. For example, we can construct ⟨1, 1⟩, ⟨0, 1⟩, and ⟨1, 0⟩ from

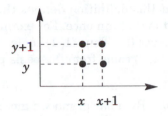

Figure 3.3

the point $\langle 0, 0 \rangle$. It seems clear that this definition will define all elements of $\mathbb{N} \times \mathbb{N}$. Let's look at two general techniques to give an inductive definition for a product $A \times B$ of two sets A and B.

The first technique can be used if both A and B are inductively defined: For the basis case, put $\langle a, b \rangle \in A \times B$ whenever a is a basis element of A and b is a basis element of B. For the inductive part, if $\langle x, y \rangle \in A \times B$ and $x' \in A$ and $y' \in B$ are elements constructed from x and y, respectively, then put the elements $\langle x, y' \rangle$, $\langle x', y \rangle$ in $A \times B$.

The second technique can be used if only one of the sets, say A, is inductively defined: For the basis case, put $\langle a, b \rangle \in A \times B$ for all basis elements $a \in A$ and all elements $b \in B$. For the induction case, if $\langle x, y \rangle \in A \times B$ and $x' \in A$ is constructed from x, then put $\langle x', y \rangle \in A \times B$. A similar definition of $A \times B$ can also be made that uses only the fact that B is inductively defined. The choice of definition usually depends on how the product set will be used. Let's look at some examples.

EXAMPLE 14. The set $\mathbb{N} \times \mathbb{N}$ can be defined inductively as follows by using the fact that the first copy of $\mathbb{N}$ is inductively defined:

> *Basis:* $\langle 0, n \rangle \in \mathbb{N} \times \mathbb{N}$ for all $n \in \mathbb{N}$.
>
> *Induction:*   If $\langle x, y \rangle \in \mathbb{N} \times \mathbb{N}$, then $\langle \mathrm{succ}(x), y \rangle \in \mathbb{N} \times \mathbb{N}$.

It's easy to see that this definition of $\mathbb{N} \times \mathbb{N}$ is correct. Just notice that for any ordered pair $\langle m, n \rangle \in \mathbb{N} \times \mathbb{N}$, either $m = 0$ or $m = \mathrm{succ}(k)$ for some $k \in \mathbb{N}$. In either case the above definition puts $\langle m, n \rangle \in \mathbb{N} \times \mathbb{N}$.   ◆

EXAMPLE 15. We can also define $\mathbb{N} \times \mathbb{N}$ by using the fact that both copies of $\mathbb{N}$ are inductively defined:

> *Basis:* $\langle 0, 0 \rangle \in \mathbb{N} \times \mathbb{N}$.
>
> *Induction:*   If $\langle x, y \rangle \in \mathbb{N} \times \mathbb{N}$, then $\langle \mathrm{succ}(x), y \rangle$, $\langle x, \mathrm{succ}(y) \rangle \in \mathbb{N} \times \mathbb{N}$.

Again, it's easy to see that this definition defines the set $\mathbb{N} \times \mathbb{N}$. Notice that some elements get defined more than once. For example, $\langle 1, 1 \rangle$ can be written as either $\langle \text{succ}(0), 1 \rangle$ or $\langle 1, \text{succ}(0) \rangle$. Thus $\langle 1, 1 \rangle$ comes from $\langle 0, 1 \rangle$ and also from $\langle 1, 0 \rangle$. Of course $\langle 0, 1 \rangle$ and $\langle 1, 0 \rangle$ come from the basis pair $\langle 0, 0 \rangle$.   ♦

EXAMPLE 16 (*A Bar Graph*). By a bar graph we mean the graph of a function whose domain is finite or at most the size of $\mathbb{N}$ and in which bars have been drawn between each point $\langle x, y \rangle$ on the graph and the point $\langle x, 0 \rangle$ on the x-axis. Most bar graphs that we see have solid bars. But from a computational point of view the solid bars are made up of discrete separated points, like pixels on a computer screen. So our bars will actually be a bunch of dots in a line, reflecting the low-level picture of things. For example, suppose we let f be the function

$$f : \{0, 1, 2, 3, 4, 5, 6\} \rightarrow \{0, 1, 2, 3, 4, 5\},$$

where $f(0) = 2$, $f(1) = 1$, $f(2) = 3$, $f(3) = 2$, $f(4) = 5$, $f(5) = 4$, and $f(6) = 2$. The diagram in Figure 3.4 shows our version of a bar graph for f.

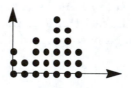

Figure 3.4

How can we compute a bar graph? Suppose we have a function $f : \mathbb{N} \rightarrow \mathbb{N}$ and we need to compute its bar graph G. In other words, we want to compute the following set:

$$G = \{\langle a, b \rangle \mid a, b \in \mathbb{N} \text{ and } 0 \leq b \leq f(a)\}.$$

Can we define G inductively? Since the domain of f is inductively defined, we can define G as follows:

Basis:	$\langle 0, f(0) \rangle \in G$.
Induction:	If $\langle x, y \rangle \in G$, then $\langle x + 1, f(x + 1) \rangle \in G$.
	If $\langle x, y \rangle \in G$ and $y > 0$, then $\langle x, y - 1 \rangle \in G$.

For example, suppose f has the definition $f(x) = x + 2$. To get the idea, we'll construct a few points in G. We start with the basis pair $\langle 0, 2 \rangle$ and follow by constructing some pairs inductively (reasons are given in parentheses):

$$\langle 0, 2 \rangle \qquad\qquad \text{(basis step)},$$
$$\langle 1, 3 \rangle, \langle 0, 1 \rangle \qquad (\langle 0, 2 \rangle \text{ and induction)},$$
$$\langle 2, 4 \rangle, \langle 1, 2 \rangle \qquad (\langle 1, 3 \rangle \text{ and induction)},$$
$$\langle 1, 3 \rangle, \langle 0, 0 \rangle \qquad (\langle 0, 1 \rangle \text{ and induction)},$$
$$\langle 3, 5 \rangle, \langle 2, 3 \rangle \qquad (\langle 2, 4 \rangle \text{ and induction)}.$$

Notice that redundant elements are constructed with this definition. In any case, each bar eventually gets filled in. The picture in Figure 3.5 shows a portion of G with the first four bars plotted. ♦

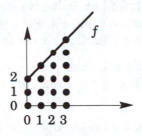

Figure 3.5

Exercises

1. Give an inductive definition for each of the following sets, under the assumption that the only known operation is the successor function, succ : $\mathbb{N} \to \mathbb{N}$.

   a. The set Odd, of odd natural numbers.

   b. The set Even, of even natural numbers.

   c. The set $S = \{4, 7, 10, 13, ...\} \cup \{3, 6, 9, 12, ...\}$.

2. Use the inductive definition of Nat, given in the second robot example, to show that $4 = \{0, 1, 2, 3\}$.

3. Rewrite each of the following expressions so that the result does not contain any occurrences of cons, ::, head, and tail.

   a. cons$(\langle\,\rangle, \langle\,\rangle)$.

   b. cons$(\langle a \rangle, \langle\,\rangle)$.

   c. cons$(\langle\,\rangle, \langle a \rangle)$.

d. cons($\langle a \rangle$, $\langle a \rangle$).

e. cons($\langle a \rangle$, cons(a, $\langle b, c \rangle$)).

f. head(cons(a, cons(b, $\langle\,\rangle$))).

g. tail(cons($\langle a \rangle$, cons(b, $\langle a \rangle$))).

4. Let x be the list $\langle\langle a \rangle, \langle b \rangle\rangle$.

a. Write x in terms of cons, a, b, and $\langle\,\rangle$.

b. Write x in terms of ::, a, b, and $\langle\,\rangle$.

5. Given a nonempty set A, find an inductive definition for each of the following subsets of Lists[A].

a. The set Even of all lists that an even number of elements.

b. The set Odd of all lists that have an odd number of elements.

6. Given the set $A = \{a,\ b\}$, find an inductive definition for the set S of all lists over A that alternate a's and b's. For example, the lists $\langle\,\rangle$, $\langle a \rangle$, $\langle b \rangle$, $\langle a, b, a \rangle$, and $\langle b, a \rangle$ are in S. But $\langle a, a \rangle$ is not in S.

7. Give an inductive definition for the set B of all binary numerals containing an odd number of digits such that the only string with a leading (leftmost) 0 is 0 itself.

8. A *palindrome* is a string that reads the same left to right as right to left. For example, RADAR is a palindrome over the English alphabet. Let A be an alphabet. Give an inductive definition of the set P of all palindromes over A.

9. Write down an inductive definition for each of the following sets.

a. The set Odd of all strings over an alphabet A that have odd length.

b. The set Even of all strings over an alphabet A that have even length.

c. The set Rat of all strings of the form $a.b$, where a and b are decimal numerals.

10. Let S be the set of strings over the alphabet $\{a, b\}$ described as follows:

$$S = \{a, b, ab, ba, aab, bba, aaab, bbba, ...\}.$$

Construct an inductive definition of S that is different from the one given in Example 11.

11. Give an inductive definition for each of the following sets.

a. The set Exp of all arithmetic expressions that are constructed from decimal numerals, +, and parentheses. For example, Exp should contain objects such as 17, 2 + 3, (3 + (4 + 5)), and 5 + 9 + 20.

   b. The set MExp of arithmetic expressions that are constructed from decimal numerals, − (subtraction), and parentheses with the property that each expression has only one meaning. For example, $9 - 34 - 10$ is not allowed.

12. a. Use tuples to represent the binary tree given by the expression

$$\text{tree}(\text{tree}(\langle\,\rangle, x, \langle\,\rangle), y, \text{tree}(\langle\,\rangle, z, \text{tree}(\langle\,\rangle, w, \langle\,\rangle))).$$

   b. Use the binary tree constructors to represent the binary tree denoted by the tuple $\langle\langle\,\rangle, 3, \langle\langle\,\rangle, 4, \langle\,\rangle\rangle\rangle$.

13. Given the following inductive definition for a subset B of $\mathbb{N} \times \mathbb{N}$:

   Basis: $\langle 0, 0 \rangle \in B$.

   Induction: If $\langle x, y \rangle \in B$, then $\langle \text{succ}(x), y \rangle$, $\langle \text{succ}(x), \text{succ}(y) \rangle \in B$.

   a. Describe the set B as a set of the form $\{\langle x, y \rangle \mid \text{some property holds}\}$.

   b. Describe those elements in B that get defined in more than one way.

14. Find two inductive definitions for each product set S. The first definition should use the fact that all components of S are inductive sets. The second definition should use only one inductive component set in S.

   a. $S = \text{Lists}[A] \times \text{Lists}[A]$ for some set A.

   b. $S = A^* \times A^*$ for some finite set A.

   c. $S = \mathbb{N} \times \text{Lists}[\mathbb{N}]$.

   d. $S = \mathbb{N} \times \mathbb{N} \times \mathbb{N}$.

15. Let A be a set. Suppose O is the set of binary trees over A that contain an odd number of nodes. Similarly, let E be the set of binary trees over A that contain an even number of nodes. Find inductive definitions for O and E. *Hint:* You can use O when defining E, and you can use E when defining O.

16. Prove that a set defined by (3.1) is countable if the basis elements in Step 1 are countable, the outside elements used in Step 2 are countable, and the rules specified in Step 2 are finite.

3.2 Language Constructions

A *language* is a set of strings. If A is an alphabet, then a *language* over A is a collection of strings whose components come from A. Recall that A^* denotes the set of all strings over A. So A^* is the biggest possible language over A, and every other language over A is a subset of A^*. Four simple examples of languages over an alphabet A are the sets $\varnothing$, $\{\Lambda\}$, A, and A^*. For example, if we let $A = \{a\}$, then the four simple languages over A are

$$\emptyset, \quad \{\Lambda\}, \quad \{a\}, \quad \text{and} \quad \{\Lambda, a, aa, aaa, \ldots\}.$$

A string in a language is often called a *well-formed formula*—or *wff* for short (pronounce wff as "woof")—because the definition of the language usually allows only certain well-formed strings. For example, suppose we consider the alphabet

$$A = \{0, 1, 2, 3, 4, 5, 6, 7, 8, 9\} \cup \{+\}.$$

There are many languages over A. The set of decimal numerals is a language over A, with wffs like 249 and 1009753. The set of arithmetic expressions that use the + operation is also a language over A. Two of its wffs are 24+173+68 and 2+98. The string 4++786+ is neither a decimal numeral wff nor an arithmetic expression wff, but it is a wff in the language A^* of all possible strings over A.

Many useful languages can be defined inductively. For example, suppose we consider the set of decimal digits

$$\text{Digits} = \{0, 1, 2, 3, 4, 5, 6, 7, 8, 9\}.$$

We can define the set D of decimal numerals inductively as follows:

Basis: Digits $\subset D$.

Induction: If $d \in$ Digits and $n \in D$, then $d \cdot n \in D$.

Notice that with this definition of D we get wffs with leading 0's such as 00034. If we want to exclude leading zero digits, then the induction part of the definition must be made more restrictive. An operation that can help is the operation of appending a character from the alphabet to the right part of a string.

EXAMPLE 1. Suppose we want to define the decimal numerals that do not have leading zeros. Of course, we will keep the single digit 0. Let D denote the set we are trying to define. We'll keep the same basis case and alter the induction case as follows, where $n \cdot d$ means append d at the right end of n:

Basis: Digits $\subset D$.

Induction: If $n \in D$ and $d \in$ Digits and $n \neq 0$, then $n \cdot d \in D$.   ◆

The natural operation of *concatenation* of strings places two strings in juxtaposition. For example, if $A = \{a, b\}$, then the concatenation of the two strings aab and ba is the string $aabba$. We will use the name "cat" to explicitly denote this operation. So

$$\text{cat}(aab, ba) = aabba.$$

Notice that the operations of appending on the left and right are special cases of the cat operation. For example, if s is a string and $b \in A$, then

$$b \cdot s = \text{cat}(b, s) \quad \text{and} \quad s \cdot b = \text{cat}(s, b).$$

We can use cat with itself, or with $\cdot$ to represent certain strings. For example, if 123 and 439 are decimal numerals, then we can create the string 123+439 by concatenating the string 123 and the string obtained by appending + to the string 439. In symbols we have

$$\text{cat}(123, + \cdot 439) = \text{cat}(123, +439) = 123{+}439.$$

We could also write

$$\text{cat}(123, \text{cat}(+, 439)) = \text{cat}(123, +439) = 123{+}439.$$

EXAMPLE 2. We can define A^* inductively using the two constructors Λ and cat as follows:

Basis: $\Lambda \in A^*$.

Induction: If $a \in A$ and $s \in A^*$, then $\text{cat}(a, s) \in A^*$.

Concatenation of strings can often be used to make inductive definitions easier and thus clearer.   ◆

Combining Languages

Since languages are sets of strings, they can be combined by the usual set operations of union, intersection, difference, and complement. Another important way to combine two languages L and M is to form the set of all concatenations of strings in L with strings in M. This new language is called the *product* of L and M and is denoted by $L \cdot M$. A formal definition can be given as follows:

$$L \cdot M = \{\text{cat}(s, t) \mid s \in L \text{ and } t \in M\}.$$

For example, if $L = \{ab, ac\}$ and $M = \{a, bc, abc\}$, then the product $L \cdot M$ is the language

$$L \cdot M = \{aba, abbc, ababc, aca, acbc, acabc\}.$$

It's easy to see that the product is associative. In other words, if L, M, and N are languages, then $L \cdot (M \cdot N) = (L \cdot M) \cdot N$. Thus we can write down products without using parentheses. On the other hand, it's easy to see that the product is not commutative. In other words, we can find two languages L and M such that $L \cdot M \neq M \cdot L$.

The product is a useful tool to help define a language in terms of already known languages. For example, we can define the set of all strings of the form $a \cdot b$, where a and b are decimal numerals, by using the three languages $\{.\}$, Digits, and Digits* in the following product:

$$\text{Digits} \cdot \text{Digits*} \cdot \{.\} \cdot \text{Digits} \cdot \text{Digits*}.$$

The product operation on languages has many properties. The following basic properties are quite useful. We'll leave the proofs as exercises.

Properties of Product   (3.6)

Let L, M, and N be languages over the alphabet A. Then

a) $L \cdot \{\Lambda\} = \{\Lambda\} \cdot L = L$.

b) $L \cdot \varnothing = \varnothing \cdot L = \varnothing$.

c) $L \cdot (M \cup N) = (L \cdot M) \cup (L \cdot N)$ and $(M \cup N) \cdot L = (M \cdot L) \cup (N \cdot L)$.

d) $L \cdot (M \cap N) = (L \cdot M) \cap (L \cdot N)$ and $(M \cap N) \cdot L = (M \cdot L) \cap (N \cdot L)$.

If L is a language, then the product $L \cdot L$ is denoted by L^2. In fact, we'll define the language product L^n for every $n \in \mathbb{N}$ as follows:

$$L^0 = \{\Lambda\},$$
$$L^n = L \cdot L^{n-1} \quad \text{if } n > 0.$$

For example, suppose $L = \{a, bb\}$. Then the first few powers of L are calculated as follows:

$$L^0 = \{\Lambda\},$$
$$L^1 = L = \{a, bb\},$$
$$L^2 = L \cdot L = \{aa, abb, bba, bbbb\},$$
$$L^3 = L \cdot L^2 = \{aaa, aabb, abba, abbbb, bbaa, bbabb, bbbba, bbbbbb\}.$$

If L is a language over A (i.e., $L \subset A^*$), then the *closure* of L is the language denoted by L^* and is defined as follows:

$$L^* = L^0 \cup L^1 \cup L^2 \cup \dots .$$

The *positive closure* of L is the language denoted by L^+ and defined as follows:

$$L^+ = L^1 \cup L^2 \cup L^3 \cup \dots .$$

It follows from the definition that $L^* = L^+ \cup \{\Lambda\}$. But it's not necessarily true that $L^+ = L^* - \{\Lambda\}$. For example, if we let our alphabet be $A = \{a\}$ and our language be $L = \{\Lambda, a\}$, then $L^+ = L^*$. Can you find a condition on a language L such that $L^+ = L^* - \{\Lambda\}$?

The closure of A coincides with our original definition of A^* as the set of all strings over A. In other words, we have a nice representation of A^* as follows:

$$A^* = A^0 \cup A^1 \cup A^2 \cup \dots ,$$

where A^n is the set of all strings over A having length n.

Some basic properties of the closure operation are given next. We'll leave the proofs as exercises.

Properties of Closure (3.7)

Let L and M be languages over the alphabet A. Then

a) $\{\Lambda\}^* = \varnothing^* = \{\Lambda\}$.

b) $L^* = L^* \cdot L^* = (L^*)^*$.

c) $\Lambda \in L$ if and only if $L^+ = L^*$.

d) $(L^* \cdot M^*)^* = (L^* \cup M^*)^* = (L \cup M)^*$.

e) $L \cdot (M \cdot L)^* = (L \cdot M)^* \cdot L$.

Grammars and Derivations

Informally, a grammar is a set of rules used to define the structure of the strings in a language. Grammars are important in computer science not only to define programming languages, but also to define data sets for programs. Typical applications try to build algorithms that test whether or not an arbitrary string belongs to some language.

We can think of an English sentence as a string in the English language if we agree to let the alphabet contain the blank character, period, comma, and so on. To *parse* a sentence means to break it up into parts that conform to a given grammar.

For example, if an English sentence consists of a subject followed by a predicate, then the sentence

"The big dog chased the cat"

would be broken up into two parts, a subject and a predicate, as follows:

subject = The big dog,

predicate = chased the cat.

To denote the fact that a sentence consists of a subject followed by a predicate we'll write the following *grammar rule*:

sentence → subject predicate.

If we agree that a subject can be an article followed by either a noun or an adjective followed by a noun, then we can break up "The big dog" into smaller parts. The corresponding grammar rule can be written as follows:

subject → article adjective noun.

Similarly, if we agree that a predicate is a verb followed by an object, then we can break up "chased the cat" into smaller parts. The corresponding grammar rule can be written as follows:

predicate → verb object.

This is the kind of activity that can be used to detect whether or not a sentence is grammatically correct.

A parsed sentence is often represented as a tree, called the *parse tree* or *derivation tree*. The parse tree for "The big dog chased the cat" is pictured in Figure 3.6.

Now that we've recalled a bit of English grammar, let's describe the general structure of grammars for arbitrary languages. If L is a language over an alphabet A, then a grammar for L consists of a set of *grammar rules* of the form

$$\alpha \to \beta,$$

where α and β denote strings of symbols taken from A and from a set of grammar symbols disjoint from A.

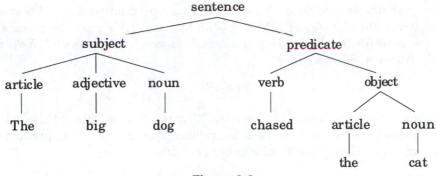

Figure 3.6

The grammar rule $\alpha \to \beta$ is often called a *production*, and it can be read in several different ways as follows:

> replace α by β,
>
> α produces β,
>
> α rewrites to β,
>
> α reduces to β.

Every grammar has a special grammar symbol called a *start symbol*, and there must be at least one production with left side consisting of only the start symbol. For example, if S is the start symbol for a grammar, then there must be at least one production of the form

$$S \to \beta.$$

Let's give an example of a grammar for a language and then discuss the process of deriving strings from the productions. Let $A = \{a, b, c\}$. Then a grammar for the language A^* can be described by the following four productions:

$$
\begin{aligned}
S &\to \Lambda \\
S &\to aS \\
S &\to bS \\
S &\to cS.
\end{aligned}
\tag{3.8}
$$

How do we know that this grammar describes the language A^*? We must be able to describe each string of the language in terms of the grammar rules. For example, let's see how we can use the productions (3.8) to show that the string *aacb* is in A^*. We'll begin with the start symbol S. Then we'll replace S

by the right side of production $S \to aS$. We chose production $S \to aS$ because *aacb* matches the right hand side of $S \to aS$ by letting $S = acb$. The process of replacing S by aS is called a *derivation*, and we say, "S derives aS." We'll denote this derivation by writing

$$S \Rightarrow aS.$$

The symbol $\Rightarrow$ means "derives in one step." The right-hand side of this derivation contains the symbol S. So we again replace S by aS using the production $S \to aS$ a second time. This results in the derivation

$$S \Rightarrow aS \Rightarrow aaS.$$

The right-hand side of this derivation contains S. In this case we'll replace S by the right side of $S \to cS$. This gives the derivation

$$S \Rightarrow aS \Rightarrow aaS \Rightarrow aacS.$$

Continuing, we replace S by the right side of $S \to bS$. This gives the derivation

$$S \Rightarrow aS \Rightarrow aaS \Rightarrow aacS \Rightarrow aacbS.$$

Since we want this derivation to produce the string *aacb*, we now replace S by the right side of $S \to \Lambda$. This gives the desired derivation of the string *aacb*:

$$S \Rightarrow aS \Rightarrow aaS \Rightarrow aacS \Rightarrow aacbS \Rightarrow aacb\Lambda = aacb.$$

Each step in a derivation corresponds to attaching a new subtree to the derivation tree whose root is the start symbol. For example, the derivation trees corresponding to the first three steps of our example are shown in Figure 3.7.

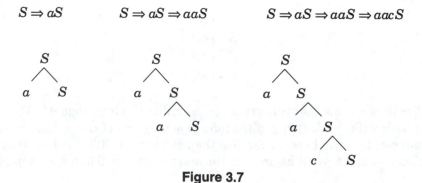

Figure 3.7

$$S \Rightarrow aS \Rightarrow aaS \Rightarrow aacS \Rightarrow aacbS \Rightarrow aacb\Lambda = aacb$$

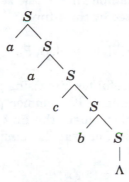

Figure 3.8

The completed derivation and derivation tree are shown in Figure 3.8.

Now that we've introduced the idea of a grammar, let's take a minute to describe the four main ingredients of any grammar.

The Four Parts of a Grammar   (3.9)

1. An alphabet N of grammar symbols called *nonterminals*.

2. An alphabet T of symbols called *terminals*. The terminals are distinct from the nonterminals.

3. A specific nonterminal called the *start* symbol.

4. A finite set of productions of the form $\alpha \to \beta$, where α and β are strings over the alphabet $N \cup T$ with the restriction that α is not the empty string. There is at least one production with only the start symbol on its left side. Each nonterminal must appear on the left side of some production.

Assumption: In this chapter, all grammar productions will have a single nonterminal on the left side. In Chapter 14 we'll see examples of grammars that allow productions to have strings of more than one symbol on the left side.

When two or more productions have the same left side, we can simplify the notation by writing one production with alternate right sides separated by the vertical line |. For example, the four productions (3.8) can be written in the following shorthand form:

$$S \to \Lambda \mid aS \mid bS \mid cS,$$

and we say, "S can be replaced by either Λ, or aS, or bS, or cS."

We can represent a grammar G as a 4-tuple $G = \langle N, T, S, P \rangle$, where P is the set of productions. For example, if P is the set of productions (3.8), then the grammar can be represented by the 4-tuple

$$\langle \{S\}, \{a, b, c\}, S, P \rangle.$$

The 4-tuple notation is useful for discussing general properties of grammars. But for a particular grammar it's common practice to only write down the productions of the grammar, where the first production listed contains the start symbol on its left side. For example, suppose we're given the following grammar:

$$S \rightarrow AB$$
$$A \rightarrow \Lambda \mid aA$$
$$B \rightarrow \Lambda \mid bB.$$

We can deduce that the nonterminals are $S, A,$ and B, the start symbol is S, and the terminals are a and b.

To discuss grammars further, we need to formalize things a bit. Suppose we're given some grammar. A string made up of terminals and/or nonterminals is called a *sentential form*. Now we can formalize the idea of a derivation. If x and y are sentential forms and $\alpha \rightarrow \beta$ is a production, then the replacement of α by β in $x\alpha y$ is called a *derivation*, and we denote it by writing

$$x\alpha y \Rightarrow x\beta y. \tag{3.10}$$

The following three symbols with their associated meanings are used quite often in discussing derivations:

$\Rightarrow$ derives in one step,

$\Rightarrow^+$ derives in one or more steps,

$\Rightarrow^*$ derives in zero or more steps.

For example, suppose we have the following grammar:

$$S \rightarrow AB$$
$$A \rightarrow \Lambda \mid aA$$
$$B \rightarrow \Lambda \mid bB.$$

Let's consider the string aab. The statement $S \Rightarrow^+ aab$ means that there exists a derivation of aab that takes one or more steps. For example, we have

$$S \Rightarrow AB \Rightarrow aAB \Rightarrow aaAB \Rightarrow aaB \Rightarrow aabB \Rightarrow aab.$$

When a grammar contains more than one nonterminal—as the preceding grammar does—it may be possible to find several different derivations of the same string. Two kinds of derivations are worthy of note. A derivation is called a *leftmost derivation* if at each step the leftmost nonterminal of the sentential form is reduced by some production. Similarly, a derivation is called a *rightmost derivation* if at each step the rightmost nonterminal of the sentential form is reduced by some production. For example, the preceding derivation of *aab* is a leftmost derivation. Here's a rightmost derivation of *aab*:

$$S \Rightarrow AB \Rightarrow AbB \Rightarrow Ab \Rightarrow aAb \Rightarrow aaAb \Rightarrow aab.$$

Grammars and Languages

Sometimes it's easy to write a grammar, and sometimes it can be quite difficult. The most important aspect of grammar writing is knowledge of the language under discussion. So we had better nail down the idea of the language associated with a grammar. If G is a grammar, then the *language of G* is the set of terminal strings derived from the start symbol of G. The language of G is denoted by

$$L(G).$$

We can also describe $L(G)$ more formally. If G is a grammar with start symbol S and set of terminals T, then the language of G is the following set:

$$L(G) = \{s \mid s \in T^* \text{ and } S \Rightarrow^+ s\}. \tag{3.11}$$

When we're trying to write a grammar for a language, we should at least check to see whether the language is finite or infinite. If the language is finite, then a grammar can consist of all productions of the form $S \rightarrow w$ for each string w in the language. For example, the language $\{a, ab\}$ can be described by the grammar $S \rightarrow a \mid ab$.

If the language is infinite, then some production or sequence of productions must be used repeatedly to construct the derivations. To see this, notice that there is no bound on the length of strings in an infinite language. Therefore there is no bound on the number of derivation steps used to derive the strings. If the grammar has n productions, then any derivation consisting of $n + 1$ steps must use some production twice (by the pigeonhole principle).

For example, the infinite language $\{a^n b \mid n \geq 0\}$ can be described by the grammar

$$S \rightarrow b \mid aS .$$

To derive the string $a^n b$, we would use the production $S \to aS$ repeatedly—n times to be exact—and then stop the derivation by using the production $S \to b$. The situation is similar to the way we make inductive definitions for sets. For example, the production $S \to aS$ allows us to make the informal statement "If S derives w, then it also derives aw."

A production is called *recursive* if its left side occurs on its right side. For example, the production $S \to aS$ is recursive. A production $S \to \alpha$ is *indirectly recursive* if S derives a sentential form that contains S. For example, suppose we have the following grammar:

$$S \to b \mid aA$$
$$A \to c \mid bS.$$

The productions $S \to aA$ and $A \to bS$ are both indirectly recursive because of the following derivations:

$$S \Rightarrow aA \Rightarrow abS,$$
$$A \Rightarrow bS \Rightarrow baA.$$

A grammar is *recursive* if it contains either a recursive production or an indirectly recursive production. So we can make the following more precise statement about grammars for infinite languages:

A grammar for an infinite language must be recursive.

Now let's look at the opposite problem of describing the language of a grammar. We know—by definition—that the language of a grammar is the set of all strings derived from the grammar. But we can also make another interesting observation about any language defined by a grammar:

Any language defined by a grammar is an inductively defined set.

Let's see why this is the case for any grammar G. We need to describe $L(G)$ by giving a basis case and an induction case. The following inductive definition does the job, where S denotes the start symbol of G:

Inductive Definition of L(G) (3.12)

Basis: For all strings w that can be derived from S without using a recursive or indirectly recursive production, put w in $L(G)$.

Induction: If $w \in L(G)$ and a derivation $S \Rightarrow^+ w$ contains a nonterminal from a recursive or indirectly recursive production, then modify the derivation by using the production to construct a new derivation $S \Rightarrow^+ x$, and put x in $L(G)$.

Proof: Let G be a grammar and let M be the inductive set defined by (3.12). We need to show that $M = L(G)$. It's clear that $M \subset L(G)$ because all strings in M are derived from the start symbol of G. Assume, by way of contradiction, that $M \neq L(G)$. In other words, we have $L(G) - M \neq \emptyset$. Since S derives all the elements of $L(G) - M$, there must be some string $w \in L(G) - M$ that has the shortest leftmost derivation among elements of $L(G) - M$. We can also assume that this derivation uses a recursive or indirectly recursive production. Otherwise, the basis case of (3.12) would force us to put $w \in M$, contrary to our assumption that $w \in L(G) - M$. So the leftmost derivation of w must have the following form, where s and t are terminal strings and α, β, and γ are sentential forms that don't include B:

$$S \Rightarrow^+ sB\gamma \Rightarrow^+ stB\beta\gamma \Rightarrow st\alpha\beta\gamma \Rightarrow^* w.$$

We can replace $sB\gamma \Rightarrow^+ stB\beta\gamma$ in this derivation with $sB\gamma \Rightarrow s\alpha\gamma$ to obtain the following derivation of a string u of terminals:

$$S \Rightarrow^+ sB\gamma \Rightarrow s\alpha\gamma \Rightarrow^* u.$$

This derivation is shorter than the derivation of w. So we must conclude that $u \in M$. Now we can apply the induction part of (3.12) to this latter derivation of u to obtain the derivation of w. This tells us that $w \in M$, contrary to our assumption that $w \notin M$. The only think left for us to conclude is that our assumption that $M \neq L(G)$ was wrong. Therefore $M = L(G)$.   QED.

Let's do a simple example to illustrate the use of (3.12).

EXAMPLE 3. Suppose we're given the following grammar G:

$$S \to \Lambda \mid aB$$
$$B \to b \mid bB.$$

We want to give an inductive definition for $L(G)$. For the basis case there are two derivations that don't contain recursive productions: $S \Rightarrow \Lambda$ and $S \Rightarrow aB \Rightarrow ab$. This gives us the basis part of the definition for $L(G)$:

Basis: $\Lambda, ab \in L(G)$.

Now let's find the induction part of the definition. The only recursive production of G is $B \to bB$. So any element of $L(G)$ whose derivation contains an occurrence of B must have the general form $S \Rightarrow aB \Rightarrow^+ ay$ for some string y. So we can use the production $B \to bB$ to add one more step to the derivation as follows:

$$S \Rightarrow aB \Rightarrow abB \Rightarrow^+ aby.$$

This gives us the induction step in the definition of $L(G)$:

> *Induction:*   If $ay \in L(G)$, then put aby in $L(G)$.

For example, the basis case tells us that $ab \in L(G)$ and the derivation $S \Rightarrow aB \Rightarrow ab$ contains an occurrence of B. So we add one more step to the derivation using the production $B \to bB$ to obtain the derivation

$$S \Rightarrow aB \Rightarrow abB \Rightarrow abb.$$

So $ab \in L(G)$ implies that $abb \in L(G)$. Now we can use the fact that $abb \in L(G)$ to put $ab^3 \in L(G)$, and so on. We can conjecture with some confidence that $L(G) = \{ab^n \mid n \in \mathbb{N}\}$.   ◆

Now let's get down to business and construct some grammars. We'll start with a few simple examples, and then we'll give some techniques for combining grammars. We should note that a language might have more than one grammar. So we shouldn't be surprised when two people come up with two different grammars for the same language.

The following list contains a few languages along with a grammar for each one. Test each grammar by constructing a few derivations for strings in the language.

Language	*Grammar*
$\{a, ab, abb, abbb\}$	$S \to a \mid ab \mid abb \mid abbb$
$\{\Lambda, a, aa, ..., a^n, ...\}$	$S \to \Lambda \mid aS$
$\{b, bbb, ..., b^{2n+1}, ...\}$	$S \to b \mid bbS$
$\{b, abc, aabcc, ..., a^nbc^n, ...\}$	$S \to b \mid aSc$
$\{ac, abc, abbc, ..., ab^nc, ...\}$	$S \to aBc$
	$B \to \Lambda \mid bB$

Sometimes a language can be written in terms of simpler languages, and a grammar can be constructed for the language in terms of the grammars for the simpler languages. We'll concentrate here on the operations of union, product, and closure.

Combining Grammars   (3.13)

Suppose M and N are languages whose grammars have disjoint sets of nonterminals. (Rename them if necessary.) Suppose also that the start symbols for the grammars of M and N are A and B, respectively. Then we have the following new languages and grammars:

Union Rule: The language $M \cup N$ starts with the two productions

$$S \rightarrow A \mid B.$$

Product Rule: The language $M \cdot N$ starts with the production

$$S \rightarrow AB.$$

Closure Rule: The language M^* starts with the production

$$S \rightarrow AS \mid \Lambda.$$

Let's see how we can use (3.13) to construct some grammars. For example, suppose we want to write a grammar for the following language:

$$L = \{\Lambda, a, b, aa, bb, ..., a^n, b^n, ...\}.$$

After a little thinking we notice that L is the union of the two languages $M = \{a^n \mid n \in \mathbb{N}\}$ and $N = \{b^n \mid n \in \mathbb{N}\}$. Thus we can write a grammar for L as follows:

$$
\begin{array}{ll}
S \rightarrow A \mid B & \text{union rule,} \\
A \rightarrow \Lambda \mid aA & \text{grammar for } M, \\
B \rightarrow \Lambda \mid bB & \text{grammar for } N.
\end{array}
$$

For another example, suppose we want to write a grammar for the following language:

$$L = \{a^m b^n \mid m, n \in \mathbb{N}\}.$$

After a little thinking we notice that L is the product of the two languages $M = \{a^m \mid m \in \mathbb{N}\}$ and $N = \{b^n \mid n \in \mathbb{N}\}$. Thus we can write a grammar for L as follows:

$$
\begin{array}{ll}
S \rightarrow AB & \text{product rule,} \\
A \rightarrow \Lambda \mid aA & \text{grammar for } M, \\
B \rightarrow \Lambda \mid bB & \text{grammar for } N.
\end{array}
$$

The closure rule in (3.13) describes the way we've been constructing grammars in some of our examples. For another example, suppose we want to construct the language L of all possible strings made up from zero or more occurrences of aa or bb. In other words, $L = \{aa, bb\}^*$. So we can write a grammar for L as follows:

$$S \rightarrow AS \mid \Lambda \qquad \text{closure rule,}$$
$$A \rightarrow aa \mid bb \qquad \text{grammar for } \{aa, bb\}.$$

We can simplify this grammar as follows:

$$S \rightarrow aaS \mid bbS \mid \Lambda.$$

Let's look at a few more examples of grammars.

EXAMPLE 4 (*Decimal Numerals*). We can find a grammar for the language of decimal numerals by observing that a decimal numeral is either a digit or a digit followed by a decimal numeral. The following grammar rules reflect this idea:

$$S \rightarrow D \mid DS$$
$$D \rightarrow 0 \mid 1 \mid 2 \mid 3 \mid 4 \mid 5 \mid 6 \mid 7 \mid 8 \mid 9.$$

We can say that S is replaced by either D or DS, and D can be replaced by any decimal digit. A derivation of the numeral 7801 can be written as follows:

$$S \Rightarrow DS \Rightarrow 7S \Rightarrow 7DS \Rightarrow 7DDS \Rightarrow 78DS \Rightarrow 780S \Rightarrow 780D \Rightarrow 7801.$$

This derivation is not unique. For example, another derivation of 7801 can we written as follows:

$$S \Rightarrow DS \Rightarrow DDS \Rightarrow D8S \Rightarrow D8DS \Rightarrow D80S \Rightarrow D80D \Rightarrow D801 \Rightarrow 7801. \quad \blacklozenge$$

EXAMPLE 5 (*Even Decimal Numerals*). We can find a grammar for the language of decimal numerals for the even natural numbers by observing that each numeral must have an even digit on its right side. In other words, either it's an even digit or it's a decimal numeral followed by an even digit. The following grammar will do the job:

$$S \rightarrow E \mid NE$$
$$N \rightarrow D \mid DN$$
$$E \rightarrow 0 \mid 2 \mid 4 \mid 6 \mid 8$$
$$D \rightarrow 0 \mid 1 \mid 2 \mid 3 \mid 4 \mid 5 \mid 6 \mid 7 \mid 8 \mid 9.$$

For example, the even numeral 136 has the derivation

$$S \Rightarrow NE \Rightarrow N6 \Rightarrow DN6 \Rightarrow DD6 \Rightarrow D36 \Rightarrow 136. \quad \blacklozenge$$

EXAMPLE 6 (*Identifiers*). Most programming languages have identifiers for names of things. Suppose we want to describe a grammar for the set of identifiers that start with a letter of the alphabet followed by zero or more letters or digits. Let Id be the start symbol. Then the grammar can be described by the following productions:

$$\text{Id} \rightarrow L \mid LA$$
$$A \rightarrow LA \mid DA \mid \Lambda$$
$$L \rightarrow a \mid b \mid \ldots \mid z$$
$$D \rightarrow 0 \mid 1 \mid \ldots \mid 9.$$

For example, we give a derivation of the string *a2b* to show that it is an identifier:

$$\text{Id} \Rightarrow LA \Rightarrow aA \Rightarrow aDA \Rightarrow a2A \Rightarrow a2LA \Rightarrow a2bA \Rightarrow a2b. \quad \blacklozenge$$

EXAMPLE 7 (*Some Rational Numerals*). Let's find a grammar for those rational numbers that have a finite decimal representation. In other words, we want to describe a grammar for the language of strings having the form *m.n* or *−m.n*, where *m* and *n* are decimal numerals. For example, 0.0 represents the number 0. Let *S* be the start symbol. We can start the grammar with the two productions

$$S \rightarrow N.N \mid -N.N.$$

To finish the job, we need to write some productions that allow *N* to derive a decimal numeral. Try out the following productions:

$$N \rightarrow D \mid DN$$
$$D \rightarrow 0 \mid 1 \mid 2 \mid 3 \mid 4 \mid 5 \mid 6 \mid 7 \mid 8 \mid 9$$

EXAMPLE 8 (*Palindromes*). We can write a grammar for the set of all palindromes over an alphabet A. Recall that a palindrome is a string that is the same when written in reverse order. For example, let $A = \{a, b, c\}$. Let P be the start symbol. Then the language of palindromes over the alphabet A has the grammar

$$P \to aPa \mid bPb \mid cPc \mid a \mid b \mid c \mid \Lambda.$$

For example, the palindrome *abcba* can be derived as follows:

$$P \Rightarrow aPa \Rightarrow abPba \Rightarrow abcba. \quad \blacklozenge$$

Meaning and Ambiguity

Most of the time we attach a meaning or value to the strings in our lives. For example, the string 3+4 means 7 to most people. The string 3–4–2 may have two distinct meanings to two different people. One person may think that 3–4–2 = (3–4)–2 = –3, while another person might think that 3–4–2 = 3–(4–2) = 1.

If we have a grammar, then we can define the *meaning* of any string in the grammar's language to be the parse tree produced by a derivation. We can often write a grammar so that each string in the grammar's language has exactly one meaning (i.e., one parse tree). A grammar is called *ambiguous* if its language contains some string that has two different parse trees. This is equivalent to saying that some string has two distinct leftmost derivations or, equivalently, some string has two distinct rightmost derivations.

To illustrate the ideas, we'll look at some grammars for simple arithmetic expressions. For example, suppose we define a set of arithmetic expressions by the grammar

$$E \to a \mid b \mid E\text{–}E.$$

The language of this grammar contains strings like a, b, b–a, a–b–a, and b–b–a–b. This grammar is ambiguous because it has a string, namely, a–b–a, that has two distinct leftmost derivations as follows:

$$E \Rightarrow E\text{–}E \Rightarrow a\text{–}E \Rightarrow a\text{–}E\text{–}E \Rightarrow a\text{–}b\text{–}E \Rightarrow a\text{–}b\text{–}a.$$
$$E \Rightarrow E\text{–}E \Rightarrow E\text{–}E\text{–}E \Rightarrow a\text{–}E\text{–}E \Rightarrow a\text{–}b\text{–}E \Rightarrow a\text{–}b\text{–}a.$$

These two derivations give us the two distinct parse trees in Figure 3.9.

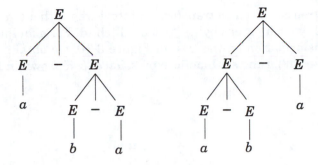

Figure 3.9

These two trees reflect the two ways we could choose to evaluate $a-b-a$. The first tree indicates the meaning

$$a-b-a = a-(b-a),$$

while the second tree indicates

$$a-b-a = (a-b)-a.$$

How can we make sure there is only one parse tree for every string in the language? We can try to find a different grammar for the same set of strings. For example, suppose we want $a-b-a$ to mean $(a-b)-a$. In other words, we want the first minus sign to be evaluated before the second minus sign. We can give the first minus sign higher precedence than the second by introducing a new nonterminal as shown in the following grammar:

$$E \rightarrow E\text{--}T \mid T$$
$$T \rightarrow a \mid b.$$

Notice that T can be replaced in a derivation only by either a, or b. Therefore every derivation of $a-b-a$ produces the unique parse tree in Figure 3.10.

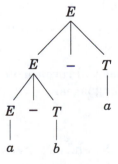

Figure 3.10

A parse tree can often be transformed into a tree without grammar symbols by replacing each parent by its "terminal" child. For example, if we perform this transformation on the tree in Figure 3.10, we obtain the familiar tree representation for the arithmetic expression *a–b–a* shown in Figure 3.11.

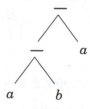

Figure 3.11

When we transform a parse tree in this manner, the resulting tree is called the *abstract syntax tree* for the string.

Note

Our notation for productions is a modification of Backus-Naur form, or BNF for short. For example, in Backus-Naur form—which is used by many authors—the production $S \to aS$ would be written as $\langle S \rangle ::= a \langle S \rangle$.

Exercises

1. Let $L = \{\Lambda, abb, b\}$ and $M = \{bba, ab, a\}$. Evaluate each of the following language expressions.

   a. $L \cdot M$.   b. $M \cdot L$.   c. L^2.

2. Use your wits to solve each of the following language equations for the unknown language.

   a. $\{\Lambda, a, ab\} \cdot L = \{b, ab, ba, aba, abb, abba\}$.

   b. $L \cdot \{a, b\} = \{a, baa, b, bab\}$.

   c. $\{a, aa, ab\} \cdot L = \{ab, aab, abb, aa, aaa, aba\}$.

   d. $L \cdot \{\Lambda, a\} = \{\Lambda, a, b, ab, ba, aba\}$

3. Let L, M, and N be languages. Prove each of the following properties of the product operation on languages.

   a. $L \cdot \{\Lambda\} = \{\Lambda\} \cdot L = L$.

   b. $L \cdot \varnothing = \varnothing \cdot L = \varnothing$.

   c. $L \cdot (M \cup N) = L \cdot M \cup L \cdot N$ and $(M \cup N) \cdot L = M \cdot L \cup N \cdot L$.

   d. $L \cdot (M \cap N) = L \cdot M \cap L \cdot N$ and $(M \cap N) \cdot L = M \cdot L \cap N \cdot L$.

4. Let L and M be languages. Prove each of the following statements about the closure of languages.

    a. $\{\Lambda\}^* = \emptyset^* = \{\Lambda\}$.

    b. $L^* = L^* \cdot L^* = (L^*)^*$.

    c. $\Lambda \in L$ if and only if $L^+ = L^*$.

    d. $L \cdot (M \cdot L)^* = (L \cdot M)^* \cdot L$.

    e. $(L^* \cdot M^*)^* = (L^* \cup M^*)^* = (L \cup M)^*$.

5. Given the following grammar with start symbol S:

$$S \to D \mid DS$$
$$D \to 0 \mid 1 \mid 2 \mid 3 \mid 4 \mid 5 \mid 6 \mid 7 \mid 8 \mid 9.$$

    a. Find the productions used for the steps of the following derivation:

$$S \Rightarrow DS \Rightarrow 7S \Rightarrow 7DS \Rightarrow 7DDS \Rightarrow 78DS \Rightarrow 780S \Rightarrow 780D \Rightarrow 7801.$$

    b. Find a leftmost derivation of the string 7801.

    c. Find a rightmost derivation of the string 7801.

6. For each grammar G, use (3.12) to find an inductive definition for $L(G)$.

    a. $S \to \Lambda \mid aaS$.

    b. $S \to a \mid aBc, B \to b \mid bB$.

7. Find a grammar for each of the following languages.

    a. $\{bb, bbbb, bbbbbb, ...\}$.

    b. $\{a, ba, bba, bbba, ...\}$.

    c. $\{\Lambda, ab, abab, ababab, ...\}$.

    d. $\{bb, bab, baab, baaab, ...\}$.

8. Find a grammar for each of the following languages.

    a. The set of decimal numerals that represent odd natural numbers.

    b. The set of binary numerals that represent odd natural numbers.

    c. The set of binary numerals that represent even natural numbers.

9. Find a grammar for each of the following languages.

    a. The set Exp of all arithmetic expressions that are constructed from decimal numerals, +, and parentheses. For example, Exp should contain objects such as 17, 2+3, (3+(4+5)), and 5+9+20.

    b. The set MExp of arithmetic expressions that are constructed from decimal numerals, − (subtraction), and parentheses with the property that each expression has only one meaning. For example, 9−34−10 is not allowed.

10. If w is a string, let w^R denote the *reverse* of w. For example, $aabc$ is the reverse of $cbaa$. Let $A = \{a, b, c\}$. Find a grammar to describe the language $\{w\, w^R \mid w \in A^*\}$.

11. Find a grammar for each of the following languages.

 a. $\{a^n b^n \mid n \geq 0\}$.

 b. $\{a^n b^m \mid n \geq 1 \text{ and } m \geq 1\}$.

 c. $\{a^n bc^n \mid n \geq 0\} \cup \{b^n a^m \mid n \geq 0 \text{ and } m \geq 0\}$.

12. Find a grammar to capture the precedence $\cdot$ over $+$ in the absence of parentheses. For example, the meaning of $a + b \cdot c$ should be $a + (b \cdot c)$.

13. The three questions below refer to the following grammar:

$$S \rightarrow S[S]S \mid \Lambda.$$

 a. Write down a sentence describing the language of this grammar.

 b. This grammar is ambiguous. Prove it.

 c. Find an unambiguous grammar that has the same language.

14. Find a grammar for the language of finite sets whose elements can be identifiers (symbolized by the letter i) or other sets.

15. For each grammar, find an equivalent grammar that has no occurrence of Λ on the right side of any rule.

 a. $S \rightarrow AB$
 $A \rightarrow Aa \mid a$
 $B \rightarrow Bb \mid \Lambda$.

 b. $S \rightarrow AcAB$
 $A \rightarrow aA \mid \Lambda$
 $B \rightarrow bB \mid b$.

3.3 Recursively Defined Functions and Procedures

Since we're going to be constructing functions and procedures in this section, we'd better agree on the idea of a procedure. From a computer science point of view a *procedure* is a subprogram that carries out one or more actions, and it can also return any number of values—including none—through its arguments. For example, a statement like print(x, y) in a program will cause the print procedure to print the values of x and y on some output device. In this case, two actions are performed, and no values are returned. For another example, a statement like sort(L) might cause the sort procedure to carry out the action of sorting the list L in place. In this case, the action of sorting L is carried out, and the sorted list is returned as L. Or there might be a statement like sort(L, M) that leaves L alone but returns its sorted version as M.

A function or a procedure is said to be *recursively defined* if it is defined in terms of itself. In other words, a function f is recursively defined if at least one value $f(x)$ is defined in terms of another value $f(y)$, where $x \neq y$. Similarly, a procedure P is recursively defined if the actions of P for some argument x are defined in terms of the actions of P for another argument y, where $x \neq y$.

Many useful recursively defined functions have domains that are inductively defined sets. Similarly, many recursively defined procedures process elements from inductively defined sets. For these cases there are very useful construction techniques. Let's describe the two techniques.

Constructing a Recursively Defined Function (3.14)

If S is an inductively defined set, we can construct a function f with domain S as follows:

Basis: For each basis element $x \in S$, specify a value for $f(x)$.

Induction: Give one or more rules that—for any inductively defined element $x \in S$—will define $f(x)$ in terms of previously defined values of f.

Any function constructed by (3.14) is recursively defined because it is defined in terms of itself by the induction part of the definition. In a similar way we can construct a recursively defined procedure to process the elements of an inductively defined set.

Constructing a Recursively Defined Procedure (3.15)

If S is an inductively defined set, we can construct a procedure P to process the elements of S as follows:

Basis: For each basis element $x \in S$, specify a set of actions for $P(x)$.

Induction: Give one or more rules that—for any inductively defined element $x \in S$—will define the actions of $P(x)$ in terms of previously defined actions of P.

In the following paragraphs we'll see how (3.14) and (3.15) can be used to construct recursively defined functions and procedures over a variety of inductively defined sets. Most of our examples will be functions. But we'll define a few procedures too. We'll also see some recursively defined stream functions that are not defined by (3.14).

Natural Numbers

Let $f : \mathbb{N} \to \mathbb{N}$ denote the function that, for each $n \in \mathbb{N}$, returns the sum of the odd numbers from 1 to $2n + 1$, as follows:

$$f(n) = 1 + 3 + \cdots + (2n + 1).$$

Notice that f(0) = 1. So we have a definition for f applied to the basis element $0 \in \mathbb{N}$. For the inductive part of the definition, notice how we can write $f(n + 1)$ in terms of $f(n)$ as follows:

$$
\begin{aligned}
f(n + 1) &= 1 + 3 + \cdots + (2n + 1) + [2(n + 1) + 1] \\
&= (1 + 3 + \cdots + 2n + 1) + 2n + 3 \\
&= f(n) + 2n + 3.
\end{aligned}
$$

This gives us the necessary ingredients for a recursive definition of f:

Basis: $f(0) = 1.$

Induction: $f(n + 1) = f(n) + 2n + 3.$

A definition like this is often called a *pattern-matching definition* because the evaluation of an expression $f(x)$ depends on $f(x)$ matching either $f(0)$ or $f(n + 1)$. For example, $f(3)$ matches $f(n + 1)$ with $n = 2$, so we would choose the second equation to evaluate $f(3)$.

An alternative form for the definition of f is the conditional form. One conditional form consists of equations with conditionals as follows:

Basis: $f(0) = 1.$

Induction: $f(n) = f(n - 1) + 2n + 1$ if $n > 0.$

A second conditional form is the familiar if-then-else equation as follows:

$$f(n) = \text{if } n = 0 \text{ then } 1 \text{ else } f(n - 1) + 2n + 1.$$

A recursively defined function can be easily evaluated by a technique called *unfolding* the definition. For example, to find the value of $f(4)$, we start by finding the appropriate expression to equate to $f(4)$ by using pattern matching or by using conditionals. Continue in this manner to unfold all expressions of the form $f(x)$ until none are left. The resulting expression can then be evaluated. Here is the sequence of unfoldings for the evaluation of $f(4)$ using the if-then-else definition:

$$
\begin{aligned}
f(4) \quad &= \quad f(3) + 2{\cdot}4{+}1 \\
&= \quad f(2) + 2{\cdot}3{+}1 + 2{\cdot}4{+}1 \\
&= \quad f(1) + 2{\cdot}2{+}1 + 2{\cdot}3{+}1 + 2{\cdot}4{+}1 \\
&= \quad f(0) + 2{\cdot}1{+}1 + 2{\cdot}2{+}1 + 2{\cdot}3{+}1 + 2{\cdot}4{+}1 \\
&= \quad 1 + 2{\cdot}1{+}1 + 2{\cdot}2{+}1 + 2{\cdot}3{+}1 + 2{\cdot}4{+}1 \\
&= \quad 1 + 3 + 5 + 7 + 9 \\
&= \quad 25.
\end{aligned}
$$

EXAMPLE 1 (*The Rabbit Problem*). The *Fibonacci numbers* are the numbers in the sequence

$$0, 1, 1, 2, 3, 5, 8, 13, \ldots$$

where each number after the first two is computed by adding the preceding two numbers. These numbers are named after the mathematician Leonardo Fibonacci, who in 1202 introduced them in his book *Liber Abaci*, in which he proposed and solved the following problem: Starting with a pair of rabbits, how many pairs of rabbits can be produced from that pair in a year if it is assumed that every month each pair produces a new pair that becomes productive after one month?

For example, if we don't count the original pair and assume that the original pair needs one month to mature and that no rabbits die, then the number of new pairs produced each month for 12 consecutive months is given by the sequence

$$0, 1, 1, 2, 3, 5, 8, 13, 21, 34, 55, 89.$$

The sum of these numbers, which is 232, is the number of pairs of rabbits produced in one year from the original pair.

Fibonacci numbers seem to occur naturally in many unrelated problems. Of course, they can also be defined recursively. For example, letting fib(n) be the nth Fibonacci number, we can define fib recursively as follows:

Basis: fib(0) = 0 and fib(1) = 1.

Induction: fib($n + 2$) = fib($n + 1$) + fib(n). ♦

From a strict point of view we should have placed the equation fib(1) = 1 in the induction part of the definition in Example 1 because 0 is the only basis element of $\mathbb{N}$ and 1 is the successor of 0. That is, we could have written the definition as follows:

Basis: fib(0) = 0.

Induction: fib(1) = 1,
                fib(n + 2) = fib(n + 1) + fib(n).

But it's common practice to place the definitions for specific elements in the basis case. We can also write the conditional form of the definition as

Basis: fib(0) = 0,
                fib(1) = 1.

Induction: fib(n) = fib(n − 1) + fib(n − 2)   if $n \geq 2$.

Or we can write the conditional form as

$$\text{fib}(n) = \text{if } n = 0 \text{ then } 0$$
$$\text{else if } n = 1 \text{ then } 1$$
$$\text{else fib}(n - 1) + \text{fib}(n - 2).$$

Sum and Product Notations

Many definitions and properties that we use without thinking are recursively defined. For example, given a sequence of numbers $\langle a_1, a_2, ..., a_n, ... \rangle$, we can represent the sum of the first n numbers of the sequence with summation notation using the symbol Σ as follows:

$$\sum_{i=1}^{n} a_i = a_1 + a_2 + ... + a_n.$$

This notation has a recursive definition, which makes the practical assumption that an empty sum is 0:

$$\sum_{i=1}^{n} a_i = \text{if } n = 0 \text{ then } 0 \text{ else } a_n + \sum_{i=1}^{n-1} a_i.$$

Similarly we can represent the product of the first n numbers in the sequence with the product notation, where the practical assumption is that an empty product is 1:

$$\prod_{i=1}^{n} a_i = \text{if } n = 0 \text{ then } 1 \text{ else } a_n \cdot \prod_{i=1}^{n-1} a_i.$$

In the special case in which $a_i = i$ this product defines the popular *factorial function*, which is denoted by $n!$ and is read "n factorial." In other words, we have

$$n! = n \cdot (n - 1) \cdots 1.$$

For example, $4! = 4 \cdot 3 \cdot 2 \cdot 1 = 24$, and $0! = 1$. So we can write $n!$ as follows:

$$n! = \text{if } n = 0 \text{ then } 1 \text{ else } n \cdot (n - 1)!.$$

The sum and product notations can be defined for any pair of indices $m \leq n$. In fact, the symbols Σ and Π can be defined as functions if we consider a sequence to be a function. For example, the sequence $\langle a_1, a_2, ..., a_n, ... \rangle$ is a listing of functional values for the function $a : \mathbb{N} \to \mathbb{N}$, where we write a_i instead of $a(i)$. Then Π is a higher-order function of three variables

$$\prod(m, n, a) = \prod_{i=m}^{n} a(i) = a(m) \cdots a(n).$$

Many definitions and laws of exponents for arithmetic can be expressed recursively. For example,

$$a^n = \text{if } n = 0 \text{ then } 1 \text{ else } a \cdot a^{n-1}.$$

The law $a^m \cdot a^n = a^{m+n}$ for multiplication can be defined recursively by the equations that follow (where n is the induction variable):

$$a^m \cdot a^0 = a^m,$$
$$a^m \cdot a^{n+1} = a^{m+1} \cdot a^n.$$

For example, we can write $a^m \cdot a^2 = a^{m+1} \cdot a^1 = a^{m+2} \cdot a^0 = a^{m+2}$.

Lists

Suppose we need to define a function $f : \mathbb{N} \to \text{Lists}[\mathbb{N}]$ that computes a backwards sequence as follows:

$$f(n) = \langle n, ..., 1, 0 \rangle.$$

Notice that the list $\langle n, n - 1, ..., 1, 0 \rangle$ can also be written

$$\text{cons}(n, \langle n - 1, ..., 1, 0 \rangle) = \text{cons}(n, f(n - 1)).$$

Therefore f can be defined recursively by

Basis: $f(0) = \langle 0 \rangle$.

Induction: $f(n) = \text{cons}(n, f(n - 1))$ if $n > 0$.

This definition can be written in if-then-else form as

$$f(n) = \text{if } n = 0 \text{ then } \langle 0 \rangle \text{ else } \text{cons}(n, f(n - 1)).$$

To see how the evaluation works, look at the unfolding that results when we evaluate $f(3)$:

$$
\begin{aligned}
f(3) &= \text{cons}(3, f(2)) \\
&= \text{cons}(3, \text{cons}(2, f(1))) \\
&= \text{cons}(3, \text{cons}(2, \text{cons}(1, f(0)))) \\
&= \text{cons}(3, \text{cons}(2, \text{cons}(1, \langle 0 \rangle))) \\
&= \text{cons}(3, \text{cons}(2, \langle 1, 0 \rangle)) \\
&= \text{cons}(3, \langle 2, 1, 0 \rangle) \\
&= \langle 3, 2, 1, 0 \rangle.
\end{aligned}
$$

We haven't given a recursively defined procedure yet. So let's give one for the problem we've been discussing. In the following procedure, $P(n)$ prints the numbers in the list $\langle n, n - 1, ..., 0 \rangle$:

$P(n)$:   **if** $n = 0$ **then** print(0)
       else
           print(n);
           $P(n - 1)$
       fi.

EXAMPLE 2 (*Length of a List*). Let S be a set and let "length" be the function of type Lists[S] $\rightarrow \mathbb{N}$, that returns the number of elements in a list. We can define "length" recursively by noticing that the length of an empty list is zero and the length of a non-empty list is one plus the length of its tail. A definition follows:

Basis: $\text{length}(\langle \, \rangle) = 0$.

Induction: $\text{length}(\text{cons}(x, t)) = 1 + \text{length}(t)$.

Recall that the infix form of cons(x, t) is $x :: t$. So we could just as well write the second equation as

Induction: length($x :: t$) = 1 + length(t).

Also, we could write the induction part of the definition with a condition as follows:

Induction: length(L) = 1 + length(tail(L))   if $L \neq \langle\,\rangle$.

In if-then-else form the definition can be written as follows:

length(L) = if $L = \langle\,\rangle$ then 0 else 1 + length(tail(L)).

The length function can be evaluated by unfolding its definition. For example, suppose we use tuples to represent lists. Then

$$
\begin{aligned}
\text{length}(\langle a, b, c \rangle) &= 1 + \text{length}(\langle b, c \rangle) \\
&= 1 + 1 + \text{length}(\langle c \rangle) \\
&= 1 + 1 + 1 + \text{length}(\langle\,\rangle) \\
&= 1 + 1 + 1 + 0 \\
&= 3. \quad \blacklozenge
\end{aligned}
$$

EXAMPLE 3 (*Distribute Function*). Suppose we want to write a recursive definition for the distribute function, which we'll denote by "dist." For example,

dist(a, $\langle b, c, d, e \rangle$) = $\langle \langle a, b \rangle, \langle a, c \rangle, \langle a, d \rangle, \langle a, e \rangle \rangle$.

Since the second argument is a list, we can use induction on that argument to define dist. For example, notice how we can write the preceding equation:

$$
\begin{aligned}
\text{dist}(a, \langle b, c, d, e \rangle) &= \langle \langle a, b \rangle, \langle a, c \rangle, \langle a, d \rangle, \langle a, e \rangle \rangle \\
&= \langle a, b \rangle :: \text{dist}(a, \langle c, d, e \rangle).
\end{aligned}
$$

That's the key to the inductive part of the definition. Since we are inducting on lists, the basis case presents us with dist(a, $\langle\,\rangle$), which we define as $\langle\,\rangle$. So the recursive definition can be written as follows:

dist(a, $\langle\,\rangle$) = $\langle\,\rangle$,

dist(a, $b :: T$) = $\langle a, b \rangle :: \text{dist}(a, T)$.

For example, let's evaluate the expression dist(3, ⟨10, 20⟩) by unfolding the above definition.

$$\begin{aligned}
\text{dist}(3, \langle 10, 20 \rangle) &= \langle 3, 10 \rangle :: \text{dist}(3, \langle 20 \rangle) \\
&= \langle 3, 10 \rangle :: \langle 3, 20 \rangle :: \text{dist}(3, \langle\ \rangle) \\
&= \langle 3, 10 \rangle :: \langle 3, 20 \rangle :: \langle\ \rangle \\
&= \langle 3, 10 \rangle :: \langle\langle 3, 20 \rangle\rangle \\
&= \langle\langle 3, 10 \rangle, \langle 3, 20 \rangle\rangle.
\end{aligned}$$

We should note that the pair ⟨a, b⟩ in the definition can be constructed by $a :: b :: \langle\ \rangle$. The if-then-else form of dist can be written as follows:

$$\begin{aligned}
\text{dist}(x, L) \quad = \quad &\text{if } L = \langle\ \rangle \text{ then } \langle\ \rangle \\
&\text{else } (x :: \text{head}(L) :: \langle\rangle) :: \text{dist}(x, \text{tail}(L)). \quad \blacklozenge
\end{aligned}$$

EXAMPLE 4 (*Pairs Function*). Recall that the "pairs" function creates a list of pairs of corresponding elements from two lists. For example,

$$\text{pairs}(\langle a, b, c \rangle, \langle 1, 2, 3 \rangle) = \langle\langle a, 1 \rangle, \langle b, 2 \rangle, \langle c, 3 \rangle\rangle.$$

The pairs function can be defined recursively by the following equations:

$$\begin{aligned}
&\text{pairs}(\langle\ \rangle, \langle\ \rangle) = \langle\ \rangle, \\
&\text{pairs}(a :: T, b :: T') = \langle a, b \rangle :: \text{pairs}(T, T').
\end{aligned}$$

For example, we'll evaluate the expression pairs(⟨a, b⟩, ⟨1, 2⟩) by unfolding the definition:

$$\begin{aligned}
\text{pairs}(\langle a, b \rangle, \langle 1, 2 \rangle) &= \langle a, 1 \rangle :: \text{pairs}(\langle b \rangle, \langle 2 \rangle) \\
&= \langle a, 1 \rangle :: \langle b, 2 \rangle :: \text{pairs}(\langle\ \rangle, \langle\ \rangle) \\
&= \langle a, 1 \rangle :: \langle b, 2 \rangle :: \langle\ \rangle \\
&= \langle a, 1 \rangle :: \langle\langle b, 2 \rangle\rangle \\
&= \langle\langle a, 1 \rangle, \langle b, 2 \rangle\rangle. \quad \blacklozenge
\end{aligned}$$

EXAMPLE 5 (*ConsRight*). Suppose we need to give a recursive definition for the sequence function. Recall, for example, that seq(4) = ⟨0, 1, 2, 3, 4⟩. Good old "cons" doesn't seem up to the task. For example, if we somehow have computed seq(3), then cons(4, seq(3)) = ⟨4, 0, 1, 2, 3⟩. It would be nice if we had a constructor to place an element on the right of a list, just as cons places an element on the left of a list. We'll write a definition for the function "consR" to do just that. For example, we want

$$\text{consR}(\langle a, b, c \rangle, d) = \langle a, b, c, d \rangle.$$

We can get an idea of how to proceed by rewriting the above equation as follows in terms of the infix form of cons:

$$
\begin{aligned}
\text{consR}(\langle a, b, c \rangle, d) \;&=\; \langle a, b, c, d \rangle \\
&=\; a :: \langle b, c, d \rangle \\
&=\; a :: \text{consR}(\langle b, c \rangle, d).
\end{aligned}
$$

So the clue is to split the list ⟨a, b, c⟩ into its head and tail. We can write the inductive definition of consR using if-then-else form as follows:

$$
\begin{aligned}
\text{consR}(L, a) \;=\; &\text{if } L = \langle\,\rangle \text{ then } \langle a \rangle \\
&\text{else head}(L) :: \text{consR}(\text{tail}(L), a).
\end{aligned}
$$

This definition can be written in pattern-matching form as follows:

$$\text{consR}(\langle\,\rangle, a) = a :: \langle\,\rangle,$$
$$\text{consR}(b :: T, a) = b :: \text{consR}(T, a).$$

For example, we can construct the list ⟨x, y⟩ with consR as follows:

$$
\begin{aligned}
\text{consR}(\text{consR}(\langle\,\rangle, x), y) \;&=\; \text{consR}(x :: \langle\,\rangle, y) \\
&=\; x :: \text{consR}(\langle\,\rangle, y) \\
&=\; x :: y :: \langle\,\rangle \\
&=\; x :: \langle y \rangle \\
&=\; \langle x, y \rangle. \quad \blacklozenge
\end{aligned}
$$

EXAMPLE 6 (*Concatenation of Lists*). An important operation on lists is the concatenation of two lists into a single list. Let "cat" denote the concatenation function. Its type is Lists[A] × Lists[A] → Lists[A]. For example,

$$\text{cat}(\langle a, b \rangle, \text{p} \langle c, d \rangle) = \langle a, b, c, d \rangle.$$

Cat can be recursively defined as follows:

$$\text{cat}(\langle\,\rangle, L) = L,$$

$$\text{cat}(a :: T, L) = a :: \text{cat}(T, L).$$

We'll unfold the definition for the expression $\text{cat}(\langle a, b \rangle, \langle c, d \rangle)$:

$$
\begin{aligned}
\text{cat}(\langle a, b \rangle, \langle d, e \rangle) &= a :: \text{cat}(\langle b \rangle, \langle d, e \rangle) \\
&= a :: b :: \text{cat}(\langle\,\rangle, \langle d, e \rangle) \\
&= a :: b :: \langle d, e \rangle \\
&= a :: \langle b, d, e \rangle \\
&= \langle a, b, d, e \rangle.
\end{aligned}
$$

We can also write cat as a recursively defined procedure that prints out the elements of the two lists:

$$\text{cat}(K, L): \quad \textbf{if } K = \langle\,\rangle \textbf{ then } \text{print}(L)$$

$$\textbf{else}$$

$$\text{print}(\text{head}(K));$$

$$\text{cat}(\text{tail}(K), L)$$

$$\textbf{fi.} \quad \blacklozenge$$

EXAMPLE 7 (*Sorting a List by Insertion*). Let's define a function to sort a list of numbers by repeatedly inserting a new number into an already sorted list of numbers. Suppose "insert" is a function that does this job. Then the sort function itself is easy. For a basis case, notice that the empty list is already sorted. For the induction case we sort the list $x :: L$ by inserting x into the list obtained by sorting L. The definition can be written as follows:

$$\text{sort}(\langle\,\rangle) = \langle\,\rangle,$$

$$\text{sort}(x :: L) = \text{insert}(x, \text{sort}(L)).$$

Everything seems to make sense as long as insert does its job. We'll assume that whenever the number to be inserted is already in the list, then a new redundant copy will be placed to the left of the one already there. Now let's define insert. Again, the basis case is easy. The empty list is sorted, and to insert x into $\langle\,\rangle$, we simply create the singleton list $\langle x \rangle$. Otherwise—if the sorted list is not empty—either x belongs on the left of the list, or it should actually be inserted somewhere else in the list. An if-then-else definition can be written as follows:

$$\text{insert}(x, S) \quad = \quad \text{if } S = \langle \, \rangle \text{ then } \langle x \rangle$$
$$\text{else if } x \leq \text{head}(S) \text{ then } x :: S$$
$$\text{else head}(S) :: \text{insert}(x, \text{tail}(S)).$$

Notice that insert works only when S is already sorted. For example, we'll unfold the definition of insert(3, $\langle 1, 2, 6, 8 \rangle$):

$$\text{insert}(3, \langle 1, 2, 6, 8 \rangle) \quad = \quad 1 :: \text{insert}(3, \langle 2, 6, 8 \rangle)$$
$$= \quad 1 :: 2 :: \text{insert}(3, \langle 6, 8 \rangle)$$
$$= \quad 1 :: 2 :: 3 :: \langle 6, 8 \rangle$$
$$= \quad \langle 1, 2, 3, 6, 8 \rangle. \quad \blacklozenge$$

EXAMPLE 8 (*Higher-Order Functions*). The function map and the selector functions can be defined recursively. For example, map has the following recursive definition, in which we use the infix expression $a :: L$ for cons(a, L):

$$\text{map}(f)(\langle \, \rangle) = \langle \, \rangle,$$
$$\text{map}(f)(a :: L) = f(a) :: \text{map}(f)(L).$$

We can unfold the expression map(f)($\langle a, b, c \rangle$) as follows:

$$\text{map}(f)(\langle a, b, c \rangle) \quad = \quad f(a) :: \text{map}(f)(\langle b, c \rangle)$$
$$= \quad f(a) :: f(b) :: \text{map}(f)(\langle c \rangle)$$
$$= \quad f(a) :: f(b) :: f(c) :: \text{map}(f)(\langle \, \rangle)$$
$$= \quad f(a) :: f(b) :: f(c) :: \langle \, \rangle$$
$$= \quad \langle f(a), f(b), f(c) \rangle. \quad \blacklozenge$$

Strings

If A is an alphabet, then the domain of the string concatenation function "cat" is $A^* \times A^*$. Since A^* is an inductively defined set, it follows that $A^* \times A^*$ is also inductively defined. We give a recursive definition of cat that uses only the fact that the first copy of A^* is inductively defined:

$$\text{cat}(\Lambda, s) = s,$$
$$\text{cat}(a \cdot t, s) = a \cdot \text{cat}(t, s).$$

The if-then-else form of the definition can be written as follows:

$$\text{cat}(x, y) = \text{if } x = \Lambda \text{ then } y \text{ else head}(x) \cdot \text{cat}(\text{tail}(x), y).$$

For example, we evaluate the expression cat(ab, cd) as follows:

$$
\begin{aligned}
\text{cat}(ab, cd) &= a \cdot \text{cat}(b, cd) \\
&= a \cdot b \cdot \text{cat}(\Lambda, cd) \\
&= a \cdot b \cdot cd \\
&= a \cdot bcd \\
&= abcd.
\end{aligned}
$$

EXAMPLE 9 (*Natural Numbers Represented as Binary Strings*). Recall that the division algorithm allows us to represent a natural number x in the form

$$x = 2 \cdot \text{floor}(x/2) + x \bmod 2.$$

For example, $27 = 2 \cdot 13 + 1$, and $48 = 2 \cdot 24 + 0$. This formula can be used to create a binary string representation of x because $x \bmod 2$ is the rightmost bit of the representation. The next bit is found by computing floor($x/2$) mod 2. The next bit is floor(floor($x/2$)/2) mod 2, and so on.

Let's try to use this idea to write a recursive definition for the function "bin" to compute the binary representation for a natural number. If $x = 0$, then x has 0 as a binary representation. So we can use this as a basis case: bin(0) = 0. If $x \neq 0$, then we should concatenate the binary string representation of floor($x/2$) with the bit $x \bmod 2$. The definition can be written in if-then-else form as follows:

$$\text{bin}(x) = \text{if } x = 0 \text{ then } 0 \text{ else cat(bin(floor}(x/2)), x \bmod 2). \tag{3.16}$$

For example, we unfold the definition to calculate the expression bin(13):

$$
\begin{aligned}
\text{bin}(13) &= \text{cat(bin(6), 1)} \\
&= \text{cat(cat(bin(3), 0), 1)} \\
&= \text{cat(cat(cat(bin(1), 1), 0), 1)} \\
&= \text{cat(cat(cat(cat(bin(0),1), 1), 0), 1)} \\
&= \text{cat(cat(cat(cat(0, 1), 1), 0), 1)} \\
&= \text{cat(cat(cat(01, 1), 0), 1)}
\end{aligned}
$$

$$= \text{cat(cat(011, 0), 1)}$$

$$= \text{cat(0110, 1)}$$

$$= 01101.$$

Notice that bin always puts a leading 0 in front of the answer. Can you find an alternative definition that leaves off the leading 0? We'll leave this as an exercise. ◆

Binary Trees

Let's look at some functions that compute properties of binary trees. To start, suppose we need to know the number of nodes in a binary tree. Since the set of binary trees over a particular set can be defined inductively, we should be able to come up with a recursively defined function that suits our needs. Let "nodes" be the function that returns the number of nodes in a binary tree. Since the empty tree has no nodes, we have nodes($\langle\,\rangle$) = 0. If the tree is not empty, then the number of nodes can be computed by adding 1 to the number of nodes in the left and right subtrees. The equational definition of nodes can be written as follows:

$$\text{nodes}(\langle\,\rangle) = 0$$

$$\text{nodes}(\text{tree}(L, a, R)) = 1 + \text{nodes}(L) + \text{nodes}(R).$$

If we want the corresponding if-then-else form of the definition, it looks like

$$\text{nodes}(T) \quad = \quad \text{if } T = \langle\,\rangle \text{ then } 0$$
$$\text{else } 1 + \text{nodes}(\text{left}(T)) + \text{nodes}(\text{right }(T)).$$

For example, we'll evaluate nodes(T) for $T = \langle\langle\langle\,\rangle, a, \langle\,\rangle\rangle, b, \langle\,\rangle\rangle$:

$$\text{nodes}(T) \quad = \quad 1 + \text{nodes}(\langle\langle\,\rangle, a, \langle\,\rangle\rangle) + \text{nodes}(\langle\,\rangle)$$

$$= \quad 1 + 1 + \text{nodes}(\langle\,\rangle) + \text{nodes}(\langle\,\rangle) + \text{nodes}(\langle\,\rangle)$$

$$= \quad 1 + 1 + 0 + 0 + 0$$

$$= \quad 2.$$

EXAMPLE 10 (*A Binary Search Tree*). Suppose we have a binary search tree whose nodes are numbers, and we want to add a new number to the tree, under the assumption that the new tree is still a binary search tree. A function to do the job needs two arguments, a number x and a binary search tree T.

Let the name of the function be "insert." The basis case is easy. If $T = \langle \rangle$, then return tree($\langle \rangle$, x, $\langle \rangle$). The induction part is straightforward. If $x < \text{root}(T)$, then we need to replace the subtree left(T) by insert(x, left(T)). Otherwise, we replace right(T) by insert(x, right(T)). Notice that redundant elements are entered to the right. If we didn't want to add redundant elements, then we could simply return T whenever $x = \text{root}(T)$. The if-then-else form of the definition looks like the following:

$$\text{insert}(x, T) \quad = \quad \text{if } T = \langle \rangle \text{ then tree}(\langle \rangle, x, \langle \rangle)$$
$$\text{else if } x < \text{root}(T) \text{ then}$$
$$\text{tree}(\text{insert}(x, \text{left}(T)), \text{root}(T), \text{right}(T))$$
$$\text{else}$$
$$\text{tree}(\text{left}(T), \text{root}(T), \text{insert}(x, \text{right}(T))).$$

Now suppose we want to build a binary search tree from a given list of numbers in which the numbers are in no particular order. We can use the insert function as the main ingredient in a recursive definition. Let "makeTree" be the name of the function. We'll use two variables to describe the function, a binary search tree T and a list of numbers L.

$$\text{makeTree}(T, L) \quad = \quad \text{if } L = \langle \rangle \text{ then } T \tag{3.17}$$
$$\text{else makeTree}(\text{insert}(\text{head}(L), T), \text{tail}(L)).$$

To construct a binary search tree with this function, we apply makeTree to the pair of arguments $(\langle \rangle, L)$. As an example, the reader should unfold the definition for the call makeTree($\langle \rangle$, $\langle 3, 2, 4 \rangle$).

The function makeTree can be defined another way. Suppose we consider the following definition for constructing a binary search tree:

$$\text{makeTree}(T, L) \quad = \quad \text{if } L = \langle \rangle \text{ then } T \tag{3.18}$$
$$\text{else insert}(\text{head}(L), \text{makeTree}(T, \text{tail}(L))).$$

You should evaluate the expression makeTree($\langle \rangle$, $\langle 3, 2, 4 \rangle$) by unfolding this alternative definition. It should help explain the difference between the two definitions. ♦

Traversing Binary Trees

There are several useful ways to list the nodes of a binary tree. The three most popular methods are called preorder, inorder, and postorder. The *preorder* listing of a binary tree has the root of the tree as its head, and its tail is the concatenation of the preorder listing of the left and right subtrees of the root, in that order. For example, the preorder listing of the nodes of the binary tree in Figure 3.12 is $\langle a, b, c, d, e \rangle$.

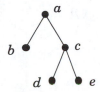

Figure 3.12

It's common practice to write the listing without any punctuation symbols as

$$a\ b\ c\ d\ e.$$

Since binary trees are inductively defined, we can easily write a recursively defined procedure to output the preorder listing of a binary tree. For example, the following recursively defined procedure prints the preorder listing of its argument T:

Preorder(T):   **if** $T \neq \langle\,\rangle$ **then**
                print(root(T));
                Preorder(left(T));
                Preorder(right(T))
        fi.

Now let's write a function to compute the preorder listing of a binary tree. Letting "preOrd" be the name of the preorder function, an equational definition can be written as follows:

$$preOrd(\langle\,\rangle) = \langle\,\rangle,$$
$$preOrd(tree(L, x, R)) = x :: cat(preOrd(L), preOrd(R)).$$

The if-then-else form of preOrdcan be written as follows:

$$preOrd(T) = \text{if } T = \langle\,\rangle \text{ then } \langle\,\rangle$$
$$\text{else root}(T) :: cat(preOrd(left(T)), preOrd(right(T))).$$

We'll evaluate the expression preOrd(T) for the tree $T = \langle\langle\langle\,\rangle, a, \langle\,\rangle\rangle, b, \langle\,\rangle\rangle$:

$$\begin{aligned}
\text{preOrd}(T) &= b :: \text{cat}(\text{preOrd}(\langle\langle\,\rangle, a, \langle\,\rangle\rangle), \text{preOrd}(\langle\,\rangle)) \\
&= b :: \text{cat}(a :: \text{cat}(\text{preOrd}(\langle\,\rangle), \text{preOrd}(\langle\,\rangle)), \text{preOrd}(\langle\,\rangle)) \\
&= b :: \text{cat}(a :: \langle\,\rangle, \langle\,\rangle) \\
&= b :: \text{cat}(\langle a\rangle, \langle\,\rangle) \\
&= b :: \langle a\rangle \\
&= \langle b, a\rangle.
\end{aligned}$$

The definitions for inorder and postorder listings are similar. The *inorder* listing of a binary tree is the concatenation of the inorder listing of the left subtree of the root with the list whose head is the root of the tree and whose tail is the inorder listing of the right subtree of the root. For example, the inorder listing of the tree in Figure 3.12 is

$$b\ a\ d\ c\ e.$$

The *postorder* listing of a binary tree is the concatenation of the postorder listings of the left and right subtrees of the root, followed lastly by the root. The postorder listing of the tree in Figure 3.12 is

$$b\ d\ e\ c\ a.$$

We'll leave the construction of the inorder and postorder procedures and functions as exercises.

The Redundant Element Problem

Suppose we want to remove redundant elements from a list. Depending on how we proceed, there might be different solutions. For example, we can remove the redundant elements from the list $\langle u, g, u, h, u\rangle$ in three ways, depending on which occurrence of u we keep: $\langle u, g, h\rangle$, $\langle g, u, h\rangle$, or $\langle g, h, u\rangle$.

We'll solve the problem by always keeping the leftmost occurrence of each element. Let "remove" be the function that takes a list L and returns the list remove(L), which has no redundant elements and contains the leftmost occurrence of each element of L.

To start things off, we can say remove($\langle\,\rangle$) = $\langle\,\rangle$. Now if $L \neq \langle\,\rangle$, then L has the form $L = b :: M$ for some list M. In this case, the head of remove(L) should be b. The tail of remove(L) can be obtained by removing all occurrences of b from M and then removing all redundant elements from the resulting list. So we need a new function to remove all occurrences of an element from a list.

Let removeAll(b, M) denote the list obtained from M by removing all occurrences of b. Now we can write an equational definition for the remove function as follows:

$$\text{remove}(\langle\,\rangle) = \langle\,\rangle,$$
$$\text{remove}(b :: M) = b :: \text{remove}(\text{removeAll}(b, M)).$$

We can rewrite the solution in if-then-else form as follows:

$$\text{remove}(L) = \;\text{if } L = \langle\,\rangle \text{ then } \langle\,\rangle$$
$$\text{else } \text{head}(L) :: \text{remove}(\text{removeAll}(\text{head}(L), \text{tail}(L))).$$

To complete the task, we need to define the removeAll function. The basis case is removeAll(b, $\langle\,\rangle$) = $\langle\,\rangle$. If $M \neq \langle\,\rangle$, then the value of removeAll(b, M) depends on head(M). If head(M) = b, then throw it away and return the value of removeAll(b, tail(M)). But if head(M) $\neq b$, then it's a keeper. So we should return the value head(M) :: removeAll(b, tail(M)). We can write the definition in if-then-else form as follows:

$$\text{removeAll}(b, M) = \;\text{if } M = \langle\,\rangle \text{ then } \langle\,\rangle$$
$$\text{else if } \text{head}(M) = b \text{ then}$$
$$\text{removeAll}(b, \text{tail}(M))$$
$$\text{else}$$
$$\text{head}(M) :: \text{removeAll}(b, \text{tail}(M)).$$

For example, we evaluate the expression removeAll(b, $\langle a, b, c, b\rangle$) by unfolding the definition:

$$
\begin{aligned}
\text{removeAll}(b, \langle a, b, c, b\rangle) \;&=\; a :: \text{removeAll}(b, \langle b, c, b\rangle)\\
&=\; a :: \text{removeAll}(b, \langle c, b\rangle)\\
&=\; a :: c :: \text{removeAll}(b, \langle b\rangle)\\
&=\; a :: c :: \text{removeAll}(b, \langle\,\rangle)\\
&=\; a :: c :: \langle\,\rangle\\
&=\; a :: \langle c\rangle\\
&=\; \langle a, c\rangle.
\end{aligned}
$$

Try to write out each unfolding step in the evaluation of the expression remove($\langle b, a, b\rangle$). Be sure to start writing at the left-hand edge of your paper.

The Power Set Problem

Suppose we want to construct the power set of a finite set. One solution uses the fact that power($\{x\} \cup T$) is the union of power(T) and the set obtained from power(T) by adding x to each of its elements. Let's see whether we can discover a solution technique by considering a small example. Let $S = \{a, b, c\}$. Then we can write power(S) as follows:

$$\text{power}(S) \ = \ \{\{\ \}, \{a\}, \{b\}, \{c\}, \{a, b\}, \{a, c\}, \{b, c\}, \{a, b, c\}\}$$
$$= \ \{\{\ \}, \{b\}, \{c\}, \{b, c\}\} \cup \{\{\underline{a}\}, \{\underline{a}, b\}, \{\underline{a}, c\}, \{\underline{a}, b, c\}\}.$$

We've written power(S) = $A \cup B$, where B is obtained from A by adding the underlined element $\underline{a}$ to each set in A. If we represent S as the list $\langle a, b, c \rangle$, then we can restate the definition for power(S) as the concatenation of the following two lists:

$$\langle\langle\ \rangle, \langle b\rangle, \langle c\rangle, \langle b, c\rangle\rangle \quad \text{and} \quad \langle\langle a\rangle, \langle a, b\rangle, \langle a, c\rangle, \langle a, b, c\rangle\rangle.$$

The first of these lists is power($\langle b, c \rangle$). The second list can be obtained from power($\langle b, c \rangle$) by working backward to the answer as follows:

$$\langle\langle a\rangle, \langle a, b\rangle, \langle a, c\rangle, \langle a, b, c\rangle\rangle \ = \ \langle a :: \langle\ \rangle, a :: \langle b\rangle, a :: \langle c\rangle, a :: \langle b, c\rangle\rangle$$
$$= \ \text{map}(::)(\langle a, \langle\ \rangle\rangle, \langle a, \langle b\rangle\rangle, \langle a, \langle c\rangle\rangle, \langle a, \langle b, c\rangle\rangle)$$
$$= \ \text{map}(::)(\text{dist}(a, \text{power}(\langle b, c\rangle))).$$

This example is the key to the induction part of the definition. Using the fact that power($\langle\ \rangle$) = $\langle\langle\ \rangle\rangle$ as the basis case, we can write down the following definition for power:

$$\text{power}(\langle\ \rangle) = \langle\langle\ \rangle\rangle,$$
$$\text{power}(a :: L) = \text{cat}(\text{power}(L), \text{map}(::)(\text{dist}(a, \text{power}(L)))).$$

The if-then-else form of the definition can be written as follows:

$$\text{power}(L) \ = \ \text{if } L = \langle\ \rangle \text{ then } \langle\langle\ \rangle\rangle \text{ else}$$
$$\text{cat}(\text{power}(\text{tail}(L)), \text{map}(::)(\text{dist}(\text{head}(L), \text{power}(\text{tail}(L))))).$$

For example, we'll evaluate the expression power($\langle a, b \rangle$) by unfolding the definition. The first step yields the equation

$$\text{power}(\langle a, b\rangle) = \text{cat}(\text{power}(\langle b\rangle), \text{map}(::)(\text{dist}(a, \text{power}(\langle b\rangle)))).$$

Next we'll evaluate power($\langle b \rangle$) and then substitute in the preceding equation:

$$power(\langle b \rangle) = cat(power(\langle\,\rangle), map(::)(dist(b, power(\langle\,\rangle))))$$
$$= cat(\langle\langle\,\rangle\rangle, map(::)(dist(b, \langle\langle\,\rangle\rangle)))$$
$$= cat(\langle\langle\,\rangle\rangle, map(::)(\langle\langle b, \langle\,\rangle\rangle\rangle))$$
$$= cat(\langle\langle\,\rangle\rangle, \langle b :: \langle\,\rangle\rangle)$$
$$= cat(\langle\langle\,\rangle\rangle, \langle\langle b\rangle\rangle)$$
$$= \langle\langle\,\rangle, \langle b\rangle\rangle.$$

Now we can continue with the evaluation of power($\langle a, b \rangle$):

$$power(\langle a, b \rangle) = cat(power(\langle b \rangle), map(::)(dist(a, power(\langle b \rangle))))$$
$$= cat(\langle\langle\,\rangle, \langle b\rangle\rangle, map(::)(dist(a, \langle\langle\,\rangle, \langle b\rangle\rangle)))$$
$$= cat(\langle\langle\,\rangle, \langle b\rangle\rangle, map(::)(\langle\langle a, \langle\,\rangle\rangle, \langle a, \langle b\rangle\rangle\rangle))$$
$$= cat(\langle\langle\,\rangle, \langle b\rangle\rangle, \langle a :: \langle\,\rangle, a :: \langle b\rangle\rangle)$$
$$= cat(\langle\langle\,\rangle, \langle b\rangle\rangle, \langle\langle a\rangle, \langle a, b\rangle\rangle)$$
$$= \langle\langle\,\rangle, \langle b\rangle, \langle a\rangle, \langle a, b\rangle\rangle.$$

Computing Streams

Recall that a stream is an infinite list. From a computational standpoint any inductively defined set that is countable can be represented by a stream. Stream building functions are easy to construct. For example, suppose the function "ints" returns the following stream for any integer x:

$$ints(x) = \langle x, x + 1, x + 2, ...\rangle.$$

We'll assume that "cons" (i.e., ::) is a stream constructor that attaches a new element to a stream. We'll also assume that "head" and "tail" work as usual. For example, the following relationships hold:

$$ints(x) = x :: ints(x + 1),$$
$$head(ints(x)) = x,$$
$$tail(ints(x)) = ints(x + 1).$$

Even though the definition of ints does not conform to (3.14), ints is still recursively defined because it's defined in terms of itself. If we executed the definition, an infinite loop would construct the stream. For example, ints(0) would construct the stream of natural numbers as follows:

$$\begin{aligned}
\text{ints}(0) &= 0 :: \text{ints}(1) \\
&= 0 :: 1 :: \text{ints}(2) \\
&= 0 :: 1 :: 2 :: \text{ints}(3) \\
&= \ldots .
\end{aligned}$$

In practice, when a stream like ints(x) is used as an argument to another function, it is evaluated only when some of its values are needed. Once the values are computed, the evaluation stops. This is an example of a technique called *lazy evaluation*. For example, we could extract the second number in the stream ints(3) as follows:

$$\begin{aligned}
\text{head}(\text{tail}(\text{ints}(3))) &= \text{head}(\text{tail}(3 :: \text{ints}(4))) \\
&= \text{head}(\text{ints}(4)) \\
&= \text{head}(4 :: \text{ints}(5)) \\
&= 4.
\end{aligned}$$

Let's try a few examples.

EXAMPLE 11 (*Summing*). Suppose we need to build a function to add up the first n elements of an arbitrary stream of integers. Letting "sum" denote the function, we can make the following general definition, where s denotes a stream of integers:

$$\text{sum}(n, s) = \text{if } n = 0 \text{ then } 0 \text{ else head}(s) + \text{sum}(n - 1, \text{tail}(s)).$$

For example, we'll unfold the definition of the sum function to add up the first three numbers in ints(4):

$$\begin{aligned}
\text{sum}(3, \text{ints}(4)) &= 4 + \text{sum}(2, \text{ints}(5)) \\
&= 4 + 5 + \text{sum}(1, \text{ints}(6)) \\
&= 4 + 5 + 6 + \text{sum}(0, \text{ints}(7)) \\
&= 4 + 5 + 6 + 0 \\
&= 15. \quad \blacklozenge
\end{aligned}$$

EXAMPLE 12 (*Skipping*). Let's construct the function "skipTwo" to build the following stream:

$$\langle x, x + 2, x + 4, \ldots \rangle$$

for any integer x. We can define skipTwo as follows:

$$\text{skipTwo}(x) = x :: \text{skipTwo}(x + 2).$$

For example, the expression skipTwo(3) evaluates to a stream as follows:

$$
\begin{aligned}
\text{skipTwo}(3) &= 3 :: \text{skipTwo}(5) \\
&= 3 :: 5 :: \text{skipTwo}(7) \\
&= \langle 3, 5, 7, \ldots \rangle.
\end{aligned}
$$

Suppose we need a more general stream-building tool. For example, suppose we want to construct a function named skip with two arguments k and x that returns the stream

$$\langle x, x + k, x + 2k, \ldots \rangle.$$

We can define skip by the following natural relationship:

$$\text{skip}(x, k) = x :: \text{skip}(x + k, k).$$

An evaluation of the expression skip(1, 3) gives the following stream:

$$
\begin{aligned}
\text{skip}(1, 3) &= 1 :: \text{skip}(4, 3) \\
&= 1 :: 4 :: \text{skip}(7, 3) \\
&= \langle 1, 4, 7, \ldots \rangle.
\end{aligned}
$$

Let's add up the first four terms of the stream skip(1, 3) using sum.

$$
\begin{aligned}
\text{sum}(4, \text{skip}(1, 3)) &= 1 + \text{sum}(3, \text{skip}(4, 3)) \\
&= 1 + 4 + \text{sum}(2, \text{skip}(7, 3)) \\
&= 1 + 4 + 7 + \text{sum}(1, \text{skip}(10, 3)) \\
&= 1 + 4 + 7 + 10 + \text{sum}(0, \text{skip}(13, 3)) \\
&= 1 + 4 + 7 + 10 + 0 \\
&= 22. \quad \blacklozenge
\end{aligned}
$$

EXAMPLE 13 (*The Sieve of Eratosthenes*). Suppose we want to build the following stream of all prime numbers:

$$\langle 2, 3, 5, 7, 11, 13, 17, ... \rangle.$$

The method of Eratosthenes (called *the sieve of Eratosthenes*) starts with the stream

$$\text{ints}(2) = \langle 2, 3, 4, 5, 6, 7, 8, 9, 10, ... \rangle,$$

and removes all multiples of 2 from its tail, to obtain the stream

$$\langle 2, 3, 5, 7, 9, 11, 13, 15, ... \rangle.$$

Next, all multiples of 3 (except 3 itself) are removed, and the process continues in this way. Let sieve denote the function to accomplish this task. Then our desired stream of primes can be constructed by evaluating the expression

$$\text{sieve}(\text{ints}(2)). \tag{3.19}$$

To define sieve, we need to think about removing multiples of an integer. Let removeM(n, t) be the stream obtained from t by removing all multiples of n from t (including n). Then we have the following relationship:

$$\text{sieve}(n :: t) = n :: \text{sieve}(\text{removeM}(n, t)).$$

We can rewrite this relationship in terms of a single argument symbol as follows:

$$\text{sieve}(s) = \text{head}(s) :: \text{sieve}(\text{removeM}(\text{head}(s), \text{tail}(s))).$$

Now, what about the removeM function? Notice first that for natural numbers m and n ($n > 0$),

$$m \text{ is a multiple of } n \text{ if and only if } m \bmod n = 0.$$

Using this fact, we can write the definition of removeM as follows:

$$\text{removeM}(n, t) = \quad \text{if head}(t) \bmod n = 0 \text{ then} \\ \text{removeM}(n, \text{tail}(t)) \\ \text{else head}(t) :: \text{removeM}(n, \text{tail}(t)).$$

For example, we'll unfold the first few steps of the definition for the expression removeM(2, ints(2)) as follows:

$$
\begin{aligned}
\text{removeM}(2,\, \text{ints}(2)) \;&=\; \text{removeM}(2,\, 2 :: \text{ints}(3)) \\
&=\; \text{removeM}(2,\, \text{ints}(3)) \\
&=\; \text{removeM}(2,\, 3 :: \text{ints}(4)) \\
&=\; 3 :: \text{removeM}(2,\, \text{ints}(4)) \\
&=\; 3 :: \text{removeM}(2,\, 4 :: \text{ints}(5)) \\
&=\; 3 :: \text{removeM}(2,\, \text{ints}(5)) \\
&=\; 3 :: \text{removeM}(2,\, 5 :: \text{ints}(6)) \\
&=\; 3 :: 5 :: \text{removeM}(2,\, \text{ints}(6)) \\
&=\; \langle 3,\, 5,\, \ldots \rangle.
\end{aligned}
$$

The following expression computes the sum of the first three prime numbers:

$$\text{sum}(3,\, \text{sieve}(\text{ints}(2))).$$

Try to evaluate this expression by unfolding all necessary definitions.   ◆

Exercises

1. Given the following definition for the nth Fibonacci number:

   fib(n) = if $n = 0$ then 0 else if n = 1 then 1 else fib($n - 1$) + fib($n - 2$).

   Write down each unfolding step in the evaluation of fib(4).

2. Given the following definition for the length of a list:

   length(L) = if $L = \langle\,\rangle$ then 0 else 1 + length(tail(L)).

   Write down each unfolding step in the evaluation of length($\langle r, s, t, u \rangle$).

3. Find a recursive definition for the function "small" to find the smallest number in a list.

4. For each of the two definitions of "makeTree" given by (3.17) and (3.18), write down all unfolding steps to evaluate makeTree($\langle\,\rangle$, $\langle 3, 2, 4 \rangle$).

5. Conway's challenge sequence is defined recursively as follows:

   Basis: $f(1) = f(2) = 1.$
   Induction: $f(n) = f(f(n-1)) + f(n - f(n-1))$ for $n > 2$.

   Calculate the first 17 elements $f(1)$, $f(2)$, ..., $f(17)$. The article by Mallows [1991] contains an account of this sequence.

6. Recall that consR attaches an element to the right of a list. For example, consR($\langle a, b \rangle, c$) = $\langle a, b, c \rangle$. Use consR to give a recursive definition for the sequence function seq. For example, seq(4) = $\langle 0, 1, 2, 3, 4 \rangle$.

7. Recall that there is an insert function that extends any binary function to take two or more arguments. For example, insert(+)(1, 4, 2, 9) = 16. Write a recursive definition for insert(f), where f is any binary function.

8. Write a recursive definition for the function eq to check two lists for equality.

9. Give a recursive definition for the function last that takes as input a nonempty string and produces as output the last element of the string. For example, last(abc) = c. Assume that the only operations available are head, tail, and $x \cdot s$, where x is a letter and s is a string of letters.

10. Write down a recursive definition for the function pal that tests a string to see whether it is a palindrome.

11. The conversion of a natural number to a binary string (3.16) placed a leading (leftmost) zero on each result. Modify the definition of the function to get rid of the leading zero.

12. Given the algebraic expression $a + (b \cdot (d + e))$, draw a picture of the binary tree representation of the expression. Then write down the preorder, inorder, and postorder listings of the tree. Are any of the listings familiar to you?

13. Write down recursive definitions for each of the following procedures to print the nodes of a binary tree.

   a. In: prints the nodes of a binary tree from an inorder traversal.

   b. Post: prints the nodes of a binary tree from a postorder traversal.

14. Write down recursive definitions for each of the following functions. Include both the equational and if-then-else forms for each definition.

   a. leaves: returns the number of leaf nodes in a binary tree.

   b. inOrd: returns the inorder listing of nodes in a binary tree.

   c. postOrd: returns the postorder listing of nodes in a binary tree.

15. Solve the redundant element problem with the restriction that we want to keep the rightmost occurrence of each redundant element. *Hint:* Invent two new tools: A tool to pick off the rightmost element of a list and a tool to pick off the leftmost sublist that excludes only the rightmost element.

16. Write a recursive definition for each of the following functions, in which the input arguments are sets represented as lists. Use the primitive operations of cons, head, and tail to build your functions (along with functions already defined):

 a. isMember. For example, isMember(a, $\langle b, a, c \rangle$) is true.

 b. isSubset. For example, isSubset($\langle a, b \rangle$, $\langle b, c, a \rangle$) is true.

 c. areEqual. For example, areEqual($\langle a, b \rangle$, $\langle b, a \rangle$) is true.

 d. union. For example, union($\langle a, b \rangle$, $\langle c, a \rangle$) = $\langle a, b, c \rangle$.

 e. intersect. For example, intersect($\langle a, b \rangle$, $\langle c, a \rangle$) = $\langle a \rangle$.

 f. difference. For example, difference($\langle a, b, c \rangle$, $\langle b, d \rangle$) = $\langle a, c \rangle$.

17. Let fib(k) denote the kth Fibonacci number, and let

$$\text{sum}(k) = 1 + 2 + \ldots + k.$$

 Write a recursive definition for the function $f : \mathbb{N} \to \mathbb{N}$ defined by $f(n) =$ sum(fib(n)). *Hint:* Write down several examples, such as $f(0)$, $f(1)$, $f(2)$, $f(3)$, $f(4)$, ... Then try to find a way to write $f(4)$ in terms of $f(3)$. This might help you discover a pattern.

18. Write a function in if-then-else form to produce the product set of two finite sets. You may assume that the sets are represented as lists.

19. The square root of a number can be approximated by the Newton-Raphson method. A stream of Newton-Raphson approximations to the square root of a number x is given as follows, where g is an initial guess at the answer:

$$\text{sqrt}(x, g) = g :: \text{sqrt}(x, (0.5)(g + (x/g))).$$

 Find the first three numbers in each of the following streams, and compare the values with the square root obtained by a calculator:

 a. sqrt(4, 1).   b. sqrt(4, 2).   c. sqrt(4, 3).

 d. sqrt(2, 1).   e. sqrt(9, 1).   f. sqrt(9, 5).

20. For each of the following problems, use the stream function (3.19) to generate the stream of prime numbers.

 a. Write a function to return the product of the first n primes.

 b. Write a function to return the list of the first n primes.

21. Find a definition for each of the following stream functions.

 a. Square: squares each element of a stream of numbers.

 b. Add: adds corresponding elements of two numeric streams.

 c. Map: applies a function to each element of a stream.

22. Suppose we define $f : \mathbb{N} \to \mathbb{N}$ by

$$f(x) = x - 10 \text{ for } x > 10 \text{ and } f(x) = f(f(x + 11)) \text{ for } 0 \leq x \leq 10.$$

This function is recursively defined even though it is not defined by (3.14). Give a simple definition of this function.

Chapter Summary

In this chapter we covered some basic construction techniques that apply to many objects of importance to computer science.

Inductively defined sets are characterized by a basis case, an induction case, and a closure case that is always assumed without comment. The constructors of an inductively defined set are the elements listed in the basis case and the rules specified in the induction case. Many sets of objects used in computer science can be defined inductively—natural numbers, lists, strings, binary trees, and products of sets.

Languages play a special role in computer science. They can be constructed not only as inductively defined sets, but also by applying operations like the product—via concatenation of strings—to existing languages. The closure operation and the usual set operations are other ways to construct languages. The most important way to construct a language is by describing a grammar. Grammar productions are used to derive strings of the language. Any grammar for an infinite language must contain at least one production that is recursive or indirectly recursive. Grammars for languages can be combined to form new grammars for unions, products, and closures of the languages. Some grammars are ambiguous.

A recursively defined function is defined in terms of itself. Most recursively defined functions have domains that are inductively defined sets. These functions are normally defined by a basis case and an induction case. The situation is similar for recursively defined procedures. Some stream functions can be defined recursively. Recursively defined functions and procedures yield powerful—but simple—programs.

Note

In this chapter we introduced some important techniques for describing sets, languages, functions, and procedures that satisfy certain properties. But once we found a description, we didn't actually prove that it satisfied the required properties. Instead, we usually checked the basis case and at least one other case to satisfy ourselves that the description was correct. In the next chapter we'll study inductive proof techniques that can be used to actually prove the correctness of claims about objects defined by the techniques of this chapter.

4

Equivalence, Order, and Inductive Proof

Good order is the foundation of all things.
—Edmund Burke (1729–1797)

Classifying things and ordering things are activities in which we all engage from time to time. Whenever we classify or order a set of things, we usually compare them in some way. That's how binary relations enter the picture.

In this chapter we'll discuss some special properties of binary relations that are useful for solving comparison problems. We'll introduce techniques to construct binary relations with the properties we need. We'll discuss the idea of equivalence by considering properties of the equality relation. We'll also study the properties of binary relations that characterize our intuitive ideas about ordering. We'll see that ordering is the fundamental ingredient that we need to discuss inductive proof techniques.

Chapter Guide

Section 4.1 introduces the basic properties of binary relations—reflexive, symmetric, transitive, irreflexive, and antisymmetric. We'll see how to construct new relations by composition and closure. We'll see how the results apply to solving path problems in graphs.

Section 4.2 concentrates on the idea of equivalence. We'll see that equivalence is closely related to partitioning of sets. We'll show how to generate equivalence relations, and we'll solve a typical equivalence problem.

Section 4.3 introduces the idea of order. We'll discuss partial orders and and how to sort them. We'll introduce well-founded orders and show some techniques for constructing them. Ordinal numbers are also introduced.

177

Section 4.4 introduces the important technique of inductive proof. After introducing inductive proof techniques for the natural numbers, we'll extend the discussion to inductive proof techniques for any well-founded set.

4.1 Properties and Tools

Recall that the statement "*R* is a binary relation over the set *A*" means that *R* relates certain pairs of elements of *A*. Thus *R* can be represented as a set of ordered pairs from *A* (i.e., *R* is a subset of $A \times A$). When $\langle a, b \rangle \in R$, we could also write $R(a, b)$, but we most often write $a R b$. Binary relations can be very useful in solving computational problems. For example, the "less" and "equal" relations defined on numbers are important tools in solving numerical problems. Binary relations often satisfy certain special properties. Many useful binary relations satisfy at most three of five special properties that we are about to discuss. So let's get to it.

The first three properties that we'll discuss are called *reflexive*, *symmetric*, and *transitive*, and they are defined as follows, where *R* is a binary relation over a set *A*:

Reflexive :	$a R a$	(for all $a \in A$).
Symmetric:	if $a R b$, then $b R a$	(for all $a, b \in A$).
Transitive:	if $a R b$ and $b R c$, then $a R c$	(for all $a, b, c \in A$).

When *R* satisfies the reflexive property, we say, "*R* is reflexive," and similarly for the other properties. Many well-known relations satisfy one or more of these three properties, and some relations don't satisfy any of them. For example, the "isParentOf" relation doesn't satisfy any of the three properties. The "isSiblingOf" relation is symmetric and transitive, and if we agree that people are self-siblings, then it's reflexive too. The relation "hasSameBirthdayAs" satisfies all three properties. The "less" relation on numbers is transitive but is neither reflexive nor symmetric. The relation $\{\langle n, n + 1 \rangle \mid n \in \mathbb{N}\}$ doesn't satisfy any of the three properties.

The simplest kind of equality on a set equates each element to itself. We'll call this *basic equality*. We can characterize basic equality on a set *A* by the relation

$$E = \{\langle x, x \rangle \mid x \in A\}.$$

Of course, *E* is often denoted by the popular symbol "=," and we write $x = x$ instead of either $x E x$ or $\langle x, x \rangle \in E$. Basic equality is reflexive, symmetric, and transitive. The reason that it's symmetric and transitive is that the if-then definitions of symmetric and transitive are vacuously satisfied.

The reflexive property has a corresponding opposite property, called *irreflexive*, which is defined as follows:

> *Irreflexive*: $a \not\mathrel{R} a$ (for all $a \in A$).

The irreflexive property says that the reflexive property does not hold for every element $a \in A$. For example, the "isAncestorOf" relation and the "less" relation are both irreflexive and transitive.

The symmetric property has a corresponding opposite property, which we can define as follows:

> *Antisymmetric*:   if $a \neq b$ and $a \mathrel{R} b$, then $b \not\mathrel{R} a$ (for all $a, b \in A$).

The antisymmetric property says that the symmetric property does not hold for every pair of elements. For example, the "isParentOf" relation is antisymmetric because if a is a parent of b, then b can't be a parent of a. We can write the antisymmetric property in an alternative form as follows:

> *Antisymmetric*:   if $a \mathrel{R} b$ and $b \mathrel{R} a$, then $a = b$ (for all $a, b \in A$).

For example, the "isParentOf" relation is antisymmetric by this definition because the hypothesis of the if-then statement is never satisfied. Therefore the antisymmetric property is vacuously true in this case.

The exercises contain many more examples of relations that satisfy some of these five properties. Next we'll discuss a few techniques to construct binary relations.

Composition

Relations can often be defined in terms of other relations. For example, we can describe the "isGrandparentOf" relation in terms of the "isParentOf" relation by saying that $\langle a, c \rangle \in$ isGrandparentOf if and only if there is some b such that $\langle a, b \rangle, \langle b, c \rangle \in$ isParentOf. This example demonstrates the fundamental idea of composing binary relations.

Although we'll be dealing with binary relations over a single set, we'll define composition for general binary relations from one set to another set. If R is a binary relation from A to B and S is a binary relation from B to C, then the *composition* of R and S is the binary relation $R \circ S$ from A to C defined as follows:

> $a \,(R \circ S)\, c$ iff $a \mathrel{R} b$ and $b \mathrel{S} c$ for some $b \in B$.

From the standpoint of ordered pairs we can write

$$\langle a, c \rangle \in R \circ S \text{ iff } \langle a, b \rangle \in R \text{ and } \langle b, c \rangle \in S \qquad \text{for some } b \in B.$$

So if $R \subset A \times B$ and $S \subset B \times C$, then we have $R \circ S \subset A \times C$.

EXAMPLE 1. We can compose the "isParentOf" relation with itself to construct the "isGrandparentOf" relation:

$$\text{isGrandparentOf} = \text{isParentOf} \circ \text{isParentOf}.$$

Similarly, we can construct the "isGreatGrandparentOf" relation by the following composition:

$$\text{isGreatGrandparentOf} = \text{isGrandparentOf} \circ \text{isParentOf}. \qquad \blacklozenge$$

If R and S are binary relations over A, then the compositions $R \circ S$ and $S \circ R$ make sense. Are they equal? In other words, is the composition of relations commutative? In general the answer is no. For example, let $R = \{\langle a, b \rangle\}$ and $S = \{\langle b, a \rangle\}$. Then $R \circ S = \{\langle a, a \rangle\}$ and $S \circ R = \{\langle b, b \rangle\}$. Let's look at another example.

EXAMPLE 2. Suppose we consider the relations "less," "greater," "equal," and "notEqual" over $\mathbb{R}$. We want to compose some of these relations and see what we get. For example, let's verify the following equality:

$$\text{greater} \circ \text{less} = \mathbb{R} \times \mathbb{R}.$$

For any pair of numbers $\langle x, y \rangle$ the definition of composition tells us that x (greater $\circ$ less)y if and only if there is some number z such that x greater z and z less y. We can write this statement more concisely as follows: $x (> \circ <)y$ if and only if there is some number z such that $x > z$ and $z < y$. We know that for any two real numbers x and y there is always another number z that is less than both. Therefore the composition must be the universe $\mathbb{R} \times \mathbb{R}$.

Many combinations are possible. For example, it's easy to verify the following two equalities:

$$\text{equal} \circ \text{notEqual} = \text{notEqual},$$

$$\text{notEqual} \circ \text{notEqual} = \mathbb{R} \times \mathbb{R}. \qquad \blacklozenge$$

Since relations are just sets (of ordered pairs), it's clear that they can also be combined by the usual set operations of union, intersection, difference, and complement. For example, if we assume that "equal" and "less" are defined over the same set of numbers, then equal $\cap$ less = $\varnothing$.

EXAMPLE 3. Suppose the underlying set is $\mathbb{R}$. The following examples show how we can combine some familiar relations. Check out each one with a few example pairs of numbers.

$$
\begin{aligned}
\text{equal} \cap \text{lessOrEqual} \quad &= \quad \text{equal,} \\
(\text{lessOrEqual})' \quad &= \quad \text{greater,} \\
\text{greaterOrEqual} - \text{equal} \quad &= \quad \text{greater,} \\
\text{equal} \cup \text{greater} \quad &= \quad \text{greaterOrEqual,} \\
\text{less} \cup \text{greater} \quad &= \quad \text{notEqual.} \quad \blacklozenge
\end{aligned}
$$

Let's list some fundamental properties of combining relations. We'll assume that the relations R, S, and T can be combined by the compositions shown.

Properties of Combining Relations (4.1)

a) $R \circ (S \circ T) = (R \circ S) \circ T$ (associativity).
b) $R \circ (S \cup T) = R \circ S \cup R \circ T$.
c) $R \circ (S \cap T) \subset R \circ S \cap R \circ T$.

We'll leave the proofs of these properties as exercises. Notice that (4.1c) is stated as a set containment rather than an equality. For example, let R, S, and T be the following relations:

$$R = \{\langle a, b \rangle, \langle a, c \rangle\}, \; S = \{\langle b, b \rangle\}, \; T = \{\langle b, c \rangle, \langle c, b \rangle\}.$$

Then $S \cap T = \varnothing$, $R \circ S = \{\langle a, b \rangle\}$, and $R \circ T = \{\langle a, c \rangle, \langle a, b \rangle\}$. Therefore

$$R \circ (S \cap T) = \varnothing \text{ and } R \circ S \cap R \circ T = \{\langle a, b \rangle\}.$$

So (4.1c) isn't always an equality. Of course, there are cases in which $R \circ (S \cap T)$ and $R \circ S \cap R \circ T$ are equal. For example, if $R = \varnothing$ or if $R = S = T$, then (4.1c) is an equality.

If R is a binary relation on A, then we'll denote the composition of R with itself n times by writing R^n. For example, if we compose isParentOf with itself, we get some familiar names as follows:

$$\text{isParentOf}^2 = \text{isGrandparentOf},$$
$$\text{isParentOf}^3 = \text{isGreatGrandparentOf}.$$

We mentioned in Chapter 1 that binary relations can be thought of as digraphs and, conversely, that digraphs can be thought of as binary relations. In other words, we can think of $\langle x, y \rangle$ as an edge from x to y in a digraph and as a member of a binary relation. So we can talk about the digraph of a binary relation. An important and useful way to think about R^n is as the digraph consisting of all edges $\langle x, y \rangle$ such that there is a path of length n from x to y. For example, if $\langle x, y \rangle \in R^2$, then we know there is some element z such that $\langle x, z \rangle, \langle z, y \rangle \in R$, which says that there is a path of length 2 from x to y in the digraph of R.

EXAMPLE 4. Let $R = \{\langle a, b \rangle, \langle b, c \rangle, \langle c, d \rangle\}$. The digraphs shown in Figure 4.1 are the digraphs for the three relations R, R^2, and R^3. ◆

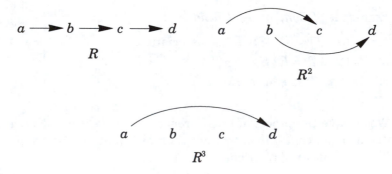

Figure 4.1

Now we'll give a more precise definition of R^n using induction. Notice the interesting choice for R^0:

$$R^0 = \{\langle a, a \rangle \mid a \in A\} \qquad \text{(basic equality)}$$
$$R^{n+1} = R^n \circ R.$$

We defined R^0 as the basic equality relation because we want to infer the equality $R^1 = R$ from the definition. To see this, notice the following evaluation of R^1:

$$R^1 = R^{0+1} = R^0 \circ R = \{\langle a, a \rangle \mid a \in A\} \circ R = R.$$

We also could have defined $R^{n+1} = R \circ R^n$ instead of $R^{n+1} = R^n \circ R$ because composition of binary operations is associative by (4.1a).

Let's note a few other interesting relationships between R and R^n. R^n inherits the reflexive, symmetric, and transitive properties from R. In other words, if R reflexive, then so is R^n. Similarly, if R is symmetric, then so is R^n. Also, if R is transitive, then so is R^n. On the other hand, if R is irreflexive, then it may not be the case that R^n is irreflexive. Similarly, if R is antisymmetric, it may not be the case that R^n is antisymmetric. We'll examine these statements in the exercises for the case of R^2.

Closures

We've seen how to construct a new relation by composing two existing relations. Let's look at another way to construct a new relation from an existing relation. Here we'll start with a binary relation R and try to construct another relation containing R that also satisfies some particular property. For example, if we have the "isParentOf" relation, then we may want to use it to construct the "isAncestorOf" relation. As always, we need to introduce some terminology.

If R is a binary relation and p is some property, then the *p closure* of R is the smallest binary relation that contains R and still satisfies property p. We'll denote the p closure of R by $p(R)$. If R already satisfies property p, then we have $R = p(R)$. We can state the relationship between R and $p(R)$ in several ways as follows: R is a *generator* of $p(R)$ or R *generates* $p(R)$ or $p(R)$ is *induced* by R.

We'll be concerned with three properties: reflexive, symmetric, and transitive. The *reflexive closure* of R is denoted by $r(R)$, the *symmetric closure* of R is denoted by $s(R)$, and the *transitive closure* of R is denoted by $t(R)$.

Our goal is to find some techniques to compute these closures. We'll start with a running example that will introduce the main ideas. Then we'll record the construction techniques. For our example we'll let $A = \{a, b, c\}$, and we'll let R be the following relation:

$$R = \{\langle a, a \rangle, \langle a, b \rangle, \langle b, a \rangle, \langle b, c \rangle\}.$$

Notice that R is neither reflexive nor symmetric nor transitive. We'll compute all three closures of R.

First we'll compute the reflexive closure $r(R)$. The two pairs missing from R are $\langle b, b \rangle$ and $\langle c, c \rangle$. So we can certainly compute $r(R)$ by forming the union of R and $\{\langle x, x \rangle \mid x \in A\}$. This gives us

$$r(R) = \{\langle a, a \rangle, \langle a, b \rangle, \langle b, a \rangle, \langle b, c \rangle, \langle b, b \rangle, \langle c, c \rangle\}.$$

Next we'll compute the symmetric closure $s(R)$. To create a symmetric relation, we need to add the pair $\langle c, b \rangle$. The *converse* of a binary relation R, which we denote by R^c, is defined as the following relation:

$$R^c = \{\langle x, y \rangle \mid \langle y, x \rangle \in R\}.$$

For example, the converse of "less" is "greater," and the converse of "equal" is itself. Notice that R is symmetric if and only if $R = R^c$. We can obtain $s(R)$ by simply forming the union of R with its converse R^c. For our example we have

$$s(R) = \{\langle a, a \rangle, \langle a, b \rangle, \langle b, a \rangle, \langle b, c \rangle, \langle c, b \rangle\}.$$

Lastly, we'll compute the transitive closure $t(R)$. For our example, R contains the pairs $\langle a, b \rangle$ and $\langle b, c \rangle$, but $\langle a, c \rangle$ is not in R. Similarly, R contains the pairs $\langle b, a \rangle$ and $\langle a, b \rangle$, but $\langle b, b \rangle$ is not in R. So $t(R)$ must contain the pairs $\langle a, c \rangle$ and $\langle b, b \rangle$. Is there some relation that we can union with R that will add the two needed pairs? The answer is yes, it's R^2. Notice that

$$R^2 = \{\langle a, a \rangle, \langle a, b \rangle, \langle b, b \rangle, \langle a, c \rangle\}.$$

It contains the two missing pairs along with two other pairs that are already in R. Thus we have

$$t(R) = R \cup R^2 = \{\langle a, a \rangle, \langle a, b \rangle, \langle b, a \rangle, \langle b, c \rangle, \langle a, c \rangle, \langle b, b \rangle\}.$$

We'll do another example so that we can get some further insight into computing the transitive closure. Let $A = \{a, b, c, d\}$, and suppose R is the following relation:

$$R = \{\langle a, b \rangle, \langle b, c \rangle, \langle c, d \rangle\}.$$

To compute $t(R)$, we need to add the three pairs $\langle a, c \rangle$, $\langle b, d \rangle$, and $\langle a, d \rangle$. In this case, $R^2 = \{\langle a, c \rangle, \langle b, d \rangle\}$. So the union of R with R^2 is missing $\langle a, d \rangle$. Can we find another relation to union with R and R^2 that will add this missing pair? Notice that $R^3 = \{\langle a, d \rangle\}$. So for this example, $t(R)$ is the union

$$t(R) = R \cup R^2 \cup R^3$$
$$= \{\langle a, b\rangle, \langle b, c\rangle, \langle c, d\rangle, \langle a, c\rangle, \langle b, d\rangle, \langle a, d\rangle\}.$$

As the examples show, $t(R)$ is a bit more difficult to construct than the other two closures. The construction techniques for all three closures are listed next.

Constructing Closures   (4.2)

If R is a binary relation over a set A, then:

a) $r(R) = R \cup R^0$ (R^0 is the equality relation).

b) $s(R) = R \cup R^c$ (R^c is the converse relation).

c) $t(R) = R \cup R^2 \cup R^3 \cup \ldots$.

d) If A is finite with n elements, then $t(R) = R \cup R^2 \cup \cdots \cup R^n$.

Part (4.2d) assures us that $t(R)$ can be calculated by taking the union of n powers of R if the cardinality of A is n. We can see this as follows: Any pair $\langle x, y\rangle \in t(R)$ represents a path from x to y in the digraph of R. Similarly, any pair $\langle x, y\rangle \in R^k$ represents a path of length k from x to y in the digraph of R. Now if $\langle x, y\rangle \in R^{n+1}$, then there is a path of length $n + 1$ from x to y in the digraph of R. Since A has n elements, it follows that some element of A occurs twice in the path from x to y. So there is a shorter path from x to y. Therefore $\langle x, y\rangle \in R^k$ for some $k \leq n$. So nothing new gets added to $t(R)$ by adding powers of R that are higher than the cardinality of A.

Sometimes we don't have to compute all the powers of R. For example, let $A = \{a, b, c, d, e\}$ and $R = \{\langle a, b\rangle, \langle b, c\rangle, \langle b, d\rangle, \langle d, e\rangle\}$. The digraphs of R and $t(R)$ are drawn in Figure 4.2. Convince yourself that $t(R) = R \cup R^2 \cup R^3$. In other words, the relations R^4 and R^5 don't add anything new. In fact, you should verify that $R^4 = R^5 = \varnothing$.

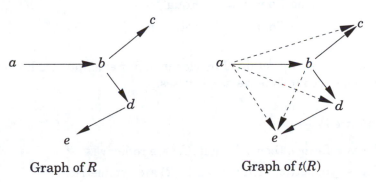

Graph of R Graph of $t(R)$

Figure 4.2

EXAMPLE 5. Let $A = \{a, b, c\}$ and $R = \{\langle a, b\rangle, \langle b, c\rangle, \langle c, a\rangle\}$. Then $R^2 = \{\langle a, c\rangle, \langle c, b\rangle, \langle b, a\rangle\}$, and $R^3 = \{\langle a, a\rangle, \langle b, b\rangle, \langle c, c\rangle\}$. Therefore we have

$$t(R) = R \cup R^2 \cup R^3 = A \times A. \quad \blacklozenge$$

EXAMPLE 6. Let $A = \{a, b, c\}$ and $R = \{\langle a, b\rangle, \langle b, c\rangle, \langle c, b\rangle\}$. Then $R^2 = \{\langle a, c\rangle, \langle b, b\rangle, \langle c, c\rangle\}$, and $R^3 = \{\langle a, b\rangle, \langle b, c\rangle, \langle c, b\rangle\}$. So we obtain

$$t(R) = \{\langle a, b\rangle, \langle b, c\rangle, \langle c, b\rangle, \langle a, c\rangle, \langle b, b\rangle, \langle c, c\rangle\}. \quad \blacklozenge$$

EXAMPLE 7. Suppose $R = \{\langle x, x + 1\rangle \mid x \in \mathbb{N}\}$. Then $R^2 = \{\langle x, x + 2\rangle \mid x \in \mathbb{N}\}$. In general, for any natural number $k > 0$ we have

$$R^k = \{\langle x, x + k\rangle \mid x \in \mathbb{N}\}.$$

Since $t(R)$ is the union of all these sets, it follows that $t(R)$ is the familiar "less" relation over $\mathbb{N}$. $\quad \blacklozenge$

EXAMPLE 8. We'll list some closures for the relations "less" and "notEqual" over the natural numbers:

$$
\begin{aligned}
r(\text{less}) &= \text{lessOrEqual}, \\
s(\text{less}) &= \text{notEqual}, \\
t(\text{less}) &= \text{less}, \\
r(\text{notEqual}) &= \mathbb{N} \times \mathbb{N}, \\
s(\text{notEqual}) &= \text{notEqual}, \\
t(\text{notEqual}) &= \mathbb{N} \times \mathbb{N}. \quad \blacklozenge
\end{aligned}
$$

Some properties are retained by closures. For example, we have the following results, which we'll leave as exercises:

Inheritance Properties (4.3)

a) If R is reflexive, then $s(R)$ and $t(R)$ are reflexive.

b) If R is symmetric, then $r(R)$ and $t(R)$ are symmetric.

c) If R is transitive, then $r(R)$ is transitive.

Notice that (4.3c) doesn't include the statement "$s(R)$ is transitive" in its conclusion. To see why, we can let $R = \{\langle a, b \rangle, \langle b, c \rangle, \langle a, c \rangle\}$. It follows that R is transitive. But $s(R)$ is not transitive because, for example, we have $\langle a, b \rangle$, $\langle b, a \rangle \in s(R)$ and $\langle a, a \rangle \notin s(R)$.

Sometimes, it's possible to take two closures of a relation and not worry about the order. Other times, we have to worry. For example, we might be interested in the double closure $r(s(R))$, which we'll denote by $rs(R)$. Do we get the same relation if we interchange r and s and compute $sr(R)$? The inheritance properties (4.3) should help us see that the answer is yes. Here are the facts:

> *Double Closure Properties* (4.4)
>
> a) $rt(R) = tr(R)$.
>
> b) $rs(R) = sr(R)$.
>
> c) $st(R) \subset ts(R)$.

Notice that (4.4c) is not an equality. To see why, let $A = \{a, b, c\}$, and consider the relation $R = \{\langle a, b \rangle, \langle b, c \rangle\}$. Then $st(R)$ and $ts(R)$ are distinct relations:

$$st(R) = \{\langle a, b \rangle, \langle b, a \rangle, \langle b, c \rangle, \langle c, b \rangle, \langle a, c \rangle, \langle c, a \rangle\}.$$

and

$$ts(R) = A \times A.$$

Therefore $st(R)$ is a proper subset of $ts(R)$. Of course, there are also situations in which $st(R) = ts(R)$. For example, if $R = \{\langle a, a \rangle, \langle b, b \rangle, \langle a, b \rangle, \langle a, c \rangle\}$, then

$$st(R) = ts(R) = \{\langle a, a \rangle, \langle b, b \rangle, \langle a, b \rangle, \langle a, c \rangle, \langle b, a \rangle, \langle c, a \rangle\}.$$

Before we close this discussion of closures, we should remark that the symbols R^+ and R^* are often used to denote the closures $t(R)$ and $rt(R)$.

Path Problems

Suppose we need to write a program that inputs two points in a city and outputs a bus route between the two points. A solution to the problem depends on the definition of "point." For example, if a point is any street intersection, then the solution may be harder than in the case in which a point is a bus stop.

This problem is an instance of a path problem. Let's consider some typical path problems in terms of a digraph.

Some Path Problems   (4.5)

Given a digraph and two of its vertices i and j.

a) Find out whether there is a path from i to j. For example, find out whether there is a bus route from i to j.

b) Find a path from i to j. For example, find a bus route from i to j.

c) Find a path from i to j with the minimum number of edges. For example, find a bus route from i to j with the minimum number of stops.

d) Find a shortest path from i to j, where each edge has a nonnegative weight. For example, find the shortest bus route from i to j, where shortest might be distance or time.

e) Find the length of a shortest path from i to j. For example, find the number of stops (or the time or miles) on the shortest bus route from i to j.

Each problem listed in (4.5) can be phrased as a question and the same question is often asked over and over again (e.g., different people asking about the same bus route). So it makes sense to get the answers in advance if possible. We'll see how to solve each of the problems in (4.5).

A useful way to represent a binary relation R over a finite set A (equivalently, a digraph with vertices A and edges R) is as a special kind of matrix called an *adjacency matrix* (or incidence matrix). For ease of notation we'll assume that $A = \{1, ..., n\}$ for some n. The adjacency matrix for R is an n by n matrix M with entries defined as follows:

$$M_{ij} = \text{if } \langle i, j \rangle \in R \text{ then 1 else 0.}$$

EXAMPLE 9. Suppose we have the relation $R = \{\langle 1, 2 \rangle, \langle 2, 3 \rangle, \langle 3, 4 \rangle, \langle 4, 3 \rangle\}$ over the set $A = \{1, 2, 3, 4\}$. The digraph for R and the adjacency matrix M for R are shown in Figure 4.3.   ♦

$$1 \longrightarrow 2 \longrightarrow 3 \longrightarrow 4 \qquad M = \begin{bmatrix} 0 & 1 & 0 & 0 \\ 0 & 0 & 1 & 0 \\ 0 & 0 & 0 & 1 \\ 0 & 0 & 1 & 0 \end{bmatrix}$$

Figure 4.3

If we look at the digraph in Figure 4.3, it's easy to see that R is neither reflexive, symmetric, nor transitive. We can see from the matrix M in Figure 4.3 that R is not reflexive because there is at least one zero on the main diagonal formed by the elements M_{ii}. Similarly, R is not symmetric because a reflection on the main diagonal is not the same as the original matrix. In other words, there are indices i and j such that $M_{ij} \neq M_{ji}$. R is not transitive, but there isn't any visual pattern in M that corresponds to transitivity.

It's an easy task to construct the adjacency matrix for $r(R)$: Just place 1's on the main diagonal of the adjacency matrix. It's also an easy task to construct the adjacency matrix for $s(R)$. We'll leave this one as an exercise. Let's look at an interesting algorithm to construct the adjacency matrix for $t(R)$. The idea, of course, is to repeat the following process until no new edges can be added to the adjacency matrix: If $\langle i, k \rangle$ and $\langle k, j \rangle$ are edges, then construct a new edge $\langle i, j \rangle$. The following algorithm to accomplish this feat with three for-loops is due to Warshall [1962].

Warshall's Algorithm for Transitive Closure (4.6)

Let M be the adjacency matrix for a relation R over $\{1, ..., n\}$. The algorithm replaces M with the adjacency matrix for $t(R)$.

```
for k := 1 to n do
    for i := 1 to n do
        for j := 1 to n do
            if (M_ik = M_kj = 1) then M_ij := 1
        od
    od
od
```

EXAMPLE 10. We'll apply Warshall's algorithm to find the transitive closure of the relation R given in Example 9. So the input to the algorithm will be the adjacency matrix M for R shown in Figure 4.3. The four matrices in Figure 4.4 show how Warshall's algorithm transforms M into the adjacency matrix for $t(R)$. Each matrix represents the value of M for the given value of k after the inner i and j loops have executed.

$$
k = 1 \qquad\qquad k = 2 \qquad\qquad k = 3 \qquad\qquad k = 4
$$

$$
M \rightarrow
\begin{bmatrix}
0 & 1 & 0 & 0 \\
0 & 0 & 1 & 0 \\
0 & 0 & 0 & 1 \\
0 & 0 & 1 & 0
\end{bmatrix}
\rightarrow
\begin{bmatrix}
0 & 1 & 1 & 0 \\
0 & 0 & 1 & 0 \\
0 & 0 & 0 & 1 \\
0 & 0 & 1 & 0
\end{bmatrix}
\rightarrow
\begin{bmatrix}
0 & 1 & 1 & 1 \\
0 & 0 & 1 & 1 \\
0 & 0 & 0 & 1 \\
0 & 0 & 1 & 1
\end{bmatrix}
\rightarrow
\begin{bmatrix}
0 & 1 & 1 & 1 \\
0 & 0 & 1 & 1 \\
0 & 0 & 1 & 1 \\
0 & 0 & 1 & 1
\end{bmatrix}
$$

Figure 4.4

To get some insight into how Warshall's algorithm works, draw the four digraphs for the adjacency matrices in Figure 4.4.   ◆

Now we have an easy way find out whether there is a path from i to j in a digraph. Let R be the set of edges in the digraph. First we represent R as an adjacency matrix. Then we apply Warshall's algorithm to construct the adjacency matrix for $t(R)$. Now we can check to see whether there is a path from i to j in the original digraph by checking M_{ij} in the adjacency matrix M for $t(R)$. So we have all the solutions to problem (4.5a).

Let's look at problem (4.5e). Can we compute the length of a shortest path in a weighted digraph? Sure. Let R denote the set of edges in the digraph. We'll represent the digraph as a *weighted adjacency matrix M* as follows: First of all, we set $M_{ij} = 0$ for $1 \le i \le n$ because we're not interested in the shortest path from i to itself. Next, for each edge $\langle i, j \rangle \in R$ with $i \ne j$, we set M_{ij} to be the nonnegative weight for that edge. Lastly, if $\langle i, j \rangle \notin R$ with $i \ne j$, then we set $M_{ij} = \infty$, where ∞ represents some number that is larger than the sum of all the weights on all the edges of the digraph.

EXAMPLE 11. The diagram in Table 4.1 represents the weighted adjacency matrix M for a weighted digraph over the vertex set {1, 2, 3, 4, 5, 6}.   ◆

	1	2	3	4	5	6
1	0	10	10	∞	20	10
2	∞	0	∞	30	∞	∞
3	∞	∞	0	30	∞	∞
4	∞	∞	∞	0	∞	∞
5	∞	∞	∞	40	0	∞
6	∞	∞	∞	∞	5	0

Table 4.1

Now we can present an algorithm to compute the shortest distances between vertices in a weighted digraph. The algorithm, due to Floyd [1962], modifies the weighted adjacency matrix M so that M_{ij} is the shortest distance between distinct vertices i and j. For example, if there are two paths from i to j, then the entry M_{ij} denotes the smaller of the two path weights. So again, transitive closure comes into play. Here's the algorithm.

Floyd's Algorithm for Shortest Distances (4.7)

Let M be the weighted adjacency matrix for a weighted digraph over the set $\{1, ..., n\}$. The algorithm replaces M with a weighted adjacency matrix that represents the shortest distances between distinct vertices.

> **for** $k := 1$ **to** n **do**
> > **for** $i := 1$ **to** n **do**
> > > **for** $j := 1$ **to** n **do**
> > > > $M_{ij} := \min\{M_{ij}, M_{ik} + M_{kj}\}$
> > > **od**
> > **od**
> **od**

EXAMPLE 12. We'll apply Floyd's algorithm to the weighted adjacency matrix in Table 4.1. The result is given in Table 4.2. The entries M_{ij} that are not zero and not ∞ represent the minimum distances (weights) required to travel from i to j in the original digraph. ◆

	1	2	3	4	5	6
1	0	10	10	40	15	10
2	∞	0	∞	30	∞	∞
3	∞	∞	0	30	∞	∞
4	∞	∞	∞	0	∞	∞
5	∞	∞	∞	40	0	∞
6	∞	∞	∞	45	5	0

Table 4.2

Let's summarize our results so far. Algorithm (4.7) creates a matrix M that allows us to easily answer two questions: Is there a path from i to j for distinct vertices i and j? Yes, if $M_{ij} \neq \infty$. What is the distance of a shortest path from i to j? It's M_{ij} if $M_{ij} \neq \infty$.

Now let's try to find a shortest path. We can make a slight modification to (4.7) to compute a "path" matrix P, which will hold the key to finding a shortest path. We'll initialize P to be all zeros. The algorithm will modify P so that $P_{ij} = 0$ means that the shortest path from i to j is the edge from i to j and $P_{ij} = k$ means that a shortest path from i to j goes through k. The modified algorithm, which computes M and P, is stated as follows:

Shortest Distances and Shortest Paths Algorithm   (4.8)

Let M be the weighted adjacency matrix for a weighted digraph over the set $\{1, ..., n\}$. Let P be the n by n matrix of zeros. The algorithm replaces M by a matrix of shortest distances and it replaces P by a path matrix.

> **for** $k := 1$ **to** n **do**
> > **for** $i := 1$ **to** n **do**
> > > **for** $j := 1$ **to** n **do**
> > > > **if** $M_{ik} + M_{kj} < M_{ij}$ **then**
> > > > > $M_{ij} := M_{ik} + M_{kj}$;
> > > > > $P_{ij} := k$
> > > > **fi**
> > > **od**
> > **od**
> **od**

EXAMPLE 13. We'll apply (4.8) to the weighted adjacency matrix in Table 4.1. The algorithm produces the matrix M in Table 4.2, and it produces the path matrix P given in Table 4.3.

	1	2	3	4	5	6
1	0	0	0	2	6	0
2	0	0	0	0	0	0
3	0	0	0	0	0	0
4	0	0	0	0	0	0
5	0	0	0	0	0	0
6	0	0	0	5	0	0

Table 4.3

For example, the shortest path between 1 and 4 passes through 2 because P_{14} = 2. Since $P_{12} = 0$ and $P_{24} = 0$, the shortest path between 1 and 4 consists of the sequence 1, 2, 4. Similarly, the shortest path between 1 and 5 is the sequence 1, 6, 5 and the shortest path between 6 and 4 is the sequence 6, 5, 4. So once we have matrix P from (4.8), it's an easy matter to compute a shortest path between two points. We'll leave this as an exercise.   ◆

Let's make a few observations about Example 13. We should note that there is another shortest path from 1 to 4, namely, 1, 3, 4. The algorithm picked 2 as the intermediate point of the shortest path because the outer index k increments from 1 to n. When the computation got to $k = 3$, the value M_{14} had already been set to the minimal value, and P_{24} had been set to 2. So the condition of the if-then statement was false and no changes were made. Therefore P_{ij} gets the value of k closest to 1 whenever there are two or more values of k that give the same value to the expression $M_{ik} + M_{kj}$, and that value is less than M_{ij}.

Before we finish with this topic, let's make a couple of comments. If we have a digraph that is not weighted, then we can still find shortest distances and shortest paths with (4.7) and (4.8). Just let each edge have weight 1. Then the matrix M produced by either (4.7) or (4.8) will give us the length of a shortest path and the matrix P produced by (4.8) will allow us to find a path of shortest length.

If we have a weighted graph that is not directed, then we can still use (4.7) and (4.8) to find shortest distances and shortest paths. Just modify the weighted adjacency matrix M as follows: For each edge between i and j having weight d, set $M_{ij} = M_{ji} = d$.

Exercises

1. Write down all of the properties that each of the following binary relations satisfies from among the three properties reflexive, symmetric, and transitive.

   a. The similarity relation on the set of triangles.

   b. The congruence relation on the set of triangles.

   c. The relation on the set of people that pairs two people if they both have the same parents. Is this the same as the "areSiblings" relation?

   d. R, where R is your favorite binary relation.

   e. The if and only if relation on the set of statements that may be true or false.

   f. The equality relation on the set of people that pairs two people if both have bachelor's degrees in computer science.

   g. "isBrotherOf" on the set of people.

   h. "hasACommonNationalLanguageWith" on the set of countries.

   i. "speaksThePrimaryLanguageOf" on the set of people.

   j. "isFatherOf" on the set of people.

2. Write down all of the properties that each of the following relations satisfies from among the properties reflexive, symmetric, and transitive.

a. The relation R over the real numbers $R = \{\langle a, b \rangle \mid a^2 + b^2 = 1\}$.

b. The relation R over the real numbers $R = \{\langle a, b \rangle \mid a^2 = b^2\}$.

c. The relation $R = \{\langle x, y \rangle \mid x \bmod y = 0 \text{ and } x, y \in \{1, 2, 3, 4\}\}$.

3. Explain why the empty relation $\varnothing$ is symmetric and transitive.

4. Write down suitable names for each of the following compositions.

   a. isChildOf ∘ isChildOf.

   b. isSisterOf ∘ isParentOf.

   c. isSonOf ∘ isSisterOf.

   d. isChildOf ∘ isParentOf.

5. Suppose we define $x\,R\,y$ to mean "x is the father of y and y has a brother." Write R as the composition of two well-known relations.

6. For each of the following conditions, find the smallest relation over the set $A = \{a, b, c\}$ satisfying the stated properties. Express your answers as graphs.

   a. Reflexive but not symmetric and not transitive.

   b. Symmetric but not reflexive and not transitive.

   c. Transitive but not reflexive and not symmetric.

   d. Reflexive and symmetric but not transitive.

   e. Reflexive and transitive but not symmetric.

   f. Symmetric and transitive but not reflexive.

   g. Reflexive, symmetric, and transitive.

7. For each of the following properties, show that if R has the property, then so does R^2.

   a. Reflexive.

   b. Symmetric.

   c. Transitive.

8. For each of the following properties, find a binary relation R such that R has the property but R^2 does not.

   a. Irreflexive.

   b. Antisymmetric.

9. Given the relation "less" over the natural numbers $\mathbb{N}$, describe each of the following compositions as a set of the form $\{\langle x, y \rangle \mid \text{property}\}$.

   a. less ∘ less.

   b. less ∘ less ∘ less.

10. Given the three relations "less," "greater," and "notEqual" over the natural numbers $\mathbb{N}$, find each of the following compositions.

    a. less ∘ greater.
    b. greater ∘ less.
    c. notEqual ∘ less.
    d. greater ∘ notEqual.

11. Prove each of the following statements about binary relations.

    a. $R \circ (S \circ T) = (R \circ S) \circ T$ (associativity).
    b. $R \circ (S \cup T) = R \circ S \cup R \circ T$.
    c. $R \circ (S \cap T) \subset R \circ S \cap R \circ T$.

12. Let A be a set, R any binary relation on A, and E the equality relation on A. Show that $E \circ R = R \circ E = R$.

13. What is the reflexive closure of the empty binary relation $\varnothing$ over a set A?

14. Find the symmetric closure of each of the following relations over the set $\{a, b, c\}$.

    a. $\varnothing$.
    b. $\{\langle a, b \rangle, \langle b, a \rangle\}$.
    c. $\{\langle a, b \rangle, \langle b, c \rangle\}$.
    d. $\{\langle a, a \rangle, \langle a, b \rangle, \langle c, b \rangle, \langle c, a \rangle\}$.

15. Find the transitive closure of each of the following relations over the set $\{a, b, c, d\}$.

    a. $\varnothing$.
    b. $\{\langle a, b \rangle, \langle a, c \rangle, \langle b, c \rangle\}$.
    c. $\{\langle a, b \rangle, \langle b, a \rangle\}$.
    d. $\{\langle a, b \rangle, \langle b, c \rangle, \langle c, d \rangle, \langle d, a \rangle\}$.

16. Find an appropriate name for the transitive closure of each of the following relations.

    a. IsParentOf.
    b. IsChildOf.
    c. $\{\langle x + 1, x \rangle \mid x \in \mathbb{N}\}$.

17. Given the relation "less" over $\mathbb{N}$, show that $st(\text{less})$ is not equal to $ts(\text{less})$.

18. Prove each of the following statements about a binary relation R over a set A.

    a. If R is reflexive, then $s(R)$ and $t(R)$ are reflexive.

    b. If R is symmetric, then $r(R)$ and $t(R)$ are symmetric.

    c. If R is transitive, then $r(R)$ is transitive.

19. Prove each of the following statements about a binary relation R over a set A.

    a. $rt(R) = tr(R)$.

    b. $rs(R) = sr(R)$.

    c. $st(R) \subset ts(R)$.

20. Write algorithms to perform each of the following actions, given a binary relation R over a finite set. Assume that R is represented as an adjacency matrix.

    a. Check R for reflexivity.

    b. Check R for symmetry.

    c. Check R for transitivity.

    d. Compute $r(R)$.

    e. Compute $s(R)$.

    f. Compute $t(R)$.

21. Suppose G is the following weighted digraph, where the triple $\langle i, j, d \rangle$ represents edge $\langle i, j \rangle$ with distance d:

$$\{\langle 1, 2, 20 \rangle, \langle 1, 4, 5 \rangle, \langle 2, 3, 10 \rangle, \langle 3, 4, 10 \rangle, \langle 4, 3, 5 \rangle, \langle 4, 2, 10 \rangle\}.$$

    a. Draw the weighted adjacency matrix for G.

    b. Use (4.8) to compute the two matrices representing the shortest distances and the shortest paths in G.

22. Write an algorithm to compute the shortest path between two points of a weighted digraph from the matrix P produced by (4.8).

4.2 Equivalence Relations

The word "equivalent" is used in many ways. For example, we've all seen statements like "Two triangles are equivalent if their corresponding angles are equal." We want to find some general properties that describe the idea of "equivalence." We'll start by discussing the idea of "equality" because, to most people, "equal" things are examples of "equivalent" things, whatever meaning is attached to the word "equivalent." Let's consider the following problem:

The Equality Problem

Write a computer program to implement the concept of equality on the elements of a set.

What is equality? Does it depend on the elements of the set? Why is equality important? What are some properties of equality? We all have an intuitive notion of what equality is because we use it all the time. Equality is important in computer science because programs use equality tests on data. If a programming language doesn't provide an equality test for certain data, then the programmer may need to implement such a test.

The simplest equality on a set A is basic equality: $\{\langle x, x\rangle \mid x \in A\}$. But most of the time we use the word "equality" in a much broader context. For example, suppose A is the set of arithmetic expressions made from natural numbers and the symbol +. Thus A contains expressions like $3 + 7$, 8, and $9 + 3 + 78$. Most of us already have a pretty good idea of what equality means for these expressions. For example, we probably agree that $3 + 2$ and $2 + 1 + 2$ are equal. In other words, two expressions (syntactic objects) are equal if they have the same value (meaning or semantics), which is obtained by evaluating all + operations.

Are there some fundamental properties that hold for any definition of equality on a set A? Certainly we want to have $x = x$ for each element x in A (the basic equality on A). Also, whenever $x = y$, it ought to follow that $y = x$. Lastly, if $x = y$ and $y = z$, then $x = z$ should hold. Of course, these are the three properties reflexive, symmetric, and transitive.

Most equalities are more than just basic equality. That is, they equate different syntactic objects that have the same meaning. In these cases the symmetric and transitive properties are needed to convey our intuitive notion of equality. For example, the following statements are true if we let "=" mean "has the same value as":

> If $2 + 3 = 1 + 4$, then $1 + 4 = 2 + 3$.
>
> If $2 + 5 = 1 + 6$ and $1 + 6 = 3 + 4$, then $2 + 5 = 3 + 4$.

Now we're ready to define equivalence. Any binary relation that is reflexive, symmetric, and transitive is called an *equivalence* relation. Sometimes people refer to an equivalence relation as an RST relation in order to remember the three properties. Equivalence relations are all around us. Of course, the basic equality relation on any set is an equivalence relation. Similarly, the notion of equivalent triangles is an equivalence relation.

For another example, suppose we relate two books in the Library of Congress if their call numbers start with the same letter. (This is an instance in which it seems to be official policy to have a number start with a letter.)

This relation is clearly an equivalence relation. Each book is related to itself (reflexive). If book b and book c have call numbers that begin with the same letter, then so do books c and b (symmetric). If books b and c have call numbers beginning with the same letter and books c and d have call numbers beginning with the same letter, then so do books b and d (transitive).

We can generalize the equality problem to the more realistic problem of equivalence:

The Equivalence Problem

Write a computer program to implement an equivalence relation on the elements of a set.

EXAMPLE 1 (*Binary Trees with the Same Structure*). Suppose we want to define a relation on binary trees that relates two binary trees whenever they have the same structure regardless of the values of the nodes. Let ~ denote the relation. We'll define $\langle \rangle \sim \langle \rangle$ as the basis case because two empty trees have the same structure. For the recursive case we'll define tree(l, x, r) ~ tree(l', y, r') if $l \sim l'$ and $r \sim r'$. It's easy to see that ~ is an equivalence relation. From a programming point of view it's easy to implement ~. For example, the function "equiv" computes ~ as follows:

$$
\begin{aligned}
\text{equiv}(s, t) \;=\; &\text{if } s = \langle \rangle \text{ then} \\
&\quad \text{if } t = \langle \rangle \text{ then true else false} \\
&\text{else if } t = \langle \rangle \text{ then false} \\
&\text{else if equiv(left}(s), \text{left}(t)) \text{ then} \\
&\quad \text{equiv(right}(s), \text{right}(t)) \\
&\text{else false.} \quad \blacklozenge
\end{aligned}
$$

Equivalence and Partitioning

An important property of any equivalence relation on a set is that it induces a partitioning of the set into a collection of subsets, each subset containing elements that are equivalent to each other. By a *partition* of a set we mean a collection of subsets that are disjoint from each other and whose union is the whole set. For example, the equality relation on the set $A = \{a, b, c\}$ induces a partitioning of A into the three singleton subsets $\{a\}$, $\{b\}$, and $\{c\}$. We write the partition as a set of sets as follows:

$$\{\{a\}, \{b\}, \{c\}\}.$$

For another example, the relation used in the Library of Congress example partitions the set of all the books into 26 subsets, one subset for each letter of the alphabet.

On the other hand, if we start with a partition of a set, then we can define an equivalence relation on the set. We simply agree to say that two elements are equivalent whenever they are in the same subset of the partition. For example, suppose we have the following partition of the set $\{a, b, c, d, e\}$:

$$\{\{a, b\}, \{c, d, e\}\}.$$

This partition defines an equivalence relation in which each element is equivalent to itself, a is equivalent to b, and the three elements c, d, and e are equivalent to each other.

We can state the important connection between equivalence relations and partitions as follows:

Equivalence Relations and Partitions (4.9)

If R is an equivalence relation on the set S, then R induces a partition of S. Conversely, if P is a partition of a set S, then P induces an equivalence relation on S.

Let's introduce some notation for partitions and equivalence relations. Let R be an equivalence relation on a set S. If $x \in S$, then we use the symbol $[x]$ to denote the subset of S consisting of all elements that are equivalent to x. So $[x]$ is the following set:

$$[x] = \{y \mid y \in S \text{ and } x R y\}.$$

The set $[x]$ is called an *equivalence class*. We say, "the equivalence class of x," or simply "bracket x." Of course, we know that $x \in [x]$ because $x R x$. Notice also that any element of $[x]$ can be used to represent the set. For example, suppose we have an equivalence class $[x] = \{x, a, y, m\}$. Then we can also represent this set in any of the following ways:

$$[a] = \{x, a, y, m\},$$
$$[y] = \{x, a, y, m\},$$
$$[m] = \{x, a, y, m\}.$$

The partition of S consisting of the collection of all such equivalence classes is called the *partition* of S by R and is denoted by S/R. We also can say

"*S* mod *R*," or "the quotient of *S* by *R*." We can write the partition *S/R* as the following set of sets:

$$S/R = \{[x] \mid x \in S\}.$$

Partitions help us simplify our thinking about sets of individuals by partitioning them into groups that are often easier to think about. For example, let *S* denote the set of all students at some university, and let *M* be the relation on *S* that pairs two students if they have the same major. (Assume here that every student has exactly one major.) *M* is clearly an equivalence relation on *S*, and it follows that *S/M* is the collection of sets of people sharing the same major. For example, one equivalence class in *S/M* is the set of computer science majors. So the partition *S/M* has the following general form:

$$S/M = \{\text{computer science majors, math majors, ...}\}.$$

The partition *S/M* can also be pictured by a Venn diagram as shown in Figure 4.5.

We'll give a few more examples to illustrate the idea of equivalence relations and partitions.

EXAMPLE 2 (*Program Testing*). If the input data set for a program is infinite, then the program can't be tested on every input. However, every program has a finite number of instructions. So we should be able to find a finite data set to cause all instructions of the program to be executed. For example, suppose *p* is the following program, where *x* is an integer and *q*, *r*, and *s* represent other parts of the program:

$$p(x): \quad \textbf{if } x > 0 \textbf{ then } q(x)$$
$$\textbf{else if } x \text{ is even } \textbf{then } r(x)$$
$$\textbf{else } s(x)$$
$$\textbf{fi}$$
$$\textbf{fi}$$

The condition "*x* > 0" causes a natural partition of the integers into the positives and the nonpositives. The condition "*x* is even" causes a natural partition of the nonpositives into the even nonpositives and the odd nonpositives. So we have the following partition of the integers:

$$\{1, 2, 3, ...\}, \{0, -2, -4, ...\}, \{-1, -3, -5, ...\}.$$

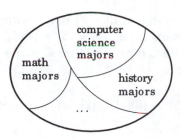

Figure 4.5

Now we can test the instructions in q, r, and s by picking three numbers, one from each set of the partition. For example, $p(1)$, $p(0)$, and $p(-1)$ will do the job. Of course, further partitioning may be necessary if q, r, or s contains further conditional statements. The equivalence relation induced by the partition relates two integers x and y if and only if $p(x)$ and $p(y)$ execute the same set of instructions. ◆

EXAMPLE 3. Let R be the relation on the real numbers that relates two numbers when they have the same absolute value. Then we can write

$$R = \{\langle a, b \rangle \mid a, b \in \mathbb{R} \text{ and } |a| = |b|\}.$$

R is reflexive because $|a| = |a|$ for all $a \in \mathbb{R}$. R is symmetric because if $|a| = |b|$, then $|b| = |a|$. And R is transitive because if $|a| = |b|$ and $|b| = |c|$, then $|a| = |c|$. For any number x we have $[x] = \{x, -x\}$. So each equivalence class has two elements, except $[0] = \{0\}$. We obtain the following partition of $\mathbb{R}$:

$$\mathbb{R}/R = \{\{x, -x\} \mid x \geq 0\}. \quad ◆$$

EXAMPLE 4. Let R be the relation on the set of integers $\mathbb{Z}$ defined by $a \, R \, b$ if and only if $(a - b) \bmod 5 = 0$. It's easy to see that R is an equivalence relation on $\mathbb{Z}$. Notice also that the partition $\mathbb{Z}/R$ consists of five equivalence classes

$$
\begin{aligned}
[0] &= \{\ldots -10, -5, 0, 5, 10, \ldots\}, \\
[1] &= \{\ldots -9, -4, 1, 6, 11, \ldots\}, \\
[2] &= \{\ldots -8, -3, 2, 7, 12, \ldots\}, \\
[3] &= \{\ldots -7, -2, 3, 8, 13, \ldots\}, \\
[4] &= \{\ldots -6, -1, 4, 9, 14, \ldots\}.
\end{aligned}
$$

This follows from the fact that it doesn't matter which element of a class is used to represent it. For example, $[0] = [5] = [-15]$. It is clear that the five classes are disjoint from each other and that $\mathbb{Z}$ is the union of the five classes. So we have the partition

$$\mathbb{Z}/R = \{[0], [1], [2], [3], [4]\}. \quad \blacklozenge$$

EXAMPLE 5 (*Rational Numbers*). When we first discussed the idea of a set representation for the rational numbers $\mathbb{Q}$, we looked at the fractions

$$F = \left\{ \frac{p}{q} \mid p, q \in \mathbb{Z} \text{ and } q \neq 0 \right\}.$$

One problem with this set is there are lots of fractions for each rational number. We then put the restriction that fractions be in lowest terms and considered the set

$$\left\{ \frac{p}{q} \mid p, q \in \mathbb{Z} \text{ and } q \neq 0 \text{ and } \frac{p}{q} \text{ is in lowest terms} \right\}.$$

There is a bijection between this set and the rational numbers. Now let's consider another way to think about the rationals. We will define a relation on the set F of fractions as follows (most of us probably already think of this relation as equality):

$$\frac{a}{b} \sim \frac{c}{d} \quad \text{if and only if} \quad ad = bc.$$

It's easy to see that $\sim$ is an equivalence relation on F. Therefore $\sim$ induces a partition of F, and the equivalence classes are none other than the sets of fractions that represent the same rational number. For example, we have

$$\left[\frac{1}{2} \right] = \left\{ \dots, \frac{-2}{-4}, \frac{-1}{-2}, \frac{1}{2}, \frac{2}{4}, \frac{3}{6}, \dots \right\}.$$

Thus each equivalence class $\left[\frac{p}{q} \right]$ in $F/\sim$ represents a unique rational number. So another representation of the rational numbers is the partition

$$F/\sim = \left\{ \left[\frac{p}{q} \right] \mid p, q \in \mathbb{Z} \text{ and } q \neq 0 \right\}. \quad \blacklozenge$$

Refinements

Suppose that P and Q are two partitions of a set A. If each equivalence class of P is contained in some equivalence class of Q, then P is said to be a *refinement* of Q. P is also said to be *finer* than Q, and Q is said to be *coarser* than P. For example, let $A = \{a, b, c, d\}$. If $P = \{\{a, b\}, \{c\}, \{d\}\}$ and $Q = \{\{a, b\}, \{c, d\}\}$, then P is a refinement of Q. We can find the two extremes by considering the equality relation E and the universal relation U on a set A. Then A/E is a refinement of A/U. For example, if $A = \{a, b, c\}$, then

$$A/E = \{\{a\}, \{b\}, \{c\}\} \quad \text{and} \quad A/U = \{\{a, b, c\}\}.$$

If A is any set and P is any partition of A, then we can say

$$A/E \text{ is a refinement of } P \text{ and } P \text{ is a refinement of } A/U.$$

In other words, we can say that the partition A/E is the finest partition of A and the partition A/U is the coarsest partition of A. This is easy to see when we notice that $A/E = \{\{a\} \mid a \in A\}$ and $A/U = \{A\}$. Thus any other partition of A must be coarser than A/E and finer than A/U. For example, the following four partitions of A are successive refinements from coarsest to finest:

$$\{\{a, b, c, d\}\}$$
$$\{\{a, b\}, \{c, d\}\}$$
$$\{\{a, b\}, \{c\}, \{d\}\}$$
$$\{\{a\}, \{b\}, \{c\}, \{d\}\}.$$

EXAMPLE 6 (*Partitions and the Mod Function*). Let R be the relation over $\mathbb{N}$ defined by $a\,R\,b$ iff $a \bmod 4 = b \bmod 4$. Then R is an equivalence relation, and the corresponding partition $\mathbb{N}/R$ consists of the four subsets

$$[0] = \{0, 4, 8, 12, ...\},$$
$$[1] = \{1, 5, 9, 13, ...\},$$
$$[2] = \{2, 6, 10, 14, ...\},$$
$$[3] = \{3, 7, 11, 15, ...\}.$$

Can we find a refinement of this partition? The relation T defined by $a\,T\,b$ iff $a \bmod 8 = b \bmod 8$ induces a partition $\mathbb{N}/T$ that is a refinement of the partition $\mathbb{N}/R$. For this relation we get eight equivalence classes:

$$[0] \;=\; \{0,\, 8,\, 16,\, \ldots\},$$
$$[1] \;=\; \{1,\, 9,\, 17,\, \ldots\},$$
$$[2] \;=\; \{2,\, 10,\, 18,\, \ldots\},$$
$$[3] \;=\; \{3,\, 11,\, 19,\, \ldots\},$$
$$[4] \;=\; \{4,\, 12,\, 20,\, \ldots\},$$
$$[5] \;=\; \{5,\, 13,\, 21,\, \ldots\},$$
$$[6] \;=\; \{6,\, 14,\, 22,\, \ldots\},$$
$$[7] \;=\; \{7,\, 15,\, 23,\, \ldots\}.$$

In fact, we can find many refinements of $\mathbb{N}/R$. Just define T by

$$a\,T\,b \quad \text{iff} \quad a \bmod k = b \bmod k, \text{ where } k \text{ is a multiple of } 4.$$

Thus there is an infinite nesting of refinements, one for each value of k in the set $\{8,\, 12,\, 16,\, \ldots\}$. Similarly, the partition $\mathbb{N}/R$ is a refinement of the coarser partition generated by the relation T defined by $a\,T\,b$ iff $a \bmod 2 = b \bmod 2$. In this case the partition consists of the two sets of odd and even natural numbers.

What can you say about a relation R defined by $a\,R\,b$ iff $a \bmod p = b \bmod p$, where p is a prime number? Can $\mathbb{N}/R$ be the refinement of a modular relation other than $\mathbb{N}/U$, where U is the universal relation on $\mathbb{N}$?   ◆

Generating Equivalence Relations

We're going to discuss two techniques for generating an equivalence relation. The first technique starts with a binary relation and adds just enough pairs to make an equivalence relation. The second technique shows how a function defines a natural equivalence relation on its domain.

What is the smallest equivalence relation containing a relation R? Do we just take the reflexive closure of R, then the symmetric closure of $r(R)$, and finally the transitive closure of $sr(R)$, resulting in $tsr(R)$? Is the result an equivalence relation? As we shall see, the answer is yes to all these questions. Does it make any difference if we apply closures in another order? For example, what about $str(R)$?

An example will suffice to show that $str(R)$ need not be an equivalence relation. Let $A = \{a, b, c\}$ and $R = \{\langle a, b\rangle, \langle a, c\rangle, \langle b, b\rangle\}$. Then

$$str(R) = \{\langle a, a\rangle,\, \langle b, b\rangle,\, \langle c, c\rangle,\, \langle a, b\rangle,\, \langle b, a\rangle,\, \langle a, c\rangle,\, \langle c, a\rangle\}.$$

This relation is reflexive and symmetric, but it's not transitive. On the other hand, we have $tsr(R) = A \times A$, which is an equivalence relation. For any relation R, can it be that $tsr(R)$ is an equivalence relation? The answer is yes, and the proof follows from the properties (4.3).

Proof: Part (a) of (4.3) implies that $sr(R)$ is reflexive, and, of course, it's symmetric. Parts (a) and (b) of (4.3) imply that $tsr(R) = t(sr(R))$ is reflexive and symmetric, and, of course it's transitive.   QED.

Since we now know that $tsr(R)$ is an equivalence relation, can it be the smallest equivalence relation containing R? The answer is yes, and we'll state the result as follows:

The Smallest Equivalence Relation   (4.10)

If R is a binary relation over A, then $tsr(R)$ is the smallest equivalence relation that contains R.

Proof: We already know that $tsr(R)$ is an equivalence relation. To see that it's the smallest equivalence relation containing R, we'll let T be an arbitrary equivalence relation containing R. So $R \subset T$. Apply tsr to both R and T to obtain $tsr(R) \subset tsr(T)$. But T is an equivalence relation. Therefore $tsr(T) = T$. Thus we have $tsr(R) \subset T$. So $tsr(R)$ is contained in every equivalence relation that contains R. Thus it's the smallest equivalence relation containing R. QED.

It can happen that $tsr(R)$ turns out to be the universal relation, as the next two examples show.

EXAMPLE 7. We'll find the equivalence relation over $\mathbb{N}$ that is induced by the following relation:

$$R = \{\langle a, b \rangle \mid b \neq 0 \text{ and } a \bmod b = 0\}.$$

First, notice that R is not reflexive because $\langle 0, 0 \rangle \notin R$. However, R is pretty close to being reflexive because $\langle a, a \rangle \in R$ if $a \neq 0$. Notice also that R is not symmetric, since $\langle 2, 1 \rangle \in R$ and $\langle 1, 2 \rangle \notin R$. But R is transitive.

Proof: If $a\,R\,b$ and $b\,R\,c$, then $a = b\,q$ and $b = c\,q'$ for two integers q and q'. Therefore $a = c\,q\,q'$. So $a\,R\,c$.   QED.

We'll build $tsr(R)$ incrementally as follows:

$$r(R) \;=\; \{\langle a, b\rangle \mid a = b = 0 \ \text{ or } \ a \bmod b = 0\}.$$

$$sr(R) \;=\; r(R) \cup r(R)^c$$
$$\;=\; \{\langle a, b\rangle \mid a = b = 0 \ \text{or } a \bmod b = 0 \ \text{or } b \bmod a = 0\}.$$

Is $sr(R)$ transitive? To see that it's not, just notice that $\langle 4, 2\rangle \in sr(R)$ and $\langle 2, 6\rangle \in sr(R)$, but $\langle 4, 6\rangle \notin sr(R)$. So how do we find $tsr(R)$? Notice that for any natural numbers n and m we have $\langle n, 1\rangle \in sr(R)$ and $\langle 1, m\rangle \in sr(R)$. Therefore the pair $\langle n, m\rangle$ must be in the transitive closure of $sr(R)$ for arbitrary values n and m. Thus $tsr(R)$ is the universal relation

$$tsr(R) = \mathbb{N} \times \mathbb{N}. \quad \blacklozenge$$

EXAMPLE 8. Let's try to find $tsr(R)$ for the relation R defined as follows:

$$R = \{\langle a, b\rangle \mid a, b \in \mathbb{Z} \ \text{and} \ 10x \le a \le b \le 10(x + 1) \ \text{for some } x \in \mathbb{Z}\}.$$

To see that R is reflexive, let $a \in \mathbb{Z}$. Then it's easy to find an integer x such that $10x \le a \le a \le 10(x + 1)$. Therefore $a \, R \, a$. What about symmetry? R is not symmetric because $1 \, R \, 2$ but not $2 \, R \, 1$. What about transitivity? R is not transitive, since $9 \, R \, 10$ and $10 \, R \, 11$ but not $9 \, R \, 11$.

Now let's find $tsr(R)$. Since R is reflexive, we already have $r(R) = R$. Since R is not symmetric, we need to compute $s(R) = R \cup R^c$. For a pair $\langle a, b\rangle \in s(R)$ either $a \le b$ or $b \le a$. So we can drop the restriction $a \le b$:

$$sr(R) = s(R) = \{\langle a, b\rangle \mid a, b \in \{10x, 10x + 1, ..., 10(x + 1)\} \ \text{for some } x \in \mathbb{Z}\}.$$

What about the transitivity of $sr(R)$? We can use the same numbers 9, 10, and 11 as before to show that $sr(R)$ is not transitive. So we've got to compute some powers of $sr(R)$. It's easy to see that

$$sr(R)^2 = \{\langle a, b\rangle \mid a, b \in \{10x, 10x + 1, ..., 10(x + 2)\} \ \text{for some } x \in \mathbb{Z}\}.$$

In fact, it's easy to see that for any integer $k \ge 1$ we have

$$sr(R)^k = \{\langle a, b\rangle \mid a, b \in \{10x, 10x + 1, ..., 10(x + k)\} \ \text{for some } x \in \mathbb{Z}\}.$$

Notice that each relation $sr(R)^k$ is a proper subset of $sr(R)^{k+1}$. So for any pair of integers $\langle a, b\rangle$ there is some k such that $\langle a, b\rangle \in sr(R)^k$. Since $tsr(R)$ is the infinite union of these relations, it follows that $tsr(R)$ is the universal relation on $\mathbb{Z}$. In other words, $tsr(R) = \mathbb{Z} \times \mathbb{Z}$. ◆

Kernel Relations

Any function $f : A \to B$ can be used as a starting point to define a natural equivalence relation on its domain A. We simply relate two elements x and y if $f(x) = f(y)$. The resulting relation is called the *kernel relation* of f on A, and we'll denote it by K_f. So we have $x\,K_f\,y$ if and only if $f(x) = f(y)$. As a set of ordered pairs we have

$$K_f = \{\langle x, y \rangle \mid f(x) = f(y)\}.$$

It's easy to see that K_f is reflexive, symmetric, and transitive. Therefore K_f is an equivalence relation. So K_f induces a partition of A by (4.9). The resulting partition A/K_f is called the *kernel partition* of A by f.

Let's do an example. Suppose we let f be the function that takes an English word and returns the set of letters in the word. So f has the set of English words for its domain and the power set of the English alphabet as its codomain. For example,

$$f(\text{hello}) = \{h, e, l, l, o\} = \{h, e, l, o\}.$$

The kernel relation induced by f relates two words if they use the same letters. Notice, for example, that $f(\text{too}) = \{t, o, o\} = \{t, o\} = f(\text{to})$. So words like

$$\text{too, to, toot, otto}$$

all belong to the same equivalence class. That is,

$$[\text{too}] = \{\text{too, to, toot, otto, ...}\}.$$

As an alternative example, let g be the function that takes an English word and returns the bag of letters in the word. So we have $g(\text{too}) = [t, o, o]$ and $g(\text{to}) = [t, o]$. Recall that $[t, o, o] \neq [t, o]$. Therefore the words "to" and "too" belong to two separate equivalence classes. Notice that the words "toot" and "otto" belong to the same equivalence class. So we get a much finer partition of the set of English words with the kernel of g. Can you see why the partition induced by K_g is a refinement of the partition induced by K_f?

Sometimes it's possible to show that a relation is an equivalence relation by rewriting its definition as a functional equality. For example, suppose we're given the relation ~ defined on the set of integers as follows:

$$x \sim y \text{ if and only if } x - y \text{ is an even integer.}$$

Notice that $x - y$ is even if and only if x and y are both odd or both even. Using this fact, we can define a function $f : \mathbb{Z} \to \{0, 1\}$ by setting $f(x) =$ if x is even then 1 else 0. So we can redefine ~ as the following functional equality:

$$x \sim y \quad \text{if and only if} \quad f(x) = f(y).$$

We can now immediately conclude that ~ is an equivalence relation because it's the kernel relation of the function f. Let's look at a couple more examples.

EXAMPLE 9 (*Solving the Equality Problem*). If we want to define an equality relation on a set S of objects and the objects do not have any established meaning, then we can use the basic equality relation $\{\langle x, x \rangle \mid x \in S\}$. On the other hand, suppose a meaning has been assigned to each element of S. We can represent the meaning by a mapping m from S to a set of values V. In other words, we have a function $m : S \to V$. It's natural to define two elements of S to be equal if they have the same meaning. That is, we define $x = y$ if and only if $m(x) = m(y)$. This equality relation is just the kernel relation K_m.

For example, let S denote the set of arithmetic expressions made from nonempty unary strings and the symbol +. For example, some typical expressions in S are 1, 11, 111, 1+1, 11+111+1. Now let's assign a meaning to each expression in S. Let $m(1^n) = n$ for each positive natural number n. If $e + e'$ is an expression of S, we define $m(e + e') = m(e) + m(e')$. We'll assume that + is applied left to right. For example, the value of the expression $1 + 111 + 11$ can be calculated as follows:

$$
\begin{aligned}
m(1 + 111 + 11) &= m((1 + 111) + 11) \\
&= m(1 + 111) + m(11) \\
&= m(1) + m(111) + 2 \\
&= 1 + 3 + 2 \\
&= 6.
\end{aligned}
$$

If we define two expressions of S to be equal when they have the same meaning, then the desired equality relation on S is the kernel relation

$$K_m = \{\langle e, d \rangle \mid m(e) = m(d)\}.$$

The partition S/K_m consists of the sets of expressions with equal values. For example, the equivalence class [1111] contains the eight expressions

1+1+1+1, 1+1+11, 1+11+1, 11+1+1, 11+11, 1+111, 111+1, 1111.   ◆

EXAMPLE 10 (*Factoring a Function*). If we have a function $f : A \to B$, then we can use the kernel partition A/K_f to help factor f into the composition of two special functions, one an injection and one a surjection. The result can be stated as follows:

Kernel Factorization (4.11)

Any function $f : A \to B$ can be factored into the composition of two functions, $f = i \circ s$, where $s : A \to A/K_f$ is a surjection defined by $s(a) = [a]$ and $i : A/K_f \to B$ is an injection defined by $i([a]) = f(a)$. The relationship $f = i \circ s$ is pictured in Figure 4.6.

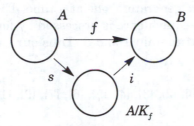

Figure 4.6

As an example, let's continue our discussion of the function f that maps an English word to its set of letters. If we factor f by kernel factorization, then s maps an English word to the set of all words that can be formed by the letters of the word, and i maps the set of all such words to the set of letters used in the words. For example, we have

$$f(\text{too}) = \{t, o\}$$
$$s(\text{too}) = \{\text{too, toot, otto, to, tot, ...}\}$$
$$i(s(\text{too})) = i(\{\text{too, toot, otto, to, tot, ...}\}) = \{t, o\}.$$

Therefore $f(\text{too}) = i(s(\text{too})) = (i \circ s)(\text{too})$. ◆

An Equivalence Problem

Suppose we have an equivalence relation over a set S that is generated by a given set of equivalences. Can we represent the generators in such a way that we can find out whether two arbitrary elements of S are equivalent? If two elements are equivalent, can we find a sequence of generators to confirm the fact? The answer to both questions is yes. We'll present a solution due to Galler and Fischer [1964], which uses a special kind of tree structure to represent the equivalence classes.

The idea is to use the generating equivalences to build a partition of S. For example, let $S = \{1, 2, ..., 10\}$, let ~ denote the equivalence relation on S, and let the generators of ~ be as follows:

$$1 \sim 8, \quad 4 \sim 5, \quad 9 \sim 2, \quad 4 \sim 10, \quad 3 \sim 7, \quad 6 \sim 3, \quad 4 \sim 9.$$

We start the construction process by building the following ten singletons, which represent the partition of S caused by the reflexive property of ~:

$$\{1\}, \{2\}, \{3\}, \{4\}, \{5\}, \{6\}, \{7\}, \{8\}, \{9\}, \{10\}.$$

Now we process the generators, one at a time. The first generator is the equivalence $1 \sim 8$. This equivalence is processed by forming the union of the equivalence classes that contain 1 and 8. Thus our partition now looks like the following:

$$\{1, 8\}, \{2\}, \{3\}, \{4\}, \{5\}, \{6\}, \{7\}, \{9\}, \{10\}.$$

Continuing in this manner to process the other generators, we eventually obtain the partition of S consisting of the following three equivalence classes:

$$\{1, 8\}, \{2, 4, 5, 9, 10\}, \{3, 6, 7\}.$$

We can represent the equivalence classes of the partition as a set of trees, where the generator $a \sim b$ will be processed by creating the branch "a is the parent of b." For our example, if we process the generators in the order in which they are written, then we obtain the three trees in Figure 4.7, which represent the three equivalence classes.

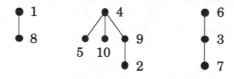

Figure 4.7

A simple way to represent these trees is with a 10-tuple (a 1-dimensional array of size 10) named p, where $p[i]$ denotes the parent of i. We'll let $p[i] = 0$ mean that i is a root. The three equivalence classes are represented by the table for p (Table 4.4).

Now it's easy to answer the question "Is $a \sim b$?" Just find the roots of the trees to which a and b belong. If the roots are the same, the answer is yes. If

i	1	2	3	4	5	6	7	8	9	10
$p[i]$	0	9	6	0	4	0	3	1	4	4

Table 4.4

the answer is yes, then there is another question, "Can you find a sequence of equivalences to show that $a \sim b$?" One way to do this is to locate one of the numbers, say b, and rearrange the tree to which b belongs so that b becomes the root. This can be done easily by reversing the links from b to the root. Once we have b at the root, it's an easy matter to read off the equivalences from a to b. We'll leave it as an exercise to construct an algorithm to do the reversing.

For example, if we ask, "Is $5 \sim 2$?", we find that 5 and 2 belong to the same tree. So the answer is yes. To find a set of equivalences to prove that $5 \sim 2$, we can reverse the links from 2 to the root of the tree. The before and after pictures are given in Figure 4.8.

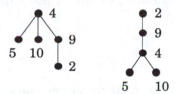

Figure 4.8

Now it's an easy computation to traverse the tree from 5 to the root 2, reading off the following equivalences: $5 \sim 4$, $4 \sim 9$, $9 \sim 2$.

EXAMPLE 11 (*Kruskal's Algorithm*). In Chapter 1 we presented Prim's algorithm to find a minimal spanning tree for a connected weighted undirected graph. Let's look an another such algorithm, due to Kruskal [1956]. The algorithm constructs a minimal spanning tree as follows: Starting with an empty tree, an edge $\{a, b\}$ of smallest weight is chosen from the graph. If there is no path in the tree from a to b, then the edge $\{a, b\}$ is added to the tree. This process is repeated with the remaining edges of the graph until the tree contains all vertices of the graph.

At any point in the algorithm, the edges in the spanning tree define an equivalence relation on the set of vertices of the graph. Two vertices a and b are equivalent iff there is a path between a and b in the tree. Whenever an edge $\{a, b\}$ is added to the spanning tree, the equivalence relation is modified by creating the equivalence class $[a] \cup [b]$. The algorithm ends when there is exactly one equivalence class consisting of all the vertices of the graph. Here's the algorithm:

1. Sort the edges of the graph by weight, and let L be the sorted list.

2. Let T be the minimal spanning tree and initialize $T := \varnothing$.

3. For each vertex v of the graph, create the equivalence class $[v] = \{v\}$.

4. **while** there are 2 or more equivalence classes **do**
   Let $\{a, b\}$ be the edge at the head of L;
   $L := \mathrm{tail}(L)$;
   if $[a] \neq [b]$ **then**
       $T := T \cup \{\{a, b\}\}$;
       Replace the equivalence classes $[a]$ and $[b]$ by $[a] \cup [b]$
   fi
   od

To implement the algorithm, we must find a representation for the equivalence classes. For example, we might use a parent array like the one we've been discussing.   ♦

Exercises

1. For each of the following relations, either prove that it is an equivalence relation or prove that it is not an equivalence relation.

   a. $a \sim b$ iff $a + b$ is even, over the set of integers.

   b. $a \sim b$ iff $a + b$ is odd, over the set of integers.

   c. $a \sim b$ iff $a \cdot b > 0$, over the set of nonzero rational numbers.

   d. $a \sim b$ iff a/b is an integer, over the set of nonzero rational numbers.

   e. $a \sim b$ iff $a - b$ is an integer, over the set of rational numbers.

   f. $a \sim b$ iff $|a - b| \leq 2$, over the set of natural numbers.

2. Let R be the relation on $\mathbb{N}$ defined by $a\,R\,b$ iff $a \bmod 4 = b \bmod 4$ and $a \bmod 6 = b \bmod 6$. Show that R is an equivalence relation, and describe the partition $\mathbb{N}/R$.

3. Let R be the relation on $\mathbb{N}$ defined by $a\,R\,b$ iff either $a \bmod 4 = b \bmod 4$ or $a \bmod 6 = b \bmod 6$. Show that R is not an equivalence relation on $\mathbb{N}$ and thus there is no partition of the form $\mathbb{N}/R$.

4. For each of the following relations, find which of the properties reflexive, symmetric, transitive hold. Also find $tsr(R)$.

   a. $R = \{\langle a, b \rangle \mid$ either $a \geq 0 \ \& \ b > 0$ or $a < 0 \ \& \ b \leq 0\}$ over the integers $\mathbb{Z}$.

    b. $R = \{\langle a, b \rangle \mid$ there is an integer x such that $10x < a < 10(x + 1)$ and $10x \leq b < 10(x + 1)\}$ over the integers $\mathbb{Z}$.

5. Write down which of the following six relations are equal to each other: $tsr(R)$, $trs(R)$, $str(R)$, $srt(R)$, $rst(R)$, and $rts(R)$.

6. Let $f : A \to B$ be a function. Show that the kernel relation K_f is an equivalence relation on A.

7. Let $f : \mathbb{Z} \to \mathbb{N}$ be defined by $f(x) = |x|$. Describe the kernel relation K_f as a particular set of ordered pairs. Also describe the partition $\mathbb{Z}/K_f$.

8. Let $f : \mathbb{R} \to \mathbb{Z}$ be defined by $f(x) = \text{floor}(x)$. Describe the kernel relation K_f as a particular set of ordered pairs. Also describe the partition $\mathbb{R}/K_f$.

9. Let $f : \mathbb{N} \to \mathbb{N}$ be defined by $f(x) = $ if $0 \leq x \leq 10$ then 10 else $x - 1$. Describe the kernel relation K_f as a particular set of ordered pairs. Also describe the partition $\mathbb{N}/K_f$.

10. Prove the equivalence and partitions theorem (4.9).

11. Prove the kernel factorization theorem (4.11).

12. In the equivalence problem we represented equivalence classes as a set of trees, where the nodes of the trees are the numbers 1, 2, ..., n. Suppose the trees are represented by an array $p[1]$, ..., $p[n]$, where $p[i]$ is the parent of i. Suppose also that $p[i] = 0$ when i is a root. Write a procedure that takes a node i and rearranges the tree that i belongs to so that i is the root, by reversing the links from the root to i.

13. Use Kruskal's algorithm to find a minimal spanning tree for the graph in Figure 4.9.

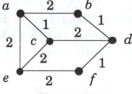

Figure 4.9

4.3 Order Relations

Each day we see the idea of "order" used in many different ways. For example, we might encounter the expression $1 < 2$. We might notice that someone is older than someone else. We might be interested in the third component of the tuple $\langle x, d, c, m \rangle$. We might try to follow a recipe. Or we might see that the word "aardvark" resides at a certain place in the dictionary. The concept of order occurs in many different forms.

Let's try to formally describe the concept of order. To have an ordering, we need a set of elements together with a binary relation having certain properties. What are these properties? Well, our intuition tells us that an ordering should be transitive. For example, if a, b, and c are natural numbers and $a < b$ and $b < c$, then we have $a < c$. Our intuition also tells us that an ordering should be antisymmetric because we don't want distinct elements "preceding" each other. For example, if $a \leq b$ and $b \leq a$, we certainly want $a = b$.

Some orders are reflexive, and some are not. For example, over the natural numbers we recognize that the relations $<$ and $\leq$ are orders because they are both transitive and antisymmetric, even though $<$ is irreflexive and $\leq$ is reflexive. So the two essential properties of any kind of order are antisymmetric and transitive.

Let's look at how different orderings can occur in trying to perform the tasks of a recipe.

EXAMPLE 1 (*Pancake Recipe*). Suppose we have the following recipe for making pancakes:

1. Mix the dry ingredients (flour, sugar, baking powder) in a bowl.
2. Mix the wet ingredients (milk, eggs) in a bowl.
3. Mix the wet and dry ingredients together.
4. Oil the pan. (It's an old pan.)
5. Heat the pan.
6. Make a test pancake and throw it away.
7. Make pancakes.

Steps 1 through 7 indicate an ordering for the steps of the recipe. But the steps could also be done in some other order. To help us discover some other orders, let's define a relation R on the seven steps of the pancake recipe as follows:

$$i \, R \, j \text{ means that step } i \text{ must be done before step } j.$$

Notice that R is antisymmetric and transitive. We can picture R as the digraph (without the transitive arrows) in Figure 4.10. The graph helps us pick out different orders for the steps of the recipe. For example, the following ordering of steps will produce pancakes just as well:

$$4, 5, 2, 1, 3, 6, 7.$$

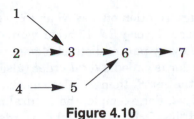

Figure 4.10

So there are several ways to perform the recipe. For example, three people could work in parallel doing tasks 1, 2, and 4 at the same time. ♦

This example demonstrates that different orderings for time-oriented tasks are possible whenever some tasks can be done at different times without changing the outcome. The orderings can be discovered by modeling the tasks by a binary relation R defined by

$i R j$ means that step i must be done before step j.

Notice that R is irreflexive because time-oriented tasks can't be done before themselves. If there are at least two tasks that are not related by R, as in Example 1, then there will be at least two different orderings of the tasks.

Now let's get down to business and discuss the basic ideas and techniques of ordering.

Partial Orders

A binary relation is called a *partial order* if it is antisymmetric and transitive. The set over which a partial order is defined is called a *partially ordered set*—or *poset* for short. If we want to emphasize the fact that R is the partial order that makes S a poset, we'll write the pair $\langle S, R \rangle$ and call it a poset. For example, in our pancake example we defined a partial order R on the set of recipe steps {1, 2, 3, 4, 5, 6, 7}. So we can say that $\langle \{1, 2, 3, 4, 5, 6, 7\}, R \rangle$ is a poset. There are many more examples of partial orders. For example, $\langle \mathbb{N}, < \rangle$ and $\langle \mathbb{N}, \leq \rangle$ are posets because the relations $<$ and $\leq$ are both antisymmetric and transitive.

The word "partial" is used in the definition because we include the possibility that some elements may not be related to each other, as in the pancake recipe example. For another example, consider the subset relation on power($\{a, b, c\}$). Certainly the subset relation is antisymmetric and transitive. So we can say that $\langle \text{power}(\{a, b, c\}), \subset \rangle$ is a poset. Notice that there are some subsets that are not related. For example, $\{a, b\}$ and $\{a, c\}$ are not related by the relation $\subset$.

Suppose R is a binary relation on a set S and $x, y \in S$. We say that x and y are *comparable* if either $x R y$ or $y R x$. In other words, elements that are related are comparable. If every pair of distinct elements in a partial order are comparable, then the order is called a *total* order (also called a *linear* order). If R is a total order on the set S, then we also say that S is a *totally ordered set* or a *linearly ordered set*. For example, the natural numbers are totally ordered by both "less" and "lessOrEqual." In other words, $\langle \mathbb{N}, < \rangle$ and $\langle \mathbb{N}, \leq \rangle$ are totally ordered sets.

EXAMPLE 2 (*The Divides Relation*). Let's look at some interesting posets that can be defined by the divides relation, $|$. First we'll consider the set $\mathbb{N}$. If $a \,|\, b$ and $b \,|\, c$, then $a \,|\, c$. Thus $|$ is transitive. Also, if $a \,|\, b$ and $b \,|\, a$, then it must be the case that $a = b$. So $|$ is antisymmetric. Therefore

$$\langle \mathbb{N}, | \rangle \text{ is a poset.}$$

But $\langle \mathbb{N}, | \rangle$ is not totally ordered because, for example, 2 and 3 are not comparable. To obtain a total order, we need to consider subsets of $\mathbb{N}$. For example, it's easy to see that for any m and n, either $2^m \,|\, 2^n$ or $2^n \,|\, 2^m$. Therefore

$$\langle \{2^n \mid n \in \mathbb{N}\}, | \rangle \text{ is a totally ordered set.}$$

Finally, let's consider some finite subsets of $\mathbb{N}$. For example, it's easy to see that

$$\langle \{1, 3, 9, 45\}, | \rangle \text{ is a totally ordered set.}$$

It's also easy to see that

$$\langle \{1, 2, 3, 4\}, | \rangle \text{ is a poset that is not totally ordered}$$

because 3 can't be compared to either 2 or 4.   ◆

We should note that the literature contains two different definitions of partial order. All definitions require the antisymmetric and transitive properties, but some authors also require the reflexive property. Since we require only the antisymmetric and transitive properties, if a partial order is reflexive and we wish to emphasize it, we'll call it a *reflexive partial order*—or a RAT to remember the properties reflexive, antisymmetric, and transitive. For example, $\leq$ is a reflexive partial order on the integers. If a partial order is irreflexive and we wish to emphasize it, we'll call it an *irreflexive partial order*.

For example, $<$ is an irreflexive partial order on the integers. When authors define partial order to mean RAT, they normally use the term quasi-order to mean irreflexive partial order.

When talking about partial orders, we'll often use the symbols $\prec$ and $\preceq$ to stand for an irreflexive partial order and a reflexive partial order, respectively. We can read $a \prec b$ as "a is less than b," and we can read $a \preceq b$ as "a is less than or equal to b." The two symbols can be defined in terms of each other. For example, if $\langle A, \prec \rangle$ is a poset, then we can define the relation $\preceq$ in terms of $\prec$ by writing

$$\preceq \ = \ \prec \ \cup \ \{\langle x, x \rangle \mid x \in A\}.$$

In other words, $\preceq$ is the reflexive closure of $\prec$. So $x \preceq y$ always means $x \prec y$ or $x = y$. Similarly, if $\langle B, \preceq \rangle$ is a poset, then we can define the relation $\prec$ in terms of $\preceq$ by writing

$$\prec \ = \ \preceq \ - \ \{\langle x, x \rangle \mid x \in B\}.$$

Therefore $x \prec y$ always means $x \preceq y$ and $x \neq y$. We also write the expression $y \succ x$ to mean the same thing as $x \prec y$.

A set of elements in a poset is called a *chain* if all the elements are comparable—linked—to each other. For example, any totally ordered set is itself a chain. A sequence of elements $x_1, x_2, x_3, \ldots$ in a poset is said to be *descending chain* if $x_i \succ x_{i+1}$ for each $i \geq 1$. We can write the descending chain in the following familiar form:

$$x_1 \ \succ \ x_2 \ \succ \ x_3 \ \succ \ \cdots .$$

For example, $4 > 2 > 0 > -2 > -4 > -6 > \ldots$ is a descending chain in $\langle \mathbb{Z}, < \rangle$. For another example, $\{a, b, c\} \supset \{a, b\} \supset \{a\} \supset \varnothing$ is a finite descending chain in $\langle \text{power}(\{a, b, c\}, \subset \rangle$. We can define an *ascending chain* of elements in a similar way. For example, $1 \mid 2 \mid 4 \mid \ldots \mid 2^n \mid \ldots$ is an ascending chain in the poset $\langle \mathbb{N}, \mid \rangle$.

If $x \prec y$, then we say that x is a *predecessor* of y, or y is a *successor* of x. Suppose that $x \prec y$ and there are no elements between x and y. In other words, suppose we have the following situation:

$$\{z \in A \mid x \prec z \prec y\} = \varnothing.$$

When this is the case, we say that x is an *immediate predecessor* of y, or y is an *immediate successor* of x. In a finite poset an element with a successor has an immediate successor. Some infinite posets also have this property. For example, every natural number x has an immediate successor $x + 1$ with respect to the "less" relation. But no rational number has an immediate successor with respect to the "less" relation.

A poset can be represented by a special graph called a *poset diagram* or a *Hasse diagram*—after the mathematician Helmut Hasse (1898–1979). Whenever $x \prec y$ and x is an immediate predecessor of y, then place an edge $\langle x, y \rangle$ in the poset diagram with x at a lower level than y. A poset diagram can often help us observe certain properties of a poset. For example, the two poset diagrams in Figure 4.11 represent the pancake recipe poset from Example 1 and the poset $\langle \{2, 3, 4, 12\}, | \rangle$.

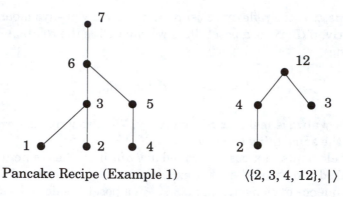

Pancake Recipe (Example 1) $\langle \{2, 3, 4, 12\}, | \rangle$

Figure 4.11

The three poset diagrams shown in Figure 4.12 are for the natural numbers and the integers with their usual orderings and for power($\{a, b\}$) with the subset relation.

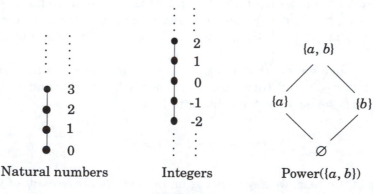

Natural numbers   Integers   Power($\{a, b\}$)

Figure 4.12

Bounds

When we have a partially ordered set, it's natural to use words like "minimal," "least," "maximal," and "greatest." Let's give these words some formal definitions.

Suppose that S is any nonempty subset of a poset P. An element $x \in S$ is called a *minimal element* of S if x has no predecessors in S. An element $x \in S$ is called the *least element* of S if x is minimal and $x \preceq y$ for all $y \in S$. For example, let's consider the poset $\langle \mathbb{N}, \, | \, \rangle$.

The subset $\{2, 4, 5, 10\}$ has two minimal elements, 2 and 5.

The subset $\{2, 4, 12\}$ has least element 2.

The set $\mathbb{N}$ has least element 1 because $1 \, | \, x$ for all $x \in \mathbb{N}$.

For another example, let's consider the poset $\langle \text{power}(\{a, b, c\}), \subset \rangle$. The subset $\{\{a, b\}, \{a\}, \{b\}\}$ has two minimal elements, $\{a\}$ and $\{b\}$. The power set itself has least element $\varnothing$.

In a similar way we can define *maximal elements* and the *greatest element* of a subset of a poset. For example, let's consider the poset $\langle \mathbb{N}, \, | \, \rangle$.

The subset $\{2, 4, 5, 10\}$ has two maximal elements, 4 and 10.

The subset $\{2, 4, 12\}$ has greatest element 12.

The set $\mathbb{N}$ itself has greatest element 0 because $x \, | \, 0$ for all $x \in \mathbb{N}$.

For another example, let's consider the poset $\langle \text{power}(\{a, b, c\}), \subset \rangle$. The subset $\{\varnothing, \{a\}, \{b\}\}$ has two maximal elements, $\{a\}$ and $\{b\}$. The power set itself has greatest element $\{a, b, c\}$.

Some sets may not have any minimal elements yet still be bounded below by some element. For example, the set of positive rational numbers has no least element yet is bounded below by the number 0. Let's introduce some standard terminology that can be used to discuss ideas like this.

If S is a nonempty subset of a poset P, an element $x \in P$ is called a *lower bound* of S if $x \preceq y$ for all $y \in S$. An element $x \in P$ is called the *greatest lower bound* (or *glb*) of S if x is a lower bound and $z \preceq x$ for all lower bounds z of S. The expression $\text{glb}(S)$ denotes the greatest lower bound of S, if it exists. For example, if we let $\mathbb{Q}^+$ denote the set of positive rational numbers, then over the poset $\langle \mathbb{Q}, \leq \rangle$ we have $\text{glb}(\mathbb{Q}^+) = 0$.

In a similar way we define upper bounds for a subset S of the poset P. An element $x \in P$ is called an *upper bound* of S if $y \preceq x$ for all $y \in S$. An element $x \in P$ is called the *least upper bound* (or *lub*) of S if x is an upper bound and $x \preceq z$ for all upper bounds z of S. The expression $\text{lub}(S)$ denotes the least upper bound of S, if it exists. For example, $\text{lub}(\mathbb{Q}^+)$ does not exist in $\langle \mathbb{Q}, \leq \rangle$.

For another example, in the poset $\langle \mathbb{N}, \leq \rangle$, every finite subset has a glb—the least element—and a lub—the greatest element. Every infinite subset has a glb but no upper bound.

Can subsets have upper bounds without having a least upper bound? Sure. Here's an example.

EXAMPLE 3. Suppose the set {1, 2, 3, 4, 5, 6} represents six time-oriented tasks. You can think of the numbers as chapters in a book, as processes to be executed on a computer, or as the steps in a recipe for making ice cream. In any case, suppose the tasks are partially ordered according to the poset diagram in Figure 4.13.

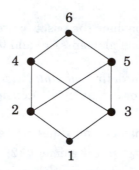

Figure 4.13

The subset {2, 3} is bounded above, but it has no least upper bound. Notice that 4, 5, and 6 are all upper bounds of {2, 3}, but none of them is a least upper bound.   ◆

A *lattice* is a poset with the property that every pair of elements has a glb and a lub. So the poset of Example 3 is not a lattice. For example, $\langle \mathbb{N}, \leq \rangle$ is a lattice in which the glb of two elements is their minimum and the lub is their maximum. For another example, if A is any set, then $\langle \text{power}(A), \subset \rangle$ is a lattice, where $\text{glb}(X, Y) = X \cap Y$ and $\text{lub}(X, Y) = X \cup Y$. The word "lattice" is used because lattices that aren't totally ordered often have poset diagrams that look like "latticeworks" or "trellisworks."

For example, the two poset diagrams in Figure 4.14 represent lattices. These two poset diagrams can represent many different lattices. For example, the poset diagram on the left represents the lattice whose elements are the positive divisors of 36, ordered by the divides relation. In other words, it represents the lattice $\langle \{1, 2, 3, 4, 6, 9, 12, 18, 36\}, \mid \rangle$. See whether you can label the poset diagram with these numbers. The diagram on the right represents the lattice $\langle \mathrm{power}(\{a, b, c\}, \subset\rangle$. It also represents the lattice whose elements are the positive divisors of 70, ordered by the divides relation. See whether you can label the poset diagram with both of these lattices. We'll give some more examples in the exercises.

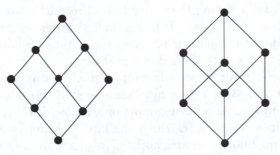

Figure 4.14

Sorting Problems

A typical computing task is to sort a list of elements taken from a totally ordered set. Here's the problem statement:

The Sorting Problem

Find an algorithm to sort a list of elements from a totally ordered set.

For example, suppose we're given the list $\langle x_1, x_2, ..., x_n \rangle$, where the elements of the list are related by a total order relation R. We might sort the list by a program "sort," which we could call as follows:

$$\mathrm{sort}(R, \langle x_1, x_2, .., x_n \rangle).$$

For example, we should be able to obtain the following results with sort:

$$\mathrm{sort}(<, \langle 8, 3, 10, 5 \rangle) = \langle 3, 5, 8, 10 \rangle,$$
$$\mathrm{sort}(>, \langle 8, 3, 10, 5 \rangle) = \langle 10, 8, 5, 3 \rangle.$$

Programming languages normally come equipped with several totally ordered sets. If a total order R is not part of the language, then R must be implemented as a relational test, which can then be called into action whenever a comparison is required in the sorting algorithm.

Can a partially ordered set be sorted? The answer is yes if we broaden our idea of what sorting means. Here's the problem statement.

The Topological Sorting Problem.

Find an algorithm to sort a list of elements from a partially ordered set.

How can we "sort" a list when some elements may not be comparable? Well, we try to find a listing that maintains the partial ordering, as in the pancake recipe from Example 1. If R is a partial order on a set, then a list of elements from the set is *topologically sorted* if, whenever two elements in the list satisfy $a\,R\,b$, then a is to the left of b in the list.

The ordering of a set of tasks is a topological sorting problem. For example, the list $\langle 4, 5, 2, 1, 3, 6, 7 \rangle$ is a topological sort of the steps in the pancake recipe from Example 1. Another example of a topological sort is the ordering of the chapters in a textbook in which the partial order is defined to be the dependence of one chapter upon another. In other words, we hope that we don't have to read some chapter farther on in the book to understand what we're reading now.

Is there a technique to do topological sorting? Yes. Suppose R is a partial order on a finite set A. For each element $y \in A$, let $P(y)$ be the number of immediate predecessors of y, and let $S(y)$ be the set of immediate successors of y. Let "Sources" be the set of sources—minimal elements—in A. Therefore y is a source if and only if $P(y) = 0$. A topological sort algorithm goes something like the following:

While the set of sources is not empty, do the following steps: (4.12)

1. Output a source y.
2. For all z in $S(y)$, decrement $P(z)$; if $P(z) = 0$, then add z to Sources.

A more formal description of the algorithm can be given as follows:

```
while Sources ≠ ∅ do
    Pick a source y from Sources;
    Output y;
    for each z in S(y) do
        P(z) := P(z) − 1;
        if P(z) = 0 then Sources := Sources ∪ {z}
    od;
    Sources := Sources − {y};
od
```

Let's do an example that includes some details on how the data for the algorithm might be represented.

EXAMPLE 4. We'll consider the steps of the pancake recipe from Example 1. Figure 4.15 shows the poset diagram for the steps of the recipe.

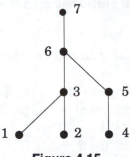

Figure 4.15

The initial set of sources is $\{1, 2, 4\}$. Letting P be an array of integers, we get the following initial table of predecessor counts:

i	1	2	3	4	5	6	7
$P(i)$	0	0	2	0	1	2	1

The following table is the initial table of successor sets S:

i	1	2	3	4	5	6	7
$S(i)$	{3}	{3}	{6}	{5}	{6}	{7}	∅

You should trace the algorithm for these data representations.   ◆

There is a very interesting and efficient implementation of algorithm (4.12) in Knuth [1968]. It involves the construction of a novel data structure to represent the set of sources, the sets $S(y)$ for each y, and the numbers $P(z)$ for each z.

Well-Founded Orders

Let's look at a special property of the natural numbers. Suppose we're given a descending chain of natural numbers, starting at x_1, as follows:

$$x_1 > x_2 > x_3 > \cdots .$$

Can this descending chain continue forever? Of course not. We know that 0 is the least natural number, so the given chain must stop after only a finite number of terms. For example, if $x_1 = 4$, then there are at most five terms in any such chain. This is not an earthshaking discovery, but we'll state it anyway:

Every descending chain of natural numbers is finite in length.

Not all posets have such a property. The integers don't satisfy such a property, nor do the positive rational numbers, as the following examples show:

$$2 > 0 > -2 > -4 > \cdots$$

$$\frac{1}{2} > \frac{1}{3} > \frac{1}{4} > \frac{1}{5} > \cdots .$$

We're going to consider posets with the property that every descending chain of elements is finite. So we'll give these posets a name. A poset is called a *well-founded set* if every descending chain of elements is finite. If a poset is well-founded, its partial order is called a *well-founded order*. Thus $\mathbb{N}$ is a well-founded set with respect to the usual less than relation, $<$. Similarly, the power set of a finite set is well-founded by the subset relation, $\subset$. This is easy to see because any finite set with n elements can start a descending chain of at most $n + 1$ subsets. For example, the following expression displays a longest descending chain starting with the set $\{a, b, c\}$:

$$\{a, b, c\} \supset \{b, c\} \supset \{c\} \supset \varnothing.$$

Notice, however, that the power set of an infinite set is not well-founded by $\subset$. For example, we can construct an infinite descending chain of elements in power($\mathbb{N}$) as follows: Let $S_k = \mathbb{N} - \{0, 1, ..., k\}$. Then we have an infinite descending chain

$$S_0 \supset S_1 \supset S_2 \supset \cdots \supset S_k \supset \cdots .$$

Many posets are not well-founded. If we consider the usual ordering "less," then the integers and the positive rationals are not well-founded because they have infinite descending chains.

Notice that any set of integers with a least element is well-founded by the "less" relation. For example, the following three sets of integers are well-founded:

$$\{1, 2, 3, 4, ...\}, \quad \{m \mid m \geq -3\}, \quad \text{and} \quad \{5, 9, 13, 17, ...\}.$$

Is a well-founded set good for anything? The answer is yes. We'll see in the next section that well-founded sets are basic tools that are used in inductive proofs. So we should get familiar with them. We'll do this by looking at another property that well-founded sets possess.

Does every subset of $\mathbb{N}$ have a least element? A quick-witted person might say, "Yes," and then think a minute and say, "Except that the empty set doesn't have any elements, so it can't have a least element." If the question is modified to "Does every nonempty subset of $\mathbb{N}$ have a least element?", then a bit of thought will convince most of us that the answer is yes.

We might reason as follows: Suppose S is some nonempty subset of $\mathbb{N}$ and x_1 is some element of S. If x_1 is the least element of S, then we are done. So assume that x_1 is not the least element of S. Then x_1 must have a predecessor x_2 in S —otherwise, x_1 would be the least element of S. If x_2 is the least element of S, then we are done. If x_2 is not the least element of S, then it has a predecessor x_3 in S, and so on. If we continue in this manner, we will obtain a descending chain of distinct elements in S:

$$x_1 > x_2 > x_3 > \cdots .$$

This looks familiar. We already know that this chain of natural numbers can't be infinite. So it stops at some value, which must be the least element of S. So we have another fundamental property of $\mathbb{N}$:

Every nonempty subset of the natural numbers has a least element.

Of course, this property is not true for all posets. For example, the set of integers has no least element. The open interval of real numbers $(0, 1)$ has no least element. Also the power set of a finite set can have collections of subsets that have no least element.

Notice however that every collection of subsets of a finite set does contain a minimal element. For example, the collection $\{\{a\}, \{b\}, \{a, b\}\}$ has two minimal elements $\{a\}$ and $\{b\}$. Remember, the property that we are looking for must be true for all well-founded sets. So the existence of least elements is out; it's too restrictive. But what about the existence of minimal elements for nonempty subsets of a well-founded set? This property is true for the natural numbers. (Least elements are certainly minimal.) It's also true for power sets of finite sets. In fact, this property is true for all well-founded sets, and we can state the result as follows:

Descending Chains and Minimality   (4.13)

If A is a well-founded set, then every nonempty subset of A has a minimal element. Conversely, if every nonempty subset of A has a minimal element, then A is well-founded.

It follows from (4.13) that the property of finite descending chains is equivalent to the property of nonempty subsets having minimal elements. In other words, if a poset has one of the properties, then it also has the other property. Thus it is also correct to define a well-founded set to be a poset with the property that every nonempty subset has a minimal element. We will call this latter property the *minimum condition* on a poset.[†]

Whenever a well-founded set is totally ordered, then each nonempty subset has a single minimal element, the least element. Such a set is called a *well-ordered set*. So a well-ordered set is a totally ordered set such that every nonempty subset has a least element. For example, $\mathbb{N}$ is well-ordered by the "less" relation. Let's examine a few more total orderings to see whether they are well-ordered.

Lexicographic and Standard Orderings

Let's look at two classic types of orderings that allow us to order things other than numbers. For example, the linear ordering < on $\mathbb{N}$ can be used to create the *lexicographic* order on $\mathbb{N}^k$, which is defined as follows:

$$\langle x_1, ..., x_k \rangle \prec \langle y_1, ..., y_k \rangle$$

if and only if there is an index $j \geq 1$ such that $x_j < y_j$ and for each $i < j$, $x_i = y_i$. This ordering is a total ordering on $\mathbb{N}^k$. It's also a well-ordering. For example, the lexicographic order on $\mathbb{N} \times \mathbb{N}$ has least element $\langle 0, 0 \rangle$. Every nonempty subset of $\mathbb{N} \times \mathbb{N}$ has a least element, namely, the pair $\langle x, y \rangle$ with the smallest value of x, where y is the smallest value among second components of pairs with x as the first component. For example, $\langle 0, 10 \rangle$ is the least element in the set $\{\langle 0, 10 \rangle, \langle 0, 11 \rangle, \langle 1, 0 \rangle\}$. Notice that $\langle 1, 0 \rangle$ has infinitely many predecessors of the form $\langle 0, y \rangle$, but $\langle 1, 0 \rangle$ has no immediate predecessors.

Let A be a finite alphabet with some agreed-upon linear ordering. Then the *lexicographic* order on A^* is defined as follows for any $x, y \in A^*$:

$x \prec_L y$ iff either x is a prefix of y or x and y have a longest common prefix u such that $x = uv, y = uw$, and head(v) $\prec_L$ head(w).

The lexicographic ordering on A^* is often called the *dictionary ordering* because it corresponds to the ordering of words that occur in a dictionary. It's clear that $\prec_L$ is a total order on A^*. For example, let $A = \{a, b\}$, where we agree that $a \prec_L b$. Then every string beginning with the letter a precedes b, but b has no immediate predecessor.

[†] Other names for well-founded set are *poset with minimum condition, poset with descending chain condition*, and *Artinian poset,* after Emil Artin, who studied algebraic structures with the descending chain condition. Some people use the term *Noetherian*, after Emmy Noether, who studied algebraic structures with the ascending chain condition.

If A is an alphabet with two or more elements, then the lexicographic ordering on A^* is NOT well-ordered. For example, let $A = \{a, b\}$, where we suppose also that $a \prec_L b$. Then the elements in the set $\{a^n b \mid n \in \mathbb{N}\}$ form the following infinite descending chain:

$$b \succ ab \succ aab \succ aaab \succ \cdots \succ a^k b \succ \cdots.$$

Now let's look at an ordering that is well-ordered. The *standard* ordering on strings uses a combination of length and the lexicographic ordering. Again assume that A is a finite alphabet with some agreed-upon linear ordering. Then the standard ordering on A^* is defined as follows: If $x, y \in A^*$, then

$$x \prec_A y \text{ iff either length}(x) < \text{length}(y), \text{ or length}(x) = \text{length}(y) \text{ and } x \prec_L y.$$

It's easy to see that $\prec_A$ is a total order on A^* and every string has an immediate successor. The standard ordering on A^* is also well-ordered because each string has a finite number of predecessors. For example, let $A = \{a, b\}$, and suppose we agree that $a \prec_L b$. Then the first few elements in the standard order of A^* are given as follows:

$$\Lambda, a, b, aa, ab, ba, bb, aaa, aab, aba, abb, baa, bab, bba, bbb, \dots.$$

Constructing Well-Founded Orderings

Collections of strings, lists, trees, graphs, or other structures that programs process can usually be made into well-founded sets by defining an appropriate order relation. For example, any finite set can be made into a well-founded set—actually a well-ordered set—by simply listing its elements in any order we wish, letting the leftmost element be the least element.

Let's look at some ways to build well-founded orderings for infinite sets. Suppose we want to define a well-founded order on some infinite set S. A simple and useful technique is to associate each element of S with some element in an existing well-founded set. For example, the natural numbers are well-founded by $<$. So we can take any function $f : S \to \mathbb{N}$ and use it to define a well-founded ordering $\prec$ on the set S by relating any two elements $x, y \in S$ in terms of the elements $f(x), f(y) \in \mathbb{N}$ as follows:

$$x \prec y \quad \text{means} \quad f(x) < f(y). \tag{4.14}$$

Does the new relation $\prec$ make S into a well-founded set? Sure. Suppose we have a descending chain of elements in S as follows:

$$x_1 \succ x_2 \succ x_3 \succ \cdots.$$

The chain must stop because $x \succ y$ is defined to mean $f(x) > f(y)$, and we know that any descending chain of natural numbers must stop. Let's look at a few more examples.

EXAMPLE 5 (*Some Well-Founded Orderings*).

a) Any set of lists is well-founded as follows: If L and M are lists, let $L \prec M$ mean length(L) < length(M).

b) Any set of strings is well-founded as follows: If s and t are strings, let $s \prec t$ mean length(s) < length(t).

c) Any set of trees is well-founded as follows: If B and C are trees, let $B \prec C$ mean nodes(B) < nodes(C), where nodes is the function that counts the number of nodes in a tree.

d) Another well-founded ordering on trees can be defined as follows: If B and C are trees, define $B \prec C$ to mean leaves(B) < leaves(C), where leaves is the function that returns the number of leaves in a tree.

e) A well-founded ordering on nonempty trees is defined as follows: For nonempty trees B and C, let $B \prec C$ mean depth(B) < depth(C).

f) The set of all people can be well-founded. Let the age of a person be the floor of the number of years they are old. Then let $A \prec B$ if age(A) < age(B) What are the minimal elements?

g) The set $\{ \ldots, -3, -2, -1 \}$ of negative integers is well-founded if we let $x \prec y$ mean $x > y$.   ◆

As the examples show, it's sometimes quite easy to find a well-founded ordering for a set. The next example constructs a finite, hence well-founded lexicographic order.

EXAMPLE 6 (*A Finite Lexicographic Order*). Let $S = \{0, 1, 2, \ldots, m\}$. Then we can define a lexicographic ordering on the set S^k in a natural way. Since S is finite, it follows that the lexicographic ordering on S^k is well-founded. The least element is $\langle 0, \ldots, 0 \rangle$, and the greatest element is $\langle m, \ldots, m \rangle$. The immediate successor of any element can be defined as follows (let $k = 3$):

$$\mathrm{succ}(\langle x, y, z \rangle) \;=\; \text{if } z < m \text{ then } \langle x, y, z + 1 \rangle$$
$$\text{else if } y < m \text{ then } \langle x, y + 1, z \rangle$$
$$\text{else if } x < m \text{ then } \langle x + 1, y, z \rangle$$
$$\text{else ? (You finish the job.)}   \blacklozenge$$

Inductively Defined Sets Are Well-Founded

It's easy to make an inductively defined set W into a well-founded set. We'll give two methods. Both methods let the basis elements of W be the minimal elements of the well-founded order.

Method 1: Define a function $f : W \to \mathbb{N}$ as follows: (4.15)

Basis: $f(c) = 0$ for all basis elements c of W.

Induction: If $x \in W$ and x is constructed from elements $y_1, y_2, ..., y_n$ in W, then define $f(x) = 1 + \max\{f(y_1), f(y_2), ..., f(y_n)\}$.

Let $x \prec y$ mean $f(x) < f(y)$.

Since 0 is the least element of $\mathbb{N}$ and $f(c) = 0$ for all basis elements c of W, it follows that the basis elements of W are minimal elements under the ordering defined by (4.15). For example, if c is a basis element of W and if $x \prec c$, then $f(x) < f(c) = 0$, which can't happen with natural numbers. Therefore c is a minimal element of W.

Let's do an example. Let W be the set of all nonempty lists over $\{a, b\}$. First we'll give an inductive definition of W. The lists $\langle a \rangle$ and $\langle b \rangle$ are the basis elements of W. For the induction case, if $L \in W$, then the lists $\mathrm{cons}(a, L)$ and $\mathrm{cons}(b, L)$ are in W. Now we'll use (4.15) to make W into a well-founded set. The function f of (4.15) turns out to be $f(L) = \mathrm{length}(L) - 1$. So for any lists L and M in W we define $L \prec M$ to mean $f(L) < f(M)$, which means $\mathrm{length}(L) - 1 < \mathrm{length}(M) - 1$, which also means $\mathrm{length}(L) < \mathrm{length}(M)$. The diagram in Figure 4.16 shows the bottom two layers of a poset diagram for W with its two minimal lists $\langle a \rangle$ and $\langle b \rangle$.

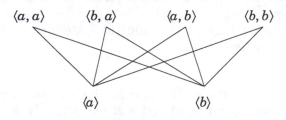

Figure 4.16

If we draw the diagram up to the next level containing triples like $\langle a, a, b \rangle$ and $\langle b, b, a \rangle$, we would have drawn 32 more lines from the two element lists up to the three element lists. So Method 1 relates many elements.

Sometimes it isn't necessary to have an ordering that relates so many elements. This brings us to the second method for defining a well-founded ordering on an inductively defined set W:

Method 2: The ordering $\prec$ is defined as follows: (4.16)

Basis: Let the basis elements of W be minimal elements.

Induction: If $x \in W$ and x is constructed from elements $y_1, y_2, ..., y_n$
               in W, then define $y_i \prec x$ for each $i = 1, ..., n$.

The actual ordering is the transitive closure of $\prec$.

The ordering of (4.16) is well-founded because any x can be constructed
from basis elements with finitely many constructions. Therefore there can be
no infinite descending chain starting at x. With this ordering, there can be
many pairs that are not related.

For example, we'll use the preceding example of nonempty lists over the
set $\{a, b\}$. The picture in Figure 4.17 shows the bottom two levels of the poset
diagram for the well-founded ordering constructed by (4.16).

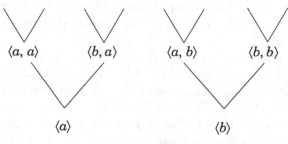

Figure 4.17

Notice that each list has only two immediate successors. For example, the two
successors of $\langle a \rangle$ are $\mathrm{cons}(a, \langle a \rangle) = \langle a, a \rangle$ and $\mathrm{cons}(b, \langle a \rangle) = \langle b, a \rangle$. The two successors of $\langle b, a \rangle$ are $\langle a, b, a \rangle$ and $\langle b, b, a \rangle$. This is much simpler than the or-
dering we got using (4.15).

Let's look at some examples of inductively defined sets that are well-
founded sets by the method of (4.16).

EXAMPLE 7. Let's define the set $\mathbb{N} \times \mathbb{N}$ inductively by using the first copy of
$\mathbb{N}$. For the basis case we put $\langle 0, n \rangle \in \mathbb{N} \times \mathbb{N}$ for all $n \in \mathbb{N}$. For the induction
case, if $\langle m, n \rangle \in \mathbb{N} \times \mathbb{N}$, then we put $\langle \mathrm{succ}(m), n \rangle \in \mathbb{N} \times \mathbb{N}$. The relation on $\mathbb{N} \times$
$\mathbb{N}$ induced by this inductive definition and (4.16) is not linearly ordered. For
example, $\langle 0, 0 \rangle$ and $\langle 0, 1 \rangle$ are not related because they are both basis elements. Notice that any pair $\langle m, n \rangle$ is the beginning of a descending chain containing at most $m + 1$ pairs. For example, the following chain is the longest
descending chain that starts with $\langle 3, 17 \rangle$:

$$\langle 3, 17 \rangle, \quad \langle 2, 17 \rangle, \quad \langle 1, 17 \rangle, \quad \langle 0, 17 \rangle. \quad \blacklozenge$$

EXAMPLE 8. Let's define the set $\mathbb{N} \times \mathbb{N}$ inductively by using both copies of $\mathbb{N}$. The single basis element is $\langle 0, 0 \rangle$. For the induction case, if $\langle m, n \rangle \in \mathbb{N} \times \mathbb{N}$, then put the three pairs $\langle \text{succ}(m), n \rangle$, $\langle m, \text{succ}(n) \rangle$, and $\langle \text{succ}(m), \text{succ}(n) \rangle \in \mathbb{N} \times \mathbb{N}$. Notice that each pair with both components nonzero is defined three times by this definition. The relation induced by this definition and (4.16) is nonlinear. For example, the two pairs $\langle 2, 1 \rangle$ and $\langle 1, 2 \rangle$ are not related. Any pair $\langle m, n \rangle$ is the beginning of a descending chain of at most $m + n + 1$ pairs. For example, the following descending chain has maximum length among the descending chains that start at the pair $\langle 2, 3 \rangle$:

$$\langle 2, 3 \rangle, \quad \langle 2, 2 \rangle, \quad \langle 1, 2 \rangle, \quad \langle 1, 1 \rangle, \quad \langle 0, 1 \rangle, \quad \langle 0, 0 \rangle.$$

Can you find a different chain of the same length starting at $\langle 2, 3 \rangle$?   ♦

Ordinal Numbers

We'll finish our discussion of order by introducing the *ordinal numbers*. These numbers are ordered, and they can be used to count things. An ordinal number is actually a set with certain properties. For example, any ordinal number x has an immediate successor defined by $\text{succ}(x) = x \cup \{x\}$. The expression $x + 1$ is also used to denote $\text{succ}(x)$. The natural numbers denote ordinal numbers when we define $0 = \varnothing$ and interpret $+$ as addition, in which case it's easy to see that

$$x + 1 = \{0, ..., x\}.$$

For example, $1 = \{0\}$, $2 = \{0, 1\}$, and $5 = \{0, 1, 2, 3, 4\}$. In this way, each natural number is an ordinal number, called a *finite ordinal*.

Now let's define some *infinite ordinals*. The first infinite ordinal is

$$\omega = \{0, 1, 2, ...\},$$

the set of natural numbers. The next infinite ordinal is

$$\omega + 1 = \text{succ}(\omega) = \omega \cup \{\omega\} = \{\omega, 0, 1, ...\}.$$

If α is an ordinal number, we'll write $\alpha + n$ in place of $\text{succ}^n(\alpha)$. So the first four infinite ordinals are ω, $\omega + 1$, $\omega + 2$, and $\omega + 3$. The infinite ordinals continue in this fashion. To get beyond this sequence of ordinals, we need to make a definition similar to the one for ω. The main idea is that any ordinal number is the union of all its predecessors. For example, we define $\omega 2 = \omega \cup \{\omega, \omega + 1, ... \}$. The ordinals continue with $\omega 2 + 1$, $\omega 2 + 2$, and so on. Of course, we can continue and define $\omega 3 = \omega 2 \cup \{\omega 2, \omega 2 + 1, ...\}$. After ω, $\omega 2$, $\omega 3$, ... comes the ordinal ω^2. Then we get $\omega^2 + 1$, $\omega^2 + 2$, ..., and we eventually get $\omega^2 + \omega$. Of course, the process goes on forever.

We can order the ordinal numbers by defining $\alpha < \beta$ iff $\alpha \in \beta$. For example, we have $x < x + 1$ for any ordinal x because $x \in \text{succ}(x) = x + 1$. So we get the familiar ordering $0 < 1 < 2 < ...$ for the finite ordinals. For any finite ordinal n we have $n < \omega$ because $n \in \omega$. Similarly, we have $\omega < \omega + 1$, and for any finite ordinal n we have $\omega + n < \omega 2$. So it goes. There are also uncountable ordinals, the least of which is denoted by Ω. And the ordinals continue on after this too.

Although every ordinal number has an immediate successor, there are some ordinals that don't have any immediate predecessors. These ordinals are called *limit ordinals* because they are defined as "limits" or unions of all their predecessors. The limit ordinals that we've seen are 0, ω, $\omega 2$, $\omega 3$, ..., ω^2, ..., Ω, ...

An interesting fact about ordinal numbers states that for any set S there is a bijection between S and some ordinal number. For example, the set $\{a, b, c\}$ is bijective with the ordinal number $3 = \{0, 1, 2\}$. For another example there are bijections between the set $\mathbb{N}$ of natural numbers and each of the ordinals ω, $\omega + 1$, $\omega + 2$, ... Some people define the cardinality of a set to be the least ordinal number that is bijective to the set. So we have $|\{a, b, c\}| = 3$ and $|\mathbb{N}| = \omega$.

More information about ordinal numbers—including ordinal arithmetic—can be found in the excellent book by Halmos [1960].

Exercises

1. Sometimes our intuition about a symbol can be challenged. For example, suppose we define the relation $\prec$ on the integers by saying that $x \prec y$ means $|x| < |y|$. Assign the value true or false to each of the following statements.

    a. $-7 \prec 7$.   b. $-7 \prec -6$.   c. $6 \prec -7$.   d. $-6 \prec 2$.

2. State whether each of the following relations is a partial order.

    a. IsFatherOf.   b. IsAncestorOf.   c. IsOlderThan.

    d. IsSisterOf   e. $\{\langle a, b \rangle, \langle a, a \rangle, \langle b, a \rangle\}$.   f. $\{\langle 2, 1 \rangle, \langle 1, 3 \rangle, \langle 2, 3 \rangle\}$.

3. Draw a poset diagram for each of the following partially ordered relations.

    a. $\{\langle a, a \rangle, \langle a, b \rangle, \langle b, c \rangle, \langle a, c \rangle, \langle a, d \rangle\}$.

    b. power($\{a, b, c\}$), with the subset relation.

    c. Lists[$\{a, b\}$], where $L \prec M$ if length(L) < length(M).

d. The set of all binary trees over the set {a, b} that contain either one or two nodes. Let $s \prec t$ mean that s is either the left or right subtree of t.

4. Suppose we wish to evaluate the following expression as a set of time-oriented tasks:

$$(f(x) + g(x))(f(x)g(x)).$$

We'll order the subexpressions by data dependency. In other words, an expression can't be evaluated until its data are available. So the subexpressions that occur in the evaluation process are

$$x, f(x), g(x), f(x) + g(x), f(x)g(x), \text{ and } (f(x) + g(x))(f(x)g(x)).$$

Draw the poset diagram for the set of subexpressions. Is the poset a lattice?

5. For any positive integer n, let D_n be the set of positive divisors of n. The poset $\langle D_n, | \rangle$ is a lattice. Describe the glb and lub for any pair of elements.

6. Why is it true that every partially ordered relation over a finite set is well-founded?

7. For each set S, show that the given partial order on S is well-founded.

   a. Let S be a set of trees. Let $s \prec t$ mean that s has fewer nodes than t.
   b. Let S be a set of trees. Let $s \prec t$ mean that s has fewer leaves than t.
   c. Let S be a set of lists. Let $L \prec M$ mean that $\text{length}(L) < \text{length}(M)$.

8. Example 8 discussed a well-founded ordering for the set $\mathbb{N} \times \mathbb{N}$. Use this ordering to construct two distinct descending chains that start at the pair $\langle 4, 3 \rangle$, both of which have maximum length.

9. Suppose we define the relation R on $\mathbb{N} \times \mathbb{N}$ as follows:

$$\langle a, b \rangle \prec \langle c, d \rangle \quad \text{if and only if} \quad \max\{a, b\} < \max\{c, d\}.$$

Is $\mathbb{N} \times \mathbb{N}$ well-founded with respect to $\prec$?

10. Show that the two properties irreflexive and transitive imply the antisymmetric property. So an irreflexive partial order can be defined by just the two properties irreflexive and transitive.

11. Trace the topological sort algorithm (4.12) for the pancake recipe in Example 1 by starting with the source 1. There are several possible answers because any source can be output by the algorithm.

12. Describe a way to perform a topological sort that uses an adjacency matrix to represent the partial order.

13. Prove the two statements of (4.13).

14. For a poset P a function $f : P \to P$ is said to be *monotonic* if $x \preceq y$ implies $f(x) \preceq f(y)$ for all $x, y \in P$. For each poset and function definition, determine whether the function is monotonic.

    a. $\langle \mathbb{N}, < \rangle$, $f(x) = 2x + 3$.   b. $\langle \mathbb{N}, < \rangle$, $f(x) = x^2$.

    c. $\langle \mathbb{Z}, < \rangle$, $f(x) = x^2$.   d. $\langle \mathbb{N}, \,|\, \rangle$, $f(x) = 2x + 3$.

    e. $\langle \mathbb{N}, \,|\, \rangle$, $f(x) = x^2$. f. $\langle \mathbb{N}, \,|\, \rangle$, $f(x) = x \bmod 5$.

    g. $\langle \text{power}(A), \subset \rangle$ for some set A, $f(X) = A - X$.

    h. $\langle \text{power}(\mathbb{N}), \subset \rangle$, $f(X) = \{n \in \mathbb{N} \mid n \,|\, x \text{ for some } x \in X\}$.

4.4 Inductive Proof

In computer science we deal not only with numbers, but also with structures such as strings, lists, trees, graphs, programs, and more complicated structures constructed from them. Do the objects that we construct have the properties that we expect? Does a program halt when it's supposed to halt and give the proper answer? These are major questions in computer science.

To answer these questions, we must find ways to reason about the objects that we construct. This section concentrates on a powerful proof technique that can be used to prove properties of objects constructed by the techniques that we introduced in the last chapter.

The Idea of Induction

Suppose we want to find the sum of numbers $1 + 2 + \cdots + n$ for any natural number n. Consider the following two programs written by two different students to calculate this sum:

$$f(n) = \text{if } n = 0 \text{ then } 0 \text{ else } n + f(n - 1),$$

$$g(n) = \frac{n\,(n + 1)}{2}.$$

Are these programs correct? That is, do they both compute the correct value of the sum $1 + 2 + \cdots + n$? We can test a few cases such as $n = 0, n = 1, n = 2$ until we feel confident that the programs are correct. Or maybe we just can't get any feeling of confidence in these programs. Is there a way to prove, once and for all, that these programs are correct for all natural numbers n? Let's look at the second program. If it's correct, then the following equation must be true for all natural numbers n:

$$1 + 2 + \cdots + n = \frac{n(n+1)}{2}.$$

Certainly we don't have the time to check it for the infinity of natural numbers. Is there some other way to prove it? Happily, we will be able to prove the infinitely many cases in just two steps with a technique called proof by induction, which we discuss next. If you don't want to see why it works, you can skip ahead to (4.18).

Interestingly, the technique that we present is based on the fact that any nonempty subset of the natural numbers has a least element. Recall that this is the same as saying that any descending chain of natural numbers is finite. In fact, this is just a statement that $\mathbb{N}$ is a well-founded set. In fact we can generalize a bit. Let m be an integer, and let W be the following set:

$$W = \{n \mid n \text{ is an integer and } n \geq m\}.$$

Every nonempty subset of W has a least element. Let's see whether this property can help us find a tool to prove infinitely many things in just two steps. First, we state the following result, which forms a basis for the inductive proof technique:

A Basis of Mathematical Induction (4.17)

Let $m \in \mathbb{Z}$ and $W = \{n \mid n \in \mathbb{Z} \text{ and } n \geq m\}$. Let S be a nonemptysubset of W such that the following two conditions hold:

1. $m \in S$.
2. Whenever $k \in S$, then $k + 1 \in S$.

Then $S = W$.

Proof: We'll prove $S = W$ by contradiction. Suppose $S \neq W$. Then $W - S$ has a least element x because every nonempty subset of W has a least element. The first condition of (4.17) tells us that $m \in S$. So it follows that $x > m$. Thus $x - 1 \geq m$, and it follows that $x - 1 \in S$. Thus we can apply the second condition to obtain $(x - 1) + 1 \in S$. In other words, we are forced to conclude that $x \in S$. This is a contradiction, since we can't have both $x \in S$ and $x \in W - S$ at the same time. Therefore $S = W$.   QED.

We should note that there is an alternative way to think about (4.17). First, notice that W is an inductively defined set. The basis case is $m \in W$. The inductive step states that whenever $k \in W$, then $k + 1 \in W$. Now we can appeal to the closure part of an inductive definition, which can be stated as follows: If S is a subset of W and S satisfies the basis and inductive steps for W, then $S = W$. From this point of view, (4.17) is just a restatement of the closure part of the inductive definition of W.

Let's put (4.17) into a practical form that can be used as a proof technique for proving that infinitely many cases of a statement are true. The technique is called the principle of mathematical induction, which we state as follows:

The Principle of Mathematical Induction   (4.18)

Let $m \in \mathbb{Z}$. To prove that $P(n)$ is true for all integers $n \geq m$, perform the following two steps:

1. Prove that $P(m)$ is true.
2. Assume that $P(k)$ is true for an arbitrary $k \geq m$. Prove that $P(k + 1)$ is true.

Proof: Let $W = \{n \mid n \geq m\}$, and let $S = \{n \mid n \geq m \text{ and } P(n) = \text{true}\}$. Assume that we have performed the two steps of (4.18). Then S satisfies the hypothesis of (4.17). Therefore $S = W$. So $P(n)$ is true for all $n \geq m$.   QED.

The principle of mathematical induction contains a technique to prove that infinitely many statements are true in just two steps. Quite a savings in time. Let's look at an example. This proof technique is just what we need to prove our opening example about computing the sum of the first n natural numbers.

EXAMPLE 1 (*A Correct Closed Form*). Let's prove once and for all that the following equation is true for all natural numbers n:

$$1 + 2 + \cdots + n = \frac{n(n+1)}{2}.$$ (4.19)

To see how to use (4.18), we can let $P(n)$ denote the above equation. Now we need to perform two steps. First, we have to show that $P(0)$ is true. Second, we have to assume that $P(k)$ is true and then prove that $P(k + 1)$ is true. When $n = 0$, equation (4.19) becomes the true statement

$$0 = \frac{0(0+1)}{2}.$$

Therefore $P(0)$ is true. Now assume that $P(k)$ is true. This means that we assume that the following equation is true:

$$1 + 2 + \cdots + k = \frac{k\,(k+1)}{2}.$$

To prove that $P(k+1)$ is true, start on the left side of the equation for the expression $P(k+1)$:

$$1 + 2 + \cdots + k + (k+1) = (1 + \cdots + k) + (k+1) \qquad \text{(associate)}$$

$$= \frac{k\,(k+1)}{2} + (k+1) \qquad \text{(assumption)}$$

$$= \frac{(k+1)\,((k+1)+1)}{2} \qquad \text{(arithmetic)}.$$

The last term is the right-hand side of $P(k+1)$. Therefore $P(k+1)$ is true. So we have performed both steps of (4.18). It follows that $P(n)$ is true for all $n \in \mathbb{N}$. In other words, equation (4.19) is true for all natural numbers $n \geq 0$. QED. ◆

EXAMPLE 2 (*A Correct Summation Program*). We show that the following if-then-else program computes the sum $1 + 2 + \cdots + n$ for all natural numbers n:

$$f(n) = \text{if } n = 0 \text{ then } 0 \text{ else } f(n-1) + n.$$

Proof: For each $n \in \mathbb{N}$, let $P(n) = $ "$f(n) = 1 + 2 + \cdots + n$." We want to show that $P(n)$ is true for all $n \in \mathbb{N}$. To start, notice that $f(0) = 0$. Thus $P(0)$ is true. Now assume that $P(k)$ is true for some $k \in \mathbb{N}$. Now we must furnish a proof that $P(k+1)$ is true. Starting on the left side of the statement for $P(k+1)$, we can proceed as follows:

$$
\begin{aligned}
f(k+1) &= f(k+1-1) + (k+1) &&\text{(definition of } f) \\
&= f(k) + (k+1) \\
&= (1 + 2 + \cdots + k) + (k+1) &&\text{(induction assumption)} \\
&= 1 + 2 + \cdots + (k+1).
\end{aligned}
$$

So if $P(k)$ is true, then $P(k+1)$ is true. Therefore by (4.18), $P(n)$ is true for all $n \in \mathbb{N}$. In other words, $f(n) = 1 + 2 + \cdots + n$ for all $n \in \mathbb{N}$.   QED.   ◆

There are usually two parts to solving a problem. The first is to guess at a solution, and the second is to verify that the guess is correct. For example, when we write programs, we are often guessing at solutions to problems.

Sometimes, the way we program (i.e., the way we express our guesses) influences whether it's easy to verify that our programs (guesses) are correct.

An important part of computer science is analyzing the efficiency of programs. We often run into summations when trying to find the number of operations performed by the execution of a program. Sometimes we don't know (or just plain forgot) a closed form for a particular sum. If we can't find the answer anywhere, then our only recourse is to guess at a closed form and then attempt to prove that it's a correct guess. Let's look at some methods to find closed forms for two well-known sums and then prove the results by induction.

When Gauss—mathematician Karl Friedrich Gauss (1777–1855)—was a 10-year-old boy, his schoolmaster, Buttner, gave the class an arithmetic progression of numbers to keep them busy. Gauss wrote down the answer just after Buttner finished writing the problem. Although the formula was known to Buttner, no boy of 10 had ever discovered it. For example, suppose we want to add up the numbers

$$21 + 36 + 51 + \cdots + 216 + 231 + 246,$$

where each number differs from its successor by a constant (15 in this case). The trick is to notice that the sum of the first and last numbers, which is 267, is the same as the sum of the second and next to last numbers, and so on:

$$(21 + 246) = (36 + 231) = (51 + 216) = \cdots = 267.$$

Then notice that there are 16 numbers. So there are eight pairs to add up:

$$8(267) = 2136.$$

We can formulate the general sum of an *arithmetic progression* of n numbers by the following formula:

$$a_1 + a_2 + \cdots + a_n = \frac{n}{2}(a_1 + a_n). \tag{4.20}$$

Do you believe this formula? Do we always get a whole number from the formula on the right? To see that the formula is correct, we can try to prove it by induction. But what set do we induct on? The subscripts give the answer. They take values in the set $\{1, 2, 3, \ldots\}$ of positive natural numbers. Let $P(n)$ denote equation (4.20). We want to show that $P(n)$ is true for all natural numbers $n \geq 1$.

Proof: We need to show that $P(1)$ is true. Then we'll assume that $P(n)$ is true and show that $P(n + 1)$ is true. Starting with $P(1)$, we obtain the statement

$$a_1 = \frac{1}{2}(a_1 + a_1).$$

Since this equation is true, we have $P(1)$ is true. Next assume that $P(n)$ is true, as stated in (4.20). Then try to prove the statement $P(n + 1)$ below:

$$a_1 + a_2 + \cdots + a_n + a_{n+1} = \frac{n+1}{2}(a_1 + a_{n+1}).$$

Starting with the left-hand side of the equation, we obtain

$$a_1 + a_2 + \cdots + a_n + a_{n+1} = (a_1 + a_2 + \cdots + a_n) + a_{n+1}$$
$$= \frac{n}{2}(a_1 + a_n) + a_{n+1}$$
$$= \frac{n}{2}a_1 + \frac{n}{2}a_n + a_{n+1}$$
$$= ?$$

What now? Somehow we would like to continue and wind up with the expression $\frac{n+1}{2}(a_1 + a_{n+1})$. To do this we can rewrite the term $\frac{n}{2}a_n$ in terms of a_1 and a_{n+1}. Since the progression is arithmetic, we can write $a_n = a_1 + (n-1)c$, where c is the constant difference between successive terms. So we can substitute $c = a_{n+1} - a_n$ to obtain the equation $a_n = a_1 + (n-1)(a_{n+1} - a_n)$. Divide this equation by 2 and collect terms in a_n to obtain the equation

$$\frac{n}{2}a_n = \frac{1}{2}a_1 + \frac{n-1}{2}a_{n+1}.$$

Now we can continue where we left off at the question mark. Using the preceding equation, we can substitute for $\frac{n}{2}a_n$ to obtain the desired expression,

$$\frac{n+1}{2}(a_1 + a_{n+1}).$$

So we have performed the two steps of (4.18), and it follows that equation (4.20) is correct for all arithmetic progressions of n numbers where $n \geq 1$. QED.

Of course, an important special case of an arithmetic progression is the sum of the first n natural numbers (4.19), which we have already discussed. Another important sum is a *geometric progression* of numbers of the following form

$$1 + x + x^2 + \cdots + x^n,$$

where x is any number and n is a natural number. A formula for this sum can be found by multiplying the given expression by the term $x - 1$ to obtain the equation

$$(x - 1)(1 + x + x^2 + \ldots + x^n) = x^{n+1} - 1.$$

Now divide both sides by $x - 1$ to obtain the formula

$$1 + x + x^2 + \ldots + x^n = \frac{x^{n+1} - 1}{x - 1}. \tag{4.21}$$

The formula works for all $x \neq 1$. An induction proof of (4.21) is also instructive.

Proof: If $n = 0$, then both sides are 1. So assume that (4.21) is true for n, and prove that it is true for $n + 1$. Starting with the left-hand side, we have

$$1 + x + x^2 + \cdots + x^n + x^{n+1} = (1 + x + x^2 + \cdots + x^n) + x^{n+1}$$

$$= \frac{x^{n+1} - 1}{x - 1} + x^{n+1}$$

$$= \frac{x^{n+1} - 1 + (x - 1)x^{n+1}}{x - 1}$$

$$= \frac{x^{(n+1)+1} - 1}{x - 1}.$$

Thus by (4.18) the formula (4.21) is true for all natural numbers k.   QED.

Sometimes, (4.18) does not have enough horsepower to do the job. For example, we might need to assume more than (4.18) will allow us, or we might be dealing with structures that are not numbers, such as lists, strings, or binary trees, and there may be no easy way to apply (4.18). The solution to many of these problems is a stronger version of induction based on well-founded sets. That's next.

Well-Founded Induction

Let's extend the idea of induction to well-founded sets. Recall that a well-founded set is a poset whose nonempty subsets have minimal elements or, equivalently, every descending chain of elements is finite. We'll start by noticing an easy extension of (4.17) to the case of well-founded sets. If you aren't interested in why the method works, you can skip ahead to (4.23).

The Basis of Well-Founded Induction   (4.22)

Let W be a well-founded set, and let S be a nonempty subset of W satisfying the following two conditions:

1. S contains all the minimal elements of W.
2. Whenever an element $x \in W$ has the property that all its predecessors are elements of S, then $x \in S$.

Then $S = W$.

Proof: The proof is by contradiction. Suppose $S \neq W$. Then $W - S$ has a minimal element x. Since x is a minimal element of $W - S$, each predecessor of x cannot be in $W - S$. In other words, each predecessor of x must be in S. The second condition in the hypothesis of the theorem now forces us to conclude that $x \in S$. This is a contradiction, since we can't have both $x \in S$ and $x \in W - S$ at the same time. Therefore $S = W$.   QED.

You might notice that condition 1 of (4.22) was not used in the proof. This is because it's a consequence of condition 2 of (4.22). We'll leave this as an exercise (something about an element that doesn't have any predecessors). Condition 1 is stated explicitly because it indicates the first thing that must be done in an inductive proof.

Practical Techniques

Let's find a more practical form of (4.22) that gives us a technique for proving a collection of statements of the form $P(x)$ for each x in a well-founded set W. The technique is called well-founded induction.

Well-Founded Induction   (4.23)

Suppose $P(x)$ is a statement for each x in the well-founded set W. To prove that $P(x)$ is true for all x in W, perform the following two steps:

1. Prove that $P(m)$ is true for all minimal elements $m \in W$.
2. Let x be an arbitrary element of W, and assume that $P(y)$ is true for all elements y that are predecessors of x. Prove that $P(x)$ is true.

Proof: Let $S = \{x \mid x \in W$ and $P(x) = $ true$\}$. Assume that we have performed the two steps of (4.23). Then S satisfies the hypothesis of (4.22). Therefore $S = W$. In other words, $P(x)$ is true for all $x \in W$.   QED.

Now we can state a corollary of (4.23), which lets us make a bigger assumption than we were allowed in (4.18):

Second Principle of Mathematical Induction   (4.24)

Let $m \in \mathbb{Z}$. To prove that $P(n)$ is true for all integers $n \geq m$, perform the following two steps:

1. Prove that $P(m)$ is true.
2. Assume that n is an arbitrary integer $n > m$, and assume that $P(k)$ is true for all k, $m \leq k < n$. Prove that $P(n)$ is true.

Proof: Let $W = \{n \mid n \geq m\}$. Notice that W is a well-founded set (actually well-ordered) whose least element is m. Let $S = \{n \mid n \in W$ and $P(n)$ is true$\}$. Assume that steps 1 and 2 have been performed. Then $m \in S$, and if $n > m$ and all predecessors of n are in S, then $n \in S$. Therefore $S = W$, by (4.23).   QED.

As an example, let's prove that every natural number greater than 1 is a product of prime numbers (1.2a).

EXAMPLE 3 (*Prime Number Theorem*). We want to prove that every natural number n $\geq$ 2 is a product of prime numbers.

Proof: For $n \geq 2$, let $P(n)$ be the statement "n is a product of prime numbers." We need to show that $P(n)$ is true for all $n \geq 2$. Since 2 is prime, it follows that $P(2)$ is true. So Step 1 of (4.24) is finished. For Step 2 we'll assume that $n > 2$ and $P(k)$ is true for $2 \leq k < n$. With this assumption we must show that $P(n)$ is true. If n is prime, then $P(n)$ is true. So assume that n is not prime. Then $n = xy$, where $2 \leq x < n$ and $2 \leq y < n$. By our assumption, $P(x)$ and $P(y)$ are both true, which means that x and y are products of primes. Therefore n is a product of primes. So $P(n)$ is true. Now (4.24) implies that $P(n)$ is true for all $n \geq 2$. *Note*: We can't use (4.18) for this proof because its induction assumption is the single statement that $P(n - 1)$ is true. We need the stronger assumption that $P(k)$ is true for $2 \leq k < n$ to allow us to say that $P(x)$ and $P(y)$ are true.   QED.   ◆

Let's pause and make a few comments about inductive proof. Remember, when you are going to prove something with an inductive proof technique, there are always two distinct steps to be performed. First prove the basis case, showing that the statement is true for each minimal element. Now comes the second step. The most important part about this step is making an assumption. Let's write it down for emphasis:

You are required to make an assumption in the inductive step of a proof.

Some people find it hard to make assumptions. But inductive proof techniques require it. So if you find yourself wondering about what to do in an inductive proof, here are two questions to ask yourself: "Have I made an induction assumption?" If the answer is yes, ask the question, "Have I used the induction assumption in my proof?" Let's write it down for emphasis:

In the inductive step, MAKE AN ASSUMPTION and then USE IT.

Look at the previous examples, and find the places where the basis case was proved, where the assumption was made, and where the assumption was used. Do the same thing as you read through the remaining examples.

Now let's do some examples that do not involve numbers. Thus we'll be using well-founded induction (4.23). We should note that some people refer to well-founded induction as "structural induction" because well-founded sets can contain structures other than numbers, such as lists, strings, binary trees, and products of sets. Whatever it's called, let's see how to use it.

EXAMPLE 4 (*Correctness of MakeSet*). The following function is supposed to take any list K as input and return the list obtained by removing all redundant elements from K:

$$\text{makeSet}(\langle\,\rangle) \quad = \quad \langle\,\rangle,$$
$$\text{makeSet}(a :: L) \quad = \quad \text{if isMember}(a, L) \text{ then makeSet}(L)$$
$$\text{else } a :: \text{makeSet}(L).$$

We'll assume that isMember correctly checks whether an element is a member of a list. Let $P(K)$ be the statement "makeSet(K) is a list obtained from K by removing its redundant elements." Now we'll prove that $P(K)$ is true for any list K.

Proof: We need a well-founded ordering for lists. For any lists K and M we'll define $K \prec M$ to mean length(K) < length(M). So the basis element is $\langle\,\rangle$. The definition of makeSet tells us that makeSet($\langle\,\rangle$) = $\langle\,\rangle$. Thus $P(\langle\,\rangle)$ is true. Next, we'll let K be an arbitrary nonempty list and assume that $P(L)$ is true for all lists $L \prec K$. In other words, we're assuming that makeSet(L) has no redundant elements for all lists $L \prec K$. We need to show that $P(K)$ is true. In other words, we need to show that makeSet(K) has no redundant elements. Since K is nonempty, we can write $K = a :: L$. There are two cases to consider. If isMember(a, L) is true, then the definition of makeSet gives

$$\text{makeSet}(K) = \text{makeSet}(a :: L) = \text{makeSet}(L).$$

Since $L \prec K$, it follows that $P(L)$ is true. Therefore $P(K)$ is true. If isMember(a, L) is false, then the definition of makeSet gives

$$\text{makeSet}(K) = \text{makeSet}(a :: L) = a :: \text{makeSet}(L).$$

Since $L \prec K$, it follows that $P(L)$ is true. Since isMember(a, L) is false, it follows that the list $a :: \text{makeSet}(L)$ has no redundant elements. Thus $P(K)$ is true. Therefore (4.23) implies that $P(K)$ is true for all lists K.   QED.   ♦

Multiple-Variable Induction

Up to this point, all our inductive proofs have involved claims and formulas with only a single variable. Often the claims that we wish to prove involve two or more variables. For example, suppose we need to show that $P(x, y)$ is true for all $\langle x, y \rangle \in A \times B$. We saw in Chapter 3 that $A \times B$ can be defined inductively in different ways. For example, it may be possible to emphasize only the inductive nature of the set A by defining $A \times B$ in terms of A. To show that $P(x, y)$ is true for all $\langle x, y \rangle$ in $A \times B$, we can perform the following steps (where y denotes an arbitrary element in B):

1. Show that $P(m, y)$ is true for minimal elements $m \in A$.
2. Assume that $P(a, y)$ is true for all predecessors a of x. Then show that $P(x, y)$ is true.

This technique is called "inducting on a single variable." The form of the statement $P(x, y)$ often gives us a clue to whether we can induct on a single variable. Here are some examples.

EXAMPLE 5. Suppose we want to prove that the following function computes the number y^{x+1} for any natural numbers x and y:

$$f(x, y) = \text{if } x = 0 \text{ then } y \text{ else } f(x - 1, y) * y.$$

In other words, we want to prove that $f(x, y) = y^{x+1}$ for all $\langle x, y \rangle$ in $\mathbb{N} \times \mathbb{N}$. We'll induct on the variable x because it's changing in the definition of f.

Proof: For the basis case the definition of f gives $f(0, y) = y = y^{0+1}$. So the basis case is proved. For the induction case, assume that $x > 0$ and $f(n, y) = y^{n+1}$ for $n < x$. We must show that $f(x, y) = y^{x+1}$. The definition of f and the induction assumption give us the following equation:

$$\begin{aligned} f(x, y) &= f(x - 1, y) * y && \text{(definition of } f) \\ &= y^{x-1+1} * y && \text{(induction assumption)} \\ &= y^{x+1}. \end{aligned}$$

The result now follows from (4.23).   QED.   ◆

EXAMPLE 6 (*Inserting an Element in a Binary Search Tree*). Let's prove that the following "insert" function does its job. Given a number x and a binary search tree T, the function returns a binary search tree obtained by inserting x in T.

$$\begin{aligned} \text{insert}(x, T) \quad = \quad &\text{if } T = \langle \rangle \text{ then tree}(\langle \rangle, x, \langle \rangle) \\ &\text{else if } x < \text{root}(T) \text{ then} \\ &\quad \text{tree}(\text{insert}(x, \text{left}(T)), \text{root}(T), \text{right}(T)) \\ &\text{else} \\ &\quad \text{tree}(\text{left}(T), \text{root}(T), \text{insert}(x, \text{right}(T))). \end{aligned}$$

The claim that we wish to prove is

insert(x, T) is a binary search tree for all binary search trees T.

Proof: We'll induct on the binary tree variable. Our ordering of binary search trees will be based on the number of nodes in a tree. For the basis case we must show that insert$(x, \langle \rangle)$ is a binary search tree. Since insert$(x, \langle \rangle)$ = tree($\langle \rangle, x, \langle \rangle$) and a single node tree is a binary search tree, the basis case is true. Next, let T = tree(L, y, R) be a binary search tree, and assume that insert(x, L) and insert(x, R) are binary search trees. Then we must show that insert(x, T) is a binary search tree. There are two cases to consider, depending on whether $x < y$. First, suppose $x < y$. Then we have

insert(x, T) = tree(insert$(x, L), y, R$).

By the induction assumption it follows that insert(x, L) is a binary search tree. Thus insert(x, T) is a binary search tree. We obtain a similar result if $x \geq y$. It follows from (4.23) that insert(x, T) is a binary search tree for all binary search trees T.   QED.   ◆

We often see induction proofs that don't mention the word "well-founded." For example, we might see a statement such as "We will induct on the depth of the trees." In such a case the induction assumption might be stated something like "Assume that $P(T)$ is true for all trees T with depth less than n." Then a proof is given that uses the assumption to prove that $P(T)$ is true for an arbitrary tree of depth n. Even though the term "well-founded" may not be mentioned in a proof, there is always a well-founded ordering lurking underneath the surface.

Before we leave the subject of inductive proof, let's discuss how we can use inductive proof to help us tell whether inductive definitions of sets are correct.

Proofs About Inductively Defined Sets

Recall that a set S is inductively defined by a basis case, an inductive case, and a closure case (which we never state explicitly). The closure case says that S is the smallest set satisfying the basis and inductive cases. The closure case can also be stated in practical terms as follows:

Closure Property of Inductive Definitions   (4.25)

If S is an inductively defined set and T is a set that also satisfies the basis and inductive cases for the definition of S, and if $T \subset S$, then it must be the case that $T = S$.

We can use this closure property to see whether an inductive definition correctly defines a given set. For example, suppose we have an inductive definition for a set named S, we have some other description of a set named T, and we wish to prove that T and S are the same set. Then we must prove three things:

1. Prove that T satisfies the basis case of the inductive definition.
2. Prove that T satisfies the inductive case of the inductive definition.
3. Prove that $T \subset S$. This can often be accomplished with an induction proof.

Let's do an example. Suppose we write down the following inductive definition for a set S:

Basis: $1 \in S$.

Induction: If $x \in S$, then $x + 2 \in S$.

This gives us a pretty good description of S. For example, suppose someone tells us that $S = \{2k + 1 \mid k \in \mathbb{N}\}$. It seems reasonable. Can we prove it? Let's give it a try. To clarify the situation, we'll let $T = \{2k + 1 \mid k \in \mathbb{N}\}$ and prove that $T = S$. We'll be done if we can show that T satisfies the basis and induction cases for S and that $T \subset S$. Then the closure property of inductive definitions will tell us that $T = S$.

Proof: The basis case of the inductive definition holds for T because 1 can be written as $1 = 2 \cdot 0 + 1 \in T$. For the induction case, assume that $x = 2k + 1 \in T$. Then we can write $x + 2 = 2(k + 1) + 1 \in T$. Therefore T satisfies the basis and induction cases of the inductive definition. Now we must prove that $T \subset S$. For this proof we'll use an induction proof. In other words, we'll prove that $2k + 1 \in S$ for all $k \in \mathbb{N}$. For $k = 0$ we have $2 \cdot 0 + 1 = 1 \in S$. Now we'll assume that $2k + 1 \in S$ and try to prove that $2(k + 1) + 1 \in S$. Since $2k + 1 \in S$, the definition of S tells us that $(2k + 1) + 2 \in S$. But we can write $2(k + 1) + 1 = (2k + 1) + 2 \in S$. Therefore $2k + 1 \in S$ for all $k \in \mathbb{N}$, which proves that $T \subset S$. So we've proven the three things that allow us to conclude—by the closure property of inductive definitions—that $T = S$.   QED.

Here's an example involving languages and grammars.

EXAMPLE 7 (*A Correct Grammar*). Suppose we're asked to find a grammar for the language $\{ab^n \mid n \in \mathbb{N}\}$, and we come up with the grammar G:

$$S \rightarrow a \mid Sb.$$

This grammar seems to do the job. But how do we know for sure? One way is to use (3.12) to create an inductive definition for $L(G)$, the language of G. Then we can try to prove that $L(G) = \{ab^n \mid n \in \mathbb{N}\}$. Using (3.12) we see that the basis case is $a \in L(G)$ because of the derivation $S \Rightarrow a$. For the induction case, if there is a string $x \in L(G)$ with derivation $S \Rightarrow^+ x$, then we can add one step to the derivation by using the recursive production $S \rightarrow Sb$ to obtain the derivation $S \Rightarrow Sb \Rightarrow^+ xb$. So we obtain the following inductive definition for $L(G)$:

Basis: $a \in L(G)$.

Induction: If $x \in L(G)$, then put xb in $L(G)$.

Since we have an inductive definition for $L(G)$, we can try to prove that $\{ab^n \mid n \in \mathbb{N}\} = L(G)$. For ease of notation we'll let $M = \{ab^n \mid n \in \mathbb{N}\}$. So we'll try to prove that $M = L(G)$. By (4.25) we must show that M satisfies the basis and induction cases and that $M \subset L(G)$. Then we can apply the closure property of inductive definitions to conclude that $M = L(G)$.

Proof: The basis case of the inductive definition holds for M because $a = ab^0 \in M$. For the induction case, let $x \in M$. Then $x = ab^n$ for some $n \in \mathbb{N}$. Thus $xb = ab^{n+1} \in M$. Therefore M satisfies the basis and induction cases of the inductive definition. Now we need to show that $M \subset L(G)$. We can do this with an induction proof. In other words, we'll show that $ab^n \in L(G)$ for all $n \in \mathbb{N}$. For $n = 0$ we have $ab^0 = a \in L(G)$. Now assume that $ab^n \in L(G)$. Then the induction part of the definition of $L(G)$ tells us that $ab^n b \in L(G)$. But $ab^{n+1} = ab^n b$. So $ab^{n+1} \in L(G)$. Therefore we've proven by induction on n that $M \subset L(G)$. Now we can conclude—by the closure property of inductive definitions—that $M = L(G)$.   QED.   ♦

Exercises

1. Find the sum of the numbers 12, 26, 40, 54, 68, ..., 278.

2. Use induction to prove each of the following equations for all natural numbers $n \geq 1$.

   a. $1^2 + 2^2 + \cdots + n^2 = \dfrac{n\,(n+1)\,(2n+1)}{6}$.

   b. $(1 + 2 + \cdots + n)^2 = 1^3 + 2^3 + \cdots + n^3$.

   c. $1 + 3 + \cdots + (2n - 1) = n^2$.

3. The *Fibonacci numbers* are defined by $F_0 = 0$, $F_1 = 1$, and $F_n = F_{n-1} + F_{n-2}$ for $n \geq 2$. Use induction to prove the following statement for all natural numbers n: $F_0 + F_1 + \ldots + F_n = F_{n+2} - 1$.

4. The *Lucas numbers* are defined by $L_0 = 2$, $L_1 = 1$, and $L_n = L_{n-1} + L_{n-2}$ for $n \geq 2$. The sequence begins as 2, 1, 3, 4, 7, 11, 18, ... These numbers are named after the mathematician Édouard Lucas (1842–1891). Use induction to prove each of the following statements.

   a. $L_0 + L_1 + \ldots + L_n = L_{n+2} - 1$.

   b. $L_n = F_{n-1} + F_{n+1}$ for $n \geq 1$, where F_n is the nth Fibonacci number.

5. Let $\mathrm{sum}(n) = 1 + 2 + \cdots + n$ for all natural numbers n. Give an induction proof to show that the following equation is true for all natural numbers m and n: $\mathrm{sum}(m + n) = \mathrm{sum}(m) + \mathrm{sum}(n) + mn$.

6. We know that $1 + 2 = 3$, $4 + 5 + 6 = 7 + 8$, and $9 + 10 + 11 + 12 = 13 + 14 + 15$. Show that we can continue these equations forever. *Hint:* The left side of each equation starts with a number of the form n^2. Formulate a general summation for each side, and then prove that the two sums are equal.

7. Use induction to prove that a finite set with n elements has 2^n subsets.

8. Use induction to prove that the function f computes the length of a list:

$$f(L) = \text{if } L = \langle\,\rangle \text{ then } 0 \text{ else } 1 + f(\text{tail}(L)).$$

9. Use induction to prove that each function performs its stated task.

    a. The function g computes the number of nodes in a binary tree:

    $$g(T) \quad = \quad \text{if } T = \langle\,\rangle \text{ then } 0$$
    $$\text{else } 1 + g(\text{left}(T)) + g(\text{right}(T)).$$

    b. The function h computes the number of leaves in a binary tree:

    $$h(T) \quad = \quad \text{if } T = \langle\,\rangle \text{ then } 0$$
    $$\text{else if } T = \text{tree}(\langle\,\rangle, x, \langle\,\rangle) \text{ then } 1$$
    $$\text{else } h(\text{left}(T)) + h(\text{right}(T)).$$

10. Suppose we have the following two procedures to write out the elements of a list. One claims to write the elements in the order listed, and one writes out the elements in reverse order. Prove that each is correct.

    a. forward(L): if $L \neq \langle\,\rangle$ then {print(head(L)); forward(tail(L))}.

    b. back(L): if $L \neq \langle\,\rangle$ then {back(tail(L)); print(head(L))}.

11. The following function "sort" takes a list of numbers and returns a sorted version of the list (from lowest to highest), where "insert" places an element correctly into a sorted list:

    $$\text{sort}(\langle\,\rangle) = \langle\,\rangle,$$
    $$\text{sort}(x :: L) = \text{insert}(x, \text{sort}(L)).$$

    a. Assume that the function insert is correct. That is, if S is sorted, then insert(x, S) is also sorted. Prove that sort is correct.

    b. Prove that the following definition for insert is correct. That is, prove that insert(x, S) is sorted for all sorted lists S.

    $$\text{insert}(x, S) \quad = \quad \text{if } S = \langle\,\rangle \text{ then } \langle x \rangle$$
    $$\text{else if } x \leq \text{head}(S) \text{ then } x :: S$$
    $$\text{else head}(S) :: \text{insert}(x, \text{tail}(S)).$$

12. Show that the following function g correctly computes the greatest common divisor for each pair of positive integers x and y: *Hint:* (2.1b) might be useful.

    $$g(x, y) \quad = \quad \text{if } x = y \text{ then } x$$
    $$\text{else if } x > y \text{ then } g(x - y, y)$$
    $$\text{else } g(x, y - x).$$

13. The following program is supposed to input a list of numbers L and output a binary search tree containing the numbers in L:

$$f(L) = \text{if } L = \langle\,\rangle \text{ then } \langle\,\rangle$$
$$\text{else insert(head}(L), f(\text{tail}(L))).$$

Assume that insert(x, T) correctly returns the binary search tree obtained by inserting the number x in the binary search tree T. Prove the following claim: $f(M)$ is a binary search tree for all lists M.

14. The following program is supposed to return the list obtained by removing the first occurrence of x from the list L:

$$\text{delete}(x, L) = \text{if } L = \langle\,\rangle \text{ then } \langle\,\rangle$$
$$\text{else if } x = \text{head}(L) \text{ then tail}(L)$$
$$\text{else head}(L) :: \text{delete}(x, \text{tail}(L)).$$

Prove that delete performs as expected.

15. The following function claims to remove all occurrences of an element from a list:

$$\text{removeAll}(a, L) = \text{if } L = \langle\,\rangle \text{ then } L$$
$$\text{else if } a = \text{head}(L) \text{ then removeAll}(a, \text{tail}(L))$$
$$\text{else head}(L) :: \text{removeAll}(a, \text{tail}(L)).$$

Prove that removeAll satisfies the claim.

16. Let r stand for the "removeAll" function from Exercise 15. Prove the following property of r for all elements a, b and all lists L:

$$r(a, r(b, L)) = r(b, r(a, L)).$$

17. The following program computes a well-known function called *Ackermann's function*. *Note:* If you try out this function, don't let x and y get too big.

$$f(x, y) = \text{if } x = 0 \text{ then } y + 1$$
$$\text{else if } y = 0 \text{ then } f(x - 1, 1)$$
$$\text{else } f(x - 1, f(x, y - 1)).$$

Prove that f is defined for all pairs $\langle x, y \rangle$ in $\mathbb{N} \times \mathbb{N}$. *Hint:* Use the lexicographic ordering on $\mathbb{N} \times \mathbb{N}$. This gives the single basis element $\langle 0, 0 \rangle$. For the induction assumption, assume that $f(x', y')$ is defined for all $\langle x', y' \rangle$ such that $\langle x', y' \rangle \prec \langle x, y \rangle$. Then show that $f(x, y)$ is defined.

18. Let the function "isMember" be defined as follows for any list L:

$$\text{isMember}(a, L) = \text{if } L = \langle\,\rangle \text{ then False}$$
$$\text{else if } a = \text{head}(L) \text{ then True}$$
$$\text{else isMember}(a, \text{tail}(L)).$$

a. Prove that isMember is correct. That is, show that isMember(a, L) is true if and only if a occurs as an element of L.

b. Prove that the following equation is true for all lists L when $a \neq b$:

$$\text{isMember}(a, \text{removeAll}(b, L)) = \text{isMember}(a, L).$$

19. Use induction to prove that the following concatenation function is associative.

$$\text{cat}(x, y) = \quad \text{if } x = \langle\,\rangle \text{ then } y$$
$$\text{else head}(x) :: \text{cat}(\text{tail}(x), y).$$

In other words, show that $\text{cat}(x, \text{cat}(y, z)) = \text{cat}(\text{cat}(x, y), z)$ for all lists x, y, and z.

20. Two students came up with the following two solutions to a problem. Both students used the "removeAll" function from Exercise 15, which we abbreviate to r.

    Student A: $f(L) = \quad \text{if } L = \langle\,\rangle \text{ then } \langle\,\rangle$
    $$\text{else head}(L) :: r(\text{head}(L), f(\text{tail}(L))).$$

    Student B: $g(L) = \quad \text{if } L = \langle\,\rangle \text{ then } \langle\,\rangle$
    $$\text{else head}(L) :: g(r(\text{head}(L), \text{tail}(L))).$$

    a. Prove that $r(a, g(L)) = g(r(a, L))$ for all elements a and all lists L. *Hint:* Exercise 16 might be useful in the proof.

    b. Prove that $f(L) = g(L)$ for all lists L. *Hint:* Part (a) could be helpful.

    c. Can you find an appropriate name for f and g? Can you prove that the name you choose is correct?

21. Prove that condition 1 of (4.22) is a consequence of condition 2 of (4.22).

22. Let G be the grammar $S \to a \mid abS$, and let $M = \{(ab)^n a \mid n \in \mathbb{N}\}$. Use (3.12) to construct an inductive definition for $L(G)$. Then use (4.25) to prove that $M = L(G)$.

23. A useful technique for recursively defined functions involves keeping—or accumulating—the results of function calls in *accumulating parameters*: The values in the accumulating parameters can then be used to compute subsequent values of the function that are then used to replace the old values in the accumulating parameters. We call the function by giving initial values to the accumulating parameters. Often these initial values are basis values for an inductively defined set of elements.

    For example, suppose we define the function f as follows:

    $$f(n, u, v) = \text{if } n = 0 \text{ then } u \text{ else } f(n - 1, v, u + v).$$

    The second and third arguments to f are accumulating parameters because they always hold two possible values of the function. Prove each of the following statements.

a. $f(n, 0, 1) = F_n$, the nth Fibonacci number.

b. $f(n, 2, 1) = L_n$, the nth Lucas number.

Hint: For part (a), show that $f(n, 0, 1) = f(k, F_{n-k}, F_{n-k+1})$ for $0 \le k \le n$. A similar hint applies to part (b).

Chapter Summary

The binary relation is the common denominator for describing the ideas of equivalence, order, and inductive proof. The basic properties that a binary relation may or may not possess are reflexive, symmetric, transitive, irreflexive, and antisymmetric. Binary relations can be constructed from other binary relations by composition and closure, and by the usual set operations. Transitive closure plays an important part in algorithms for solving path problems—Warshall's algorithm, Floyd's algorithm, and the modification of Floyd's algorithm to find shortest paths.

Equivalence relations are characterized by being reflexive, symmetric, and transitive. These relations generalize the idea of basic equality by partitioning a set into classes of equivalent elements. Any set has a hierarchy of partitions ranging from fine to coarse. Equivalence relations can be generated from other relations by taking the transitive symmetric reflexive closure. They can also be generated from functions by the kernel relation. The equivalence problem can be solved by a novel tree structure. Kruskal's algorithm uses an equivalence relation to find a minimal spanning tree for a weighted undirected graph.

Order relations are characterized by being transitive and antisymmetric. Sets with these properties are called posets—for partially ordered sets—because it may be the case that not all pairs of elements are related. The ideas of successor and predecessor apply to posets. Posets can also be "topologically" sorted. A well-founded poset is characterized by the condition that no descending chain of elements can go on forever. This is equivalent to the condition that any nonempty subset has a minimal element. Well-founded sets can be constructed by mapping objects into a known well-founded set such as the natural numbers. Inductively defined sets are well-founded.

Inductive proof is a powerful technique that can be used to prove infinitely many statements. The most basic inductive proof technique is the principle of mathematical induction. A more useful inductive proof technique is well-founded induction. The important thing to remember about applying inductive proof techniques is to *make an assumption* and then *use the assumption* that you made. Inductive proof techniques can be used to prove properties of recursively defined functions and inductively defined sets.

5

Analysis Techniques

Remember that time is money.
—Benjamin Franklin (1706–1790)

Time and space are important words in computer science because we want fast algorithms and we want algorithms that don't use a lot of memory. The purpose of this chapter is to study some fundamental techniques and tools that can be used to analyze algorithms for the time and space that they require. Although the study of algorithm analysis is beyond our scope, we'll give some examples to show how the process works. After introducing the idea of an optimal algorithm and some examples, we'll concentrate on techniques for counting (permutations, combinations, and finite probability), solving recurrences, and comparing the growth rates of functions.

Chapter Guide

Section 5.1 introduces the optimal algorithm problem. We'll define the worst case performance of an algorithm, and we'll analyze a few example algorithms.

Section 5.2 introduces the basic counting techniques for permutations and combinations. We'll introduce finite probability, and we'll discuss the average case performance of algorithms.

Section 5.3 introduces some techniques for solving recurrences that crop up in the analysis of algorithms. We'll include a short discussion of the powerful technique of generating functions.

Section 5.4 presents some techniques for comparing the rates of growth of functions. We'll apply the results to those functions that describe the approximate running time of algorithms.

5.1 Optimal Algorithms

An important question of computer science is: Can you convince another person that your algorithm is efficient? This takes some discussion. Let's start by stating the following problem.

The Optimal Algorithm Problem

Suppose algorithm A solves problem P. Is A the best solution to P?

What does "best" mean? Two typical meanings are *least time* and *least space*. In either case we still need to clarify what it means for an algorithm to solve a problem in the least time or the least space. For example, an algorithm running on two different machines may take different amounts of time. Do we have to compare A to every possible solution of P on every type of machine? This is impossible. So we need to make a few assumptions in order to discuss the optimal algorithm problem. We'll concentrate on "least time" as the meaning of "best" because time is the most important factor in most computations.

Instead of executing an algorithm on a real machine to find its running time, we'll analyze the algorithm by counting the number of certain operations that it will perform when executed on a real machine. In this way we can compare two algorithms by simply comparing the number of operations of the same type that each performs. If we make a good choice of the type of operations to count, we should get a good measure of an algorithm's performance. For example, we might count addition operations and multiplication operations for a numerical problem. On the other hand, we might choose to count comparison operations for a sorting problem.

The number of operations performed by an algorithm usually depends on the size or structure of the input. The size of the input again depends on the problem. For example, for a sorting problem, "size" usually means the number of items to be sorted. Sometimes inputs of the same size can have different structures that affect the number of operations performed. For example, some sorting algorithms perform very well on an input data set that is all mixed up but perform badly on an input set that is already sorted!

Because of these observations we need to define the idea of a worst case input for an algorithm A. An input of size n is a *worst case input* if, when compared to all other inputs of size n, it causes A to execute the largest number of operations. Now let's get down to business. For any input I we'll denote its size by size(I), and we'll let time(I) denote the number of operations executed by A on I. Then the *worst case function* for A is defined as follows:

$$W_A(n) = \max\{\text{time}(I) \mid I \text{ is an input and size}(I) = n\}.$$

Now let's discuss comparing different algorithms that solve the same problem P. We'll always assume that the algorithms we compare use certain specified operations that we intend to count. If A and B are algorithms that solve P and if $W_A(n) \le W_B(n)$ for all $n > 0$, then we know algorithm A has worst case performance that is better than or equal to that of algorithm B. An algorithm A is *optimal in the worst case* for problem P if, for any algorithm B that exists or ever will exist, the following relationship holds:

$$W_A(n) \le W_B(n) \quad \text{for all } n > 0.$$

How in the world can we ever find an algorithm that is optimal in the worst case for a problem P? The answer involves the following three steps:

1. (Find an algorithm) Find or design an algorithm A to solve P. Then do an analysis of A to find the worst case function W_A.
2. (Find a lower bound) Find a function F such that $F(n) \le W_B(n)$ for all $n > 0$ and for all algorithms B that solve P.
3. Compare F and W_A. If $F = W_A$, then A is optimal in the worst case.

An interesting situation occurs when $F \ne W_A$ in Step 3. This means that $F(n) < W_A(n)$ for some n. In this case there are two possible courses of action to consider:

Try to find a better algorithm than A. In other words, try to find an algorithm C such that $W_C(n) \le W_A(n)$ for all $n > 0$.

Try to find a "better" lower bound than F. In other words, try to find a new function G such that $F(n) \le G(n) \le W_B(n)$ for all $n > 0$ and for all algorithms B that solve P.

We should note that the zero function is always a lower bound, but it's not very interesting because most algorithms take more than zero time. A few problems have optimal algorithms. For the vast majority of problems that have solutions, optimal algorithms have not yet been found. The examples contain both kinds of problems.

EXAMPLE 1 (*Matrix Multiplication*). We can "multiply" two n by n matrices A and B to obtain the product AB, which is the n by n matrix defined by letting the element in the ith row and jth column of AB be the value of the expression $\sum_{k=1}^{n} A_{ik} B_{kj}$. In other words, for $1 \le i \le n$ and $1 \le j \le n$ we have

$$(AB)_{ij} = \sum_{k=1}^{n} A_{ik} B_{kj}.$$

For example, let A and B be the following 2 by 2 matrices:

$$A = \begin{bmatrix} a & b \\ c & d \end{bmatrix}, \qquad B = \begin{bmatrix} e & f \\ g & h \end{bmatrix}.$$

The product AB is given by the following 2 by 2 matrix:

$$AB = \begin{bmatrix} ae + bg & af + bh \\ ce + dg & cf + dh \end{bmatrix}.$$

Notice that the computation of AB takes eight multiplications and four additions. The definition of matrix multiplication of two n by n matrices uses n^3 multiplication operations and $n^2(n - 1)$ addition operations.

A known lower bound for the number of multiplication operations needed to multiply two n by n matrices is n^2. Strassen [1969] showed how to multiply two matrices with about $n^{2.81}$ multiplication operations. The number 2.81 is an approximation to the value of $\log_2(7)$. It stems from the fact that a pair of 2 by 2 matrices can be multiplied by using seven multiplication operations. Multiplication of larger-size matrices is broken down into multiplying many 2 by 2 matrices. Therefore the number of multiplication operations becomes less than n^3. This revelation got research going in two camps. One camp is trying to find a better algorithm. The other camp is trying to raise the lower bound above n^2. In recent years, algorithms have been found with still lower numbers. Pan [1978] gave an algorithm to multiply two 70×70 matrices using 143,640 multiplications, which is less than $70^{2.81}$ multiplication operations. Coppersmith and Winograd [1987] gave an algorithm that, for large values of n, uses $n^{2.376}$ multiplication operations. So it goes.   ◆

EXAMPLE 2 (*Finding the Minimum*). Let's look at an example of an optimal algorithm to find the minimum number in an unsorted list of n numbers. We'll count the number of comparison operations that an algorithm makes between elements of the list. To find the minimum number in a list of n numbers, the minimum number must be compared with the other $n - 1$ numbers. So $n - 1$ is a lower bound on the number of comparisons needed to find the minimum number in a list of n numbers. If we represent the list as an array a indexed from 1 to n, then the following algorithm is optimal because the operation $\leq$ is executed exactly $n - 1$ times.

$$m := a[1];$$
for $i := 2$ **to** n **do**
$$m := \text{if } m \leq a[i] \text{ then } m \text{ else } a[i]$$
od   ◆

EXAMPLE 3 (*Simple Sort*). In this example we'll construct a simple sorting algorithm and analyze it to find the number of comparison operations. We'll sort an array a of numbers indexed from 1 to n as follows: Find the smallest element in a, and exchange it with the first element. Then find the smallest element in positions 2 through n, and exchange it with the element in position 2. Continue in this manner to obtain a sorted array. To write the algorithm, we'll use a function "min" and a procedure "exchange," which are defined as follows:

min(a, i, n) is the index of the minimum number among the elements $a[i]$, $a[i + 1]$, ..., $a[n]$. We can easily modify the algorithm in Example 2 to accomplish this task with with $n - i$ comparisons.

exchange($a[i]$, $a[j]$) represents the usual operation of swapping elements and does not use any comparisons.

We can write the sorting algorithm as follows:

> **for** $i := 1$ **to** $n - 1$ **do**
> $\quad j := \min(a, i, n)$;
> $\quad$ exchange($a[i]$, $a[j]$)
> **od**

Now let's compute the number of comparison operations. The algorithm for min(a, i, n) makes $n - i$ comparisons. So as i moves from 1 to $n - 1$, the number of comparison operations moves from $n - 1$ to $n - (n - 1)$. Adding these comparisons gives the arithmetic expression

$$(n - 1) + (n - 2) + \cdots + 1 = \frac{n\,(n - 1)}{2}.$$

The algorithm makes the same number of comparisons no matter what the form of the input array, even if it is sorted to begin with. So any arrangement of numbers is a worst case input. For example, to sort 1000 items it would take 499,500 comparisons, no matter how the items are arranged to begin with.

There are many faster sorting algorithms. For example, an algorithm called "heapsort" takes no more than $2n \log_2 n$ comparisons for its worst case performance. So for 1000 items, heapsort would take a maximum of 20,000 comparisons—quite an improvement over our simple sort algorithm. In Section 5.2 we'll discover a good lower bound for the worst case performance of comparison sorting algorithms. ◆

Decision Trees

We can often use a tree to represent the decision processes that take place in an algorithm. A *decision tree* for an algorithm is a tree whose nodes represent decision points in the algorithm and whose leaves represent possible outcomes. Decision trees can be useful in trying to construct an algorithm or trying to find properties of an algorithm. For example, lower bounds may equate to the depth of a decision tree.

If an algorithm makes decisions based on the comparison of two objects, then it can be represented by a binary decision tree. Each node in the tree represents a pair of objects to be compared, and each branch from that node represents a path taken by the algorithm based on the comparison. Each leaf can represent an outcome of the algorithm. Many sorting and searching algorithms can be analyzed with decision trees because they perform comparisons. Let's look at some examples to illustrate the idea.

EXAMPLE 4 (*Binary Search*). Suppose we search a sorted list in a binary fashion. That is, we check the middle element of the list to see whether it's the key we are looking for. If not, then we perform the same operation on either the left half or the right half of the list, depending on the value of the key. This algorithm has a nice representation as a decision tree. For example, suppose we have the following sorted list that contains the first fifteen prime numbers:

$$2, 3, 5, 7, 11, 13, 17, 19, 23, 29, 31, 37, 41, 43, 47.$$

Suppose we're given a key K, and we must find whether it is in the list.

The decision tree for a binary search of the list of primes has the number 19 at its root. This represents the comparison of K with 19. If $K = 19$, then we are successful in one comparison. If $K < 19$, then we go to the left child of 19; otherwise we go to the right child of 19. The result is a ternary decision tree in which the leaves are labeled with either S, for successful search, or U, for unsuccessful search. The tree is pictured in Figure 5.1.

It's easy to see in this case that there will be at most four comparisons to find whether K is in the list. So a worst case lower bound for the number of comparisons is 4, which is 1 plus the depth of the binary tree whose nodes are the numbered nodes in Figure 5.1. We know that the minimum depth of a binary tree with n nodes is $\lfloor \log_2 n \rfloor$. So the lower bound for the worst case of a binary search algorithm on a sorted input list of n elements is

$$1 + \lfloor \log_2 n \rfloor. \quad \blacklozenge$$

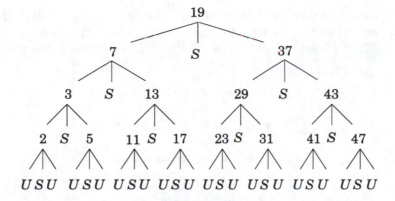

Figure 5.1

EXAMPLE 5 (*Weighing Things*). Suppose that we are given eight coins and told to find the heavy coin among the eight with the assumption that they all look alike, and the other seven all have the same weight, and we must use a pan balance. There are two ways to proceed, depending on whether or not we want to consider the possibility that the balance may balance. If the pan never balances, then we will obtain a binary decision tree. Otherwise, we get a ternary decision tree.

Solution 1 (binary tree): Let each internal node of the tree represent the pan balance, with an equal number of coins on each side. If the left side goes down, then the heavy coin is on the left side of the balance. Otherwise, the heavy coin is on the right side of the balance. Each leaf represents one coin that is the heavy coin. Suppose we label the coins with the numbers 1, 2, ..., 8. One algorithm's decision tree is pictured in Figure 5.2, where the numbers on either side of a nonleaf node represent the coins on either side of the pan balance. This algorithm finds the heavy coin in three weighings. Can we do any better?

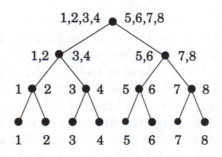

Figure 5.2

Solution 2 (ternary tree): Here we allow for the third possibility that the two pans are balanced. So we don't have to use all eight coins on the first weighing. The decision tree in Figure 5.3 shows one solution to the problem.

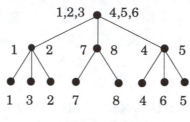

Figure 5.3

Notice that there is no middle branch on the middle subtree, since at this point, one of the coins 7 or 8 must be the heavy one. This algorithm finds the heavy coin in two weighings.

The second solution is an optimal pan balance algorithm for this problem, where we are counting the number of weighings to find the heavy coin. To see this, notice that any one of the eight coins could be the heavy one. Therefore there must be at least eight leaves on any algorithm's decision tree. But a binary tree of depth k can have 2^k possible leaves. So to get eight leaves, we must have $2^k \geq 8$. This implies that $k \geq 3$. But a ternary tree of depth k can have 3^k possible leaves. So to get eight leaves, we must have $3^k \geq 8$, or $k \geq 2$. Therefore 2 is a lower bound for the number of weighings. Since the second solution solves the problem in two weighings, it is optimal.   ◆

EXAMPLE 6. Suppose we have a set of 13 coins in which at most one coin is bad and a bad coin may be heavier or lighter than the other coins. The problem is to use a pan balance to find the bad coin if it exists and say whether it is heavy or light. We'll find a lower bound on the heights of decision trees for pan balance algorithms to solve the problem.

Any solution must tell whether a bad coin is heavy or light. Thus there are 27 possible conditions: no bad coin and the 13 pairs of conditions (ith coin light, ith coin heavy). Therefore any decision tree for the problem must have at least 27 leaves. So a ternary decision tree of depth k must satisfy $3^k \geq 27$, or $k \geq 3$. This gives us a lower bound of 3. Now the big question: Is there an algorithm to solve the problem, where the decision tree of the algorithm has depth 3? The answer is no. Just look at the cases of different initial weighings, and note in each case that the remaining possible conditions cannot be distinguished with just two more weighings. Thus any decision tree for this problem must have depth 4 or more.   ◆

Exercises

1. For the following algorithm, answer each question by giving a formula in terms of n:

$$
\begin{aligned}
&\textbf{for } i := 1 \textbf{ to } n \textbf{ do} \\
&\quad \textbf{for } j := i \textbf{ downto } 1 \textbf{ do } x := x + f(x) \textbf{ od}; \\
&\quad x := x + g(x) \\
&\textbf{od}
\end{aligned}
$$

   a. Find the number of times the assignment statement (:=) is executed during the running of the program. Notice that an assignment statement is found at four places in the program.

   b. Find the number of times the addition operation (+) is executed during the running of the program.

2. Draw a picture of the decision tree for an optimal algorithm to find the maximum number in the list $\langle x_1, x_2, x_3, x_4 \rangle$.

3. Suppose there are 95 possible answers to some problem. For each of the following types of decision tree, find a reasonable lower bound for the number of decisions necessary to solve the problem.

   a. Binary tree.   b. Ternary tree.   c. Four-way tree.

4. Find a nonzero lower bound on the number of weighings necessary for any ternary pan balance algorithm to solve the following problem: A set of 30 coins contains at most one bad coin, which may be heavy or light. Is there a bad coin? If so, state whether it's heavy or light.

5. Find an optimal pan balance algorithm to find a bad coin, if it exists, from 12 coins, where at most one coin is bad (i.e., heavier or lighter than the others). *Hint:* Once you've decided on the coins to weigh for the root of the tree, then the coins that you choose at the second level should be the same coins for all three branches of the tree.

5.2 Elementary Counting Techniques

In this section we'll discuss some elementary principles of counting things that may be ordered or unordered. We'll also discuss some elementary aspects of finite probability.

Permutations (Order Is Important)

In how many different ways can we arrange the elements of a set S? If S has n elements, then there are n choices for the first element. For each of these choices there are $n - 1$ choices for the second element. Continuing in this way, we obtain $n! = n \cdot (n - 1) \cdots 2 \cdot 1$ different arrangements of n elements. Any arrangement of n distinct objects is called a *permutation* of the objects. We'll write down the rule for future use:

Permutations

There are $n!$ permutations of an n-element set.   (5.1)

For example, if $S = \{a, b, c\}$, then the six possible permutations of S, written as strings, are listed as follows:

$$abc,\ acb,\ bac,\ bca,\ cab,\ cba.$$

Now suppose we want to count the number of permutations of r elements chosen from an n-element set, where $1 \leq r \leq n$. There are n choices for the first element. For each of these choices there are $n - 1$ choices for the second element. We continue this process r times to obtain the answer,

$$n\ (n - 1) \cdots (n - r + 1).$$

This number is denoted by the symbol $P(n, r)$ and is read "the number of permutations of n objects taken r at a time." We should emphasize here that we are counting r distinct objects. So we have the formula

$$P(n, r) = n\ (n - 1) \cdots (n - r + 1),$$ (5.2)

which can also be written,

$$P(n, r) = \frac{n!}{(n - r)!}.$$ (5.3)

Notice that $P(n, 1) = n$ and $P(n, n) = n!$. If $S = \{a, b, c, d\}$, then there are 12 permutations of two elements from S, given by the formula $P(4, 2) = \frac{4!}{2!} = 12$. The permutations are listed as follows:

$$ab,\ ba,\ ac,\ ca,\ ad,\ da,\ bc,\ cb,\ bd,\ db,\ cd,\ dc.$$

Permutations can be thought of as arrangements of objects selected from a set *without replacement*. In other words, we can't pick an element from the set more than once. If we can pick an element more than once, then the objects are said to be selected *with replacement*. In this case the number of arrangements of r objects from an n-element set is just n^r. We can state this idea in terms of bags as follows: The number of distinct permutations of r objects taken from a bag containing n distinct objects, each occurring r times, is n^r. For example, consider the bag $B = [a, a, b, b, c, c]$. Then the number of distinct permutations of 2 objects chosen from B is 3^2, and they can be listed as follows:

aa, ab, ac, ba, bb, bc, ca, cb, cc.

Let's look now at permutations of all the elements in a bag. For example, suppose we have the bag $B = [a, a, b, b, b]$. We can write down the distinct permutations of B as follows:

aabbb, ababb, abbab, abbba, baabb, babab, babba, bbaab, bbaba, bbbaa.

There are 10 strings. Let's see how to compute the number 10 from the information we have about the bag B. One way to proceed is to place subscripts on the elements in the bag, obtaining the five distinct elements a_1, a_2, b_1, b_2, b_3. Then we get $5! = 120$ permutations of the five distinct elements. Now we remove all the subscripts on the elements, and we find that there are many redundant strings among the original 120 strings.

For example, suppose we remove the subscripts from the two strings,

$$a_1 b_1 b_2 a_2 b_3 \quad \text{and} \quad a_2 b_1 b_3 a_1 b_2 .$$

Then we obtain two occurrences of the string *abbab*. If we wrote them all down, we would find 12 strings, all of which reduce to the string *abbab* when subscripts are removed. This is because there are $2!$ permutations of the letters a_1 and a_2, and there are $3!$ permutations of the letters b_1, b_2, and b_3. So there are $2! 3! = 12$ distinct ways to write the string *abbab* when we use subscripts. Of course, the number is the same for any string of two a's and three b's. Therefore the number of distinct strings of two a's and three b's is found by dividing the total number of subscripted strings by $2! 3!$ to obtain $\frac{5!}{2! \, 3!} = 10$. This argument generalizes to obtain the following result about permutations that can contain redundant elements.

Permutations of a Bag

Let B be an n-element bag with k distinct elements, where each of the numbers $m_1, ..., m_k$ denotes the number of occurrences of each element. Then the number of permutations of the n elements of B is

$$\frac{n!}{m_1! \cdots m_k!} . \tag{5.4}$$

Now let's look at a few examples to see how permutations (5.1)–(5.4) can be used to solve a variety of problems. We'll start with an important result about sorting.

EXAMPLE 1 (*A Worst Case Lower Bound for Comparison Sorting*). Let's find a lower bound for the number of comparison operations performed by any sorting algorithm that sorts by comparing elements in the list to be sorted. Assume that we have a set of n distinct numbers. Since there are $n!$ possible arrangements of these numbers, it follows that any algorithm to sort a list of n numbers has $n!$ possible input arrangements. Therefore any decision tree for a comparison sorting algorithm must contain at least $n!$ leaves, one leaf for each possible outcome of sorting one arrangement. We know that a binary tree of depth d has at most 2^d leaves. So the depth d of the decision tree for any comparison sort of n items must satisfy the inequality

$$n! \leq 2^d.$$

We can solve this inequality for the natural number d as follows:

$$\log_2 n! \leq d$$
$$\lceil \log_2 n! \rceil \leq d.$$

In other words, $\lceil \log_2 n! \rceil$ is a worst case lower bound for the number of comparisons to sort n items. We'll state it for the record:

Any algorithm that sorts by comparing elements must use at least $\lceil \log_2 n! \rceil$ comparisons in the worst case to sort n items.

The number $\lceil \log_2 n! \rceil$ is hard to calculate for large values of n. We'll see in Section 5.4 that it is "approximately" the same as $n \log_2 n$.   ◆

EXAMPLE 2. In how many ways can 20 people be arranged in a circle if we don't count a rotation of the circle as a different arrangement? There are 20! arrangements of 20 people in a line. We can form a circle by joining the two ends of a line. Since there are 20 distinct rotations of the same circle of people, it follows that there are $\frac{20!}{20} = 19!$ distinct arrangements of 20 people in a circle. Another way to proceed is to put one person in a certain fixed position of the circle. Then fill in the remaining 19 people in all possible ways to get 19! arrangements.   ◆

EXAMPLE 3. How many distinct strings can be made by rearranging the letters of the word *banana*? One letter is repeated twice, one letter is repeated three times, and one letter stands by itself. So we can answer the question by finding the number of permutations of the bag of letters $[b, a, n, a, n, a]$. Therefore (5.4) gives us the result

$$\frac{6!}{1! \, 2! \, 3!} = 60. \quad \blacklozenge$$

EXAMPLE 4. How many distinct strings of length 10 can be constructed from the digits 0 and 1 with the restriction that five characters must be 0 and five must be 1? The answer is

$$\frac{10!}{5! \, 5!} = 252$$

because we are looking for the number of permutations from a 10-element bag with five 1's and five 0's. $\quad \blacklozenge$

EXAMPLE 5. Suppose we want to build a code to represent each of 29 distinct objects with a binary string having the same minimal length n, where each string has the same number of 0's and 1's. Somehow we need to solve an inequality like

$$\frac{n!}{k! \, k!} \geq 29,$$

where $k = n/2$. We find by trial and error that $n = 8$. Try it. $\quad \blacklozenge$

Combinations (Order Is Not Important)

Suppose we want to count the number of r-element subsets in an n-element set. For example, if $S = \{a, b, c, d\}$, then there are four 3-element subsets of S: $\{a, b, c\}$, $\{a, b, d\}$, $\{a, c, d\}$, and $\{b, c, d\}$. Is there a formula for the general case? The answer is yes. An easy way to see this is to first count the number of r-element permutations of the n elements, which is given by the formula

$$P(n, r) = \frac{n!}{(n - r)!} \, .$$

Now each r-element subset has $r!$ distinct r-element permutations, which we have included in our count $P(n, r)$. How do we remove the redundant permutations from the count? Let $C(n, r)$ denote the number of r-element subsets of an n-element set. Since each of the r-element subsets has $r!$ distinct permutations, it follows that $r! \cdot C(n, r) = P(n, r)$. Now divide both sides by $r!$ to obtain the formula

$$C(n, r) = \frac{P(n, r)}{r!} = \frac{n!}{r! \, (n - r)!}. \tag{5.5}$$

EXAMPLE 6. If $S = \{a, b, c, d, e\}$, then all the three-element subsets of S are listed as follows:

$$\{a, b, c\}, \{a, b, d\}, \{a, b, e\}, \{a, c, d\}, \{a, c, e\},$$

$$\{a, d, e\}, \{b, c, d\}, \{b, c, e\}, \{b, d, e\}, \{c, d, e\}.$$

There are 10 such subsets, and we can verify the number by the calculation

$$C(5, 3) = \tfrac{5!}{3! \, 2!} = 10. \quad \blacklozenge$$

The expression $C(n, r)$ is usually said to represent the number of *combinations* of n things taken r at a time. With combinations, the order in which the objects appear is not important. We count only the different sets of objects. The expression $C(n, r)$ is often read "n choose r."

Notice how $C(n, r)$ crops up in the following binomial expansion of the expression $(x + y)^4$:

$$(x + y)^4 = x^4 + 4x^3y + 6x^2y^2 + 4xy^3 + y^4$$

$$= C(4,0)x^4 + C(4,1)x^3y + C(4,2)x^2y^2 + C(4,3)xy^3 + C(4,4)y^4.$$

Another useful way to represent $C(n, r)$ is with the *binomial coefficient symbol*:

$$\binom{n}{r} = C(n, r).$$

Using this symbol, we can write the expansion for $(x + y)^4$ as follows:

$$(x + y)^4 = x^4 + 4x^3y + 6x^2y^2 + xy^3 + y^4$$

$$= \binom{4}{0}x^4 + \binom{4}{1}x^3y + \binom{4}{2}x^2y^2 + \binom{4}{3}xy^3 + \binom{4}{4}y^4.$$

This is an instance of a well-known formula called the *binomial theorem*, which can be written as follows, where n is a natural number:

$$(x+y)^n = \sum_{i=0}^{n} \binom{n}{i} x^{n-i} y^i. \tag{5.6}$$

The binomial coefficients for the expansion of $(x + y)^n$ can be read from the nth row of Table 5.1. The table is called *Pascal's triangle*—after the philosopher and mathematician Blaise Pascal (1623–1662). However, the triangle was known prior to the time of Pascal in China, India, the Middle East, and Europe. Notice that any interior element is the sum of the two elements above and to its left.

n	0	1	2	3	4	5	6	7	8	9	10
0	1										
1	1	1									
2	1	2	1								
3	1	3	3	1							
4	1	4	6	4	1						
5	1	5	10	10	5	1					
6	1	6	15	20	15	6	1				
7	1	7	21	35	35	21	7	1			
8	1	8	28	56	70	56	28	8	1		
9	1	9	36	84	126	126	84	36	9	1	
10	1	10	45	120	210	252	210	120	45	10	1

Table 5.1

But how do we really know that the following statement is correct?

The element in the nth row and kth column of the triangle is $\binom{n}{k}$. (5.7)

Proof: For convenience we will designate a position in the triangle by an ordered pair of the form ⟨row, column⟩. Notice that the edge elements of the triangle are all 1, and they occur at positions ⟨n, 0⟩ or ⟨n, n⟩. Notice also that $\binom{n}{0}$ = 1 = $\binom{n}{n}$. So (5.7) is true when $k = 0$ or $k = n$. Next, we need to consider the interior elements of the triangle. So let $n > 1$ and $0 < k < n$. We want to show that the element in position ⟨n, k⟩ is $\binom{n}{k}$. To do this, we need the following useful result about binomial coefficients:

$$\binom{n}{k} = \binom{n-1}{k} + \binom{n-1}{k-1}. \tag{5.8}$$

To prove (5.8), just expand each of the three terms and simplify. Continuing with the proof of (5.7), we'll use well-founded induction. To do this, we need to define a well-founded order on something. For our purposes we will let the something be the set of positions in the triangle. We agree that any position in row $n-1$ precedes any position in row n. In other words, if $n' < n$, then $\langle n', k' \rangle$ precedes $\langle n, k \rangle$ for any values of k' and k. Now we can use well-founded induction. We pick position $\langle n, k \rangle$ and assume that (5.7) is true for all pairs in row $n-1$. In particular, we can assume that the elements in positions $\langle n-1, k \rangle$ and $\langle n-1, k-1 \rangle$ have values

$$\binom{n-1}{k} \text{ and } \binom{n-1}{k-1}.$$

Now we use this assumption along with (5.8) to tell us that the value of the element in position $\langle n, k \rangle$ is $\binom{n}{k}$.   QED.

Can you find some other interesting patterns in Pascal's triangle? There are lots of them. For example, look down the column labeled 2 and notice that, for each $n \geq 2$, the element in position $\langle n, 2 \rangle$ is the value of the arithmetic sum $1 + 2 + \cdots + (n-1)$. In other words, we have the formula

$$\binom{n}{2} = \frac{n(n-1)}{2}.$$

Let's continue our discussion about combinations by counting bags of things rather than sets of things. Suppose we have the set $A = \{a, b, c\}$. How many three-element bags can we construct from the elements of A? We can list them as follows:

$$[a, a, a], \quad [a, a, b], \quad [a, a, c], \quad [a, b, c], \quad [a, b, b],$$

$$[a, c, c], \quad [b, b, b], \quad [b, b, c], \quad [b, c, c], \quad [c, c, c].$$

So there are 10 three-element bags constructed from the elements of $\{a, b, c\}$.

Let's see if we can find a general formula for the number of k-element bags that can be constructed from an n-element set. For convenience, we'll assume that the n-element set is $A = \{1, 2, ..., n\}$. Suppose that $b = [x_1, x_2, x_3, ..., x_k]$ is some k-element bag with elements chosen from A, where the elements of b are written so that $x_1 \leq x_2 \leq ... \leq x_k$. This allows us to construct the following k-element set:

$$B = \{x_1, x_2 + 1, x_3 + 2, ..., x_k + (k-1)\}.$$

The numbers $x_i + (i-1)$ are used to ensure that the elements of B are distinct elements in the set $C = \{1, 2, ..., n + (k-1)\}$. So we've associated each k-element bag b over A with a k-element subset B of C. Conversely, suppose that $\{y_1, y_2, y_3, ..., y_k\}$ is some k-element subset of C, where the elements are written so that $y_1 \leq y_2 \leq ... \leq y_k$. This allows us to construct the k-element bag

$$[y_1, y_2 - 1, y_3 - 2, ..., y_k - (k-1)],$$

whose elements come from the set A. So we've associated each k-element subset of C with a k-element bag over A.

Therefore the number of k-element bags over an n-element set is exactly the same as the number of k-element subsets of a set with $n + (k-1)$ elements. This gives us the following result:

Bag Combinations

The number of k-element bags whose distinct elements are chosen from an n-element set is given by the following formula, where k and n are positive:

$$\binom{n+k-1}{k}. \tag{5.9}$$

EXAMPLE 7. In how many ways can four coins be selected from a collection of pennies, nickels, and dimes? Let $S = \{\text{penny, nickel, dime}\}$. Then we need the number of four-element bags chosen from S. The answer is

$$\binom{3+4-1}{4} = \binom{6}{4} = 15. \quad \blacklozenge$$

EXAMPLE 8. In how many ways can five people be selected from a collection of Democrats, Republicans, and Independents? Here we are choosing five-element bags from a set of three characteristics $\{$Democrat, Republican, Independent$\}$. The answer is

$$\binom{3+5-1}{5} = \binom{7}{5} = 21. \quad \blacklozenge$$

Finite Probability

Often we're concerned with the average behavior of an algorithm. That is, instead of the worst case performance, we might be interested in the average case performance. This can get a bit tricky because it usually forces us to make one or two assumptions. Some people hate to make assumptions. But it's not so bad. Let's do an example.

Suppose we have a sorted list of the first 15 prime numbers, and we want to know the average number of comparisons needed to find a number in the list, using a binary search. The decision tree for a binary search of the list is pictured in Figure 5.4.

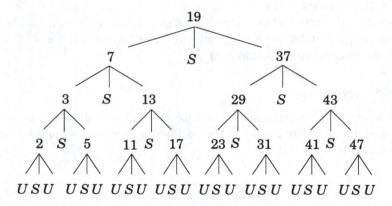

Figure 5.4

After some thought, it might seem reasonable to add up all the path lengths from the root to a leaf marked with an S (for successful search) and divide by the number of S leaves, which is 15. In this case there are eight paths of length 4, four paths of length 3, two paths of length 2, and one path of length 1. So we get

$$\text{Average path length} = \frac{32 + 12 + 4 + 1}{15} = \frac{49}{15} \approx 3.27.$$

This gives us the average number of comparisons needed to find a number in the list. Or does it? Have we made any assumptions here? Sure. We assumed that each path in the tree has the same chance of being traversed as any other path. Of course, this might not be the case. For example, suppose that we always wanted to look up the number 37. Then the average number of comparisons would be two. So our calculation was made under the assumption that each of the 15 numbers had the same chance of being picked.

Let's pause here and introduce some notions and notation. If some operation or experiment has n possible outcomes and each outcome has the same chance of occurring, then we say that each outcome has *probability* $1/n$. In the preceding example we assumed that each number had probability 1/15 of being picked. As another example, let's consider the coin-flipping problem. If we flip a fair coin, then there are two possible outcomes, assuming that the coin does not land on its edge. Thus the probability of a head is 1/2, and the probability of a tail is 1/2. If we toss the coin 1000 times, we should expect about 500 heads and 500 tails. So probability has something to do with expectation.

Now for some terminology. The set of all possible outcomes of an experiment is called a *sample space*. The elements in a sample space are called *sample points* or simply *points*. Further, any subset of a sample space is called an *event*. For example, suppose we toss two coins and are interested in the set of possible outcomes. Let H and T mean head and tail, respectively, and let the string HT mean that the first coin lands H and the second coin lands T. Then the sample space for this experiment is the set

$$\{HH,\ HT,\ TH,\ TT\}.$$

For example, the event that one coin lands as a head and the other coin lands as a tail can be represented by the subset $\{HT,\ TH\}$.

A *probability distribution* on a sample space S is an assignment of probabilities to the points of S such that the sum of all the probabilities is 1. We can say this more precisely as follows: Let $S = \{x_1, x_2, ..., x_n\}$ be a sample space (we'll discuss only finite sample spaces). A probability distribution p on S is a function

$$p : S \to [0, 1]$$

such that

$$p(x_1) + p(x_2) + \cdots + p(x_n) = 1.$$

For example, in the two-coin-toss experiment it makes sense to define the following probability distribution on the sample space $S = \{HH, HT, TH, TT\}$:

$$p(HH) = p(HT) = p(TH) = p(TT) = \frac{1}{4}.$$

Once we have a probability distribution p defined on the points of a sample space S, we can use p to define the probability of any event E in S. The *probability of* E is denoted by $P(E)$ and is defined by

$$P(E) = \sum_{x \in E} p(x).$$

In particular, we have $P(S) = 1$ and $P(\emptyset) = 0$. If E' is the complement of E in S, then we have the equation

$$P(E') = 1 - P(E).$$

Of course, we also have $P(E) = 1 - P(E')$.

In our two-coin-toss example, let E be the event that at least one coin is a tail. Then $E = \{HT, TH, TT\}$. We can calculate $P(E)$ as follows:

$$P(E) = P(\{HT, TH, TT\}) = p(HT) + p(TH) + p(TT) = \frac{1}{4} + \frac{1}{4} + \frac{1}{4} = \frac{3}{4}.$$

Let's look at some examples to see how probability is used.

EXAMPLE 9 (*The Birthday Problem*). Suppose we ask 25 people, chosen at random, their birthday (month and day). Would you bet that they all have different birthdays? It seems a likely bet that no two have the same birthday since there are 365 birthdays in the year. But in fact the probability that two out of 25 people have the same birthday is greater than $\frac{1}{2}$. Again, we're assuming some things here, which we'll get to shortly. Let's see why this is the case. The question we want to ask is:

> Given n people in a room, what is the probability that at least two of the people have the same birthday (month and day)?

We'll neglect leap year and assume that there are 365 days in the year. So there are 365^n possible n-tuples of birthdays for n people. This set of n-tuples is our sample space S. We'll also assume that birthdays are equally distributed throughout the year. So for any n-tuple $\langle b_1, ..., b_n \rangle$ of birthdays, we have $p(\langle b_1, ..., b_n \rangle) = \frac{1}{365^n}$. The event E that we are concerned with is the subset of S consisting of all n-tuples that contain two or more equal entries. So our question can be written as follows:

> What is $P(E)$?

To answer the question, let's use the negation technique. That is, we'll compute the probability of the event $E' = S - E$, consisting of all n-tuples that have distinct entries. In other words, no two of the n people have the same birthday. Then the probability that we want is $P(E) = 1 - P(E')$. So let's concentrate on E'.

An n-tuple is in E' exactly when all its components are distinct. The cardinality of E' can be found in several ways. For example, there are 365 possible values for the first element of an n-tuple in E'. For each of these 365 values there are 364 values for the second element of an n-tuple in E'. Thus we obtain

$$365 \cdot 364 \cdot 363 \cdots (365 - n + 1)$$

n-tuples in E'. Of course, this is also the number $P(365, n)$ of permutations of 365 things taken n at a time. Since each n-tuple of E' is equally likely with probability $\frac{1}{365^n}$, it follows that

$$P(E') = \frac{365 \cdot 364 \cdot 363 \cdots (365 - n + 1)}{365^n}.$$

Thus the probability that we desire is

$$P(E) = 1 - P(E') = 1 - \frac{365 \cdot 364 \cdot 363 \cdots (365 - n + 1)}{365^n}.$$

Table 5.2 gives a few calculations for different values of n. Notice the case when $n = 23$. The probability is better than 0.5 that two people have the same birthday.

n	$P(E)$
10	0.117
20	0.411
23	0.507
30	0.706
40	0.891

Table 5.2

Try this out next time you're in a room full of people. It always seems like magic when two people have the same birthday. ◆

EXAMPLE 10 (*Switching Pays*). Suppose there is a set of three numbers. One of the three numbers will be chosen as the winner of a three-number lottery. We pick one of the three numbers. Later, we are told that one of the two remaining numbers is not a winner, and we are given the chance to keep the number that we picked or to switch and choose the remaining number. What should we do? We should switch.

To see this, notice that once we pick a number, the probability that we did not pick the winner is $\frac{2}{3}$. In other words, it is more likely that one of the other two numbers is a winner. So when we are told that one of the other numbers is not the winner, it follows that the remaining other number has probability 2/3 of being the winner. So go ahead and switch. Try this experiment a few times with a friend to see that in the long run it's better to switch.

Another way to see that switching is the best policy is to modify the problem to a set of 50 numbers and a 50-number lottery. If we pick a number, then the probability that we did not pick a winner is $\frac{49}{50}$. Later we are told that 48 of the remaining numbers are not winners, but we are given the chance to keep the number we picked or switch and choose the remaining number. What should we do? We should switch because the chance that the remaining number is the winner is $\frac{49}{50}$. ◆

Expectation = Average Behavior

Let's get back to talking about averages and expectations. We all know that the average of a bunch of numbers is the sum of the numbers divided by the number of numbers. So what's the big deal? The deal is that we often assign numbers to each outcome in a sample space. For example, we assigned a path length to each of the first 15 prime numbers. We added up the 15 path lengths and divided by 15 to get the average. Makes sense, doesn't it? But remember, we assumed that each number was equally likely to occur. This is not always the case. So we also have to consider the probabilities assigned to the points in the sample space.

Let's look at another example. Suppose we agree to flip a coin. If the coin comes up heads, we agree to pay 4 dollars; if it comes up tails, we agree to accept 5 dollars. Notice here that we have assigned a number to each of the two possible outcomes of this experiment. What is our expected take from this experiment? It depends on the coin. Suppose the coin is fair. After one toss we are either 4 dollars poorer or 5 dollars richer. Suppose we play the game 10 times. What then? Well, since the coin is fair, it seems likely that we can expect to win five times and lose five times. So we can expect to pay 20 dollars and receive 25 dollars. Thus our expectation from 10 tosses is 5 dollars.

Suppose we knew that the coin was biased with p(head) = 2/5 and p(tail) = 3/5. What would our expectation be? Again, we can't say much for just one toss. But for 10 tosses we can expect about four heads and six tails. Thus we can expect to pay out 16 dollars and receive 30 dollars, for a net profit of 14 dollars. An equation to represent our reasoning follows:

$$10\,p(\text{head})(-4) + 10\,p(\text{tail})(5) = 10\left(\frac{2}{5}\right)(-4) + 10\left(\frac{3}{5}\right)(5) = \frac{70}{5} = 14.$$

Can we learn anything from this equation? Yes we can. The 14 dollars represents our take over 10 tosses. What's the average profit? Just divide by 10 to get $1.40. This can be expressed by the following equation:

$$p(\text{head})(-4) + p(\text{tail})(5) = \left(\frac{2}{5}\right)(-4) + \left(\frac{3}{5}\right)(5) = \frac{7}{5} = 1.4.$$

So we can compute the average profit per toss without using the number of coin tosses. The average profit per toss is $1.40 no matter how many tosses there are. That's what probability gives us. It's called *expectation*, and we'll generalize from this example to define expectation for any sample space having an assignment of numbers to the sample points. Let S be a sample space, p a probability distribution on S, and $v : S \to \mathbb{R}$ an assignment of numbers to the points of S. Suppose $S = \{x_1, x_2, ..., x_n\}$. Then the *expected value* (or *expectation*) of v is defined by the following formula:

$$v(x_1)p(x_1) + v(x_2)p(x_2) + \cdots + v(x_n)p(x_n).$$

So when we want the average behavior, we're really asking for the expectation.

Average Performance of an Algorithm

To compute the average performance of an algorithm A, we must do several things: First, we must decide on a sample space to represent the possible inputs of size n. Suppose our sample space is $S = \{I_1, I_2, ..., I_k\}$. Second, we must define a probability distribution p on S that represents our idea of how likely it is that the inputs will occur. Third, we must count the number of operations required by A to process each sample point. We'll denote this count by the function $v : S \to \mathbb{N}$. Lastly, we can compute the average number of operations to execute A as a function of input size n by calculating the following expectation:

$$\text{Avg}_A(n) = v(I_1)p(I_1) + v(I_2)p(I_2) + \cdots + v(I_k)p(I_k).$$

To show that an algorithm A is *optimal in the average case* for a problem P, we need to specify a particular sample space and probability distribution. Then we need to show that $\text{Avg}_A(n) \leq \text{Avg}_B(n)$ for all $n > 0$ and for all algorithms B that solve P. The problem of finding lower bounds for the average case is just as difficult as finding lower bounds for the worst case. So we're often content to just compare known algorithms to find the best of the bunch.

We'll finish the section with an example showing an average case analysis of a simple algorithm.

EXAMPLE 11 (*Analysis of Sequential Search*). Suppose we have the following algorithm to search for an element X in an array L, indexed from 1 to n. If X is in L, the algorithm returns the index of the rightmost occurrence of X. The index 0 is returned if X is not in L:

$$i := n;$$
$$\textbf{while } i \geq 1 \textbf{ and } X \neq L[i] \textbf{ do}$$
$$i := i - 1$$
$$\textbf{od}$$

We'll count the average number of comparisons $X \neq L[i]$ performed by the algorithm. Frst we need a sample space. Suppose we let I_i denote the input case where the rightmost occurrence of X is at the ith position of L. Let I_{n+1} denote the case in which X is not in L. So the sample space is the set

$$\{I_1, I_2, ..., I_{n+1}\}.$$

Let $v(I)$ denote the number of comparisons made by the algorithm when the input has the form I. Looking at the algorithm, we obtain the following values:

$$v(I_i) = n - i + 1 \quad \text{for } 1 \leq i \leq n,$$
$$v(I_{n+1}) = n.$$

Suppose we let q be the probability that X is in L. Thus $1 - q$ is the probability that X is not in L. Let's also assume that whenever X is in L, its position is random. This gives us the following probability distribution p over the sample space:

$$p(I_i) = \frac{q}{n} \quad \text{for } 1 \leq i \leq n,$$
$$p(I_{n+1}) = 1 - q.$$

Therefore the expected number of comparisons made by the algorithm for this probability distribution is given by the expected value of v:

$$
\begin{aligned}
\text{Avg}_A(n) &= v(I_1)p(I_1) + \cdots + v(I_{n+1})p(I_{n+1}) \\
&= \frac{q}{n}(n + (n-1) + \cdots + 1) + (1-q)\,n \\
&= q\left(\frac{n+1}{2}\right) + (1-q)\,n.
\end{aligned}
$$

Let's observe a few things about the expected number of comparisons. If we know that X is in L, then $q = 1$. So the expectation is $\frac{n+1}{2}$ comparisons. If we know that X is not in L, then $q = 0$, and the expectation is n comparisons. If X is in L and it occurs at the first position, then the algorithm takes n comparisons. So the worst case occurs for the two input cases I_{n+1} and I_1, and we have $W_A(n) = n$. ◆

Exercises

1. Evaluate each of the following expressions.

   a. $P(6, 6)$.   b. $P(6, 0)$.   c. $P(6, 2)$.

   d. $P(10, 4)$.   e. $C(5, 2)$.   f. $C(10, 4)$.

2. Let $S = \{a, b, c\}$. Write down the objects satisfying each of the following descriptions.

   a. All permutations of the three letters in S.

   b. All permutations consisting of two letters from S.

   c. All combinations of the three letters in S.

   d. All combinations consisting of two letters from S.

   e. All bag combinations consisting of two letters from S.

3. For each part of Exercise 2, write down the formula, in terms of P or C, for the number of objects requested.

4. Given the bag $B = [a, a, b, b]$, write down all the bag permutations of B, and verify with a formula that you wrote down the correct number.

5. Find the number of ways to arrange the letters in each of the following words. Assume all letters are lowercase.

   a. Computer.   b. Radar.   c. States.

   d. Mississippi.   e. Tennessee.

6. A *derangement* of a string is a permutation of the letters such that each letter changes its position. For example, a derangement of the string ABC is BCA. But ACB is not a derangement of ABC, since A does not change position. Write down all derangements for each of the following strings.

   a. A.   b. AB.   c. ABC.   d. $ABCD$.

7. Suppose we want to build a code to represent 29 objects in which each object is represented as a binary string of length n, which consists of k 0's and m 1's, and $n = k + m$. Find n, k, and m, where n has the smallest possible value.

8. We wish to form a committee of seven people chosen from five Democrats, four Republicans, and six Independents. The committee will contain two Democrats, two Republicans, and three Independents. In how many ways can we choose the committee?

9. Each row of Pascal's triangle (Table 5.1) has a largest number. Find a formula to describe which column contains the largest number in row n.

10. Suppose we have an algorithm that must perform 2,000 operations as follows: The first 1,000 operations are performed by a processor with a capacity of 100,000 operations per second. Then the second 1,000 operations are performed by a processor with a capacity of 200,000 operations per second. Find the average number of operations per second performed by the two processors to execute the 2,000 operations.

11. Find the chances of winning a lottery that allows you to pick six numbers from the set $\{1, 2, ..., 44\}$.

12. Suppose three coins are tossed. Find the probability for each of the following events.

    a. Exactly one coin is a head.   b. Exactly two coins are tails.

    c. At least one coin is a head.   d. At most two coins are tails.

13. Suppose a pair of dice are tossed. Find the probability for each of the following events.

    a. The sum of the dots is 7.

    b. The sum of the dots is even.

    c. The sum of the dots is either 7 or 11.

    d. The sum of the dots is at least 5.

14. For each of the following problems, compute the expected value.

    a. The expected number of dots that show when a die is tossed.

    b. The expected score obtained by guessing all 100 questions of a true-false exam in which a correct answer is worth 1 point and an incorrect answer is worth $-\frac{1}{2}$ point.

15. Test the birthday problem on a group of people.

16. Suppose an operating system must schedule the execution of n processes, where each process consists of k separate actions that must be done in order. Assume that any action of one process may run before or after any action of another process. How many execution schedules are possible?

17. Count the number of strings consisting of n 0's and n 1's such that each string is subject to the following restriction: As we scan a string from left

to right, the number of 0's is never greater than the number of 1's. For example, the string 110010 is OK, but the string 100110 is not. *Hint:* Count the total number of strings of length $2n$ with n 0's and n 1's. Then try to count the number that are not OK, and subtract this number from the total number.

18. Given a nonempty finite set S with n elements, prove that there are $n!$ bijections from S to S.

5.3 Solving Recurrences

Suppose we write down the following sequence of three numbers:

$$2, 4, 8.$$

What is the next number in the sequence? The problem below might make you think about your answer.

The n-ovals problem

Suppose that n ovals (an oval is a closed curve that does not cross over itself) are drawn on the plane such that no three ovals meet in a point and each pair of ovals intersects in exactly two points. How many distinct regions of the plane are created by n ovals?

For example, the diagrams in Figure 5.5 show the cases for one, two, and three ovals.

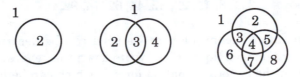

Figure 5.5

If we let r_n denote the number of distinct regions of the plane for n ovals, then it's clear that the first three values are

$$r_1 = 2,$$
$$r_2 = 4,$$
$$r_3 = 8.$$

What is the value of r_4? Is it 16? Check it out. To find r_n, consider the following description: $n - 1$ ovals divide the region into r_{n-1} regions. The nth oval will meet each of the previous $n - 1$ ovals in $2(n - 1)$ points. So the nth oval will itself be divided into $2(n - 1)$ arcs. Each of these $2(n - 1)$ arcs splits some region in two. Therefore we add $2(n - 1)$ regions to r_{n-1} to obtain r_n. This gives us the following recursive definition for r, which is called a *recurrence*:

$$\begin{aligned}
r_1 &= 2, \\
r_n &= r_{n-1} + 2(n - 1).
\end{aligned}$$

This recurrence fits a well-known pattern that makes it easy to solve. We start with r_n and keep substituting for r on the right side until there aren't any occurrences of r:

$$\begin{aligned}
r_n &= r_{n-1} + 2(n - 1) \\
&= r_{n-2} + 2(n - 2) + 2(n - 1) \\
&= r_{n-3} + 2(n - 3) + 2(n - 2) + 2(n - 1) \\
&\ \ \vdots \\
&= 2 + 2(1) + 2(2) + \cdots + 2(n - 3) + 2(n - 2) + 2(n - 1) \\
&= 2 + 2(1 + 2 + \cdots + (n - 1)) \\
&= 2 + n(n - 1) \qquad \text{(arithmetic progression)} \\
&= n^2 - n + 2.
\end{aligned}$$

So we can use this formula to calculate $r_4 = 14$. Therefore the sequence of numbers 2, 4, 8 could very well be the first three numbers in the following sequence for the n ovals problem:

$$2, 4, 8, 14, 22, 32, 44, 62, 74, 92, \dots .$$

The substitution technique that we used can be described in another way as a *cancellation* technique. First, write down the general equation for r_n. Then write down the general equation for r_{n-1}, which can be found by replacing n by $n - 1$ in the equation for r_n. Continue the process until you see a pattern emerging. Then write down a last equation that contains the base case r_0 on the right-hand side. For our example we get the following equations:

$$\begin{aligned}
r_n &= r_{n-1} + 2(n - 1) \\
r_{n-1} &= r_{n-2} + 2(n - 2) \\
&\ \ \vdots \\
r_2 &= r_1 + 2(1).
\end{aligned}$$

Next, we add up all the equations, noting that cancellation takes place, to yield the resulting equation:

$$\begin{aligned} r_n &= r_1 + 2(1 + 2 + \ldots + (n-1)) \\ &= 2 + 2\,\frac{(n-1)\,n}{2} \\ &= 2 + n(n-1). \end{aligned}$$

Therefore we have our result $r_n = n(n-1) + 2$ for all $n > 0$.

This cancellation technique works on recurrence systems that have a more general form, as follows, where a_n and b_n denote either constants or expressions involving n but not involving the recurrence r:

$$\begin{aligned} r_0 &= b_0, \\ r_n &= a_n\,r_{n-1} + b_n. \end{aligned} \tag{5.10}$$

Let's look at an example to get the idea.

EXAMPLE 1. Suppose we are given the following recurrence:

$$\begin{aligned} r_0 &= 1, \\ r_n &= 2r_{n-1} + n. \end{aligned}$$

This recurrence fits the form of (5.10), so we can solve it by cancellation. Starting with the general term, we obtain the following sequence of equations (remember that the term on the left side of a new equation is always the term that contains r from the right side of the preceding equation):

$$\begin{aligned} r_n &= 2r_{n-1} + n \\ 2r_{n-1} &= 2(2\,r_{n-2} + (n-1)) = 2^2\,r_{n-2} + 2\,(n-1) \\ 2^2\,r_{n-2} &= 2^2\,(2\,r_{n-3} + (n-2)) = 2^3\,r_{n-3} + 2^2\,(n-2) \\ &\;\;\vdots \\ 2^{n-1}\,r_1 &= 2^n\,r_0 + 2^{n-1}\,(1). \end{aligned}$$

Now add up all the equations, cancel the like terms, and replace r_0 by its value, to get the following equation:

$$r_n = 2^n + n + 2(n-1) + \cdots + 2^{n-1}\,(1). \quad \blacklozenge$$

The expression on right side of the preceding equation contains an ellipsis. We would like to simplify expressions like this so that they don't contain an ellipsis. That's the next topic.

Finding Closed Forms for Sums

Often an expression involving a sum of terms can be simplified into a form that can be easily computed with familiar operations, without using loops, and without using recursion. Such a form is often called a *closed form*. We can describe a closed form as an expression that can be computed by applying a fixed number of familiar operations to the arguments. A closed form can't have an ellipsis because the number of operations to evaluate the form would not be fixed.

Let's start by reviewing a few important facts about summation notation and the indexes used for summing things. We can use summation notation to represent a sum like $a_1 + a_2 + \cdots + a_n$ as follows:

$$\sum_{i=1}^{n} a_i = a_1 + a_2 + \cdots + a_n.$$

Many problems can be solved by the simple manipulation of sums. So we'll begin by listing some useful facts about sums, which are easily verified.

Summation Facts (5.11)

a) $\displaystyle\sum_{i=m}^{n} c = (n - m + 1)c.$

b) $\displaystyle\sum_{i=m}^{n} (a_i + b_i) = \sum_{i=m}^{n} a_i + \sum_{i=m}^{n} b_i.$

c) $\displaystyle\sum_{i=m}^{n} c\, a_i = c \sum_{i=m}^{n} a_i.$

d) $\displaystyle\sum_{i=m}^{n} a_{i+k} = \sum_{i=m+k}^{n+k} a_i$ (k is any integer).

e) $\displaystyle\sum_{i=m}^{n} a_i x^{i+k} = x^k \sum_{i=m}^{n} a_i x^i$ (k is any integer).

Now let's look at closed forms for some elementary sums. We already know some of them, and we'll discuss the others.

Closed Forms of Elementary Finite Sums   (5.12)

a) $\displaystyle\sum_{i=1}^{n} i \;=\; \frac{n(n+1)}{2}.$

b) $\displaystyle\sum_{i=1}^{n} i^2 \;=\; \frac{n(n+1)(2n+1)}{6}.$

c) $\displaystyle\sum_{i=0}^{n} a^i \;=\; \frac{a^{n+1}-1}{a-1} \qquad (a \neq 1).$

d) $\displaystyle\sum_{i=1}^{n} i a^i \;=\; \frac{a-(n+1)a^{n+1}+na^{n+2}}{(a-1)^2} \qquad (a \neq 1).$

Sums like (5.12) are useful because they pop up in many situations in trying to count the number of operations performed by an algorithm. Are closed forms easy to find? Sometimes yes, sometimes no. Here's an example.

EXAMPLE 2. For example, suppose we need a closed form for the sum of the odd natural numbers up to a certain point:

$$1 + 3 + 5 + \ldots + (2n + 1).$$

We can write this using summation notation as follows:

$$\sum_{i=0}^{n} (2i + 1) \;=\; 1 + 3 + 5 + \ldots + (2n + 1).$$

Now, let's manipulate the sum to find a closed form. Make sure you can supply the reason for each step.

$$
\begin{aligned}
\sum_{i=0}^{n} (2i+1) \;&=\; \sum_{i=0}^{n} 2i + \sum_{i=0}^{n} 1 \\[2mm]
&=\; 2\sum_{i=0}^{n} i + \sum_{i=0}^{n} 1 \\[2mm]
&=\; 2\frac{n(n+1)}{2} + (n+1) = (n+1)^2. \qquad \blacklozenge
\end{aligned}
$$

There are several techniques that can be used to find closed forms. Let's look at a couple that can be used to derive some of the closed forms in (5.12).

Technique One

With this method we let S_n denote the sum that we wish to find. Usually, we try to create an equation with S_n on both sides but with different coefficients so that we can solve for S_n.

EXAMPLE 3. For example, suppose we let S_n be the geometric progression

$$S_n = 1 + a + a^2 + \cdots + a^n.$$

We can add the next term a^{n+1} to both sides of the sum and then manipulate the right side as follows:

$$
\begin{aligned}
S_n + a^{n+1} &= 1 + a + a^2 + \cdots + a^{n+1} \\
&= 1 + a(1 + a + \cdots + a^n) \\
&= 1 + a\, S_n.
\end{aligned}
$$

So when $a \neq 1$, we can solve the equation for S_n to obtain the well-known formula (5.12c) for a geometric progression:

$$1 + a + a^2 + \cdots + a^n = \frac{a^{n+1} - 1}{a - 1}.$$

We can verify the answer by mathematical induction on n or by multiplying both sides of the answer by $a - 1$ to get an equality as the result.   ◆

EXAMPLE 4. Let's try to derive the closed formula given in (5.12d) for the sum $\sum_{i=1}^{n} ia^i$. Suppose we let $S_n = \sum_{i=1}^{n} ia^i$. We'll add the term $(n + 1)a^{n+1}$ to S_n, and then proceed as follows to get S_n on the right side of an equation:

$$
\begin{aligned}
S_n + (n + 1)a^{n+1} &= \sum_{i=1}^{n+1} ia^i \\
&= \sum_{i=0}^{n} (i+1)a^{i+1} \quad \text{(change the index range)} \\
&= \sum_{i=0}^{n} i\, a^{i+1} + \sum_{i=0}^{n} a^{i+1} \\
&= a\sum_{i=0}^{n} i\, a^i + a\sum_{i=0}^{n} a^i \\
&= aS_n + a\left(\frac{a^{n+1} - 1}{a - 1}\right).
\end{aligned}
$$

Since S_n appears on both sides of the equation, with different coefficients, we can solve for S_n to get the closed formula (5.12d):

$$\sum_{i=1}^{n} ia^i = \frac{a - (n+1)a^{n+1} + na^{n+2}}{(a-1)^2}. \qquad \blacklozenge$$

Technique Two

Another technique that sometimes works is to replace some occurrence of a variable by a more complicated expression.

EXAMPLE 5. Let's again consider the sum (5.12d). We'll replace the coefficient i by $i - 1 + 1$. This yields the following equations:

$$S_n = \sum_{i=1}^{n} ia^i = \sum_{i=1}^{n} (i - 1 + 1)a^i$$

$$= \sum_{i=1}^{n} (i-1)a^i + \sum_{i=1}^{n} a^i$$

$$= \sum_{i=0}^{n-1} ia^{i+1} + \sum_{i=1}^{n} a^i.$$

See whether you can manipulate these sums to find S_n on the right side. Then solve for S_n to find the closed formula. $\blacklozenge$

EXAMPLE 6. A sum like $\sum_{i=1}^{n} i^3$ can be solved in terms of the two sums

$$\sum_{i=1}^{n} i \quad \text{and} \quad \sum_{i=1}^{n} i^2.$$

We'll start by adding the term $(n+1)^4$ to $\sum_{i=1}^{n} i^4$. Then we obtain the following equations:

$$\sum_{i=1}^{n} i^4 + (n+1)^4 = \sum_{i=0}^{n} (i+1)^4$$

$$= \sum_{i=0}^{n} (i^4 + 4i^3 + 6i^2 + 4i + 1)$$

$$= \sum_{i=1}^{n} i^4 + 4\sum_{i=1}^{n} i^3 + 6\sum_{i=1}^{n} i^2 + 4\sum_{i=1}^{n} i + \sum_{i=0}^{n} 1.$$

Now cancel the term $\sum_{i=1}^{n} i^4$ from both sides of the above equation to obtain the following equation:

$$(n+1)^4 \;=\; 4\sum_{i=1}^{n} i^3 + 6\sum_{i=1}^{n} i^2 + 4\sum_{i=1}^{n} i + \sum_{i=0}^{n} 1. \tag{5.13}$$

Since we know the closed forms for the latter three sums on the right side of the equation, we can solve for $\sum_{i=1}^{n} i^3$ to find its closed form. We'll leave this as an exercise. We can use the same method to find a closed form for the expression $\sum_{i=1}^{n} i^k$ for any natural number k. ◆

For some problems we need to find new techniques. For example, suppose we wish to find a closed form for the nth Fibonacci number F_n, which is defined by the recurrence system

$$
\begin{aligned}
F_0 &= 0, \\
F_1 &= 1, \\
F_n &= F_{n-1} + F_{n-2} \quad (n \geq 2).
\end{aligned}
$$

We can't use the cancellation technique with this system because F occurs twice on the right side of the general equation. This problem belongs to a large class of problems that need a more powerful technique. That's next.

Generating Functions

As we have seen, recurrences cannot always be solved by cancellation. We need a more powerful technique. The technique that we present comes from the simple idea of equating the coefficients of two polynomials. For example, suppose we have the following equation:

$$a + bx + cx^2 \;=\; 4 + 7x^2.$$

We can solve for a, b, and c by equating coefficients to yield $a = 4$, $b = 0$, and $c = 7$. We'll extend this idea to expressions that have infinitely many terms of the form $a_n x^n$ for each natural number n.

Let's get to the definition. Suppose we have an infinite sequence of numbers

$$a_0, a_1, \dots, a_n, \dots.$$

The *generating function* for this sequence is the following infinite expression, which is also called a formal power series or an infinite polynomial:

$$A(x) = a_0 + a_1 x + a_2 x^2 + \cdots + a_n x^n + \cdots$$

$$= \sum_{n=0}^{\infty} a_n x^n.$$

Two generating functions may be added by adding the corresponding co-efficients. Similarly, two generating functions may be multiplied by extending the rule for multiplying regular polynomials. In other words, multiply each term of one generating function by every term of the other generating function, and then add up all the results. Two generating functions are equal if their corresponding coefficients are equal.

We'll be interested in those generating functions that have closed forms. For example, let's consider the following generating function for the infinite sequence 1, 1, ..., 1, ...:

$$\sum_{n=0}^{\infty} x^n.$$

This generating function is often called a *geometric series*, and its closed form is given by the following formula:

$$\frac{1}{1-x} = \sum_{n=0}^{\infty} x^n. \tag{5.14}$$

To justify equation (5.14), just multiply both sides by the term $1 - x$.

But how can we use this formula to solve recurrences? The idea, as we shall see, is to create an equation in which $A(x)$ is the unknown, solve for $A(x)$, and hope that our solution has a nice closed form. For example, if we find that $A(x) = \frac{1}{1-2x}$, then we can rewrite it using (5.14) as follows:

$$A(x) = \frac{1}{1-2x} = \frac{1}{1-(2x)} = \sum_{n=0}^{\infty} (2x)^n = \sum_{n=0}^{\infty} 2^n x^n.$$

Now we can equate coefficients to obtain the solution $a_n = 2^n$. In other words, the solution sequence is 1, 2, 4, ..., 2^n, ...

The Method

How do we obtain the closed form for $A(x)$? It's a four-step process, and we'll present it with an example. Suppose we want to solve the following recurrence:

$$a_0 = 0, \tag{5.15}$$

$$a_1 = 1,$$

$$a_n = 5a_{n-1} - 6a_{n-2} \quad (n \geq 2).$$

Step 1

Use the general equation in the recurrence to write an infinite polynomial with coefficients a_n. We start the index of summation at 2 because the general equation in (5.15) holds for $n \geq 2$. Thus we obtain the following equation:

$$\sum_{n=2}^{\infty} a_n x^n = \sum_{n=2}^{\infty} (5a_{n-1} - 6a_{n-2}) x^n$$

$$= \sum_{n=2}^{\infty} 5a_{n-1} x^n - \sum_{n=2}^{\infty} 6a_{n-2} x^n \qquad (5.16)$$

$$= 5\sum_{n=2}^{\infty} a_{n-1} x^n - 6\sum_{n=2}^{\infty} a_{n-2} x^n$$

We want to solve for $A(x)$ from this equation. Therefore we need to transform each infinite polynomial in (5.16) into an expression containing $A(x)$. To do this, notice that the left-hand side of (5.16) can be written as follows:

$$\sum_{n=2}^{\infty} a_n x^n = A(x) - a_0 - a_1 x$$

$$= A(x) - x \qquad \text{(substitute for } a_0 \text{ and } a_1\text{)}.$$

The first infinite polynomial on the right side of (5.16) can be written as follows:

$$\sum_{n=2}^{\infty} a_{n-1} x^n = \sum_{n=1}^{\infty} a_n x^{n+1} \qquad \text{(by a change of indices)}$$

$$= x\sum_{n=1}^{\infty} a_n x^n$$

$$= x(A(x) - a_0)$$

$$= xA(x).$$

The second infinite polynomial on the right side of (5.16) can be written as follows:

$$\sum_{n=2}^{\infty} a_{n-2} x^n = \sum_{n=0}^{\infty} a_n x^{n+2} \qquad \text{(by a change of indices)}$$

$$= x^2 \sum_{n=0}^{\infty} a_n x^n$$

$$= x^2 A(x).$$

Thus equation (5.16) can be rewritten in terms of $A(x)$ as follows:

$$A(x) - x = 5 x A(x) - 6 x^2 A(x). \tag{5.17}$$

Step 1 can often be done equationally by starting with the definition of $A(x)$ and continuing until an equation involving $A(x)$ is obtained. For this example the process goes as follows:

$$
\begin{aligned}
A(x) &= \sum_{n=0}^{\infty} a_n x^n \\[2mm]
&= a_0 + a_1 x + \sum_{n=2}^{\infty} a_n x^n \\[2mm]
&= x + \sum_{n=2}^{\infty} a_n x^n \\[2mm]
&= x + \sum_{n=2}^{\infty} (5 a_{n-1} - 6 a_{n-2}) x^n \\[2mm]
&= x + \sum_{n=2}^{\infty} 5 a_{n-1} x^n - \sum_{n=2}^{\infty} 6 a_{n-2} x^n \\[2mm]
&= x + 5 \sum_{n=2}^{\infty} a_{n-1} x^n - 6 \sum_{n=2}^{\infty} a_{n-2} x^n \\[2mm]
&= x + 5x\, (A(x) - a_0) - 6x^2 A(x) \\[2mm]
&= x + 5x\, A(x) - 6x^2 A(x).
\end{aligned}
$$

Step 2

Solve the equation for $A(x)$ and try to transform the result into an expression containing closed forms of known generating functions. We solve equation (5.17) by isolating $A(x)$ as follows:

$$A(x)(1 - 5x + 6x^2) = x.$$

Therefore we can solve for $A(x)$ and try to obtain known closed forms, which can then be replaced by generating functions:

$$A(x) \;=\; \frac{x}{1 - 5x + 6x^2}$$

$$=\; \frac{x}{(2x - 1)(3x - 1)}$$

$$=\; \frac{1}{2x - 1} - \frac{1}{3x - 1} \qquad \text{(partial fractions)}$$

$$=\; -\frac{1}{1 - 2x} + \frac{1}{1 - 3x} \qquad \left(\text{put into the form } \frac{1}{1 - t}\,\right)$$

$$=\; -\sum_{n=0}^{\infty} (2x)^n + \sum_{n=0}^{\infty} (3x)^n$$

$$=\; -\sum_{n=0}^{\infty} 2^n x^n + \sum_{n=0}^{\infty} 3^n x^n$$

$$=\; \sum_{n=0}^{\infty} (-2^n + 3^n) x^n.$$

Step 3

Equate coefficients, and obtain the result. In other words, we equate the original definition for $A(x)$ and the form of $A(x)$ obtained in Step 2:

$$\sum_{n=0}^{\infty} a_n x^n \;=\; \sum_{n=0}^{\infty} (-2^n + 3^n) x^n.$$

These two infinite polynomials are equal if and only if the corresponding coefficients are equal. Equating the coefficients, we obtain the following closed form for a_n:

$$a_n = 3^n - 2^n \quad \text{for } n \ge 0. \tag{5.18}$$

Step 4 (Check the answer)

One way to do this is to use induction to prove that (5.18) is the correct answer to the recurrence (5.15). Since the recurrence has two basis cases, we need to start our induction by verifying the special cases $n = 0$ and $n = 1$. These cases are verified below:

$$a_0 = 3^0 - 2^0 = 0,$$
$$a_1 = 3^1 - 2^1 = 1.$$

Now assume that $n \geq 2$ and (5.15) is true for all $k < n$. Then, starting on the right side of (5.15), we have

$$
\begin{aligned}
5a_{n-1} - 6a_{n-2} &= 5\left(3^{n-1} - 2^{n-1}\right) - 6\left(3^{n-2} - 2^{n-2}\right) \quad &\text{(induction)} \\
&= 3^n - 2^n \quad &\text{(simplification)} \\
&= a_n.
\end{aligned}
$$

An Aside on Partial Fractions

Let's recall a few facts about partial fractions. Suppose we are given the following quotient of two polynomials $p(x)$ and $q(x)$:

$$
\frac{p(x)}{q(x)},
$$

where the degree of $p(x)$ is less than the degree of $q(x)$. The first thing to do is factor $q(x)$ into a product of linear and/or quadratic polynomials that can't be factored further (say over the real numbers). Therefore each factor of $q(x)$ has one of the following forms:

$$
ax + b \quad \text{or} \quad cx^2 + dx + e.
$$

The *partial fraction* representation of

$$
\frac{p(x)}{q(x)}
$$

is a sum of terms, where each term in the sum is a quotient as follows:

1. If the linear polynomial $ax + b$ is repeated k times as a factor of $q(x)$, then add the following terms to the partial fraction representation, where A_1, ..., A_k are constants to be determined:

$$
\frac{A_1}{ax + b} + \frac{A_2}{(ax + b)^2} + \cdots + \frac{A_k}{(ax + b)^k}.
$$

2. If the quadratic polynomial $cx^2 + dx + e$ is repeated k times as a factor of $q(x)$, then add the following terms to the partial fraction representation, where A_i and B_i are constants to be determined:

$$
\frac{A_1 x + B_1}{cx^2 + dx + e} + \frac{A_2 x + B_2}{(cx^2 + dx + e)^2} + \cdots + \frac{A_k x + B_k}{(cx^2 + dx + e)^k}.
$$

EXAMPLE 7. Here are a few samples of partial fractions that can be obtained from the two rules:

$$\frac{x-1}{x\,(x-2)\,(x+1)} = \frac{A}{x} + \frac{B}{x-2} + \frac{C}{x+1},$$

$$\frac{x^3-1}{x^2\,(x-2)^3} = \frac{A}{x} + \frac{B}{x^2} + \frac{C}{x-2} + \frac{D}{(x-2)^2} + \frac{E}{(x-2)^3},$$

$$\frac{x^2}{(x-1)\,(x^2+2x+1)} = \frac{A}{x-1} + \frac{Bx+C}{x^2+2x+1},$$

$$\frac{x}{(x-1)\,(x^2+1)^2} = \frac{A}{x-1} + \frac{Bx+C}{x^2+1} + \frac{Dx+E}{(x^2+1)^2}. \quad \blacklozenge$$

To determine the constants in a partial fraction representation, we can solve simultaneous equations. Suppose there are n constants to be found. Then we need to create n equations. To create an equation, pick some value for x, with the restriction that the value for x does not make any denominator zero. Do this for n distinct values for x. Then solve the resulting n equations. For example, in Step 2 of the generating function example we wrote down the following equalities:

$$
\begin{aligned}
A(x) &= \frac{x}{1-5x+6x^2} \\[2mm]
&= \frac{x}{(2x-1)\,(3x-1)} \\[2mm]
&= \frac{1}{2x-1} - \frac{1}{3x-1}.
\end{aligned}
$$

The last equality is the result of partial fractions. First we write the partial fraction representation

$$\frac{x}{(2x-1)\,(3x-1)} = \frac{A}{2x-1} + \frac{B}{3x-1}.$$

Then we create two equations in A and B by letting $x = 0$ and $x = 1$:

$$0 = -A - B$$

$$\frac{1}{2} = A + \frac{1}{2}B$$

Solving for A and B, we get $A = 1$ and $B = -1$. This yields the desired equality

$$\frac{x}{(2x-1)(3x-1)} = \frac{1}{2x-1} - \frac{1}{3x-1}.$$

A Final Note on Partial Fractions

If the degree of the numerator $p(x)$ is greater than or equal to the degree of $q(x)$, then a simple division of $p(x)$ by $q(x)$ will yield an equation of the form

$$\frac{p(x)}{q(x)} = s(x) + \frac{p'(x)}{q'(x)},$$

where the degree of $p'(x)$ is less than the degree of $q'(x)$. Then we can apply partial fractions to the quotient

$$\frac{p'(x)}{q'(x)}.$$

More Generating Functions

There are many useful generating functions. Since our treatment is not intended to be exhaustive, we'll settle for listing two more generating functions that have many applications.

Two More Useful Generating Functions

$$\frac{1}{(1-x)^{k+1}} = \sum_{n=0}^{\infty} \binom{k+n}{n} x^n \qquad \text{for } k \in \mathbb{N}. \tag{5.19}$$

$$(1+x)^r = \sum_{n=0}^{\infty} \left(\frac{r(r-1) \ldots (r-n+1)}{n!} \right) x^n \qquad \text{for } r \in \mathbb{R}. \tag{5.20}$$

The numerator of the coefficient expression for the nth term in (5.20) contains a product of n numbers. When $n = 0$, we use the convention that a vacuous product—of zero numbers—has the value 1. Therefore the 0th term of (5.20) is $\frac{1}{0!} = 1$. So the first few terms of (5.20) look like the following:

$$(1+x)^r = 1 + rx + \frac{r(r-1)}{2}x^2 + \frac{r(r-1)(r-2)}{6}x^3 + \cdots.$$

Let's finish things off with another complete example.

EXAMPLE 8 (*Parentheses*). Suppose we want to find the number of ways to parenthesize the expression

$$t_1 + t_2 + \cdots + t_{n-1} + t_n \qquad (5.21)$$

so that a parenthesized form of the expression reflects the process of adding two terms. For example, the expression $t_1 + t_2 + t_3 + t_4$ has several different forms, as follows:

$$((t_1 + t_2) + (t_3 + t_4))$$

$$(t_1 + (t_2 + (t_3 + t_4)))$$

$$(t_1 + ((t_2 + t_3) + t_4))$$

$$\vdots$$

To solve the problem, we'll let b_n denote the total number of possible parenthesizations for an n-term expression. Notice that if $1 \le k \le n - 1$, then we can split the expression (5.21) into two subexpressions as follows:

$$t_1 + \cdots + t_{n-k} \quad \text{and} \quad t_{n-k+1} + \cdots + t_n. \qquad (5.22)$$

So there are $b_{n-k}b_k$ ways to parenthesize the expression (5.21) if the final + is placed between the two subexpressions (5.22). If we let k range from 1 to $k - 1$, we obtain the following formula for b_n when $n \ge 2$:

$$b_n = b_{n-1}b_1 + b_{n-2}b_2 + \cdots + b_2 b_{n-2} + b_1 b_{n-1}. \qquad (5.23)$$

But we need $b_1 = 1$ for (5.23) to make sense. It's OK to make this assumption because we're concerned only about expressions that contain at least two terms. Similarly, we can let $b_0 = 0$. So we can write down the recurrence to describe the solution as follows:

$$
\begin{aligned}
b_0 &= 0, \\
b_1 &= 1, \\
b_n &= b_n b_0 + b_{n-1}b_1 + \cdots + b_1 b_{n-1} + b_0 b_n \quad (n \ge 2).
\end{aligned}
\qquad (5.24)
$$

Notice that this system cannot be solved by the cancellation method. Let's try generating functions. Let $B(x)$ be the generating function for the sequence

$$b_0, b_1, \ldots, b_n, \ldots \ .$$

So $B(x) = \sum_{n=0}^{\infty} b_n x^n$. Now let's try to apply the four-step procedure for generating functions. First we use the general equation in the recurrence to introduce the partial (since $n \geq 2$) generating function

$$\sum_{n=2}^{\infty} b_n x^n \;=\; \sum_{n=2}^{\infty} (b_n b_0 + b_{n-1} b_1 + \cdots + b_1 b_{n-1} + b_0 b_n) x^n. \qquad (5.25)$$

Now the left-hand side of (5.25) can be written in terms of $B(x)$:

$$\sum_{n=2}^{\infty} b_n x^n \;=\; B(x) - b_1 x - b_0$$

$$\;=\; B(x) - x \qquad \text{(since } b_0 = 0 \text{ and } b_1 = 1\text{).}$$

Before we discuss the right hand-side of equation (5.25), notice that we can write the product

$$B(x)\,B(x) \;=\; \left(\sum_{n=0}^{\infty} b_n x^n \right)\!\left(\sum_{n=0}^{\infty} b_n x^n \right) \qquad (5.26)$$

$$\;=\; \sum_{n=0}^{\infty} c_n x^n,$$

where $c_0 = b_0 b_0$ and for $n > 0$,

$$c_n \;=\; b_n b_0 + b_{n-1} b_1 + \cdots + b_1 b_{n-1} + b_0 b_n.$$

So the right-hand side of equation (5.25) can be written as

$$\sum_{n=2}^{\infty} (b_n b_0 + b_{n-1} b_1 + \cdots + b_1 b_{n-1} + b_0 b_n) x^n$$

$$\;=\; B(x)\,B(x) - b_0 b_0 - (b_1 b_0 + b_0 b_1)$$

$$\;=\; B(x)\,B(x) \qquad \text{(since } b_0 = 0\text{).}$$

Now equation (5.25) can be written in simplified form as

$$B(x) - x = B(x)\,B(x) \quad \text{or} \quad B(x)^2 - B(x) + x = 0.$$

Now, thinking of $B(x)$ as the unknown, the equation is a quadratic equation with two solutions:

$$B(x) \;=\; \frac{1 \pm \sqrt{1 - 4x}}{2}.$$

Notice that $\sqrt{1-4x}$ is the closed form for a binomial generating function obtained from (5.20), where $r = \frac{1}{2}$. Thus we can write

$$\sqrt{1-4x} \;=\; (1+(-4x))^{\frac{1}{2}}$$

$$=\; \sum_{n=0}^{\infty} \frac{\frac{1}{2}\left(\frac{1}{2}-1\right)\left(\frac{1}{2}-2\right)\cdots\left(\frac{1}{2}-n+1\right)}{n!}(-4x)^n$$

$$=\; \sum_{n=0}^{\infty} \frac{\frac{1}{2}\left(-\frac{1}{2}\right)\left(-\frac{3}{2}\right)\cdots\left(-\frac{2n-3}{2}\right)}{n!}(-2)^n\,2^n\,x^n$$

$$=\; 1+\sum_{n=1}^{\infty} \frac{(-1)(1)(3)\cdots(2n-3)}{n!}\,2^n\,x^n$$

$$=\; 1+\sum_{n=1}^{\infty}\left(-\frac{2}{n}\right)\binom{2n-2}{n-1}x^n.$$

The last equality is left as an exercise. Notice that for $n \geq 1$ the coefficient of x^n is negative in this generating function. In other words, the nth term ($n \geq 1$) of the generating function for $\sqrt{1-4x}$ always has a negative coefficient. Since we need positive values for b_n, we must choose the following solution of our quadratic equation:

$$B(x) \;=\; \frac{1}{2}-\frac{1}{2}\sqrt{1-4x}.$$

Putting things together, we can write our desired generating function as follows:

$$\sum_{n=0}^{\infty} b_n x^n \;=\; B(x) \;=\; \frac{1}{2}-\frac{1}{2}\sqrt{1-4x}$$

$$=\; \frac{1}{2}-\frac{1}{2}\left\{1+\sum_{n=1}^{\infty}\left(-\frac{2}{n}\right)\binom{2n-2}{n-1}x^n\right\}$$

$$=\; 0+\sum_{n=1}^{\infty}\frac{1}{n}\binom{2n-2}{n-1}x^n.$$

Now we can finish the job by equating coefficients to obtain the following solution:

$$b_n \;=\; \text{if } n = 0 \text{ then } 0 \text{ else } \frac{1}{n}\binom{2n-2}{n-1}. \quad \blacklozenge$$

Exercises

1. Solve each of the following recurrences by the cancellation technique. Put each answer in closed form (no ellipsis allowed).

   a. $a_1 = 0$,
      $a_n = a_{n-1} + 4$.

   b. $a_1 = 0$,
      $a_n = a_{n-1} + 2n$.

   c. $a_0 = 1$,
      $a_n = 2a_{n-1} + 3$.

2. Find a closed form for each of the following sums.

   a. $3 + 2 \cdot 3^2 + 3 \cdot 3^3 + 4 \cdot 3^4 + \cdots + n3^n$.

   b. $n + 2(n-1) + 2^2(n-2) + \cdots + 2^{n-1}$.

3. Solve equation (5.13) to find a closed form for the expression $\sum_{i=1}^{n} i^3$.

4. The *Tower of Hanoi* puzzle was invented by Lucas in 1883. It consists of three stationary pegs with one peg containing a stack of n disks that form a tower (each disk has a hole in the center for the peg) in which each disk has a smaller diameter than the disk below it. The problem is to move the tower to one of the other pegs by transferring one disk at a time from one peg to another peg, no disk ever being placed on a smaller disk. Find the minimum number of moves H_n to do the job.

   Hint: It takes 0 moves to transfer a tower of 0 disks and 1 move to transfer a tower of 1 disk. So $H_0 = 0$ and $H_1 = 1$. Try it out for $n = 2$ and $n = 3$ to get the idea. Then try to find a recurrence relation for the general term H_n as follows: Move the tower consisting of the top $n - 1$ disks to the nonchosen peg; then move the bottom disk to the chosen peg; then move the tower of $n - 1$ disks onto the chosen peg.

5. A diagonal in a polygon is a line from one vertex to another nonadjacent vertex. For example, a triangle doesn't have any diagonals because each vertex is adjacent to the other vertices. Find the number of diagonals in an n-sided polygon where $n \geq 3$.

6. Given the generating function $A(x) = \sum_{n=0}^{\infty} a_n x^n$, for each of the following representations of $A(x)$, write down the closed form for the general term a_n.

   a. $A(x) = \dfrac{1}{x-2} - \dfrac{2}{3x+1}$.

   b. $A(x) = \dfrac{1}{2x+1} + \dfrac{3}{x+6}$.

   c. $A(x) = \dfrac{1}{3x-2} - \dfrac{1}{(1-x)^2}$.

7. Use generating functions to solve each of the following recurrences.

   a. $a_0 = 0$,
      $a_1 = 4$,
      $a_n = 2a_{n-1} + 3a_{n-2}$ $(n \geq 2)$.

b. $a_0 = 0$,
   $a_1 = 1$,
   $a_n = 7a_{n-1} - 12a_{n-2}$ $(n \geq 2)$.

c. $a_0 = 0$,
   $a_1 = 1$,
   $a_2 = 1$,
   $a_n = 2a_{n-1} + a_{n-2} - 2a_{n-3}$ $(n \geq 3)$.

8. Use generating functions to solve each recurrence in Exercise 1. For those recurrences that do not have an a_0 term, assume that $a_0 = 0$.

9. For each of the following functions, find a recurrence to describe the number of times the cons operation :: is called. Solve each recurrence.

   a. $\mathrm{cat}(L, M) = $ if $L = \langle\,\rangle$ then M else $\mathrm{head}(L) :: \mathrm{cat}(\mathrm{tail}(L), M)$.

   b. $\mathrm{dist}(x, L) = $ if $L = \langle\,\rangle$ then $\langle\,\rangle$
      else $(x :: \mathrm{head}(L) :: \langle\,\rangle) :: \mathrm{dist}(x, \mathrm{tail}(L))$.

   c. $\mathrm{power}(L) = $ **if** $L = \langle\,\rangle$ **then** return $\langle\,\rangle :: \langle\,\rangle$
      else
         $A := \mathrm{power}(\mathrm{tail}(L))$;
         $B := \mathrm{dist}(\mathrm{head}(L), A)$;
         $C := \mathrm{map}(::)\,(B)$;
         return $\mathrm{cat}(A, C)$
      fi

10. Prove that the following equation holds for all positive integers n, in two different ways, as indicated:

$$\frac{(1)(1)(3)\cdots(2n-3)}{n!}\,2^n = \frac{2}{n}\binom{2n-2}{n-1}.$$

   a. Use induction.

   b. Transform the left side into the right side by "inserting" the missing even numbers in the numerator.

11. Find a closed form for the number of structurally distinct binary trees with n nodes, where $n \geq 0$.

12. Recall that a derangement of a string is a permutation of the letters of the string such that no letter remains in the same position. In terms of bijections a derangement of a set S is a bijection f on S such that $f(x) \neq x$ for all x in S. The number of derangements of an n-element set can be given by the following recurrence:

$$d_1 = 0,$$
$$d_2 = 1,$$
$$d_n = (n-1)(d_{n-1} + d_{n-2}) \qquad (n \geq 3).$$

Solve this recurrence. *Hint:* One way is to show that the general term d_n can be rewritten as follows: $d_n = nd_{n-1} + (-1)^n$. Then solve this recurrence by cancellation.

13. Find a closed form for the nth Fibonacci number defined by the following recurrence system:

$$F_0 = 0,$$
$$F_1 = 1,$$
$$F_n = F_{n-1} + F_{n-2} \qquad (n \geq 2).$$

5.4 Comparing Rates of Growth

Sometimes it makes sense to approximate the number of steps required to execute an algorithm because of the difficulty involved in finding a closed form for an expression or the difficulty in evaluating an expression. To approximate one function with another function, we need some way to compare them. That's where "rate of growth" comes in. We want to give some meaning to statements like "f has the same growth rate as g" and "f has a lower growth rate than g."

For our purposes we will consider functions whose domains and codomains are subsets of the real numbers. We'll examine the asymptotic behavior of two functions f and g by comparing $f(n)$ and $g(n)$ for large positive values of n (i.e., as n approaches infinity).

Big Theta

Let's begin by discussing the meaning of the statement "f has the same growth rate as g." We say that f has the *same growth rate* as g, or f has the *same order* as g, if we can find a number m and two nonzero numbers c and d such that

$$cg(n) \leq f(n) \leq dg(n) \qquad \text{for all } n \geq m. \tag{5.27}$$

In this case we write

$$f(n) = \Theta(g(n)),$$

which is read, "$f(n)$ is *big theta* of $g(n)$." It's easy to verify that the relation "has the same growth rate as" is an equivalence relation. In other words, the following three properties hold for all functions:

$$f(n) = \Theta(f(n)).$$

If $f(n) = \Theta(g(n))$, then $g(n) = \Theta(f(n))$.

If $f(n) = \Theta(g(n))$ and $g(n) = \Theta(h(n))$, then $f(n) = \Theta(h(n))$.

If $f(n) = \Theta(g(n))$ and we also know that $g(n)$ is positive for all $n \geq m$, then we can divide the inequality (5.27) by $g(n)$ to obtain

$$c \leq \frac{f(n)}{g(n)} \leq d \quad \text{for all } n \geq m.$$

This inequality gives us a better way to think about "having the same growth rate." It tells us that the ratio of the two functions is always within a fixed bound beyond some point. We can always take this point of view for functions that count the steps of algorithms because they are positive valued.

Now let's see whether we can find some functions that have the same growth rate. To start things off, suppose f and g are proportional. This means that there is a nonzero constant c such that $f(n) = cg(n)$ for all n. In this case, definition (5.27) is satisfied by letting $d = c$. Thus we have the following statement:

If two functions f and g are proportional, then $f(n) = \Theta(g(n))$.   (5.28)

EXAMPLE 1. Recall that log functions with different bases are proportional. In other words, if we have two bases $a > 1$ and $b > 1$, then

$$\log_a n = (\log_a b)(\log_b n) \quad \text{for all } n > 0.$$

So we can disregard the base of the log function when considering rates of growth. In other words, we have

$$\log_a n = \Theta(\log_b n). \quad \blacklozenge \qquad (5.29)$$

It's interesting to note that two functions can have the same growth rate without being proportional. Here's an example.

EXAMPLE 2. Let's show that $n^2 + n$ and n^2 have the same growth rate. The following inequality is true for all $n \geq 1$:

$$1n^2 \leq n^2 + n \leq 2n^2.$$

Therefore $n^2 + n = \Theta(n^2)$.   ◆

The following theorem gives us a nice tool for showing that two functions have the same growth rate:

$$\text{If } \lim_{n \to \infty} \frac{f(n)}{g(n)} = c \ \text{ where } c \neq 0 \text{ and } c \neq \infty, \text{ then } f(n) = \Theta(g(n)). \qquad (5.30)$$

For example, the quotient $(25n^2 + n)/n^2$ approaches 25 as n approaches infinity. Therefore $25n^2 + n = \Theta(n^2)$.

We should note that the limit in (5.30) is not a necessary condition for $f(n) = \Theta(g(n))$. For example, suppose we let f and g be the functions defined as follows:

$$f(n) = \text{if } n \text{ is odd then } 2 \text{ else } 4,$$

$$g(n) = 2.$$

We can write $1g(n) \leq f(n) \leq 2g(n)$ for all $n \geq 1$. Therefore $f(n) = \Theta(g(n))$. But the quotient $f(n)/g(n)$ alternates between the two values 1 and 2. Therefore the limit of the quotient does not exist. Still the limit test (5.30) will work for the majority of functions that occur in analyzing algorithms.

Approximations can be quite useful for those of us who can't remember formulas that we don't use all the time. For example, we can write the sums from (5.12) in terms of Θ as follows:

$$\sum_{i=1}^{n} i = \Theta(n^2). \qquad (5.31)$$

$$\sum_{i=1}^{n} i^2 = \Theta(n^3). \qquad (5.32)$$

$$\text{If } a \neq 1, \text{ then } \sum_{i=0}^{n} a^i = \Theta(a^{n+1}). \qquad (5.33)$$

$$\text{If } a \neq 1, \text{ then } \sum_{i=0}^{n} ia^i = \Theta(na^{n+1}). \qquad (5.34)$$

The first two sums (5.31) and (5.32) are special cases of the following general result:

$$\sum_{i=1}^{n} i^k = \Theta(n^{k+1}).\qquad(5.35)$$

EXAMPLE 3. Let's clarify a statement that we made in Example 1 of Section 5.2. We showed that $\lceil \log_2 n! \rceil$ is the worst case lower bound for comparison sorting algorithms. But $\log n!$ is hard to calculate for even modest values of n. We stated that $\lceil \log_2 n! \rceil$ is "approximately" equal to $n \log_2 n$. Now we can make the following statement:

$$\log n! = \Theta(n \log n).\qquad(5.36)$$

To prove this statement, we'll find some bounds on $\log n!$ as follows:

$$
\begin{aligned}
\log n! &= \log n + \log(n-1) + ... + \log 1 \\
&\leq \log n + \log n + ... + \log n \qquad (n \text{ terms}) \\
&= n \log n.
\end{aligned}
$$

$$
\begin{aligned}
\log n! &= \log n + \log(n-1) + ... + \log 1 \\
&\geq \log \mathrm{n} + \log(n-1) + ... + \log(\lceil n/2 \rceil) \quad (\lceil n/2 \rceil \text{ terms}) \\
&\geq \log \lceil n/2 \rceil + ... + \log \lceil n/2 \rceil \qquad\qquad (\lceil n/2 \rceil \text{ terms}) \\
&= \lceil n/2 \rceil \log \lceil n/2 \rceil \\
&\geq (n/2) \log(n/2).
\end{aligned}
$$

So we have the inequality:

$$(n/2) \log(n/2) \leq \log n! \leq n \log n.$$

It's easy to see (i.e., as an exercise) that if $n > 4$, then $(1/2) \log n < \log(n/2)$. Therefore we have the following inequality for $n > 4$:

$$(1/2)(n \log n) \leq (n/2) \log(n/2) \leq \log n! \leq n \log n.$$

So there are nonzero constants $1/2$ and 1 and the number 4 such that

$$(1/2)(n \log n) \leq \log n! \leq (1)(n \log n) \quad \text{for all } n > 4.$$

This tells us that $\log n! = \Theta(n \log n)$.   ◆

An important approximation to $n!$ is *Stirling's formula*—named for the mathematician James Stirling (1692–1770)—which is written as follows:

$$n! = \Theta\left(\sqrt{2\pi n}\left(\frac{n}{e}\right)^n\right). \tag{5.37}$$

Let's see how we can use big theta to discuss the approximate performance of algorithms. For example, the worst case performance of the binary search algorithm is $\Theta(\log n)$ because the actual value is $1 + \lfloor \log_2 n \rfloor$. Both the average and worst case performances of a linear sequential search are $\Theta(n)$ because the average number of comparisons is $\frac{n+1}{2}$ and the worst case number of comparisons is n.

For sorting algorithms that sort by comparison the worst case lower bound is $\lceil \log_2 n! \rceil = \Theta(n \log n)$. Many sorting algorithms, like the simple sort algorithm in Section 5.1, have worst case performance of $\Theta(n^2)$. The "dumbSort" algorithm, which constructs a permutation of the given list and then checks to see whether it is sorted, may have to construct all possible permutations before it gets the right one. Thus dumbSort has worst case performance of $\Theta(n!)$. An algorithm called "heapsort" will sort any list of n items using at most $2n \log_2 n$ comparisons. So heapsort is a $\Theta(n \log n)$ algorithm in the worst case.

Little Oh

Now let's discuss the meaning of the statement "f has a lower growth rate than g." We say that f has a *lower growth rate* than g or f has *lower order* than g if

$$\lim_{n \to \infty} \frac{f(n)}{g(n)} = 0. \tag{5.38}$$

In this case we write

$$f(n) = o(g(n)),$$

which can be read, "f is *little oh* of g." When you say, "little oh," think of "lower order."

For example, the quotient n/n^2 approaches 0 as n goes to infinity. Therefore $n = o(n^2)$, and we can say that n has lower order than n^2. For another example, if a and b are positive numbers such that $a < b$, then $a^n = o(b^n)$. To see this, notice that the quotient $a^n/b^n = (a/b)^n$ approaches 0 as n approaches infinity because $0 < a/b < 1$.

For those readers familiar with derivatives, the evaluation of limits can often be accomplished by using L'Hôpital's rule:

$$\text{If } \lim_{n \to \infty} f(n) = \lim_{n \to \infty} g(n) = \infty \quad \text{or} \quad \lim_{n \to \infty} f(n) = \lim_{n \to \infty} g(n) = 0 \qquad (5.39)$$

and f and g are differentiable beyond some point, then

$$\lim_{n \to \infty} \frac{f(n)}{g(n)} = \lim_{n \to \infty} \frac{f'(n)}{g'(n)}.$$

EXAMPLE 4. We'll show that $\log n = o(n)$. Since both n and $\log n$ approach infinity as n approaches infinity, we can apply (5.39) to $(\log n)/n$. Since we can write $\log n = (\log e)(\log_e n)$, it follows that the derivative of $\log n$ is $(\log e)(1/n)$. Therefore we obtain the following equations:

$$\lim_{n \to \infty} \frac{\log n}{n} = \lim_{n \to \infty} \frac{(\log e)(1/n)}{1} = 0.$$

So $\log n$ has lower order than n, and we can write $\log n = o(n)$.   ◆

Let's list a hierarchy of some familiar functions according to their growth rates, where $f(n) \prec g(n)$ means that $f(n) = o(g(n))$:

$$1 \prec \log n \prec n \prec n \log n \prec n^2 \prec n^3 \prec 2^n \prec 3^n \prec n! \prec n^n. \qquad (5.40)$$

This hierarchy can help us compare different algorithms. For example, we would certainly choose an algorithm with running time $\Theta(\log n)$ over an algorithm with running time $\Theta(n)$.

Big Oh and Big Omega

It is often useful to express the fact that f has order less than or equal to g. In terms of Θ and o this means either $f(n) = \Theta(g(n))$ or $f(n) = o(g(n))$. The standard notation for this either-or situation is

$$f(n) = O(g(n)), \qquad (5.41)$$

which we read, "$f(n)$ is *big oh* of $g(n)$." For example, we have $2n + 1 = O(n^2)$ and $300n^2 + n = O(n^2)$. We can define $f(n) = O(g(n))$ without regard to big theta and little oh as follows:

$f(n) = O(g(n))$ means that there are positive numbers c and m (5.42)
such that $|f(n)| \le c\,|g(n)|$ for all $n \ge m$.

EXAMPLE 5. We'll show that $n^2 = O(n^3)$ and $5n^3 + 2n^2 = O(n^3)$. Since $n^2 \le 1n^3$ for all $n \ge 1$, it follows that $n^2 = O(n^3)$. Since $5n^3 + 2n^2 \le 7n^3$ for all $n \ge 1$, it follows that $5n^3 + 2n^2 = O(n^3)$.   ◆

Now let's go the other way. Suppose we want to express the fact that f has order greater than or equal to g. In terms of Θ and o this means that either

$$f(n) = \Theta(g(n)) \quad \text{or} \quad g(n) = o(f(n)).$$

The standard notation for this either-or situation is

$$f(n) = \Omega(g(n)), \tag{5.43}$$

which we can read, "$f(n)$ is *big omega* of $g(n)$." For example, $2n^3 = \Omega(n^2)$ and $300n^2 = \Omega(n^2)$. We can define $f(n) = \Omega(g(n))$ without regard to big theta and little oh as follows:

$f(n) = \Omega(g(n))$ means that there are positive numbers c and m (5.44)
such that $|f(n)| \ge c\,|g(n)|$ for all $n \ge m$.

EXAMPLE 6. We'll show that $n^3 = \Omega(n^2)$ and $3n^2 + 2n = \Omega(n^2)$. Since $n^3 \ge 1n^2$ for all $n \ge 1$, it follows that $n^3 = \Omega(n^2)$. Since $3n^2 + 2n \ge 1n^2$ for all $n \ge 1$, it follows that $3n^2 + 2n = \Omega(n^2)$.

Let's see how we can use the terms that we've defined so far to discuss algorithms. For example, suppose we have constructed an algorithm A to solve some problem P. Suppose further that we've analyzed A and found that it takes $5n^2$ operations in the worst case for an input of size n. This allows us to make a few general statements. First, we can say that the worst case performance of A is $\Theta(n^2)$. Second, we can say that an optimal algorithm for P, if one exists, must have a worst case performance of $O(n^2)$. In other words, an optimal algorithm for P must do no worse than our algorithm A.

Continuing with our example, suppose some good soul has computed a worst case theoretical lower bound of $\Theta(n \log n)$ operations for any algorithm that solves P. Then we can say that an optimal algorithm, if one exists, must have a worst case performance of $\Omega(n \log n)$. In other words, an optimal algorithm for P can do no better than the given lower bound of $\Theta(n \log n)$.

Before we leave our discussion of approximate optimality, let's look at some other ways to use the symbols. The four symbols Θ, o, O, and Ω can also be used to represent terms within an expression. For example, the equation

$$h(n) = 4n^3 + O(n^2)$$

means that $h(n)$ equals $4n^3$ plus a term of order at most n^2. When used as part of an expression, big oh is the most popular of the four symbols because it gives a nice way to concentrate on those terms that contribute the most muscle.

We should also note that the four symbols Θ, o, O, and Ω can be formally defined to represent sets of functions. In other words, for a function g we define the following four sets:

$\Theta(g)$ is the set of functions with the same order as g;

$o(g)$ is the set of functions with lower order than g;

$O(g)$ is the set of functions with the same order or lower order than g;

$\Omega(g)$ is the set of function with the same order or higher order than g.

When set representations are used, we can use an expression like $f(n) \in \Theta(g(n))$ to mean that f has the same order as g. The set representations also give some nice relationships. For example, we have the following set equalities:

$$O(g(n)) = \Theta(g(n)) \cup o(g(n)),$$
$$\Theta(g(n)) = O(g(n)) \cap \Omega(g(n)),$$
$$o(g(n)) = O(g(n)) - \Theta(g(n)).$$

Exercises

1. Prove that the binary relation on functions defined by $f(n) = \Theta(g(n))$ is an equivalence relation.

2. For any constant $k > 0$, prove each of the following statements.
   a. $\log(kn) = \Theta(\log n)$.
   b. $\log(k + n) = \Theta(\log n)$.

3. Find an example of an increasing function f such that $f(n) = \Theta(1)$.

4. Prove the following sequence of orders: $n \prec n \log n \prec n^2$.

5. Find a place to insert the function $\log \log n$ in the sequence (5.40).

6. For each each of the following functions f, find an appropriate place in the sequence (5.40).

    a. $f(n) = \log 1 + \log 2 + \log 3 + \cdots + \log n$.

    b. $f(n) = \log 1 + \log 2 + \log 4 + \cdots + \log 2^n$.

7. For any constant k, show that n^k has lower order than 2^n.

8. For each of the following values of n, calculate the following three numbers: the exact value of $n!$, Stirling's approximation (5.37) for the value of $n!$, and the difference between the two values.

    a. $n = 5$.   b. $n = 10$.

9. Prove the following sequence of orders: $2^n \prec n! \prec n^n$.

10. Let $f(n) = O(h(n))$ and $g(n) = O(h(n))$. Prove each of the following statements.

    a. $af(n) = O(h(n))$ for any real number a.

    b. $f(n) + g(n) = O(h(n))$.

Chapter Summary

This chapter introduces some basic tools and techniques that are used to analyze algorithms. The idea of an optimal algorithm is introduced by comparing algorithms by the number of operations they perform in the worst case. A lower bound is a value that can't be beat by any algorithm in a particular class. An optimal algorithm's performance matches the lower bound.

Two useful things to count are permutations, in which order is important, and combinations, in which order is not important. Pascal's triangle contains formulas for combinations, which are the same as binomial coefficients. There are formulas to count permutations and combinations of bags, which allow redundant elements. Finite probability—with finite sample spaces—gives us the tools to define the average case performance of an algorithm.

Counting problems can give rise to recurrences that need to be solved. Two techniques to solve recurrences are cancellation and the use of generating functions. These solution techniques often give rise to finite sums that need closed form solutions.

Often it makes sense to find approximations for functions that describe the number of operations performed by an algorithm. The rates of growth of two functions can be compared in various ways—big theta, little oh, big oh, and big omega.

Notes

In this chapter we've just scratched the surface of techniques for manipulating expressions that crop up in counting things while analyzing algorithms. The book by Knuth [1968] contains the first account of a collection of techniques for the analysis of algorithms. The book by Graham, Knuth, and Patashnik [1989] contains a host of techniques, formulas, anecdotes, and further references to the literature. The book also introduces an alternative notation for working with sums, which often makes it easier to manipulate them without having to change the expressions for the upper and lower limits of summation. The notation is called Iverson's convention, and it is also described in the article by Knuth [1992].

6

Elementary Logic

> *... if it was so, it might be; and if it were so, it would be: but as it isn't, it ain't. That's logic.*
> —Tweedledee in *Through the Looking-Glass*
> by Lewis Carroll (1832–1898)

Why is it important to study logic? Two things that we continually try to accomplish are to understand and to be understood. We attempt to understand an argument given by someone so that we can agree with the conclusion or, possibly, so that we can say that the reasoning does not make sense. We also attempt to express arguments to others without making a mistake. A formal study of logic will help improve these fundamental communication skills.

Why should a student of computer science study logic? A computer scientist needs logical skills to argue whether or not a problem can be solved on a machine, to transform logical statements from everyday language to a variety of computer languages, to argue that a program is correct, and to argue that a program is efficient. Computers are constructed from logic devices and are programmed in a logical fashion. Computer scientists must be able to understand and apply new ideas and techniques for programming, many of which require a knowledge of the formal aspects of logic.

In this chapter we'll discuss the formal character of sentences that contain words like "and," "or," and "not" or a phrase like "if A then B."

Chapter Guide

Section 6.1 starts our study of logic with the philosophical question "How do we reason?" We'll discuss some common things that we all do when we reason, and we'll introduce the general idea of a "calculus" as a thing with which to study logic.

Section 6.2 introduces the basic notions and notations of propositional calculus. We'll discuss the properties of tautology, contradiction, and contingency. We'll introduce the idea of equivalence, and we'll use it to find disjunctive and conjunctive normal forms for formulas.

Section 6.3 introduces the basic techniques for formal reasoning in the propositional calculus. We'll introduce some fundamental inference rules. We'll give a formal definition of "proof," and we'll introduce two basic proof techniques: conditional proof and indirect proof.

6.1 How Do We Reason?

How do we reason with each other in our daily lives? We probably state some facts and then state a conclusion based on the facts. For example, the words and the phrase in the following list are often used to indicate that some kind of conclusion is being made:

therefore, thus, whence, so, ergo, hence, it follows that.

When we state a conclusion of some kind, we are applying a rule of logic called an *inference rule*.

The most common rule of inference is called *modus ponens,* and it works like this: Suppose *A* and *B* are two statements and we assume that *A* and "If *A* then *B*" are both true. We can then infer that *B* is true. A typical example of inference by modus ponens is given by the following three sentences:

If it is raining, then there are clouds in the sky.
It is raining.
Therefore there are clouds in the sky.

We use the modus ponens inference rule without thinking about it. We certainly learned it when we were children, probably by testing a parent. For example, if a child receives a hug from a parent after performing some action, it might dawn on the child that the hug follows after the action. The parent might reinforce the situation by saying, "If you do that again, then you will be rewarded." Parents often make statements such as "If you touch that stove burner, then you will burn your finger." After touching the burner, the child probably knows a little bit more about modus ponens. A parent might say, "If you do that again, then you are going to be punished." The normal child probably will do it again and notice that punishment follows. Eventually, in the child's mind, the statement "If ... then ... punishment" is accepted as a true statement, and the modus ponens rule has taken root.

Another inference rule, which we also learned when we were children, is called *modus tollens*, and it works like this: Suppose *A* and *B* are any two statements. If the statement "If A then B" is true and the statement B is false, then we infer the falsity of statement A. A child might learn this rule initially by thinking, "If I'm not being punished, then I must not be doing anything wrong."

Most of us are also familiar with the false reasoning exhibited by the *non sequitur*, which means "It does not follow." For example, someone might make several true statements and then conclude that some other statement is true, even though it has nothing to do with the assumptions. The hope is that we can recognize this kind of false reasoning so that we never use it. For example, the following three sentences form a non sequitur:

> Io is a moon of Jupiter.
> Titan is a moon of Saturn.
> Therefore Earth is the third planet from the sun.

Here's another example of a non sequitur:

> You squandered the money entrusted to you.
> You did not keep required records.
> You incurred more debt than your department is worth.
> Therefore you deserve a promotion.

So we reason by applying inference rules to sentences that we assume are true, obtaining new sentences that we conclude are true. Each of us has our own personal reasoning system in which the assumptions are those English sentences that we assume are true and the inference rules are all the rules that we personally use to convince other people that something is true. But there's a problem.

When two people disagree on what they assume to be true or on how they reason about things, then they have problems trying to reason with each other. Some people call this "lack of communication." Other people call it something worse, especially when things like non sequiturs are part of a person's reasoning system. Can common ground be found? Are there any reasoning systems that are, or should be, contained in everyone's personal reasoning system? The answer is yes. The study of logic helps us understand and describe the fundamental parts of all reasoning systems.

What Is a Calculus?

The Romans used small beads called "calculi" to perform counting tasks. The word "calculi" is the plural of the word "calculus." So it makes sense to think that "calculus" has something to do with calculating. Since there are many kinds of calculation, it shouldn't surprise us that "calculus" is used in many different contexts. Let's give a definition.

A *calculus* is a language of expressions of some kind, with definite rules for forming the expressions. There are values, or meanings, associated with the expressions, and there are definite rules to transform one expression into another expression having the same value.

The English language is something like a calculus, where the expressions are sentences formed by English grammar rules. Certainly, we associate meanings with English sentences. But there are no definite rules for transforming one sentence into another. So our definition of a calculus is not quite satisfied. Let's try again with a programming language X. We'll let the expressions be the programs written in the X language. Is this a calculus? Well, there are certainly rules for forming the expressions, and the expressions certainly have meaning. Are there definite rules for transforming one X language program into another X language program? For most modern programming languages the answer is no. So we don't quite have a calculus. We should note that compilers transform X language programs into Y language programs, where X and Y are different languages. Thus a compiler does not qualify as a calculus transformation rule.

In mathematics the word "calculus" usually means the calculus of real functions. For example, the two expressions

$$D_x[f(x)g(x)] \quad \text{and} \quad f(x)D_x g(x) + g(x)D_x f(x)$$

are equivalent in this calculus. The calculus of real functions satisfies our definition of a calculus because there are definite rules for forming the expressions and there are definite rules for transforming expressions into equivalent expressions.

We'll be studying some different kinds of "logical" calculi. In a logical calculus the expressions are defined by rules, the values of the expressions are related to the concepts of true and false, and there are rules for transforming one expression into another. We'll start with a question.

How Can We Tell Whether Something Is a Proof?

When we reason with each other, we normally use informal proof techniques from our personal reasoning systems. This brings up a few questions:

> What is an informal proof?
> What is necessary to call something a proof?
> How can I tell whether an informal proof is correct?
> Is there a proof system to learn for each subject of discussion?
> Can I live my life without all this?

A formal study of logic will provide us with some answers to these questions. We'll find general methods for reasoning that can be applied informally in many different situations. We'll introduce a precise language for expressing arguments formally, and we'll discuss ways to translate an informal argument into a formal argument. This is especially important in computer science, in which formal solutions (programs) are required for informally stated problems.

6.2   Propositional Calculus

To discuss reasoning, we need to agree on some rules and notation about the truth of sentences. A sentence that is either true or false is called a *proposition*. For example, each of the following lines contains a proposition:

> Winter begins in June in the Southern Hemisphere.
>
> $2 + 2 = 4$.
>
> If it is raining, then there are clouds in the sky.
>
> I may or may not go to a movie tonight.
>
> All integers are even.
>
> There is a prime number greater than a googol.

For this discussion we'll denote propositions by the letters P, Q, and R, possibly subscripted. Propositions can be combined to form more complicated propositions, just the way we combine sentences, using the words "not," "and," "or," and the phrase "if... then..." These combining operations are often called *connectives*. We'll denote them by the following symbols and words:

> $\neg$ not, negation.
>
> $\wedge$ and, conjunction.
>
> $\vee$ or, disjunction.
>
> $\rightarrow$ conditional, implication.

Some common ways to read the expression $P \rightarrow Q$ are "if P then Q," "Q if P," "P implies Q," "P is a sufficient condition for Q," and "Q is a necessary condition for P." P is called the *antecedent*, *premise*, or *hypothesis*, and Q is called the *consequent* or *conclusion* of $P \rightarrow Q$.

Now that we have some symbols, we can denote propositions in symbolic form. For example, if P denotes the proposition "It is raining" and Q denotes the proposition "There are clouds in the sky," then $P \rightarrow Q$ denotes the proposition "If it is raining, then there are clouds in the sky." Similarly, $\neg P$ denotes the proposition "It is not raining."

The four logical operators are defined to reflect their usage in everyday English. Table 6.1, a *truth table*, defines the operators for all possible truth values of their operands.

P	Q	$\neg P$	$P \vee Q$	$P \wedge Q$	$P \to Q$
true	true	false	true	true	true
true	false	false	true	false	false
false	true	true	true	false	true
false	false	true	false	false	true

Table 6.1

Well-Formed Formulas and Semantics

Like any programming language or any natural language, whenever we deal with symbols, at least two questions always arise. The first deals with syntax: Is an expression grammatically (or syntactically) correct? The second deals with semantics: What is the meaning of an expression? Let's look at the first question first.

A grammatically correct expression is called a *well-formed formula*, or wff for short, which can be pronounced "woof." To decide whether an expression is a wff, we need to precisely define the syntax (or grammar) rules for the formation of wffs in our language. So let's do it.

Syntax

As with any language, we must agree on a set of symbols to use as the alphabet. For our discussion we will use the following sets of symbols:

Truth symbols: true, false
Connectives: $\neg, \to, \wedge, \vee$
Propositional variables: Capital letters like P, Q, and R
Punctuation symbols: (,).

Next we need to define those expressions (strings) that form the wffs of our language. We do this by giving the following informal inductive definition for the set of propositional wffs:

A wff is either a truth symbol, or a propositional letter, or the negation of a wff, or the conjunction of two wffs, or the disjunction of two wffs, or the implication of one wff from another, or a wff surrounded by parentheses.

For example, the following expressions are wffs:

$$\text{true, false, } P, \neg Q, P \wedge Q, P \to Q, (P \vee Q) \wedge R, P \wedge Q \to R.$$

If push comes to shove and we must justify that some expression is a wff, then we can apply the inductive definition. Let's look at an example.

EXAMPLE 1. We'll show that the expression $P \wedge Q \vee R$ is a wff. First, we know that P, Q, and R are wffs because they are propositional letters. Therefore $Q \vee R$ is a wff because it's a disjunction of two wffs. It follows that $P \wedge Q \vee R$ is a wff because it's a conjunction of two wffs. We could have arrived at the same conclusion by saying that $P \wedge Q$ is a wff and then stating that $P \wedge Q \vee R$ is a wff, since it is the disjunction of two wffs.   ♦

Can we associate a truth table with each wff? Yes we can, once we agree on a hierarchy of precedence among the connectives. For example, $P \wedge Q \vee R$ is a perfectly good wff. But to find a truth table, we need to agree on which connective to evaluate first. We will define the following hierarchy of evaluation for the connectives of the propositional calculus:

$$\neg \qquad \text{(highest, do first)}$$
$$\wedge$$
$$\vee$$
$$\to \qquad \text{(lowest, do last)}$$

We also agree that the operations $\wedge$, $\vee$, and $\to$ are left associative. In other words, if the same operation occurs two or more times in succession, without parentheses, then evaluate the operations from left to right. Be sure you can tell the reason for each of the following lines, where each line contains a wff together with a parenthesized wff with the same meaning:

$P \vee Q \wedge R$	means	$P \vee (Q \wedge R)$.
$P \to Q \to R$	means	$(P \to Q) \to R$.
$\neg P \vee Q$	means	$(\neg P) \vee Q$.
$\neg P \to P \wedge Q \vee R$	means	$(\neg P) \to ((P \wedge Q) \vee R)$.
$\neg \neg P$	means	$\neg (\neg P)$.

Any wff has a natural syntax tree that clearly displays the hierarchy of the connectives. For example, the syntax tree for the wff $P \wedge (Q \vee \neg R)$ is given by the diagram in Figure 6.1.

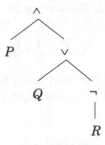

Figure 6.1

Semantics

Now we can say that any wff has a unique truth table. For example, suppose we want to find the truth table for the wff

$$\neg P \rightarrow Q \wedge R.$$

From the hierarchy of evaluation we know that this wff has the following parenthesized form:

$$(\neg P) \rightarrow (Q \wedge R).$$

So we can construct the truth table as follows: Begin by writing down all possible truth values for the three letters P, Q, and R. This gives us a table with eight lines. Next, compute a column of values for $\neg P$. Then compute a column of values for $Q \wedge R$. Finally, use these two columns to compute the column of values for $\neg P \rightarrow Q \wedge R$. Table 6.2 gives the result.

Although we've talked some about meaning, we haven't specifically defined the *meaning*, or *semantics*, of a wff. Let's do it now. We know that any

P	Q	R	$\neg P$	$Q \wedge R$	$\neg P \rightarrow Q \wedge R$
true	true	true	false	true	true
true	true	false	false	false	true
true	false	true	false	false	true
true	false	false	false	false	true
false	true	true	true	true	true
false	true	false	true	false	false
false	false	true	true	false	false
false	false	false	true	false	false

Table 6.2

wff has a unique truth table. So we'll associate each wff with its truth table:

> The meanings of the truth symbols true and false are true and false, respectively. Otherwise, the meaning of a wff is its truth table.

If all the truth table values for a wff are true, then the wff is called a *tautology*. For example, the wffs $P \vee \neg P$ and $P \rightarrow P$ are tautologies. If all the truth table values are false, then the wff is called a *contradiction*. The wff $P \wedge \neg P$ is a contradiction. If some of the truth values are true and some are false, then the wff is called a *contingency*. The wff P is a contingency.

We will often use capital letters to refer to arbitrary propositional wffs. For example, if we say, "A is a wff," we mean that A represents some arbitrary wff. We also use capital letters to denote specific propositional wffs. For example, if we want to talk about the wff $P \wedge (Q \vee \neg R)$ several times in a discussion, we might let $W = P \wedge (Q \vee \neg R)$. Then we can refer to W instead of always writing down the symbols $P \wedge (Q \vee \neg R)$.

Equivalence

Some wffs have the same meaning even though their expressions are different. For example, the wffs $P \wedge Q$ and $Q \wedge P$ have the same meaning because they have the same truth tables. Two wffs are said to be *equivalent* if they have the same meaning. In other words, two wffs are equivalent if their truth tables have the same values. If A and B are equivalent wffs, we denote this fact by writing

$$A \equiv B.$$

For example, we can write $P \wedge Q \equiv Q \wedge P$. The definition of equivalence also allows us to make the following formulation in terms of conditionals:

$$A \equiv B \quad \text{if and only if} \quad (A \rightarrow B) \wedge (B \rightarrow A) \text{ is a tautology.}$$

Before we go much further, let's list a few easy equivalences. See Table 6.3. All these equivalences are easily verified by truth tables, so we'll leave them as exercises. Can we do anything with these equivalences? Sure. We can use them to show that other wffs are equivalent without checking truth tables. But first we need to observe two general properties of equivalence.

The first thing to observe is that equivalence is an "equivalence" relation. In other words, $\equiv$ satisfies the reflexive, symmetric, and transitive properties. The transitive property is the most important property for our purposes. It can be stated as follows for any wffs W, X, and Y:

$$\text{If } W \equiv X \text{ and } X \equiv Y \text{ then } W \equiv Y.$$

Some Basic Equivalences (6.1)

Negation	Disjunction	Conjunction	Implication
$\neg\neg A \equiv A$	$A \vee \text{true} \equiv \text{true}$	$A \wedge \text{true} \equiv A$	$A \rightarrow \text{true} \equiv \text{true}$
	$A \vee \text{false} \equiv A$	$A \wedge \text{false} \equiv \text{false}$	$A \rightarrow \text{false} \equiv \neg A$
	$A \vee A \equiv A$	$A \wedge A \equiv A$	$\text{true} \rightarrow A \equiv A$
	$A \vee \neg A \equiv \text{true}$	$A \wedge \neg A \equiv \text{false}$	$\text{false} \rightarrow A \equiv \text{true}$
			$A \rightarrow A \equiv \text{true}$

Some Conversions	Absorption laws
$A \rightarrow B \equiv \neg A \vee B.$	$A \wedge (A \vee B) \equiv A$
$\neg(A \rightarrow B) \equiv A \wedge \neg B.$	$A \vee (A \wedge B) \equiv A$
$A \rightarrow B \equiv A \wedge \neg B \rightarrow \text{false}.$	$A \wedge (\neg A \vee B) \equiv A \wedge B$
$\wedge$ and $\vee$ are associative.	$A \vee (\neg A \wedge B) \equiv A \vee B$
$\wedge$ and $\vee$ are commutative.	
$\wedge$ and $\vee$ distribute over each other:	De Morgan's laws
$A \wedge (B \vee C) \equiv (A \wedge B) \vee (A \wedge C)$	$\neg(A \wedge B) \equiv \neg A \vee \neg B$
$A \vee (B \wedge C) \equiv (A \vee B) \wedge (A \vee C)$	$\neg(A \vee B) \equiv \neg A \wedge \neg B$

Table 6.3

This property allows us to write a sequence of equivalences and then conclude that the first wff is equivalent to the last wff, just the way we do it with ordinary equality of algebraic expressions.

The next thing to observe is the *replacement rule* of equivalences, which can be stated as follows:

Any subwff of a wff can be replaced by an equivalent wff without changing the truth value of the original wff.

It's just like the old phrase "Substituting equals for equals doesn't change the value of an expression." Can you see why this is OK for equivalences?

For example, suppose we want to simplify the wff $B \rightarrow (A \vee (A \wedge B))$. We might notice that one of the laws from (6.1) gives $A \vee (A \wedge B) \equiv A$. Therefore we can apply the replacement rule and write the following equivalence:

$$B \rightarrow (A \vee (A \wedge B)) \equiv B \rightarrow A.$$

Let's do an example to illustrate the process of showing that two wffs are equivalent without checking truth tables.

EXAMPLE 2. The following equivalence shows an interesting relationship involving the connective →:

$$A \to (B \to C) \equiv B \to (A \to C).$$

We'll prove it using equivalences that we already know. Make sure you can give the reason for each line of the proof.

Proof:

$$
\begin{aligned}
A \to (B \to C) &\equiv A \to (\neg B \vee C) \\
&\equiv \neg A \vee (\neg B \vee C) \\
&\equiv (\neg A \vee \neg B) \vee C \\
&\equiv (\neg B \vee \neg A) \vee C \\
&\equiv \neg B \vee (\neg A \vee C) \\
&\equiv B \to (\neg A \vee C) \\
&\equiv B \to (A \to C). \quad \text{QED.} \quad \blacklozenge
\end{aligned}
$$

This example illustrates that we can use known equivalences like (6.1) as rules to transform wffs into other wffs having the same meaning. This justifies the word "calculus" in the name "propositional calculus."

Now let's look at another application of equivalences to finding the truth value of a proposition.

Is It a Tautology, a Contradiction, or a Contingency?

Suppose our task is to classify a wff W as a tautology, a contradiction, or a contingency. If W contains n letters, then there are 2^n different assignments of truth values to the letters of W. Building a truth table with 2^n rows can be tedious when n is moderately large. Is there another way? Yes. We can use equivalences to solve the problem. If A is a letter and W is a wff, we let

$$W(A/\text{true})$$

denote the wff obtained from W by replacing all occurrences of A by true. Similarly, we define $W(A/\text{false})$ to be the wff obtained from W by replacing all occurrences of A by false. Now we come to the key observation:

W is a tautology   if and only if $W(A/\text{true})$ and $W(A/\text{false})$ are tautologies.

The idea is to simplify $W(A/\text{true})$ and $W(A/\text{false})$ by using the basic properties. After simplification we do the same thing to the resulting wffs. Enough said. It's time for an example to see what we're talking about.

EXAMPLE 3 (*Quine's Method*). Suppose we want to check the meaning of the following wff W:

$$[(A \land B \to C) \land (A \to B)] \to (A \to C).$$

First we compute the two wffs $W(A/\text{true})$ and $W(A/\text{false})$ and simplify them:

$$
\begin{aligned}
W(A/\text{true}) \;&=\; [(\text{true} \land B \to C) \land (\text{true} \to B)] \to (\text{true} \to C) \\
&\equiv\; [(B \to C) \land (\text{true} \to B)] \to (\text{true} \to C) \\
&\equiv\; [(B \to C) \land B] \to C.
\end{aligned}
$$

$$
\begin{aligned}
W(A/\text{false}) \;&=\; [(\text{false} \land B \to C) \land (\text{false} \to B)] \to (\text{false} \to C) \\
&\equiv\; [(\text{false} \to C) \land \text{true}] \to \text{true} \\
&\equiv\; \text{true}.
\end{aligned}
$$

Therefore $W(A/\text{false})$ is a tautology. Now we need to check the simplification of $W(A/\text{true})$. Call it X. We continue the process by constructing the two wffs $X(B/\text{true})$ and $X(B/\text{false})$:

$$
\begin{aligned}
X(B/\text{true}) \;&=\; [(\text{true} \to C) \land \text{true}] \to C \\
&\equiv\; [C \land \text{true}] \to C \\
&\equiv\; C \to C \\
&\equiv\; \text{true}.
\end{aligned}
$$

So $X(B/\text{true})$ is a tautology. Now let's look at $X(B/\text{false})$.

$$
\begin{aligned}
X(B/\text{false}) \;&=\; [(\text{false} \to C) \land \text{false}] \to C \\
&\equiv\; [\text{true} \land \text{false}] \to C \\
&\equiv\; \text{false} \to C \\
&\equiv\; \text{true}.
\end{aligned}
$$

So $X(B/\text{false})$ is also a tautology. Thus X is a tautology, and it follows that W is a tautology.   ◆

Quine's method can also be described graphically with a binary tree. Let W be the root. If N is any node, pick one of its variables, say V, and let the two children of N be $N(V/\text{true})$ and $N(V/\text{false})$. Each node should be simplified as much as possible. Then W is a tautology if all leaves are true, a contradiction if all leaves are false, and a contingency otherwise. Let's illustrate the idea with the wff $P \to Q \wedge P$. The binary tree in Figure 6.2 shows that the wff $P \to Q \wedge P$ is a contingency because Quine's method gives one false leaf and two true leaves.

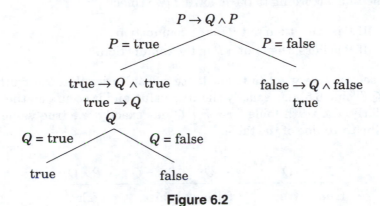

$$P \to Q \wedge P$$

P = true P = false

true $\to Q \wedge$ true   false $\to Q \wedge$ false
true $\to Q$ true
Q

Q = true Q = false

true   false

Figure 6.2

Truth Functions and Normal Forms

A *truth function* is a function whose arguments can take only the values true or false and whose values are either true or false. So any wff defines a truth function. For example, the function g defined by

$$g(P, Q) = P \wedge Q$$

is a truth function. Is the converse true? In other words, is every truth function a wff? The answer is yes. To see why this is true, we'll present a technique to construct a wff for any truth function.

For example, suppose we define the truth function f by saying that $f(P, Q)$ is true exactly when P and Q have opposite truth values. Is there a wff that has the same truth table as f? We'll introduce the technique with this example. Table 6.4 is the truth table for f. We'll explain the statements on the right side of this table.

P	Q	$f(P, Q)$	
true	true	false	
true	false	true	Create $P \wedge \neg Q$
false	true	true	Create $\neg P \wedge Q$
false	false	false	

Table 6.4

We've written the two wffs $P \wedge \neg\, Q$ and $\neg\, P \wedge Q$ on the second and third lines of the table because the values of f are true on these lines. Each wff is a conjunction of argument letters or their negations according to their values on the same line, according to the following two rules:

If P is true, then put P in the conjunction.

If P is false, then put $\neg\, P$ in the conjunction.

Let's see why we want to follow these rules. Notice that the truth table for $P \wedge \neg\, Q$ (Table 6.5) has exactly one true value and it occurs on the second line. Similarly, the truth table for $\neg\, P \wedge Q$ has exactly one true value and it occurs on the third line of the table.

P	Q	$f(P, Q)$	$P \wedge \neg\, Q$	$\neg\, P \wedge Q$
true	true	false	false	false
true	false	true	true	false
false	true	true	false	true
false	false	false	false	false

Table 6.5

Thus each of the tables for $P \wedge \neg\, Q$ and $\neg\, P \wedge Q$ has exactly one true value per column, and these true values occur on the same lines as the true values for f. Since there is one conjunctive wff for each occurrence of true in the table for f, it follows that the table for f can be obtained by taking the disjunction of the tables for $P \wedge \neg\, Q$ and $\neg\, P \wedge Q$. Thus we obtain the following equivalence:

$$f(P, Q) \equiv (P \wedge \neg\, Q) \vee (\neg\, P \wedge Q).$$

Let's do another example to get the idea. Then we will discuss the special forms that we obtain by using this technique.

EXAMPLE 4. Let f be the truth function defined as follows:

$f(P, Q, R) = \text{true}$ if and only if   either $P = Q = \text{false}$ or $Q = R = \text{true}$.

Then f is true in exactly the following four cases:

$f(\text{false}, \text{false}, \text{true}),$
$f(\text{false}, \text{false}, \text{false}),$
$f(\text{true}, \text{true}, \text{true}),$
$f(\text{false}, \text{true}, \text{true}).$

So we can construct a wff equivalent to f by taking the disjunction of the four wffs that correspond to these four cases. The disjunction follows:

$$(\neg P \wedge \neg Q \wedge R) \vee (\neg P \wedge \neg Q \wedge \neg R) \vee (P \wedge Q \wedge R) \vee (\neg P \wedge Q \wedge R). \quad \blacklozenge$$

The method we have described can be generalized to construct an equivalent wff for any truth function having at least one true value. If a truth function doesn't have any true values, then it is a contradiction and is equivalent to false. So every truth function is equivalent to some propositional wff. We'll state this as the following theorem:

Every truth function is equivalent to a propositional wff. (6.2)

Now we're going to discuss some useful forms for propositional wffs. But first we need a little terminology. A *literal* is a propositional letter or its negation. For example, P, Q, $\neg P$, and $\neg Q$ are literals.

Disjunctive Normal Form

A *fundamental conjunction* is either a literal or a conjunction of two or more literals. For example, P and $P \wedge \neg Q$ are fundamental conjunctions. A *disjunctive normal form* (DNF) is either a fundamental conjunction or a disjunction of two or more fundamental conjunctions. For example, the following wffs are DNFs:

$$P \vee (\neg P \wedge Q),$$
$$(P \wedge Q) \vee (\neg Q \wedge P),$$
$$(P \wedge Q \wedge R) \vee (\neg P \wedge Q \wedge R).$$

Sometimes the trivial cases are hardest to see. For example, try to explain why the following four wffs are DNFs: P, $\neg P$, $P \vee \neg P$, and $\neg P \wedge Q$. The propositions that we constructed for truth functions are DNFs.

It is often the case that a DNF is equivalent to a simpler DNF. For example, the DNF $P \vee (P \wedge Q)$ is equivalent to the simpler DNF P by using (6.1). For another example, consider the following DNF:

$$(P \wedge Q \wedge R) \vee (\neg P \wedge Q \wedge R) \vee (P \wedge R).$$

The first fundamental conjunction is equivalent to $(P \wedge R) \wedge Q$, which we see contains the third fundamental conjunction $P \wedge R$ as a subexpression. Thus the first term of the DNF can be absorbed by (6.1) into the third term, which gives the following simpler equivalent DNF:

$$(\neg P \wedge Q \wedge R) \vee (P \wedge R).$$

For any wff W we can always construct an equivalent DNF. If W is a contradiction, then it is equivalent to the single term DNF $P \wedge \neg P$. If W is not a contradiction, then we can write down its truth table and use the technique that we used for truth functions to construct a DNF. So we can make the following statement:

$$\textit{Every wff is equivalent to a DNF.} \qquad (6.3)$$

Another way to construct a DNF for a wff is to transform it into a DNF by using the equivalences of (6.1). In fact we'll outline a short method that will always do the job:

First, remove all occurrences (if there are any) of the connective $\rightarrow$ by using the equivalence

$$A \rightarrow B \equiv \neg A \vee B.$$

Next, move all negations inside to create literals by using De Morgan's equivalences

$$\neg (A \wedge B) \equiv \neg A \vee \neg B \quad \text{and} \quad \neg (A \vee B) \equiv \neg A \wedge \neg B.$$

Finally, apply the distributive equivalences to obtain a DNF. Let's look at an example.

EXAMPLE 5. We'll construct a DNF for the wff $((P \wedge Q) \rightarrow R) \wedge S$.

$$
\begin{aligned}
((P \wedge Q) \rightarrow R) \wedge S \ &\equiv\ (\neg (P \wedge Q) \vee R) \wedge S \\
&\equiv\ (\neg P \vee \neg Q \vee R) \wedge S \\
&\equiv\ (\neg P \wedge S) \vee (\neg Q \wedge S) \vee (R \wedge S). \quad \blacklozenge
\end{aligned}
$$

Suppose W is a wff having n distinct propositional letters. A DNF for W is called a *full disjunctive normal form* if each fundamental conjunction has exactly n literals, one for each of the n letters appearing in W. For example, the following wff is a full DNF:

$$(P \wedge Q \wedge R) \vee (\neg P \wedge Q \wedge R).$$

The wff $P \vee (\neg P \wedge Q)$ is a DNF but not a full DNF because the letter Q does not occur in the first fundamental conjunction.

The truth table technique to construct a DNF for a truth function automatically builds a full DNF because all of the letters in a wff occur in each fundamental conjunction. So we can state the following result:

Every wff that is not a contradiction is equivalent to a full DNF. (6.4)

Conjunctive Normal Form

In a manner entirely analogous to the above discussion we can define a *fundamental disjunction* to be either a literal or the disjunction of two or more literals. A *conjunctive normal form* (CNF) is either a fundamental disjunction or a conjunction of two or more fundamental disjunctions. For example, the following wffs are CNFs:

$$P \wedge (\neg P \vee Q),$$
$$(P \vee Q) \wedge (\neg Q \vee P),$$
$$(P \vee Q \vee R) \wedge (\neg P \vee Q \vee R).$$

Let's look at some trivial examples. Notice that the following four wffs are CNFs: P, $\neg P$, $P \wedge \neg P$, and $\neg P \vee Q$. As in the case for DNFs, some CNFs are equivalent to simpler CNFs. For example, the CNF $P \wedge (P \vee Q)$ is equivalent to the simpler CNF P by (6.1).

Suppose some wff W has n distinct propositional letters. A CNF for W is called a *full conjunctive normal form* if each fundamental disjunction has exactly n literals, one for each of the n letters that appear in W. For example, the following wff is a full CNF:

$$(P \vee Q \vee R) \wedge (\neg P \vee Q \vee R).$$

On the other hand, the wff $P \wedge (\neg P \vee Q)$ is a CNF but not a full CNF.

It's possible to write any truth function f that is not a tautology as a full CNF. In this case we associate a fundamental disjunction with each line of the truth table in which f has a false value, with the property that the fundamental disjunction is false on only that line. Let's return to our original example, in which $f(P, Q)$ is true exactly when P and Q have opposite truth values. In the table for the function f (Table 6.6) we have created the fundamental disjunctions on the right of the lines with false values.

P	Q	$f(P, Q)$	
true	true	false	Create $\neg P \vee \neg Q$
true	false	true	
false	true	true	
false	false	false	Create $P \vee Q$

Table 6.6

In this case, $\neg P$ is added to the disjunction if P = true, and P is added to the disjunction if P = false. Then we take the conjunction of these disjunctions to obtain the following conjunctive normal form of f:

$$f(P, Q) \equiv (\neg P \vee \neg Q) \wedge (P \vee Q).$$

Of course, any tautology is equivalent to the single term CNF $P \vee \neg P$. Now we can state the following results for CNFs, which correspond to statements (6.3) and (6.4) for DNFs:

> *Every wff is equivalent to a CNF.*   (6.5)
>
> *Every wff that is not a tautology is equivalent to a full CNF.*   (6.6)

We should note that some authors use the terms "disjunctive normal form" and "conjunctive normal form" to describe the expressions that we have called "full disjunctive normal forms" and "full conjunctive normal forms." For example, they do not consider $P \vee (\neg P \wedge Q)$ to be a DNF. We use the more general definitions of DNF and CNF because they are useful in describing methods for automatic reasoning and they are useful in describing methods for simplifying digital logic circuits.

Constructing Full Normal Forms Using Equivalences

We can construct full normal forms for wffs without resorting to truth table techniques. Let's start with the full disjunctive normal form. To find a full DNF for a wff, we first convert it to a DNF by the usual actions: eliminate conditionals, move negations inside, and distribute $\wedge$ over $\vee$. For example, the wff $P \wedge (Q \to R)$ can be converted to a DNF in two steps, as follows:

$$
\begin{aligned}
P \wedge (Q \to R) &\equiv P \wedge (\neg Q \vee R) \\
&\equiv (P \wedge \neg Q) \vee (P \wedge R).
\end{aligned}
$$

The right side of the equivalence is a DNF. However, it's not a full DNF because the two fundamental conjunctions don't contain all three letters. The trick to add the extra letters can be described as follows:

> To add a letter, say R, to a fundamental conjunction C without changing the value of C, write the following equivalences:
>
> $$C \equiv C \wedge \text{true} \equiv C \wedge (R \vee \neg R).$$

Then distribute $\wedge$ over $\vee$ to obtain a disjunction of two fundamental conjunctions.

Let's continue with our example. First, we'll add the letter R to the fundamental conjunction $P \wedge \neg Q$. Be sure to justify each step of the following calculation:

$$
\begin{aligned}
P \wedge \neg Q &\equiv (P \wedge \neg Q) \wedge \text{true} \\
&\equiv (P \wedge \neg Q) \wedge (R \vee \neg R) \\
&\equiv (P \wedge \neg Q \wedge R) \vee (P \wedge \neg Q \wedge \neg R).
\end{aligned}
$$

Next, we'll add the letter Q to the fundamental conjunction $P \wedge R$:

$$
\begin{aligned}
P \wedge R &\equiv (P \wedge R) \wedge \text{true} \\
&\equiv (P \wedge R) \wedge (Q \vee \neg Q) \\
&\equiv (P \wedge R \wedge Q) \vee (P \wedge R \wedge \neg Q).
\end{aligned}
$$

Lastly, we put the two wffs together to obtain a full DNF for $P \wedge (Q \to R)$:

$$
(P \wedge \neg Q \wedge R) \vee (P \wedge \neg Q \wedge \neg R) \vee (P \wedge R \wedge Q) \vee (P \wedge R \wedge \neg Q).
$$

EXAMPLE 6. We'll construct a full DNF for the wff $P \to Q$. Make sure to justify each line of the following calculation:

$$
\begin{aligned}
P \to Q &\equiv \neg P \vee Q \\
&\equiv (\neg P \wedge \text{true}) \vee Q \\
&\equiv (\neg P \wedge (Q \vee \neg Q)) \vee Q \\
&\equiv (\neg P \wedge Q) \vee (\neg P \wedge \neg Q) \vee Q \\
&\equiv (\neg P \wedge Q) \vee (\neg P \wedge \neg Q) \vee (Q \wedge \text{true}) \\
&\equiv (\neg P \wedge Q) \vee (\neg P \wedge \neg Q) \vee (Q \wedge (P \vee \neg P)) \\
&\equiv (\neg P \wedge Q) \vee (\neg P \wedge \neg Q) \vee (Q \wedge P) \vee (Q \wedge \neg P) \\
&\equiv (\neg P \wedge Q) \vee (\neg P \wedge \neg Q) \vee (Q \wedge P). \quad \blacklozenge
\end{aligned}
$$

We can proceed in an entirely analogous manner to find a full CNF for a wff. The trick in this case is to add letters to a fundamental disjunction without changing its truth value. It goes as follows:

To add a letter, say R, to a fundamental disjunction D without changing the value of D, write the following equivalences:

$$
D \equiv D \vee \text{false} \equiv D \vee (R \wedge \neg R).
$$

Then distribute $\vee$ over $\wedge$ to obtain a conjunction of two fundamental disjunctions.

For example, let's find a full CNF for the wff $P \wedge (P \to Q)$. To start off, we put the wff in conjunctive normal form as follows:

$$P \wedge (P \to Q) \equiv P \wedge (\neg P \vee Q).$$

The right side is not a full CNF because the letter Q does not occur in the fundamental disjunction P. So we'll apply the trick to add the letter Q. Make sure you can justify each step in the following calculation:

$$
\begin{aligned}
P \wedge (P \to Q) &\equiv P \wedge (\neg P \vee Q) \\
&\equiv (P \vee \text{false}) \wedge (\neg P \vee Q) \\
&\equiv (P \vee (Q \wedge \neg Q)) \wedge (\neg P \vee Q) \\
&\equiv (P \vee Q) \wedge (P \vee \neg Q) \wedge (\neg P \vee Q).
\end{aligned}
$$

The result is a full CNF that is equivalent to the original wff. Let's do another example.

EXAMPLE 7. We'll construct a full CNF for $(P \to (Q \vee R)) \wedge (P \vee Q)$. After converting the wff to conjunctive normal form, all we need to do is add the letter R to the fundamental disjunction $P \vee Q$. Here's the transformation:

$$
\begin{aligned}
(P \to (Q \vee R)) \wedge (P \vee Q) &\equiv (\neg P \vee Q \vee R) \wedge (P \vee Q) \\
&\equiv (\neg P \vee Q \vee R) \wedge ((P \vee Q) \vee \text{false}) \\
&\equiv (\neg P \vee Q \vee R) \wedge ((P \vee Q) \vee (R \wedge \neg R)) \\
&\equiv (\neg P \vee Q \vee R) \wedge ((P \vee Q \vee R) \wedge (P \vee Q \vee \neg R)) \\
&\equiv (\neg P \vee Q \vee R) \wedge (P \vee Q \vee R) \wedge (P \vee Q \vee \neg R). \quad \blacklozenge
\end{aligned}
$$

Complete Sets of Connectives

The four connectives in the set $\{\neg, \wedge, \vee, \to\}$ are used to form the wffs of the propositional calculus. Are there other sets of connectives that will do the same job? The answer is yes. A set of connectives is called *complete* if every wff of the propositional calculus is equivalent to a wff using connectives from the set. We've already seen that every wff has a disjunctive normal form, which uses only the connectives $\neg$, $\wedge$, and $\vee$. Therefore $\{\neg, \wedge, \vee\}$ is a complete set of connectives for the propositional calculus. Recall that we don't need implication because we have the equivalence

$$A \to B \equiv \neg A \vee B.$$

For a second example, consider the two connectives ¬ and ∨. To show that these connectives generate the propositional calculus, we need only show that statements of the form $A \wedge B$ can be written in terms of ¬ and ∨. This can be seen by the equivalence

$$A \wedge B \equiv \neg (\neg A \vee \neg B).$$

Therefore {¬, ∨} is a complete set of connectives for the propositional calculus because we know that {¬, ∧, ∨} is a complete set. Other complete sets are {¬, ∧} and {¬, →}. We'll leave these as exercises.

Are there any single connectives that are complete? The answer is yes, but we won't find one among the four basic connectives. There is a connective called the NAND operator, which is short for the "Negation of AND." We'll write NAND in functional form NAND(P, Q), since there is no well-established symbol for it. Table 6.7 is the truth table for NAND.

P	Q	NAND(P, Q)
true	true	false
true	false	true
false	true	true
false	false	true

Table 6.7

To see that NAND is complete, we have to show that the other connectives can be defined in terms of it. For example, we can write negation in terms of NAND as follows:

$$\neg P \equiv \text{NAND}(P, P).$$

We'll leave it as an exercise to show that the other connectives can be written in terms of NAND.

Another single connective that is complete for the propositional calculus is the NOR operator. NOR is short for the "Negation of OR." Table 6.8 is its truth table.

P	Q	NOR(P, Q)
true	true	false
true	false	false
false	true	false
false	false	true

Table 6.8

We'll leave it as an exercise to show that NOR is a complete connective for the propositional calculus. NAND and NOR are important because they represent the behavior of two important building blocks for logic circuits.

Exercises

1. Write down the parenthesized version of each of the following expressions.

    a. $\neg P \wedge Q \to P \vee R$.

    b. $P \vee \neg Q \wedge R \to P \vee R \to \neg Q$.

    c. $A \to B \vee \neg C \wedge D \wedge E \to F$.

2. Let A, B, and C be propositional wffs. Find a wff whose meaning is reflected by the statement "If A then B else C."

3. Remove as many parentheses as possible from each of the following wffs.

    a. $(((P \vee Q) \to (\neg R)) \vee (((\neg Q) \wedge R) \wedge P))$.

    b. $((A \to (B \vee C)) \to (A \vee (\neg (\neg B)))))$.

4. Use truth tables to verify the equivalences in (6.1).

5. Use other equivalences to prove the equivalence

$$A \to B \equiv A \wedge \neg B \to \text{false}.$$

    Hint: Start with the right side.

6. Show that $\to$ is not associative. That is, show that $(A \to B) \to C$ is not equivalent to $A \to (B \to C)$.

7. Verify each of the following equivalences by writing an equivalence proof. That is, start on one side and use known equivalences to get to the other side.

    a. $(A \to B) \wedge (A \vee B) \equiv B$.

    b. $A \wedge B \to C \equiv (A \to C) \vee (B \to C)$.

    c. $A \wedge B \to C \equiv A \to (B \to C)$.

    d. $A \vee B \to C \equiv (A \to C) \wedge (B \to C)$.

    e. $A \to B \wedge C \equiv (A \to B) \wedge (A \to C)$.

    f. $A \to B \vee C \equiv (A \to B) \vee (A \to C)$.

8. Use Quine's method to determine whether each of the following wffs is a tautology, a contradiction, or a contingency.

    a. $A \wedge B \to A$.   b. $A \vee B \to B$.

    c. $(A \to B) \vee ((C \to \neg B) \wedge \neg C)$.

9. Show that each of the following statements is not a tautology by finding truth values for the variables that make the premise true and the conclusion false.

   a. $(A \vee B) \rightarrow (C \vee A) \wedge (\neg C \vee B)$.

   b. $(A \rightarrow B) \wedge (B \rightarrow \neg A) \rightarrow A$.

   c. $(A \rightarrow B) \wedge (B \rightarrow C) \rightarrow (C \rightarrow A)$.

   d. $(A \vee B \rightarrow C) \wedge A \rightarrow (C \rightarrow B)$.

10. Use equivalences to transform each of the following wffs into a DNF.

    a. $(P \rightarrow Q) \rightarrow P$.

    b. $P \rightarrow (Q \rightarrow P)$.

    c. $Q \wedge \neg P \rightarrow P$.

    d. $(P \vee Q) \wedge R$.

    e. $P \rightarrow Q \wedge R$.

    f. $(A \vee B) \wedge (C \rightarrow D)$.

11. Use equivalences to transform each of the following wffs into a CNF.

    a. $(P \rightarrow Q) \rightarrow P$.

    b. $P \rightarrow (Q \rightarrow P)$.

    c. $Q \wedge \neg P \rightarrow P$.

    d. $(P \vee Q) \wedge R$.

    e. $P \rightarrow Q \wedge R$.

    f. $(A \wedge B) \vee E \vee F$.

    g. $(A \wedge B) \vee (C \wedge D) \vee (E \rightarrow F)$.

12. For each of the following functions, write down the full DNF and full CNF representations.

    a. $f(P, Q) =$ true if and only if P is true.

    b. $f(P, Q, R) =$ true if and only if either Q is true or R is false.

13. Transform each of the following wffs into a full DNF if possible.

    a. $(P \rightarrow Q) \rightarrow P$.

    b. $Q \wedge \neg P \rightarrow P$.

    c. $P \rightarrow (Q \rightarrow P)$.

    d. $(P \vee Q) \wedge R$.

    e. $P \rightarrow Q \wedge R$.

14. Transform each of the following wffs into a full CNF if possible.

    a. $(P \rightarrow Q) \rightarrow P$.

    b. $P \rightarrow (Q \rightarrow P)$.

c. $Q \wedge \neg P \rightarrow P$.

d. $P \rightarrow Q \wedge R$.

e. $(P \vee Q) \wedge R$.

15. Show that each of the following sets of operations is a complete set of connectives for the propositional calculus.

    a. $\{\neg, \wedge\}$.   b. $\{\neg, \rightarrow\}$.   c. $\{$false, $\rightarrow\}$.

    d. $\{$NAND$\}$.   e. $\{$NOR$\}$.

16. Show that there are no complete single binary connectives other than NAND and NOR. *Hint:* Let f be the truth function for a complete binary connective. Show that $f(\text{true, true}) = \text{false}$ and $f(\text{false, false}) = \text{true}$ because the negation operation must be represented in terms of f. Then consider the remaining cases in the truth table for f.

6.3 Formal Reasoning Systems

We have seen that truth tables are sufficient to find the truth of any proposition. However, if a proposition has three or more variables and contains several connectives, then a truth table can become quite complicated. When we use an equivalence proof, rather than truth tables, to decide the equivalence of two wffs, it seems somehow closer to the way we communicate with each other. Although there is no need to formally reason about the truth of propositions, it turns out that all other parts of logic need tools other than truth tables to reason about the truth of wffs. Thus we need to introduce the basic ideas of a formal reasoning system. We'll do it here because the techniques carry over to all logical systems. A formal reasoning system must have a set of well-formed formulas (wffs) to represent the statements of interest. But two other ingredients are required, and we'll discuss them next.

A reasoning system needs some rules to help us conclude things. An *inference rule* maps one or more wffs, called *premises, hypotheses,* or *antecedents*, to a single wff called the *conclusion*, or *consequent*. For example, the modus ponens rule maps the two wffs A and $A \rightarrow B$ to the wff B. If we let MP stand for modus ponens, then we can represent the rule by using functional notation as follows:

$$\text{MP}(A, A \rightarrow B) = B.$$

In general, if R is an inference rule and $R(P_1, ..., P_k) = C$, then we say something like, "C is inferred from P_1 and ... and P_k by R." A common way to represent an inference rule is to draw a horizontal line, place the premises above the line and place the conclusion below the line. The premises can be listed horizontally or vertically, and the conclusion is prefixed by the symbol ∴ as follows:

$$P_1, ..., P_k \over \therefore\ C \qquad \text{or} \qquad \begin{array}{c} P_1 \\ \vdots \\ P_k \\ \hline \therefore\ C \end{array}.$$

The symbol ∴ can be read as any of the following words:

therefore, thus, whence, so, ergo, hence.

We can also say, "C is a direct consequence of P_1 and ... and P_k." For example, the modus ponens rule can be written as follows:

Modus ponens (MP)

$$\frac{A \to B,\ A}{\therefore\ B} \qquad \text{or} \qquad \frac{\begin{array}{c} A \\ A \to B \end{array}}{\therefore\ B}. \tag{6.7}$$

We would like our inference rules to preserve truth. In other words, if all the premises are tautologies, then we want the conclusion to be a tautology. So an inference rule $R(P_1, ..., P_k) = C$ preserves truth if the following wff is a tautology:

$$P_1 \wedge ... \wedge P_k \to C.$$

For example, the modus ponens rule preserves truth because whenever $A \to B$ and A are both tautologies, then B is also a tautology. We can prove this by showing that the following wff is a tautology:

$$A \wedge (A \to B) \to B.$$

Any conditional tautology of the form $C \to D$ can be used as an inference rule. For example, the tautology

$$(A \to B) \wedge \neg B \to \neg A$$

gives rise to the *modus tollens* inference rule (MT) for propositions:

Modus tollens (MT)

$$\frac{A \to B,\ \neg B}{\therefore\ \neg A}. \tag{6.8}$$

Now let's list a few more useful inference rules for the propositional calculus. Each rule can be easily verified by showing that $C \to D$ is a tautology, where C is the conjunction of the premises and D is the conclusion.

Conjunction (Conj)

$$\frac{A, \; B}{\therefore \;\; A \wedge B} \,. \tag{6.9}$$

Simplification (Simp)

$$\frac{A \wedge B}{\therefore \;\; A} \,. \tag{6.10}$$

Addition (Add)

$$\frac{A}{\therefore \;\; A \vee B} \,. \tag{6.11}$$

Disjunctive syllogism (DS)

$$\frac{A \vee B, \; \neg A}{\therefore \;\; B} \,. \tag{6.12}$$

Hypothetical syllogism (HS)

$$\frac{A \to B, \; B \to C}{\therefore \;\; A \to C} \,. \tag{6.13}$$

For any reasoning system to work, it needs some fundamental truths to start the process. An *axiom* is a wff that we wish to use as a basis from which to reason. So an axiom is usually a wff that we "know to be true" from our initial investigations (e.g., a proposition that has been shown to be a tautology by a truth table). When we apply logic to a particular subject, then an axiom might also be something that we "want to be true" to start out our discussion For example, "Two points lie on one and only one line," for a geometry reasoning system.

We've introduced the three ingredients that make up any *formal reasoning system*:

A set of wffs,   a set of axioms,   and   a set of inference rules.

A formal reasoning system is often called a *formal theory*. How do we reason in such a system? Can we describe the reasoning process in some reasonable way? What is a proof? What is a theorem? Let's start off by describing a proof.

A *proof* is a finite sequence of wffs with the property that each wff in the sequence either is an axiom or can be inferred from previous wffs in the sequence. The last wff in a proof is called a *theorem*. For example, suppose the following sequence of wffs is a proof:

$$W_1, ..., W_n.$$

Then we know that W_1 is an axiom because there aren't any previous wffs in the sequence to infer it. We also know that for any $i > 1$, either W_i is an axiom or W_i is the conclusion of an inference rule, where the premises of the rule are taken from the set of wffs $\{W_1, ..., W_{i-1}\}$. We also know that W_n is a theorem. So when we say that a wff W is a theorem, we mean that there is a proof W_1, ..., W_n such that $W_n = W$.

Suppose we are unlucky enough to have a formal theory with a wff W such that both W and $\neg W$ can be proved as theorems. A formal theory exhibiting this bad behavior is said to be *inconsistent*. We probably would agree that inconsistency is a bad situation. A formal theory that doesn't possess this bad behavior is said to be *consistent*. We certainly would like our formal theories to be consistent. For the propositional calculus we'll get consistency if we choose our axioms to be tautologies and we choose our inference rules to map tautologies to tautologies. Then every theorem will have to be a tautology.

We'll write proofs in table format, where each line is numbered and contains a wff together with the reason it's there. For example, a proof sequence $W_1, ..., W_n$ will be written as follows:

Proof: 1. W_1 Reason for W_1
      2. W_2 Reason for W_2
      ⋮   ⋮   ⋮
      n. W_n Reason for W_n.

The reason column for each line always contains a short indication of why the wff is on the line. If the line depends on previous lines because of an inference rule, then we'll always include the line numbers of those previous lines.

Now let's get down to business and study some formal proof techniques for the propositional calculus. The two techniques that we will discuss are conditional proof and indirect proof.

Conditional Proof

Most statements that we want to prove either are in the form of a conditional or can be restated in the form of a conditional. For example, someone might make a statement of the form "*D* follows from *A*, *B*, and *C*." The statement can be rephrased in many ways. For example, we might say, "From the premises *A*, *B*, and *C* we can conclude *D*." We can avoid wordiness by writing the statement as the conditional

$$A \land B \land C \to D.$$

Let's discuss a rule to help us prove conditionals in a straightforward manner. The rule is called the *conditional proof rule* (CP), and we can describe it as follows:

Conditional Proof Rule (CP) (6.14)

Suppose we wish to construct a proof for a conditional of the form

$$A_1 \land A_2 \land \ ... \ \land A_n \to B.$$

Start the proof by writing each of the premises A_1, A_2, ..., A_n on a separate line with the letter *P* in the reason column. Now treat these premises as axioms, and construct a proof of *B*.

Let's look at the structure of a typical conditional proof. For example, a conditional proof of the wff $A \land B \land C \to D$ will contain lines for the three premises *A*, *B*, and *C*. It will also contain a line for the conclusion *D*. Finally, it will contain a line for $A \land B \land C \to D$, with CP listed in the reason column along with the line numbers of the premises and conclusion. The following proof structure exhibits these properties:

Proof: 1. *A*   *P*
        2. *B*   *P*
        3. *C*   *P*
       ⋮   ⋮   ⋮
       k. *D*   ...
   k+1. $A \land B \land C \to D$ 1, 2, 3, *k*, CP.

Since many conditionals are quite complicated, it may be difficult to write them on the last line of a proof. So we'll agree to write QED in place of the conditional on the last line of the proof and we'll also omit the last line number. With these changes, the form of the above proof can be abbreviated as follows:

Proof: 1. *A*   *P*
       2. *B*   *P*
       3. *C*   *P*
       ⋮ ⋮   ⋮
       k. *D*   ...
            QED   1, 2, 3, *k*, CP.

EXAMPLE 1. Let's do a real example to get the flavor of things. We'll give a conditional proof of the following statement:

$$(A \vee B) \wedge (A \vee C) \wedge \neg A \rightarrow B \wedge C.$$

The three premises are $A \vee B$, $A \vee C$, and $\neg A$. So we'll list them as premises to start the proof. Then we'll construct a proof of $B \wedge C$.

Proof: 1. $A \vee B$ *P*
       2. $A \vee C$ *P*
       3. $\neg A$ *P*
       4. *B*   1, 3, DS
       5. *C*   2, 3, DS
       6. $B \wedge C$ 4, 5, Conj
            QED   1, 2, 3, 6, CP.   ♦

Subproofs

Often a conditional proof will occur as part of another proof. We'll call a proof that is part of another proof a *subproof* of the proof. We'll always indicate a conditional subproof by indenting the wffs on its lines. We'll always write the conditional to be proved on the next line of the proof, without the indentation, because it will be needed as part of the main proof. Let's do an example to show the idea.

EXAMPLE 2. We'll give a proof of the following statement:

$$((A \vee B) \rightarrow (B \wedge C)) \rightarrow (B \rightarrow C) \vee D.$$

Notice that this wff is a conditional, where the conclusion $(B \rightarrow C) \vee D$ contains a second conditional $B \rightarrow C$. So we'll use a conditional proof that contains another conditional proof as a subproof. Here goes:

Proof: 1. $(A \lor B) \to (B \land C)$ P
        2. B P Start subproof of $B \to C$
        3. $A \lor B$ 2, Add
        4. $B \land C$ 1, 3, MP
        5. C 4, Simp
        6. $B \to C$ 2, 5, CP   Finish subproof of $B \to C$
        7. $(B \to C) \lor D$ 6, Add
            QED   1, 7, CP.   ◆

An Important Rule about Conditional Subproofs

If there is a conditional proof as a subproof within another proof, as indicated by the indented lines, then these indented lines may not be used to infer some line that occurs after the subproof is finished. The only exception to this rule is if an indented line does not depend, either directly or indirectly, on any premise of the subproof. In this case the indented line could actually be placed above the subproof, with the indentation removed.

EXAMPLE 3. The following sequence of lines indicates a general proof structure for some conditional statement having the form $A \to M$. In the reason column for each line we have listed the possible lines that might be used to infer the given line.

     1. A P
     2. B 1
     3. C P
     4. D 1, 2, 3
     5. E P
     6. F 1, 2, 3, 4, 5
     7. G 1, 2, 3, 4, 5, 6
     8. $E \to G$ 5, 7, CP
     9. H 1, 2, 3, 4, 8
    10. $C \to H$ 3, 9, CP
    11. I 1, 2, 10
    12. J P
    13. K 1, 2, 10, 11, 12
    14. L 1, 2, 10, 11, 12, 13
    15. $J \to L$ 12, 14, CP
    16. M 1, 2, 10, 11, 15
         QED   1, 16, CP.

Simplifications in Conditional Proofs

We can make some simplifications in our proofs to reflect the way informal proofs are written. If W is a theorem, then we can use it to prove other theorems. We can put W on a line of a proof and treat it as an axiom. Or, better yet, we can leave W out of the proof sequence but still use it as a reason for some line of the proof.

EXAMPLE 4. Let's prove the following statement:

$$\neg (A \wedge B) \wedge (B \vee C) \wedge (C \rightarrow D) \rightarrow (A \rightarrow D).$$

Proof:

1.	$\neg (A \wedge B)$	P
2.	$B \vee C$	P
3.	$C \rightarrow D$	P
4.	$\neg A \vee \neg B$	$1, \neg (A \wedge B) \equiv \neg A \vee \neg B$
5.	$\quad A$	P
6.	$\quad\quad \neg B$	$4, 5, \text{DS}$
7.	$\quad\quad C$	$2, 6, \text{DS}$
8.	$\quad\quad D$	$3, 7, \text{MP}$
9.	$A \rightarrow D$	$5, 8, \text{CP}$
	QED	$1, 2, 3, 9, \text{CP}.$

Line 4 of the proof is OK because of the equivalence we listed in the reason column. ◆

Instead of writing down the specific tautology or theorem in the reason column, we'll usually write the symbol

$$T$$

to indicate that a tautology or theorem is being used.

Some proofs are straightforward, while others can be brain busters. Remember, when you construct a proof, it may take several false starts before you come up with a correct proof sequence. Let's do some more examples of conditional proofs.

EXAMPLE 5. We will give a conditional proof of the following wff:

$$A \wedge ((A \rightarrow B) \vee (C \wedge D)) \rightarrow (\neg B \rightarrow C).$$

Proof: 1. A P
       2. $(A \rightarrow B) \vee (C \wedge D)$ P
       3. $\neg B$ P
       4. $A \wedge \neg B$ 1, 3, Conj
       5. $\neg\neg A \wedge \neg B$ 4, T
       6. $\neg(\neg A \vee B)$ 5, T
       7. $\neg(A \rightarrow B)$ 6, T
       8. $C \wedge D$ 2, 7, DS
       9. C 8, Simp
     10. $\neg B \rightarrow C$ 3, 9, CP
           QED   1, 2, 10, CP. ◆

EXAMPLE 6. Consider the following collection of statements:

The team wins or I am sad. If the team wins, then I go to a movie. If I am sad, then my dog barks. My dog is quiet. Therefore I go to a movie.

Let's formalize these statements. First, we'll make some simplifications as follows:

W: The team wins.
S: I am sad.
M: I go to a movie.
B: My dog barks.

Now we can symbolize the given statements by the wff

$$(W \vee S) \wedge (W \rightarrow M) \wedge (S \rightarrow B) \wedge \neg B \rightarrow M.$$

We'll show this wff is a theorem by giving a proof as follows:

Proof: 1. $W \vee S$ P
       2. $W \rightarrow M$ P
       3. $S \rightarrow B$ P
       4. $\neg B$ P
       5. $\neg S$ 3, 4, MT
       6. W 1, 5, DS
       7. M 2, 6, MP
          QED   1, 2, 3, 4, 7, CP. ◆

Indirect Proof

Suppose we want to prove a statement $A \to B$, but we just can't seem to find a way to get going using the conditional proof technique. Sometimes it may be worthwhile to try to prove the contrapositive of $A \to B$. In other words, we try a conditional proof of $\neg B \to \neg A$. Since we have the equivalence

$$A \to B \equiv \neg B \to \neg A,$$

we can start by letting $\neg B$ be the premise. Then we can try to find a proof that ends with $\neg A$.

What if we still have problems finding a proof? Then we might try another indirect method, called *proof by contradiction* or *reductio ad absurdum*. The idea is based on the following equivalence:

$$A \to B \equiv A \wedge \neg B \to \text{false}.$$

This indirect method gives us more information to work with than the contrapositive method because we can use both A and $\neg B$ as premises. We also have more freedom because all we need to do is find any kind of contradiction. You might try it whenever there doesn't seem to be enough information from the given premises or when you run out of ideas. You might also try it as your first method. Here's the formal description:

Indirect Proof Rule (IP) (6.15)

Suppose we wish to construct an indirect proof of the conditional

$$A_1 \wedge A_2 \wedge \ldots \wedge A_n \to B.$$

Start the proof by writing each of the premises A_1, A_2, ..., A_n on a separate line with the letter P in the reason column. Then place the wff $\neg B$ on the next line, and write "P for IP" in the reason column to indicate that $\neg B$ is a premise for indirect proof. Now treat these premises as axioms, and construct a proof of a false statement.

Let's look at the structure of a typical indirect proof. For example, an indirect proof of the wff $A \wedge B \wedge C \to D$ will contain lines for the three premises A, B, and C, and for the IP premise $\neg D$. It will also contain a line for a false statement. Finally, it will contain a line for $A \wedge B \wedge C \to D$, with IP listed in the reason column along with the line numbers of the premises, the IP premise, and the false statement. The following proof structure exhibits these properties:

Proof: 1. A P
         2. B P
         3. C P
         4. $\neg D$ P for IP
             $\vdots$ $\vdots$ $\vdots$
         k. false   ...
     $k + 1$. $A \wedge B \wedge C \rightarrow D$ 1, 2, 3, 4, k, IP.

As with the case for conditional proofs, we'll agree to write QED in place the conditional on the last line of the proof, and we'll also omit the last line number. So the form of the above proof can be abbreviated as follows:

Proof: 1. A P
         2. B P
         3. C P
         4. $\neg D$ P for IP
             $\vdots$ $\vdots$ $\vdots$
         k. false   ...
             QED   1, 2, 3, 4, k, IP.

EXAMPLE 7. Let's do a real example to get our feet wet. We'll give an indirect proof of the following statement that is derived from the movies problem in Example 6:

$$(W \vee S) \wedge (W \rightarrow M) \wedge (S \rightarrow B) \wedge \neg B \ \rightarrow M.$$

Proof: 1. $W \vee S$ P
          2. $W \rightarrow M$ P
          3. $S \rightarrow B$ P
          4. $\neg B$ P
          5. $\neg M$ P for IP
          6. $\neg W$ 2, 5, MT
          7. $\neg S$ 3, 4, MT
          8. $\neg W \wedge \neg S$ 6, 7, Conj
          9. $\neg (W \vee S)$ 8, T
         10. $(W \vee S) \wedge \neg (W \vee S)$ 1, 9, Conj
         11. false   10, T
              QED   1–4, 5, 11, IP.   ◆

Compare this proof to the earlier direct proof of the same statement. It's a bit longer, and it uses different rules. Sometimes a longer proof can be easier to create, using simpler steps. Just remember, there's more than one way to proceed when trying a proof.

Proof Notes

When proving something, we should always try to *tell the proof, the whole proof, and nothing but the proof*. Here are a few concrete suggestions that should make life easier for beginning provers.

Don't Use Unnecessary Premises

Sometimes beginners like to put extra premises in proofs to help get to a conclusion. But then they forget to give credit to these extra premises. For example, suppose we want to prove a conditional of the form $A \to C$. We start the proof by writing A as a premise. Suppose that along the way we decide to introduce another premise, say B, and then use A and B to infer C, either directly or indirectly. The result is not a proof of $A \to C$. Instead, we have given a proof of the statement $A \land B \to C$.

Remember: Be sure to use a premise only when it's the hypothesis of a conditional that you want to prove. Another way to say this is: If you use a premise to prove something, then the premise becomes part of the antecedent of the thing you proved. Still another way to say this is:

> *The conjunction of all the premises that you use to prove something is precisely the antecedent of the conditional that you proved.*

Don't Apply Inference Rules to Subexpressions

Beginners sometimes use an inference rule incorrectly by applying it to a subexpression of a larger wff. This violates the definition of a proof, which states that a wff in a proof either is an axiom or is inferred by previous wffs in the proof. In other words, an inference rule can be applied only to entire wffs that appear on previous lines of the proof. So

> *Don't apply inference rules to subexpressions of wffs.*

Let's do an example to see what we are talking about. Suppose we have the following wff:

$$A \land ((A \to B) \lor C) \to B.$$

This wff is not a tautology. For example, let A = true, B = false, and C = true. The following sequence attempts to show that the wff is a theorem:

$$
\begin{array}{lll}
1. & A & P \\
2. & (A \rightarrow B) \vee C & P \\
3. & B & 1, 2, \text{MP} \qquad \text{Incorrect use of MP} \\
  & \text{QED} & 1, 2, 3, \text{CP.}
\end{array}
$$

The reason that the proof is wrong is that MP is applied to the two wffs A on line 1 and $A \rightarrow B$ on line 2. But $A \rightarrow B$ does not occur on a line by itself. Rather, it's a subexpression of $(A \rightarrow B) \vee C$. Therefore MP cannot be used.

Failure to Find a Proof

Sometimes we can obtain valuable information about a wff by failing to find a proof. After a few unsuccessful attempts, it may dawn on us that the thing is not a theorem. For example, no proof exists for the following conditional wff:

$$(A \rightarrow C) \rightarrow (A \vee B \rightarrow C).$$

To see that this wff is not a tautology, let A = false, C = false, and B = true.

But remember that we cannot conclude that a wff is not a tautology just because we can't find a proof.

Reasoning Systems for Propositional Calculus

Although truth tables are sufficient to decide any question about the propositional calculus, most of us do not reason by truth tables. Instead, we use our personal reasoning systems. In the proofs presented up to now we allowed the use of any of the inference rules (6.7) to (6.13). Similarly, we allowed the use of any known tautology as an axiom. This is a loose kind of formal system for the propositional calculus.

Can we find a proof system for the propositional calculus that contains a specific set of axioms and inference rules? If we do find such a system, how do we know that it does the job? In fact, what is the job that we want done? Basically, we want two things. We want our proofs to yield theorems that are tautologies, and we want any tautology to be provable as a theorem. These properties are converses of each other, and they have the following names:

Soundness: *We want all proofs to yield theorems that are indeed tautologies.*

Completeness: *We want all tautologies to be provable as theorems. Completeness is the converse of soundness.*

Any inference rule in a theory automatically creates a conditional theorem $A \to B$ in the theory, where A is the conjunction of the premises and B is the conclusion of the rule. For example, suppose we have the following inference rule for the propositional calculus, which we'll call rule X:

$$\frac{A, B, C}{\therefore \ D}.$$

Rule X gives us the conditional theorem $A \wedge B \wedge C \to D$. We'll give a conditional proof of this theorem as follows:

Proof: 1. A P
       2. B P
       3. C P
       4. D $1, 2, 3, X$
          QED $1, 2, 3, 4,$ CP.

An important point to notice here is that the inference rules in a theory will in some sense define the concept of truth. For example, suppose we wish to make up a theory of propositions, named BAD, that has the following inference rule:

Bad Inference Rule: $\dfrac{A \vee B}{\therefore \ A}.$

This inference rule thus gives us the theorem $A \vee B \to A$. This contradicts our knowledge that $A \vee B \to A$ is not a tautology—it's a contingency.

So if we want to reason in a theory for which our theorems are actually true, then we must ensure that we use only inference rules that map true statements to true statements. Of course, we must also ensure that the axioms that we choose are true.

Is there some simple formal system for the propositional calculus that is both sound and complete? Yes, there is. In fact, there are many of them. Each one specifies a fixed set of axioms and inference rules. We'll look at a system that is similar to a system introduced by the mathematician David Hilbert (1862–1943). Hilbert's system is described in Hilbert and Ackermann [1938]. So we'll call our system *Hilbert's system*. We'll use the connectives $\neg$, $\vee$, $\wedge$, and $\to$. Hilbert's system starts with the following axioms, where A, B, and C stand for arbitrary wffs:

Hilbert's Axioms   (6.16)

1. $A \vee A \to A$.
2. $A \to A \vee B$.
3. $A \vee B \to B \vee A$.
4. $(A \to B) \to (C \vee A \to C \vee B)$.
5. $A \to B \equiv \neg A \vee B \equiv \neg (A \wedge \neg B)$.

The first four axioms may appear a bit strange, but they can be verified by truth tables. Axiom 5 relates negation with the other operations, where an equivalence $D \equiv E$ is short for the two statements $D \to E$ and $E \to D$.

We'll use the modus ponens (MP) inference rule together with the conditional proof rule (CP). That's it. Everything that we do must be based only on the five axioms (6.16), MP, and CP. The remarkable thing is that this system is sound and complete. In other words, every proof yields a theorem that is a tautology, and there is a proof for every tautology of the propositional calculus.

Let's use Hilbert's system to prove some familiar statements. The first theorem we'll prove is the tautology known as hypothetical syllogism. It's the basis for inference rule (6.13).

Theorem 1: $(A \to B) \wedge (B \to C) \to (A \to C)$.   (6.17)

Proof: 1. $A \to B$ P
         2. $B \to C$ P
         3. A P
         4. B 1, 3, MP
         5. C 2, 4, MP
         6. $A \to C$ 3, 5, CP
            QED   1, 2, 6, CP.

Now that we have our first theorem, we can use it in subsequent proofs as an inference rule. Let's continue with our examples. We know that the statement $A \to A$ is a tautology. But can we prove it in Hilbert's system? We'll do this next.

Theorem 2: $A \to A$.   (6.18)

Proof: 1. $A \to A \vee A$ Axiom 2
         2. $A \vee A \to A$ Axiom 1
         3. $A \to A$ 1, 2, Theorem 1
            QED.

We can also obtain some familiar statements about negation, which we'll list as follows:

Theorem 3: $\neg A \vee A.$ (6.19)

Theorem 4: $A \vee \neg A.$ (6.20)

Theorem 5: $\neg A \vee \neg \neg A.$ (6.21)

Theorem 6: $A \rightarrow \neg \neg A.$ (6.22)

We'll leave the proofs of these theorems as exercises. If we can prove the converse of Theorem 6, then the two statements can be combined to give us the equivalence $A \equiv \neg \neg A$. The next theorem does the job.

Theorem 7: $\neg \neg A \rightarrow A.$ (6.23)

Proof: 1. $\neg A \rightarrow \neg \neg \neg A$ Theorem 6
         2. $(\neg A \rightarrow \neg \neg \neg A) \rightarrow (A \vee \neg A \rightarrow A \vee \neg \neg \neg A)$ Axiom 4
         3. $(A \vee \neg A) \rightarrow (A \vee \neg \neg \neg A)$ 1, 2, MP
         4. $A \vee \neg A$ Theorem 4
         5. $A \vee \neg \neg \neg A$ 3, 4, MP
         6. $(A \vee \neg \neg \neg A) \rightarrow (\neg \neg \neg A \vee A)$ Axiom 3
         7. $\neg \neg \neg A \vee A$ 5, 6, MP
         8. $(\neg \neg \neg A \vee A) \rightarrow (\neg \neg A \rightarrow A)$ Part of Axiom 5
         9. $\neg \neg A \rightarrow A$ 7, 8, MP
             QED.

There are important reasons for studying small formal systems like Hilbert's system. Small systems are easier to test and easier to compare with other systems because there are only a few basic operations to worry about. For example, if we build a program to do automatic reasoning, it may be easier to implement a small set of axioms and inference rules. This also applies to computers with small instruction sets and to programming languages with a small number of basic operations.

Logic Puzzles

For most of us, logic puzzles are interesting challenges. But if a solution eludes us, it can be a frustrating experience. Sometimes a little thought and ingenuity can make the light bulb go on. For example, it might be helpful to formalize the problem with the symbols of logic and then do some formal reasoning. Another technique that often does the trick is to create a table of possibilities. The solution process often consists of eliminating possibilities by finding contradictions until a solution is found. Let's try an example.

EXAMPLE 8 (*Blind Man Sees*). The county jail is full. The sheriff, Anne Oakley, brings in a newly caught criminal and decides to make some space for the criminal by letting one of the current inmates go free. She picks prisoners *A*, *B*, and *C* to choose from. She puts blindfolds on *A* and *B* because *C* is already blind. Next she selects three hats from five hats hanging on the hat rack, two of which are red and three of which are white, and places the three hats on the prisoner's heads. She hides the remaining two hats. Then she takes the blindfolds off *A* and *B* and tells them what she has done, including the fact that there were three white hats and two red hats to choose from. Sheriff Oakley then makes the following statement:

If you can tell me the color of the hat you are wearing, without looking at your own hat, then you can go free.

The following things happen:

1. *A* says that he can't tell the color of his hat. So the sheriff has him returned to his cell.

2. Then *B* says that he can't tell the color of his hat. So he is also returned to his cell.

3. Then *C*, the blind prisoner, says that he knows the color of his hat. He tells the sheriff, and she sets him free.

What color was *C*'s hat, and how did *C* do his reasoning?

Don't read further until you have tried the problem.

C might have reasoned as follows: Since *A* doesn't know the color of his hat, it follows that the hats on *B* and *C* are not both red. After listening to *A*, *B* says that he doesn't know the color of his hat. So of course *A* and *C* don't both have red hats. But *C* can't be wearing a red hat. For if *C* were wearing a red hat, then *B* would have been wearing a white hat and would have told the sheriff this fact. Therefore *C* must be wearing a white hat. He tells this to the sheriff and is set free.   ◆

Exercises

1. Let W denote the wff $(A \rightarrow B) \rightarrow B$. It's easy to see that W is not a tautology. Just let B = false and A = false. Now, suppose someone claims that the following sequence of statements is a "proof" of W:

   1. $A \rightarrow B$ P
   2. A P
   3. B 1, 2, MP
     QED   1, 3, CP.

   a. What is wrong with the above "proof" of W?

   b. Write down the statement that the "proof" proves.

2. Let W denote the wff $(A \rightarrow (B \wedge C)) \rightarrow (A \rightarrow B) \wedge C$. It's easy to see that W is not a tautology. Suppose someone claims that the following sequence of statements is a "proof" of W:

   1. $A \rightarrow (B \wedge C)$ P
   2. A P
   3. $B \wedge C$ 1, 2, MP
   4. B 3, Simp
   5. $A \rightarrow B$ 2, 4, CP
   6. C 3, Simp
   7. $(A \rightarrow B) \wedge C$ 5, 6, Conj
     QED   1, 7, CP.

   What is wrong with this "proof" of W?

3. Find the number of premises required for a conditional proof of each of the following wffs. Assume that the letters stand for other wffs.

   a. $A \rightarrow (B \rightarrow (C \rightarrow D))$.

   b. $((A \rightarrow B) \rightarrow C) \rightarrow D$.

4. Give a formalized version of the following proof:

   If I am dancing, then I am happy. There is a mouse in the house or I am happy. I am sad. Therefore there is a mouse in the house and I am not dancing.

5. Give formal proofs for each of the following tautologies by using the CP rule.

   a. $A \rightarrow (B \rightarrow (A \wedge B))$.

   b. $A \rightarrow (\neg B \rightarrow (A \wedge \neg B))$.

   c. $(A \vee B \rightarrow C) \wedge A \rightarrow C$.

   d. $(B \rightarrow C) \rightarrow (A \wedge B \rightarrow A \wedge C)$.

e. $(A \vee B \to C \wedge D) \to (B \to D)$.

f. $(A \vee B \to C) \wedge (C \to D \wedge E) \to (A \to D)$.

g. $\neg (A \wedge B) \wedge (B \vee C) \wedge (C \to D) \to (A \to D)$.

h. $(A \to (B \to C)) \to (B \to (A \to C))$.

i. $(A \to C) \to (A \wedge B \to C)$.

j. $(A \to C) \to (A \to B \vee C)$.

6. Give formal proofs for each of the following tautologies by using the IP rule.

a. $A \to (B \to A)$.

b. $(A \to B) \wedge (A \vee B) \to B$.

c. $\neg B \to (B \to C)$.

d. $(A \to B) \to (C \vee A \to C \vee B)$.

e. $(A \to B) \to ((A \to \neg B) \to \neg A)$.

f. $(A \to B) \to ((B \to C) \to (A \vee B \to C))$.

g. $(A \to B) \wedge (B \to C) \to (A \to C)$. *Note:* This is the HS inference rule.

h. $(C \to A) \wedge (\neg C \to B) \to (A \vee B)$.

7. Give formal proofs for each of the following tautologies by using the IP rule somewhere in each proof.

a. $A \to (B \to (A \wedge B))$.

b. $A \to (\neg B \to (A \wedge \neg B))$.

c. $(A \vee B \to C) \wedge A \to C$.

d. $(B \to C) \to (A \wedge B \to A \wedge C)$.

e. $(A \vee B \to C \wedge D) \to (B \to D)$.

f. $(A \vee B \to C) \wedge (C \to D \wedge E) \to (A \to D)$.

g. $\neg (A \wedge B) \wedge (B \vee C) \wedge (C \to D) \to (A \to D)$.

h. $(A \to B) \to ((B \to C) \to (A \vee B \to C))$.

i. $(A \to (B \to C)) \to (B \to (A \to C))$.

j. $(A \to C) \to (A \wedge B \to C)$.

k. $(A \to C) \to (A \to B \vee C)$.

8. Give a formal proof of the equivalence $A \wedge B \to C \equiv A \to (B \to C)$. In other words, prove both of the following statements. Use either CP or IP.

a. $(A \wedge B \to C) \to (A \to (B \to C))$.

b. $(A \to (B \to C)) \to (A \wedge B \to C)$.

9. The following two inference rules, called *dilemmas*, can be useful tools in certain proofs. Give a formal proof for each rule, showing that it maps

tautologies to tautologies. In other words, prove that the conjunction of the premises implies the conclusion.

a. *Constructive dilemma (CD):*
$$\frac{A \vee B, \; A \to C, \; B \to D}{\therefore \;\; C \vee D}.$$

b. *Destructive dilemma (DD):*
$$\frac{\neg C \vee \neg D, \; A \to C, \; B \to D}{\therefore \;\; \neg A \vee \neg B}.$$

10. Use the Hilbert reasoning system (6.16) to prove each of the following theorems.

   a. $\neg A \vee A$. b. $A \vee \neg A$. c. $\neg A \vee \neg \neg A$. d. $A \to \neg \neg A$.

11. Use only the axioms, corollaries, and theorems of the Hilbert reasoning system to prove each of the following statements.

   a. $(A \to B) \to (\neg B \to \neg A)$.
   b. $(\neg B \to \neg A) \to (A \to B)$.

12. Four men and four women were nominated for two positions on the school board. One man and one woman were elected to the positions. Suppose the men are named $A, B, C,$ and D and the women are named $E, F, G,$ and H. Further, suppose that the following four statements are true:

   1. If neither A nor E won a position, then G won a position.
   2. If neither A nor F won a position, then B won a position.
   3. If neither B nor G won a position, then C won a position.
   4. If neither C nor F won a position, then E won a position.

   Who were the two people elected to the school board?

Chapter Summary

The propositional calculus is the basic building block of formal logic. Each wff represents a statement that can be checked by truth tables to determine whether it is a tautology, a contradiction, or a contingency. There are basic equivalences (6.1) that allow us to simplify and transform wffs into other wffs. We can use these equivalences with Quine's method to determine without truth tables whether a wff is a tautology, a contradiction, or a contingency. We can also use the equivalences to transform any wff into a DNF or a CNF. Any truth function has one of these forms.

   Propositional calculus also provides us with formal techniques for proving properties of wffs without using truth tables. A formal reasoning system has wffs, axioms, and inference rules. Some useful inference rules are modus ponens, modus tollens, conjunction, simplification, addition, disjunctive syllogism, and hypothetical syllogism. Two basic proof techniques are conditional proof and the indirect proof. When constructing proofs, remember: *Don't use unnecessary premises*, and *don't apply inference rules to subexpressions*.

We want formal reasoning systems to be sound—proofs yield theorems that are tautologies—and complete—all tautologies can be proven as theorems. The system presented in this chapter is sound and complete as long as we always use tautologies as axioms. Hilbert's system is a little example of such a system with only five axioms and one inference rule—modus ponens.

Notes on Logic

The logical symbols that we've used in this chapter are not universal. So you should be flexible in your reading of the literature. From a historical point of view, Whitehead and Russell [1910] introduced the symbols $\supset$, $\vee$, $\cdot$, $\sim$, and $\equiv$ to stand for implication, disjunction, conjunction, negation, and equivalence, respectively. A prefix notation for the logical operations was introduced by Lukasiewicz [1929], where the letters C, A, K, N, and E stand for implication, disjunction, conjunction, negation, and equivalence, respectively. So in terms of our notation we have $Cpq = p \rightarrow q$, $Apq = p \vee q$, $Kpq = p \wedge q$, $Nq = \neg q$, and $Epq = p \equiv q$. This notation is called Polish notation; and its advantage is that each expression has a unique meaning without using parentheses and precedence. For example, $(p \rightarrow q) \rightarrow r$ and $p \rightarrow (q \rightarrow r)$ are represented by the expressions $CCpqr$ and $CpCqr$, respectively. The disadvantage of the notation is that it's harder to read. For example, $CCpqKsNr = (p \rightarrow q) \rightarrow (s \wedge \neg r)$.

The fact that a wff W is a theorem is often denoted by placing a turnstile in front of it as follows:

$$\vdash W.$$

So this means that there is a proof $W_1, ..., W_n$ such that $W_n = W$. Turnstiles are also used in discussing conditionals. For example, the notation

$$A_1, A_2, ..., A_n \vdash B$$

means that there is a conditional proof of the wff $A_1 \wedge A_2 \wedge ... \wedge A_n \rightarrow B$.

We should again emphasize that the logic that we are studying in this book deals with statements that are either true or false. This is sometimes called the *Law of the Excluded Middle*: Every statement is either true or false. If our logic does not assume the law of the excluded middle, then we can no longer use indirect proof because we can't conclude that a statement is false from the assumption that it is not true. A logic called *intuitionist logic* omits this law and thus forces all proofs to be direct. Intuitionists like to construct things in a direct manner.

Logics that assume the law of the excluded middle are called *two-valued logics*. Some logics take a more general approach and consider statements that may not be true or false. For example, a *three-valued logic* assigns one of three values to each statement: 0, .5, or 1, where 0 stands for false, .5 stands for unknown, and 1 stands for true. We can build truth tables for this logic by defining $\neg A = 1 - A$, $A \vee B = \max(A, B)$, and $A \wedge B = \min(A, B)$. We still use the equivalence $A \rightarrow B \equiv \neg A \vee B$. So we can discuss three-valued logic.

In a similar manner we can discuss *n-valued logic* for any natural number $n \geq 2$, where each statement takes on one of n specific values in the range 0 to 1. Some n-valued logics assign names to the values such as "necessarily true," "probably true," "probably false," and "necessarily false." For example, there is a logic called *modal logic* that uses two extra unary operators, one to indicate that a statement is necessarily true and one to indicate that a statement is possibly true. So modal logic can represent a sentence like "If P is necessarily true, then P is true."

Some logics assign values over an infinite set. For example, the term *fuzzy logic* is used to describe a logic in which each statement is assigned some value in the closed unit interval [0, 1].

All these logics have applications in computer science but they are beyond our scope and purpose. However, it's nice to know that they all depend on a good knowledge of two-valued logic. In this chapter we've covered the fundamental parts of two-valued logic—the properties and reasoning rules of the propositional calculus. We'll see that these ideas occur in all the logics and applications that we cover in the next three chapters.

7

Predicate Logic

*Error of opinion may be tolerated where reason
is left free to combat it.*
—Thomas Jefferson (1743–1826)

We need a new logic if we want to describe arguments that deal with all cases
or with some case out of many cases. In this chapter we'll introduce the notions and notations of first-order predicate calculus. This logic will allow us to
analyze and symbolize a wider variety of statements and arguments than can
be done with propositional logic.

Chapter Guide

Section 7.1 introduces the basic syntax and semantics of the first-order predicate calculus. We'll discuss the properties of validity and satisfiability,
and we'll discuss the problem of deciding whether a formula is valid.

Section 7.2 introduces the fundamental equivalences of first-order predicate
calculus. We'll discuss the prenex normal forms, and we'll look at the
problem of formalizing English sentences.

Section 7.3 introduces the standard inference rules for formal reasoning in
first-order predicate calculus.

7.1 First-Order Predicate Calculus

The propositional calculus provides adequate tools for reasoning about propositional wffs, which are combinations of propositions. But a proposition is a
sentence taken as a whole. With this restrictive definition, propositional calculus doesn't provide the tools to do everyday reasoning. For example, in the

following argument it is impossible to find a formal way to test the correctness of the inference without further analysis of each sentence:

> All computer science majors own a personal computer.
> Socrates does not own a personal computer.
> Therefore Socrates is not a computer science major.

To discuss such an argument, we need to break up the sentences into
parts. The words in the set {All, own, not} are important to understand the
argument. Somehow we need to symbolize a sentence so that the information
needed for reasoning is characterized in some way. Therefore we will study
the inner structure of sentences.

The statement "x owns a personal computer" is not a proposition because
its truth value depends on x. If we give x a value, like x = Socrates, then the
statement becomes a proposition because it has the value true or false. From
the grammar point of view, the property "owns a personal computer" is a
predicate, where a predicate is the part of a sentence that gives a property of
the subject. A predicate usually contains a verb, like "owns" in our example.
The word predicate comes from the Latin word *praedicare*, which means to
proclaim.

From the logic point of view, a *predicate* is a relation, which of course we
can also think of as a property. For example, suppose we let $p(x)$ mean "x
owns a personal computer." Then p is a predicate that describes the relation
(i.e., property) of owning a personal computer. Sometimes it's convenient to
call $p(x)$ a predicate, although p is the actual predicate. If we replace the
variable x by some definite value such as Socrates, then we obtain the proposition p(Socrates). For another example, suppose that for any two natural
numbers x and y we let $q(x, y)$ mean "$x < y$." Then q is the predicate that we
all know of as the "less than" relation. For example, the proposition $q(1, 5)$ is
true, and the proposition $q(8, 3)$ is false.

Let $p(x)$ mean "x is an odd integer." Then the proposition $p(9)$ is true,
and the proposition $p(20)$ is false. Similarly, the following proposition is true:

$$p(2) \vee p(3) \vee p(4) \vee p(5).$$

We can describe this proposition by saying, "There exists an element x in
the set {2, 3, 4, 5} such that $p(x)$ is true." By letting D = {2, 3, 4, 5} the statement can be shortened to "There exists $x \in D$ such that $p(x)$ is true." If we
don't care about the truth of the statement, then we can still describe the preceding disjunction by saying, "There exists $x \in D$ such that $p(x)$." Still more
formally, we can write the expression

$$\exists x \in D : p(x).$$

Now if we want to consider the truth of the expression for a different set of numbers, say S, then we would write

$$\exists x \in S : p(x).$$

This expression would be true if S contained an odd integer. If we want to consider the statement without regard to any particular set of numbers, then we write

$$\exists x \; p(x).$$

This expression is not a proposition because we don't have a specific set of elements over which x can vary. Thus $\exists x \; p(x)$ cannot be given a truth value. If we don't know what p stands for, we can still say, "There exists an x such that $p(x)$." The symbol $\exists x$ is called an *existential quantifier*.

Now let's look at conjunctions rather than disjunctions. Suppose we have the following proposition:

$$p(1) \wedge p(3) \wedge p(5) \wedge p(7).$$

This conjunction can be represented by the expression $\forall x \in D : p(x)$, where $D = \{1, 3, 5, 7\}$. We say, "Every x in D is an odd integer." If we want to consider the statement without regard to any particular set of numbers, then we write

$$\forall x \; p(x).$$

The expression $\forall x \; p(x)$ is read, "For every x $p(x)$." The symbol $\forall x$ is called a *universal quantifier*.

Let's see how the quantifiers can be used together to represent certain statements. If $p(x, y)$ is a predicate and we let the variables x and y vary over the set $D = \{0, 1\}$, then the proposition

$$[p(0, 0) \vee p(0, 1)] \wedge [p(1, 0) \vee p(1, 1)]$$

can be represented by the following expression:

$$\forall x \in D : \exists y \in D : p(x, y).$$

To see this, notice that each of the two disjunctions in the proposition can be expressed in the following form, where $x \in D$:

$$p(x, 0) \vee p(x, 1) \; = \; \exists y \in D : p(x, y).$$

Now we use the universal quantifier $\forall x$ to represent the conjunction of the two disjunctions to obtain the desired expression, as follows:

$$[p(0, 0) \vee p(0, 1)] \wedge [p(1, 0) \vee p(1, 1)] \; = \; \forall x \in D : p(x, 0) \vee p(x, 1)$$
$$= \; \forall x \in D : \exists y \in D : p(x, y).$$

Now let's go the other way. We'll start with an expression containing different quantifiers and try to write it as a proposition. For example, if we use the same set of values $D = \{0, 1\}$, the expression

$$\exists y \in D : \forall x \in D : p(x, y)$$

denotes the proposition

$$[p(0, 0) \wedge p(1, 0)] \vee [p(0, 1) \wedge p(1, 1)].$$

It's easier to see why this is the case. All we need to do is evaluate the expression as follows:

$$\exists y \in D : \forall x \in D : p(x, y) \; = \; \exists y \in D : p(0, y) \wedge p(1, y)$$
$$= \; [p(0, 0) \wedge p(1, 0)] \vee [p(0, 1) \wedge p(1, 1)].$$

Of course, not every expression containing quantifiers results in a proposition. For example, if $D = \{0, 1\}$, then the expression $\forall x \in D : p(x, y)$ can be written as follows:

$$\forall x \in D : p(x, y) \; = \; p(0, y) \wedge p(1, y).$$

To obtain a proposition, each variable of the expression must be quantified or assigned some value in D. We'll discuss this shortly, when we talk about semantics.

Now let's look at two examples, which will introduce us to the important process of formalizing English sentences with quantifiers.

EXAMPLE 1. We'll formalize the three sentences about Socrates that we listed at the beginning of the section. Let P be the set of all people, let $cs(x)$ mean that x is a computer science major, and let $pc(x)$ mean that x owns a personal computer. Then the sentence "All computer science majors own a personal computer" can be formalized as follows:

$$\forall x \in P : (cs(x) \rightarrow pc(x)).$$

The sentence "Socrates does not own a personal computer" becomes

$$\neg \, \text{pc(Socrates)}.$$

The sentence "Socrates is not a computer science major" becomes

$$\neg \, \text{cs(Socrates)}. \quad \blacklozenge$$

EXAMPLE 2 (*Natural Numbers*). Suppose we consider the following two elementary facts about the natural numbers $\mathbb{N}$:

1. Every natural number has a successor.
2. There is no natural number whose successor is 0.

Let's formalize these sentences. We'll begin by writing down a semiformal version of the first sentence:

For each $x \in \mathbb{N}$ there exists $y \in \mathbb{N}$ such that the successor of x is y.

If we let $s(x, y)$ mean that the successor of x is y, then the formal version of the sentence can then be written as follows:

$$\forall x \in \mathbb{N} : \exists y \in \mathbb{N} : s(x, y).$$

Now let's look at the second sentence. It can be written in a semiformal version as follows:

There does not exist $x \in \mathbb{N}$ such that the successor of x is 0.

The formal version of this sentence is $\neg \, \exists x \in \mathbb{N} : s(x, 0)$. $\quad \blacklozenge$

These notions of quantification belong to a logic called *first-order predicate calculus*. The words "first-order" refer to the fact that quantifiers can quantify only variables that occur in predicates. In Chapter 8 we'll discuss "higher-order" logics in which quantifiers can quantify additional things. To discuss first-order predicate calculus, we need to give a precise description of its well-formed formulas and their meanings. That's the task of this section.

Well-Formed Formulas

To give a precise description of a first-order predicate calculus, we need an alphabet of symbols. For this discussion we'll use several kinds of letters and symbols, described as follows:

Individual variables:	x, y, z
Individual constants:	a, b, c
Function constants:	f, g, h
Predicate constants:	p, q, r
Connective symbols:	$\neg, \rightarrow, \wedge, \vee$
Quantifier symbols:	$\exists, \forall$
Punctuation symbols:	$(,)$

From time to time we will use other letters, or strings of letters, to denote variables or constants. We'll also allow letters to be subscripted. The number of arguments for a predicate or function will normally be clear from the context. A predicate with no arguments is considered to be a proposition.

A *term* is either a variable, a constant, or a function applied to arguments that are terms. For example, x, a, and $f(x, g(b))$ are terms. An *atomic formula* (or simply *atom*) is a predicate applied to arguments that are terms. For example, $p(x, a)$ and $q(y, f(c))$ are atoms.

We can define the wffs—the well-formed formulas—of the first-order predicate calculus inductively as follows:

Basis: Any atom is a wff.

Induction: If W and V are wffs and x is a variable, then the following expressions are also wffs:

$$(W),\ \neg\ W,\ W \vee V,\ W \wedge V,\ W \rightarrow V,\ \exists x\ W,\ \text{and}\ \forall x\ W.$$

To write formulas without too many parentheses and still maintain a unique meaning, we'll agree that the quantifiers have the same precedence as the negation symbol. We'll continue to use the same hierarchy of precedence for the operators $\neg$, $\wedge$, $\vee$, and $\rightarrow$. Therefore the hierarchy of precedence now looks like the following:

$$\neg, \exists x, \forall y \qquad \text{(highest, do first)}$$
$$\wedge$$
$$\vee$$
$$\rightarrow \qquad \text{(lowest, do last)}$$

If any of the quantifiers or the negation symbol appear next to each other, then the rightmost symbol is grouped with the smallest wff to its right. Here are a few wffs in both unparenthesized form and parenthesized form:

Unparenthesized Form	Parenthesized Form
$\forall x \neg \exists y \, \forall z \, p(x, y, z)$	$\forall x \, (\neg \, (\exists y \, (\forall z \, p(x, y, z))))$.
$\exists x \, p(x) \vee q(x)$	$(\exists x \, p(x)) \vee q(x)$.
$\forall x \, p(x) \rightarrow q(x)$	$(\forall x \, p(x)) \rightarrow q(x)$.
$\exists x \neg p(x, y) \rightarrow q(x) \wedge r(y)$	$(\exists x \, (\neg \, p(x, y))) \rightarrow (q(x) \wedge r(y))$.
$\exists x \, p(x) \rightarrow \forall x \, q(x) \vee p(x) \wedge r(x)$	$(\exists x \, p(x)) \rightarrow ((\forall x \, q(x)) \vee (p(x) \wedge r(x)))$.

Because wffs are defined inductively, there are easy techniques to check whether or not an arbitrary expression is a wff. For example, let's show that the following expression is a wff:

$$\exists x \, p(x, y) \rightarrow q(x).$$

There are two approaches: The *bottom-up* approach starts with basis case of atoms; the *top-down* approach starts with the overall structure of the wff.

Bottom-Up

Start with $p(x, y)$ and $q(x)$. These expressions are atoms. So it follows that they are wffs by the basis case. Therefore $\exists x \, p(x, y)$ is a wff by the induction case. Finally, $\exists x \, p(x, y) \rightarrow q(x)$ is a wff by the induction case.

Top-Down

Start by noticing that the expression has the form $W \rightarrow V$ where $W = \exists x \, p(x, y)$ and $V = q(x)$. So the expression is a wff if we can show that W and V are wffs. V is a wff by the basis case. W has the form $\exists x \, U$, where $U = p(x, y)$. Since U is a wff by the basis case, it follows that W is a wff by the induction case. Therefore $\exists x \, p(x, y) \rightarrow q(x)$ is a wff.

Now let's discuss the relationship between the quantifiers and the variables that appear in a wff. When a quantifier occurs in a wff, it influences some occurrences of the quantified variable. The extent of this influence is called the scope of the quantifier, which we define as follows:

In the wff $\exists x \, W$, W is the *scope* of the quantifier $\exists x$.

In the wff $\forall x \, W$, W is the *scope* of the quantifier $\forall x$.

For example, the scope of $\exists x$ in the wff

$$\exists x\, p(x, y) \to q(x)$$

is $p(x, y)$ because the parenthesized version of the wff is $(\exists x\, p(x, y)) \to q(x)$. On the other hand, the scope of $\exists x$ in the wff $\exists x\, (p(x, y) \to q(x))$ is $p(x, y) \to q(x)$.

An occurrence of the variable x in a wff is said to be *bound* if it lies within the scope of either $\exists x$ or $\forall x$ or if it's the quantifier variable x itself. Otherwise, an occurrence of x is said to be *free* in the wff. For example, consider the following wff:

$$\exists x\, p(x, y) \to q(x).$$

The first two occurrences of x are bound because the scope of $\exists x$ is $p(x, y)$. The only occurrence of y is free, and the third occurrence of x is free.

So every occurrence of a variable in a wff can be classified as either bound or free, and this classification is determined by the scope of the quantifiers in the wff. Now we're in position to discuss the meaning of wffs.

Semantics

Up to this point a wff is just a string of symbols with no meaning attached. For a wff to have a meaning, we must give an interpretation to its symbols so that the wff can be read as a statement that is true or false. For example, suppose we let $p(x)$ mean "x is an even integer" and we let x be the number 236. With this interpretation, $p(x)$ becomes the statement "236 is an even integer," which is true.

As another example, let's give an interpretation to the wff

$$\forall x\, \exists y\, s(x, y).$$

We'll let $s(x, y)$ mean that the successor of x is y, where the variables x and y take values from the set of natural numbers $\mathbb{N}$. With this interpretation the wff becomes the statement "For every natural number x there exists a natural number y such that the successor of x is y," which is true.

Before we proceed any further, we need to make a precise definition of an interpretation. Here's the definition:

Interpretations

An *interpretation* for a wff consists of a nonempty set D, called the *domain* of the interpretation, together with an assignment that associates the symbols of the wff to values in D as follows:

Each predicate letter must be assigned some relation over D. A predicate with no arguments is a proposition and must be assigned a truth value.

Each function letter must be assigned a function over D.

Each free variable must be assigned a value in D. All free occurrences of a variable x are assigned the same value in D.

Each constant must be assigned a value in D. All occurrences of the same constant are assigned the same value in D.

So there can be many interpretations for a wff. To describe the meaning of an interpreted wff, we need some notation. Suppose W is a wff, x is a free variable in W, and t is a term. Then the wff obtained from W by replacing all free occurrences of x by t is denoted by the expression

$$W(x/t).$$

The expression x/t is called a *binding* of x to t and can be read as "x gets t" or "x is bound to t" or "x has value t" or "x is replaced by t." We can read $W(x/t)$ as "W with x replaced by t" or "W with x bound to t." For example, suppose we have the following wff:

$$W = p(x) \vee \exists x \; q(x, y).$$

We can apply the binding x/a to W to obtain the wff

$$W(x/a) \; = \; p(a) \vee \exists x \; q(x, y).$$

We can apply the binding y/b to $W(x/a)$ to obtain the wff

$$W(x/a) \, (y/b) \; = \; p(a) \vee \exists x \; q(x, b).$$

We often want to emphasize the fact that a wff W contains a free variable x. In this case we'll use the notation

$$W(x).$$

When this is the case, we write $W(t)$ to denote the wff $W(x/t)$. For example, if $W = p(x) \vee \exists x \; q(x, y)$ and we want to emphasize the fact the x is free in W, we'll write

$$W(x) = p(x) \vee \exists x \; q(x, y).$$

Then we can apply the binding x/a to W by writing

$$W(a) = p(a) \vee \exists x \, q(x, y).$$

Now we have all the ingredients to define the *meaning*, or *semantics*, of wffs in first-order predicate calculus.

The Meaning of a Wff

Suppose we have an interpretation with domain D for a wff.

1. If the wff has no quantifiers, then its meaning is the truth value of the statement obtained from the wff by applying the interpretation.

2. If the wff contains quantifiers, then each quantified wff is evaluated as follows:

   $\exists x \, W$ is true if $W(x/d)$ is true for some $d \in D$. Otherwise, $\exists x \, W$ is false.

   $\forall x \, W$ is true if $W(x/d)$ is true for every $d \in D$. Otherwise, $\forall x \, W$ is false.

The meaning of a wff containing quantifiers can be computed by recursively applying this definition. For example, consider a wff of the form $\forall x \, \exists y \, W$, where W does not contain any further quantifiers. The meaning of $\forall x \, \exists y \, W$ is true if the meaning of $(\exists y \, W)(x/d)$ is true for every $d \in D$. We can write

$$(\exists y \, W)(x/d) = \exists y \, W(x/d).$$

So for each $d \in D$ we must find the meaning of $\exists y \, W(x/d)$. The meaning of $\exists y \, W(x/d)$ is true if there is some element $e \in D$ such that $W(x/d)(y/e)$ is true.

Let's look at a few examples that use actual interpretations.

EXAMPLE 3. The meaning of the wff $\forall x \, \exists y \, s(x, y)$ is true with respect to the interpretation $D = \mathbb{N}$, and $s(x, y)$ means "the successor of x is y." The interpreted wff can be restated as "Every natural number has a successor."   ◆

EXAMPLE 4. Let's consider the "isFatherOf" predicate, where isFatherOf(x, y) means "x is the father of y." The domain of our interpretation is the set of all people now living or who have lived. Assume also that Jim is the father of Andy. The following wffs are given along with their interpreted values:

$$\text{isFatherOf(Jim, Andy)} = \text{true,}$$
$$\exists x \text{ isFatherOf}(x, \text{Andy}) = \text{true ,}$$
$$\forall x \text{ isFatherOf}(x, \text{Andy}) = \text{false,}$$
$$\forall x \, \exists y \text{ isFatherOf}(x, y) = \text{false,}$$
$$\forall y \, \exists x \text{ isFatherOf}(x, y) = \text{true,}$$
$$\exists x \forall y \text{ isFatherOf}(x, y) = \text{false,}$$
$$\exists y \forall x \text{ isFatherOf}(x, y) = \text{false,}$$
$$\exists x \, \exists y \text{ isFatherOf}(x, y) = \text{true,}$$
$$\exists y \, \exists x \text{ isFatherOf}(x, y) = \text{true,}$$
$$\forall x \, \forall y \text{ isFatherOf}(x, y) = \text{false,}$$
$$\forall y \, \forall x \text{ isFatherOf}(x, y) = \text{false.} \quad \blacklozenge$$

EXAMPLE 5. Let's look at some interpretations for $W = \exists x \forall y \, (p(y) \rightarrow q(x, y))$. For each of the following interpretations we'll let $q(x, y)$ denote the equality relation "$x = y$."

   a) Let the domain $D = \{a\}$, and let $p(a) = $ true. Then W is true.

   b) Let the domain $D = \{a\}$, and let $p(a) = $ false. Then W is true.

   c) Let the domain $D = \{a, b\}$, and let $p(a) = p(b) = $ true. Then W is false.

   d) Notice that W is true for any domain D for which $p(d) = $ true for at most one element $d \in D$. $\blacklozenge$

EXAMPLE 6. Let $W = \forall x \, (p(f(x, x), x) \rightarrow p(x, y))$. One interpretation for W can be made as follows: Let $\mathbb{N}$ be the domain, let p be equality, let $y = 0$, and let f be the function defined by $f(a, b) = (a + b) \bmod 3$. With this interpretation, W can be written in more familiar notation as follows:

$$\forall x \, ((2x \bmod 3 = x) \rightarrow x = 0).$$

A bit of checking will convince us that W is true with respect to this interpretation. $\blacklozenge$

EXAMPLE 7. Let $W = \forall x \, (p(f(x, x), x) \rightarrow p(x, y))$. Let $D = \{a, b\}$ be the domain of an interpretation such that $f(a, a) = a$, $f(b, b) = b$, p is equality, and $y = a$. Then W is false with respect to this interpretation. $\blacklozenge$

An interpretation for a wff W is called a *model* for W if W is true with respect to the interpretation. Otherwise, the interpretation is a *countermodel* for W. The previous two examples gave a model and a countermodel, respectively, for the wff

$$W = \forall x \ (p(f(x, x), x) \rightarrow p(x, y)).$$

Validity

Can any wff be true for every possible interpretation? Although it may seem unlikely, this property holds for many wffs. The property is important enough to introduce some terminology. A wff is *valid* if it's true for all possible interpretations. So a wff is valid if every interpretation is a model. Otherwise, the wff is *invalid*. A wff is *unsatisfiable* if it's false for all possible interpretations. So a wff is unsatisfiable if all of its interpretations are countermodels. Otherwise, it is *satisfiable*. From these definitions we see that every wff satisfies exactly one of the following pairs of properties:

> valid and satisfiable,
>
> satisfiable and invalid,
>
> unsatisfiable and invalid.

In the propositional calculus the words *tautology*, *contingency*, and *contradiction* correspond, respectively, to the preceding three pairs of properties for the predicate calculus.

EXAMPLE 8. The wff $\exists x \ \forall y \ (p(y) \rightarrow q(x, y))$ is satisfiable and invalid. To see that the wff is satisfiable, notice that the wff is true with respect to the following interpretation: The domain is the singleton $\{3\}$, and we define $p(3) =$ true and $q(3, 3) =$ true. To see that the wff is invalid, notice that it is false with respect to the following interpretation: The domain is still the singleton $\{3\}$, but now we define $p(3) =$ true and $q(3, 3) =$ false.   ◆

In the propositional calculus we can use truth tables to decide whether any propositional wff is a tautology. But how can we show that a wff of the predicate calculus is valid? We can't check the infinitely many interpretations of the wff to see whether each one is a model. This is the same as checking infinitely many truth tables, one for each interpretation. So we are forced to use some kind of reasoning to show that a wff is valid. Here are two strategies to prove validity:

Direct Approach

If the wff has the form $A \to B$, then assume that there is an arbitrary interpretation for $A \to B$ that is a model for A. Show that the interpretation is a model for B. This proves that any interpretation for $A \to B$ is a model for $A \to B$. So $A \to B$ is valid.

Indirect Approach

Assume that the wff is invalid, and try to obtain a contradiction. Start by assuming the existence of a countermodel for the wff. Then try to argue toward a contradiction of some kind. For example, if the wff has the form $A \to B$, then a countermodel for $A \to B$ makes A true and B false. This information should be used to find a contradiction.

In the following example we'll use the direct approach and the indirect approach to prove the validity of a wff.

EXAMPLE 9. Let W denote the following wff:

$$\exists y \, \forall x \, p(x, y) \to \forall x \, \exists y \, p(x, y).$$

We'll give two proofs to show that W is valid—one direct and one indirect. In both proofs we'll let A be the antecedent and B be the consequent of W.

Direct approach: Let M be an interpretation with domain D for W such that M is a model for A. Then there is an element $d \in D$ such that $\forall x \, p(x, d)$ is true. Therefore $p(e, d)$ is true for all $e \in D$. This says that M is also a model for B. Therefore W is valid.   QED.

Indirect approach: Assume that W is invalid. Then it has a countermodel with domain D that makes A true and B false. Therefore there is an element $d \in D$ such that the wff $\exists y \, p(d, y)$ is false. Thus $p(d, e)$ is false for all $e \in D$. Now we are assuming that A is true. Therefore there is an element $c \in D$ such that $\forall x \, p(x, c)$ is true. In other words, $p(b, c)$ is true for all $b \in D$. In particular, this says that $p(d, c)$ is true. But this contradicts the fact that $p(d, e)$ is false for all elements $e \in D$. Therefore W is valid.   QED.   ◆

There are two interesting transformations that we can apply to any wff containing free variables. One is to universally quantify each free variable, and that other is to existentially quantify each free variable. It seems reasonable to expect that these transformations will change the meaning of the original wff, as the following examples show:

$p(x) \wedge \neg\, p(y)$ is satisfiable, but $\forall x\ \forall y\ (p(x) \wedge \neg\, p(y))$ is unsatisfiable.

$p(x) \to p(y)$ is invalid, but $\exists x\ \exists y\ (p(x) \to p(y))$ is valid.

The interesting thing about the process is that validity is preserved if we universally quantify the free variables and unsatisfiability is preserved if we existentially quantify the free variables. To make this more precise, we need a little terminology.

Suppose W is a wff with free variables $x_1, ..., x_n$. The *universal closure* of W is the wff

$$\forall x_1 \cdots \forall x_n\ W.$$

The *existential closure* of W is the wff

$$\exists x_1 \cdots \exists x_n\ W.$$

For example, suppose $W = \forall x\ p(x, y)$. W has y as its only free variable. So the universal closure of W is

$$\forall y\ \forall x\ p(x, y),$$

and the existential closure of W is

$$\exists y\ \forall x\ p(x, y).$$

As we have seen, the meaning of a wff may change by taking either of the closures. But there are two properties that don't change, and we'll state them for the record as follows:

Closure Properties   (7.1)

1. A wff is valid if and only if its universal closure is valid.
2. A wff is unsatisfiable if and only if its existential closure is unsatisfiable.

We'll prove property 1 and leave the proof of property 2 as an exercise.

Proof of property 1: Let W be a wff and x be the only free variable of W. Let M be a model for W with domain D. Then $W(x/d)$ is true for every element $d \in D$. In other words, M is a model for $\forall x\ W$. On the other hand, let M be a model for $\forall x\ W$, where D is the domain. Then $W(x/d)$ is true for each element $d \in D$.

Therefore M is a model for W. If there are more free variables in a wff, then apply the argument to each variable.   QED.

The Validity Problem

We'll end this section with a short discussion about deciding the validity of wffs. First we need to introduce the general notion of decidability. Any problem that can be stated as a question with a yes or no answer is called a *decision problem*. Practically every problem can be stated as a decision problem, perhaps after some work. A decision problem is called *decidable* if there is an algorithm that halts with the answer to the problem. Otherwise, the problem is called *undecidable*. A decision problem is called *partially decidable* if there is an algorithm that halts with the answer yes if the problem has a yes answer but may not halt if the problem has a no answer. The words *solvable*, *unsolvable*, and *partially solvable* are also used to mean decidable, undecidable, and partially decidable, respectively.

Now let's get back to logic. The *validity problem* for a formal theory can be stated as follows:

Given a wff, is it valid?

The validity problem for the propositional calculus can be stated as follows: Given a wff, is it a tautology? This problem is decidable by Quine's method. Another algorithm would be to build a truth table for the wff and then check it.

Although the validity problem for the first-order predicate calculus is undecidable, it is partially decidable. There are two partial decision procedures for the first-order predicate calculus that are of interest: natural deduction (due to Gentzen [1935]) and resolution (due to Robinson [1965]). Natural deduction is a formal reasoning system that models the natural way we reason about the validity of wffs by using inference rules, as we did in Chapter 6 and as we'll discuss in the last section of this chapter. Resolution is a mechanical way to reason, which is not easily adaptable to people. It is, however, adaptable to machines. Resolution is an important ingredient in logic programming and automatic reasoning, which we'll discuss in Chapter 9.

Exercises

1. Write down the proposition denoted by each of the following expressions, where variables can take values in the domain $D = \{0, 1\}$.

   a. $\exists x \in D : \forall y \in D : p(x, y)$.

   b. $\forall y \in D : \exists x \in D : p(x, y)$.

2. Write down a quantified expression over some domain to denote each of the following propositions or predicates.

    a. $q(0) \wedge q(1)$.

    b. $q(0) \vee q(1)$.

    c. $p(x, 0) \wedge p(x, 1)$.

    d. $p(0, x) \vee p(1, x)$.

    e. $p(1) \vee p(3) \vee p(5) \vee \dots$.

    f. $p(2) \wedge p(4) \wedge p(6) \wedge \dots$.

3. Write down a formal representation of each of the following statements as a quantified wff.

    a. Every natural number other than 0 has a predecessor.

    b. Any two nonzero natural numbers have a common divisor.

4. Explain why each of the following expressions is a wff.

    a. $\exists x\, p(x) \rightarrow \forall x\, p(x)$.   b. $\exists x \forall y\, (p(y) \rightarrow q(f(x), y))$.

5. Explain why the expression $\forall y\, (p(y) \rightarrow q(f(x), p(x)))$ is not a wff.

6. For each of the following wffs, label each occurrence of the variables as either bound or free.

    a. $p(x, y) \vee (\forall y\, q(y) \rightarrow \exists x\, r(x, y))$.

    b. $\forall y\, q(y) \wedge \neg p(x, y)$.

    c. $\neg\, q(x, y) \vee \exists x\, p(x, y)$.

7. Write down a single wff containing three variables $x, y,$ and z, with the following properties: x occurs twice as a bound variable; y occurs once as a free variable; z occurs three times, once as a free variable and twice as a bound variable.

8. Given the wff $W = \exists x\, p(x) \rightarrow \forall x\, p(x)$.

    a. Find all possible interpretations of W over the domain $D = \{a\}$. Also givethe truth value of W over each of the interpretations.

    b. Find all possible interpretations of W over the domain $D = \{a, b\}$. Also give the truth value of W over each of the interpretations.

9. Find a model for each of the following wffs.

    a. $p(c) \wedge \exists x\, \neg\, p(x)$.

    b. $\exists x\, p(x) \rightarrow \forall x\, p(x)$.

    c. $\exists y\, \forall x\, p(x, y) \rightarrow \forall x\, \exists y\, p(x, y)$.

    d. $\forall x\, \exists y\, p(x, y) \rightarrow \exists y\, \forall x\, p(x, y)$.

    e. $\forall x\, (p(x, f(x)) \rightarrow p(x, y))$.

10. Find a countermodel for each of the following wffs.

    a. $p(c) \wedge \exists x \neg p(x)$.
    b. $\exists x \, p(x) \rightarrow \forall x \, p(x)$.
    c. $\forall x \, (p(x) \vee q(x)) \rightarrow \forall x \, p(x) \vee \forall x \, q(x)$.
    d. $\exists x \, p(x) \wedge \exists x \, q(x) \rightarrow \exists x \, (p(x) \wedge q(x))$.
    e. $\forall x \, \exists y \, p(x, y) \rightarrow \exists y \forall x \, p(x, y)$.
    f. $\forall x \, (p(x, f(x)) \rightarrow p(x, y))$.

11. Given the wff $W = \forall x \, \forall y \, (p(x) \rightarrow p(y))$.

    a. Show that W is true for any interpretation whose domain is a singleton.
    b. Show that W is not valid.

12. Given the wff $W = \forall x \, p(x, x) \rightarrow \forall x \, \forall y \, \forall z \, (p(x, y) \vee p(x, z) \vee p(y, z))$.

    a. Show that W is true for any interpretation whose domain is a singleton.
    b. Show that W is true for any interpretation whose domain has two elements.
    c. Show that W is not valid.

13. Find an example of a wff that is true for any interpretation having a domain with three or fewer elements but is not valid. *Hint:* Look at the structure of the wff in Exercise 12.

14. Prove that each of the following wffs is valid. *Hint:* Either show that every interpretation is a model or assume that the wff is invalid and find a contradiction.

    a. $\forall x \, (p(x) \rightarrow p(x))$.
    b. $p(c) \rightarrow \exists x \, p(x)$.
    c. $\forall x \, p(x) \rightarrow \exists x \, p(x)$.
    d. $\exists x \, (A(x) \wedge B(x)) \rightarrow \exists x \, A(x) \wedge \exists x \, B(x)$.
    e. $\forall x \, A(x) \vee \forall x \, B(x) \rightarrow \forall x \, (A(x) \vee B(x))$.
    f. $\forall x \, (A(x) \rightarrow B(x)) \rightarrow (\exists x \, A(x) \rightarrow \exists x \, B(x))$.
    g. $\forall x \, (A(x) \rightarrow B(x)) \rightarrow (\forall x \, A(x) \rightarrow \exists x \, B(x))$.
    h. $\forall x \, (A(x) \rightarrow B(x)) \rightarrow (\forall x \, A(x) \rightarrow \forall x \, B(x))$.

15. Prove that each of the following wffs is unsatisfiable. *Hint:* Either show that every interpretation is a countermodel or assume that the wff is satisfiable and find a contradiction.

    a. $p(c) \wedge \neg p(c)$.   b. $\exists x \, (p(x) \wedge \neg p(x))$.   c. $\exists x \, \forall y \, (p(x, y) \wedge \neg p(x, y))$.

16. Prove that a wff is unsatisfiable if and only if its existential closure is unsatisfiable.

17. Prove that any wff of the form $A \to B$ is valid if and only if whenever A is valid, then B is valid.

7.2 Equivalent Formulas

In this section we'll discuss the important notion of equivalence for wffs of the first-order predicate calculus.

Equivalence

Two wffs A and B are *equivalent* if they both have the same truth value with respect to every interpretation of both A and B. By an interpretation of both A and B, we mean that all free variables, constants, functions, and predicates that occur in either A or B are interpreted with respect to a single domain. We denote the fact that A and B are equivalent by writing

$$A \equiv B.$$

We can describe equivalence in terms of a single valid wff as follows:

$$A \equiv B \quad \text{if and only if} \quad (A \to B) \wedge (B \to A) \text{ is valid.}$$

To start things off, let's see how propositional equivalences give rise to predicate calculus equivalences. A wff W is an *instance* of a propositional wff V if W is obtained from V by replacing each propositional letter of V by a wff, where all occurrences of each propositional letter in V are replaced by the same wff. For example, the wff

$$\forall x \, p(x) \to \forall x \, p(x) \vee q(x)$$

is an instance of $P \to P \vee Q$ because Q is replaced by $q(x)$ and both occurrences of P are replaced by $\forall x \, p(x)$.

If W is an instance of a propositional wff V, then the truth value of W for any interpretation can be obtained by assigning truth values to the letters of V. For example, suppose we define an interpretation with domain $D = \{a, b\}$ and we set $p(a) = p(b) =$ true and $q(a) = q(b) =$ false. For this interpretation, the truth value of the wff $\forall x \, p(x) \to \forall x \, p(x) \vee q(x)$ is the same as the truth value of the propositional wff $P \to P \vee Q$, where $P =$ true and $Q =$ false.

So we can say that two wffs are equivalent if they are instances of two equivalent propositional wffs, where both instances are obtained by using the same replacement of propositional letters. For example, we have

$$\forall x \, p(x) \to q(x) \equiv \neg \, \forall x \, p(x) \vee q(x)$$

because the left and right sides are instances of the left and right sides of the propositional equivalence $P \to Q \equiv \neg \, P \vee Q$, where both occurrences of P are replaced by $\forall x \, p(x)$ and both occurrences of Q are replaced by $q(x)$. We'll state the result again for emphasis:

> *Two wffs are equivalent whenever they are instances of two equivalent propositional wffs, where both instances are obtained by using the same replacement of propositional letters.*

Let's see whether we can find some more equivalences to make our logical life easier. We'll start by listing two equivalences that relate the two quantifiers by negation. For any wff W we have the two equivalences

$$\neg \, (\forall x \, W) \equiv \exists x \, \neg \, W \quad \text{and} \quad \neg \, (\exists x \, W) \equiv \forall x \, \neg \, W. \qquad (7.2)$$

It's easy to believe that these two equivalences are true. For example, we can illustrate the equivalence $\neg \, (\forall x \, W) \equiv \exists x \, \neg \, W$ by observing that the negation of the statement "Something is true for all possible cases" has the same meaning as the statement "There is some case for which the something is false." Similarly, we can illustrate the equivalence $\neg \, (\exists x \, W) \equiv \forall x \, \neg \, W$ by observing that the negation of the statement "There is some case for which something is true" has the same meaning as the statement "Every case of the something is false."

Another way to demonstrate these equivalences is to use De Morgan's laws. For example, let $W = p(x)$ and suppose that we have an interpretation with domain $D = \{0, 1, 2, 3\}$. Then no matter what values we assign to p, we can apply De Morgan's laws to obtain the following propositional equivalence:

$$
\begin{aligned}
\neg \, (\forall x \, p(x)) \quad &\equiv \quad \neg \, (p(0) \wedge p(1) \wedge p(2) \wedge p(3)) \\
&\equiv \quad \neg \, p(0) \vee \neg \, p(1) \vee \neg \, p(2) \vee \neg \, p(3) \\
&\equiv \quad \exists x \, \neg \, p(x).
\end{aligned}
$$

We also get the following equivalence:

$$\neg\,(\exists x\,p(x)) \quad \equiv \quad \neg\,(p(0) \vee p(1) \vee p(2) \vee p(3))$$
$$\equiv \quad \neg\,p(0) \wedge \neg\,p(1) \wedge \neg\,p(2) \wedge \neg\,p(3)$$
$$\equiv \quad \forall x\,\neg\,p(x).$$

These examples are nice, but they don't prove (7.2). Let's give an actual proof, using validity, of the equivalences (7.2). We'll prove the first equivalence, $\neg\,(\forall x\,W) \equiv \exists x\,\neg\,W$, and then use it to prove the second equivalence.

Proof: Let I be an interpretation with domain D for the wffs $\neg\,(\forall x\,W)$ and $\exists x\,\neg\,W$. We want to show that I is a model for one of the wffs if and only if I is a model for the other wff. The following equivalent statements do the job:

I is a model for $\neg\,(\forall x\,W)$ iff $\neg\,(\forall x\,W)$ is true for I

iff $\forall x\,W$ is false for I

iff $W(x/d)$ is false for some $d \in D$

iff $\neg\,W(x/d)$ is true for some $d \in D$

iff $\exists x\,\neg\,W$ is true for I

iff I is a model for $\exists x\,\neg\,W$.

This proves the equivalence $\neg\,(\forall x\,W) \equiv \exists x\,\neg\,W$. Now, since W is arbitrary, we can replace W by the wff $\neg\,W$ to obtain the following equivalence:

$$\neg\,(\forall x\,\neg\,W) \equiv \exists x\,\neg\,\neg\,W.$$

Now take the negation of both sides of this equivalence, and simplify the double negations to obtain the second equivalence of (7.2):

$$\forall x\,\neg\,W \equiv \neg\,(\exists x\,W). \quad \text{QED.}$$

Now let's look at two equivalences that allow us to interchange universal quantifiers if they are next to each other and similarly for existential quantifiers.

$$\forall x\,\forall y\,W \equiv \forall y\,\forall x\,W \quad \text{and} \quad \exists x\,\exists y\,W \equiv \exists y\,\exists x\,W. \tag{7.3}$$

Again, this is easy to believe. For example, suppose that $W = p(x, y)$ and we have an interpretation with domain $D = \{0, 1\}$. Then we have the following equivalences:

$$\forall x \; \forall y \; p(x, y) \;\equiv\; \forall y \; p(0, y) \land \forall y \; p(1, y)$$

$$\equiv\; (p(0, 0) \land p(0, 1)) \land (p(1, 0) \land p(1, 1))$$

$$\equiv\; (p(0, 0) \land p(1, 0)) \land (p(0, 1) \land p(1, 1))$$

$$\equiv\; \forall x \; p(x, 0) \land \forall x \; p(x, 1)$$

$$\equiv\; \forall y \; \forall x \; p(x, y).$$

We also have the following equivalences:

$$\exists x \; \exists y \; p(x, y) \;\equiv\; \exists y \; p(0, y) \lor \exists y \; p(1, y)$$

$$\equiv\; (p(0, 0) \lor p(0, 1)) \lor (p(1, 0) \lor p(1, 1))$$

$$\equiv\; (p(0, 0) \lor p(1, 0)) \lor (p(0, 1) \lor p(1, 1))$$

$$\equiv\; \exists x \; p(x, 0) \lor \exists x \; p(x, 1)$$

$$\equiv\; \exists y \; \exists x \; p(x, y).$$

We leave the proofs of equivalences (7.3) as exercises.

EXAMPLE 1. We want to prove the following equivalence:

$$\exists x \; (p(x) \to q(x)) \equiv \forall x \; p(x) \to \exists x \; q(x). \tag{7.4}$$

Proof: First, we'll prove $\exists x \; (p(x) \to q(x)) \to (\forall x \; p(x) \to \exists x \; q(x))$: Let I be a model for $\exists x \; (p(x) \to q(x))$ with domain D. Then $\exists x \; (p(x) \to q(x))$ is true for I, which means that $p(d) \to q(d)$ is true for some $d \in D$. Therefore either $p(d) =$ false or $p(d) = q(d) =$ true for some $d \in D$. If $p(d) =$ false, then $\forall x \; p(x)$ is false for I; if $p(d) = q(d) =$ true, then $\exists x \; q(x)$ is true for I. In either case we obtain $\forall x \; p(x) \to \exists x \; q(x)$ is true for I. Therefore I is a model for $\forall x \; p(x) \to \exists x \; q(x)$.

   Now we'll prove $(\forall x \; p(x) \to \exists x \; q(x)) \to \exists x \; (p(x) \to q(x))$: Let I be a model for $\forall x \; p(x) \to \exists x \; q(x)$ with domain D. Then $\forall x \; p(x) \to \exists x \; q(x)$ is true for I. Therefore either $\forall x \; p(x)$ is false for I or both $\forall x \; p(x)$ and $\exists x \; q(x)$ are true for I. If $\forall x \; p(x)$ is false for I, then $p(d)$ is false for some $d \in D$. Therefore $p(d) \to q(d)$ is true. If both $\forall x \; p(x)$ and $\exists x \; q(x)$ are true for I, then there is some $c \in D$ such that $p(c) = q(c) =$ true. Therefore $p(c) \to q(c)$ is true. So in either case, $\exists x \; (p(x) \to q(x))$ is true for I. Therefore I is a model for $\exists x \; (p(x) \to q(x))$.   QED.   ◆

Of course, once we know some equivalences, we can use them to prove other equivalences. For example, let's see how previous results can be used to prove the following equivalence:

$$\exists x\ (p(x) \lor q(x)) \equiv \exists x\ p(x) \lor \exists x\ q(x). \qquad (7.5)$$

Proof:
$$
\begin{aligned}
\exists x\ (p(x) \lor q(x)) \ &\equiv\ \exists x\ (\neg\, p(x) \to q(x)) \\
&\equiv\ \forall x\ \neg\, p(x) \to \exists x\ q(x) &&(7.4) \\
&\equiv\ \neg\, \exists x\ p(x) \to \exists x\ q(x) &&(7.2) \\
&\equiv\ \exists x\ p(x) \lor \exists x\ q(x) &&\text{QED.}
\end{aligned}
$$

Next we'll look at some equivalences that hold under certain restrictions.

Restricted Equivalences

We'll start with the simple idea of name replacement. With certain restrictions we can rename variables in a quantified wff without changing its meaning. For example, suppose we are given the wff

$$\forall x\ (p(x) \to p(w)).$$

We can replace all occurrences of x by y to obtain the equivalence

$$\forall x\ (p(x) \to p(w)) \equiv \forall y\ (p(y) \to p(w)).$$

But we can't replace all occurrences of x by w because

$$\forall x\ (p(x) \to p(w)) \text{ is not equivalent to } \forall w\ (p(w) \to p(w)).$$

The key is to choose a new variable that doesn't occur in the wff and then be consistent with the replacement. Here's the rule:

Renaming Rule $\qquad\qquad\qquad\qquad\qquad\qquad\qquad\qquad (7.6)$

If y does not occur in $W(x)$, then the following equivalences hold:

a) $\exists x\ W(x) \equiv \exists y\ W(y).$

b) $\forall x\ W(x) \equiv \forall y\ W(y).$

Remember that $W(y)$ is obtained from $W(x)$ by replacing all free occurrences of x by y.

For example, let's use (7.6) to make all the quantifier variables distinct in the following wff:

$$\forall x\ \exists y\ (p(x, y) \to \exists x\ q(x, y) \lor \forall y\ r(x, y)).$$

Since the variables u and v don't occur in this wff, we can use (7.6) to write the equivalent wff

$$\forall u \, \exists v \, (p(u, v) \to \exists x \, q(x, v) \vee \forall y \, r(u, y)).$$

Now we'll look at some restricted equivalences that allow us to move a quantifier past a wff that doesn't contain the quantified variable.

Equivalences with Restrictions

IF x does not occur in the wff C, THEN the following equivalences hold:

Disjunction (7.7)

$$\forall x \, (C \vee A(x)) \equiv C \vee \forall x \, A(x).$$
$$\exists x \, (C \vee A(x)) \equiv C \vee \exists x \, A(x).$$

Conjunction (7.8)

$$\forall x \, (C \wedge A(x)) \equiv C \wedge \forall x \, A(x).$$
$$\exists x \, (C \wedge A(x)) \equiv C \wedge \exists x \, A(x).$$

Implication (7.9)

$$\forall x \, (C \to A(x)) \equiv C \to \forall x \, A(x).$$
$$\exists x \, (C \to A(x)) \equiv C \to \exists x \, A(x).$$
$$\forall x \, (A(x) \to C) \equiv \exists x \, A(x) \to C.$$
$$\exists x \, (A(x) \to C) \equiv \forall x \, A(x) \to C.$$

The implication equivalences (7.9) are easily derived from the other equivalences. For example, the third equivalence of (7.9) can be proved as follows:

$$
\begin{aligned}
\forall x \, (A(x) \to C) &\equiv \forall x \, (\neg A(x) \vee C) \\
&\equiv \forall x \, \neg A(x) \vee C && (7.7) \\
&\equiv \neg \, \exists x \, A(x) \vee C && (7.2) \\
&\equiv \exists x \, A(x) \to C.
\end{aligned}
$$

The equivalences (7.7), (7.8), and (7.9) also hold under the weaker assumption that C does not contain a free occurrence of x. For example, if $C = \exists x \, p(x)$, then we can rename the variable x to y to obtain $\exists x \, p(x) \equiv \exists y \, p(y)$. Now x does not occur in the wff $\exists y \, p(y)$, and we can apply the rules as they are stated. For this example we can prove the first equivalence of (7.7) in the following way:

$$\forall x\,(C \vee A(x)) \quad \equiv \quad \forall x\,(\exists x\,p(x) \vee A(x))$$

$$\equiv \quad \forall x\,(\exists y\,p(y) \vee A(x)) \qquad \text{(rename)}$$

$$\equiv \quad \exists y\,p(y) \vee \forall x\,A(x) \qquad (7.7)$$

$$\equiv \quad \exists x\,p(x) \vee \forall x\,A(x) \qquad \text{(rename)}$$

$$\equiv \quad C \vee \forall x\,A(x).$$

Now that we have some equivalences on hand, we can use them to prove other equivalences. In other words, we have a set of rules to transform wffs into other wffs having the same meaning. This justifies the word "calculus" in the name "predicate calculus."

Normal Forms

In the propositional calculus we know that any wff is equivalent to a wff in conjunctive normal form and to a wff in disjunctive normal form. Let's see whether we can do something similar with the wffs of the predicate calculus. We'll start with a definition. A wff W is in *prenex normal form* if all its quantifiers are on the left of the expression. In other words, a prenex normal form looks like the following:

$$Q_1 x_1 \ldots Q_n x_n\,M,$$

where each Q_i is either $\forall$ or $\exists$, each x_i is distinct, and M is a wff without quantifiers. For example, the following wffs are in prenex normal form:

$$p(x),$$

$$\exists x\,p(x),$$

$$\forall x\,p(x, y),$$

$$\forall x\,\exists y\,(p(x, y) \rightarrow q(x)),$$

$$\forall x\,\exists y\,\forall z\,(p(x) \vee q(y) \wedge r(x, z)).$$

Is any wff equivalent to some wff in prenex normal form? Yes. In fact there's an easy algorithm to obtain the desired form. The idea is to make sure that variables have distinct names and then apply equivalences that send all quantifiers to the left end of the wff. Here's the algorithm:

Prenex Normal Form Algorithm (7.10)

Any wff *W* has an equivalent prenex normal form, which can be constructed as follows:

1. Rename the variables of W so that no quantifiers use the same variable name and such that the quantified variable names are distinct from the free variable names.

2. Move quantifiers to the left by using equivalences (7.2), (7.7), (7.8), and (7.9).

The renaming of variables is important to the success of the algorithm. For example, we can't replace $p(x) \vee \forall x\, q(x)$ by $\forall x\, (p(x) \vee q(x))$ because they aren't equivalent. But we can rename variables to obtain the following equivalence:

$$p(x) \vee \forall x\, q(x) \equiv p(x) \vee \forall y\, q(y) \equiv \forall y\, (p(x) \vee q(y)).$$

EXAMPLE 2. Suppose we are given the following wff W:

$$A(x) \wedge \forall x\, (B(x) \rightarrow \exists y\, C(x, y) \vee \neg\, \exists y\, A(y)).$$

Let's put this wff in prenex normal form. First notice that y is used in two quantifiers and x occurs both free and in a quantifier. After changing names, we obtain the following version of W:

$$A(x) \wedge \forall z\, (B(z) \rightarrow \exists y\, C(z, y) \vee \neg\, \exists w\, A(w)).$$

Now each quantified variable is distinct, and the quantified variables are distinct from the free variable x. Now we'll apply equivalences to move all the quantifiers to the left:

$$
\begin{aligned}
W \;\equiv\;& A(x) \wedge \forall z\, (B(z) \rightarrow \exists y\, C(z, y) \vee \neg\, \exists w\, A(w)) \\[2pt]
\equiv\;& \forall z\, (A(x) \wedge (B(z) \rightarrow \exists y\, C(z, y) \vee \neg\, \exists w\, A(w))) && (7.8) \\[2pt]
\equiv\;& \forall z\, (A(x) \wedge (B(z) \rightarrow \exists y\, (C(z, y) \vee \neg\, \exists w\, A(w)))) && (7.7) \\[2pt]
\equiv\;& \forall z\, (A(x) \wedge \exists y\, (B(z) \rightarrow C(z, y) \vee \neg\, \exists w\, A(w))) && (7.9) \\[2pt]
\equiv\;& \forall z\, \exists y\, (A(x) \wedge (B(z) \rightarrow C(z, y) \vee \neg\, \exists w\, A(w))) && (7.8) \\[2pt]
\equiv\;& \forall z\, \exists y\, (A(x) \wedge (B(z) \rightarrow C(z, y) \vee \forall w\, \neg\, A(w))) && (7.2) \\[2pt]
\equiv\;& \forall z\, \exists y\, (A(x) \wedge (B(z) \rightarrow \forall w\, (C(z, y) \vee \neg\, A(w)))) && (7.7) \\[2pt]
\equiv\;& \forall z\, \exists y\, (A(x) \wedge \forall w\, (B(z) \rightarrow C(z, y) \vee \neg\, A(w))) && (7.9) \\[2pt]
\equiv\;& \forall z\, \exists y\, \forall w\, (A(x) \wedge (B(z) \rightarrow C(z, y) \vee \neg\, A(w))) && (7.8).
\end{aligned}
$$

This wff is in the desired prenex normal form. ◆

There are two special prenex normal forms that correspond to the disjunctive normal form and the conjunctive normal form for propositional calculus. We define a *literal* in the predicate calculus to be an atom or the negation of an atom. For example, $p(x)$ and $\neg q(x, y)$ are literals. A prenex normal form is called a *prenex disjunctive normal form* if it has the form

$$Q_1 x_1 \dots Q_n x_n \, (D_1 \vee \dots \vee D_k),$$

where each D_i is a conjunction of one or more literals. Similarly, a prenex normal form is called a *prenex conjunctive normal form* if it has the form

$$Q_1 x_1 \dots Q_n x_n \, (C_1 \wedge \dots \wedge C_k),$$

where each C_i is a disjunction of one or more literals.

It's easy to construct either of these normal forms from a prenex normal form. Just eliminate conditionals, move $\neg$ inwards, and either distribute $\wedge$ over $\vee$ or distribute $\vee$ over $\wedge$. If we want to start with an arbitrary wff, then we can put everything together in a nice little algorithm. We can save some thinking by removing all conditionals at an early stage of the process. Then we won't have to remember the formulas (7.9). The algorithm can be stated as follows:

Prenex Disjunctive/Conjunctive Normal Form Algorithm   (7.11)

Any wff W has an equivalent prenex disjunctive/conjunctive normal form, which can be constructed as follows:

1. Rename the variables of W so that no quantifiers use the same variable name and such that the quantified variable names are distinct from the free variable names.

2. Remove implications by using the equivalence $A \rightarrow B \equiv \neg A \vee B$.

3. Move negations to the right to form literals by using the equivalences (7.2) and the equivalences $\neg (A \wedge B) \equiv \neg A \vee \neg B$, $\neg (A \vee B) \equiv \neg A \wedge \neg B$, and $\neg \neg A \equiv A$.

4. Move quantifiers to the left by using equivalences (7.7), and (7.8).

5. To obtain the disjunctive normal form, distribute $\wedge$ over $\vee$. To obtain the conjunctive normal form, distribute $\vee$ over $\wedge$.

Now let's do an example that uses (7.11) to transform a wff into a prenex normal form.

EXAMPLE 3. Suppose W is the following wff:

$$\forall x\, A(x) \vee \exists x\, B(x) \to C(x) \wedge \exists x\, C(x).$$

We'll construct the prenex disjunctive normal form for W as the following sequence of equivalences:

$$
\begin{aligned}
W ={}& \forall x\, A(x) \vee \exists x\, B(x) \to C(x) \wedge \exists x\, C(x) \\
\equiv{}& \forall y\, A(y) \vee \exists z\, B(z) \to C(x) \wedge \exists w\, C(w) & \text{(rename variables)} \\
\equiv{}& \neg\,(\forall y\, A(y) \vee \exists z\, B(z)) \vee (C(x) \wedge \exists w\, C(w)) & \text{(remove} \to) \\
\equiv{}& (\neg\, \forall y\, A(y) \wedge \neg\, \exists z\, B(z)) \vee (C(x) \wedge \exists w\, C(w)) & \text{(move negation)} \\
\equiv{}& (\exists y\, \neg\, A(y) \wedge \forall z\, \neg\, B(z)) \vee (C(x) \wedge \exists w\, C(w)) & \text{(7.2)} \\
\equiv{}& \exists y\, (\neg\, A(y) \wedge \forall z\, \neg\, B(z)) \vee (C(x) \wedge \exists w\, C(w)) & \text{(7.8)} \\
\equiv{}& \exists y\, ((\neg\, A(y) \wedge \forall z\, \neg\, B(z)) \vee (C(x) \wedge \exists w\, C(w))) & \text{(7.7)} \\
\equiv{}& \exists y\, (\forall z\, (\neg\, A(y) \wedge \neg\, B(z)) \vee (C(x) \wedge \exists w\, C(w))) & \text{(7.8)} \\
\equiv{}& \exists y\, \forall z\, ((\neg\, A(y) \wedge \neg\, B(z)) \vee (C(x) \wedge \exists w\, C(w))) & \text{(7.7)} \\
\equiv{}& \exists y\, \forall z\, ((\neg\, A(y) \wedge \neg\, B(z)) \vee \exists w\, (C(x) \wedge C(w))) & \text{(7.8)} \\
\equiv{}& \exists y\, \forall z\, \exists w\, ((\neg\, A(y) \wedge \neg\, B(z)) \vee (C(x) \wedge C(w))) & \text{(7.7).}
\end{aligned}
$$

This wff is in prenex disjunctive normal form. To obtain the prenex conjunctive normal form, we'll distribute $\vee$ over $\wedge$ to obtain the following equivalences:

$$
\begin{aligned}
\equiv{}& \exists y\, \forall z\, \exists w\, ((\neg\, A(y) \wedge \neg\, B(z)) \vee C(x)) \wedge (\neg\, A(y) \wedge \neg\, B(z)) \vee C(w))) \\
\equiv{}& \exists y\, \forall z\, \exists w\, ((\neg\, A(y) \vee C(x)) \wedge (\neg\, B(z) \vee C(x)) \\
& \qquad\qquad \wedge (\neg\, A(y) \vee C(w)) \wedge (\neg\, B(z)) \vee C(w))).
\end{aligned}
$$

This wff is in prenex conjunctive normal form.   ♦

Formalizing English Sentences

Now that we have a few tools at hand, let's see whether we can find some heuristics for formalizing English sentences. We'll look at several sentences dealing with people and the characteristics of being a politician and being crooked. Let $p(x)$ denote the statement "x is a politician," and let $q(x)$ denote the statement "x is crooked." For each of the following sentences we've listed a formalization with quantifiers. Before you look at each formalization, try to find one of your own. It may be correct, even though it doesn't look like the listed answer.

> "Some politician is crooked." $\exists x \, (p(x) \wedge q(x))$.
>
> "No politician is crooked." $\forall x \, (p(x) \rightarrow \neg \, q(x))$.
>
> "All politicians are crooked." $\forall x \, (p(x) \rightarrow q(x))$.
>
> "Not all politicians are crooked." $\exists x \, (p(x) \wedge \neg \, q(x))$.
>
> "Every politician is crooked." $\forall x \, (p(x) \rightarrow q(x))$.
>
> "There is an honest politician." $\exists x \, (p(x) \wedge \neg \, q(x))$.
>
> "No politician is honest." $\forall x \, (p(x) \rightarrow q(x))$.
>
> "All politicians are honest." $\forall x \, (p(x) \rightarrow \neg \, q(x))$.

Can we notice anything interesting about the formalizations of these sentences? Yes, we can. Notice that each formalization satisfies one of the following two properties:

> The universal quantifier $\forall x$ quantifies a conditional.
>
> The existential quantifier $\exists x$ quantifies a conjunction.

To see why this happens, let's look at the statement "Some politician is crooked." We came up with the wff $\exists x \, (p(x) \wedge q(x))$. Someone might argue that the answer could also be the wff $\exists x \, (p(x) \rightarrow q(x))$. Notice that the second wff is true even if there are no politicians, while the first wff is false in this case, as it should be. Another way to see the difference is to look at equivalent wffs. From (7.4) we have the equivalence $\exists x \, (p(x) \rightarrow q(x)) \equiv \forall x \, p(x) \rightarrow \exists x \, q(x)$. Let's see how the wff $\forall x \, p(x) \rightarrow \exists x \, q(x)$ reads when applied to our example. It says, "If everyone is a politician, then someone is crooked." This doesn't seem to convey the same thing as our original sentence.

Another thing to notice is that people come up with different answers. For example, the second sentence, "No politician is crooked," might also be written as follows:

$$\neg \, \exists x \, (p(x) \wedge q(x)).$$

It's nice to know that this answer is OK too because it's equivalent to the listed answer, $\forall x \, (p(x) \rightarrow \neg \, q(x))$. We'll prove the equivalence of the two wffs by applying (7.2) as follows:

$$\neg\, \exists x\, (p(x) \wedge q(x)) \;\equiv\; \forall x\, \neg\, (p(x) \wedge q(x))$$
$$\equiv\; \forall x\, (\neg\, p(x) \vee \neg\, q(x))$$
$$\equiv\; \forall x\, (p(x) \rightarrow \neg\, q(x)).$$

Of course, not all sentences are easy to formalize. For example, suppose we want to formalize the following sentence:

It is not the case that not every widget has no defects.

Suppose we let $w(x)$ mean "x is a widget" and let $d(x)$ mean "x has a defect." We might look at the latter portion of the sentence, which says, "every widget has no defects." We can formalize this statement as $\forall x\, (w(x) \rightarrow \neg\, d(x))$. Now the beginning part of the sentence says, "It is not the case that not." This is a double negation. So the formalization of the entire sentence is

$$\neg\, \neg\, \forall x\, (w(x) \rightarrow \neg\, d(x)),$$

which of course is equivalent to $\forall x\, (w(x) \rightarrow \neg\, d(x))$.

Let's discuss the little words "is" and "are." Their usage can lead to quite different formalizations. For example, the three statements

"4 is 2 + 2," "x is a widget," and "Widgets are defective"

have the three formalizations $4 = 2 + 2$, $w(x)$, and $\forall x\, (w(x) \rightarrow d(x))$. So we have to be careful when we try to formalize English sentences.

As a final example, which we won't discuss, consider the following sentence taken from Section 2, Article I, of the Constitution of the United States of America:

No person shall be a Representative who shall not have attained to the Age of twenty-five Years, and been seven Years a Citizen of the United States, and who shall not, when elected, be an Inhabitant of that State in which he shall be chosen.

Summary

We collect here some of the fundamental results, which include some equivalences, some restricted equivalences, and some conditionals that are not equivalences.

Some Equivalences

1. $\neg\, \forall x\ W(x) \equiv \exists x \neg W(x).$ (7.2)

2. $\neg\, \exists x\ W(x) \equiv \forall x \neg W(x).$ (7.2)

3. $\exists x\ (A(x) \vee B(x)) \equiv \exists x\ A(x) \vee \exists x\ B(x).$ (7.5)

4. $\forall x\ (A(x) \wedge B(x)) \equiv \forall x\ A(x) \wedge \forall x\ B(x).$

5. $\exists x\ (A(x) \to B(x)) \equiv \forall x\ A(x) \to \exists x\ B(x).$ (7.4)

6. $\forall x\ \forall y\ W(x, y) \equiv \forall y\ \forall x\ W(x, y).$ (7.3)

7. $\exists x\ \exists y\ W(x, y) \equiv \exists y\ \exists x\ W(x, y).$ (7.3)

Some Restricted Equivalences

The following equivalences hold if x does not occur in the wff C:

Disjunction   (7.7)

$$\forall x\ (C \vee A(x)) \equiv C \vee \forall x\ A(x).$$

$$\exists x\ (C \vee A(x)) \equiv C \vee \exists x\ A(x).$$

Conjunction   (7.8)

$$\forall x\ (C \wedge A(x)) \equiv C \wedge \forall x\ A(x).$$

$$\exists x\ (C \wedge A(x)) \equiv C \wedge \exists x\ A(x).$$

Implication   (7.9)

$$\forall x\ (C \to A(x)) \equiv C \to \forall x\ A(x).$$

$$\exists x\ (C \to A(x)) \equiv C \to \exists x\ A(x).$$

$$\forall x\ (A(x) \to C) \equiv \exists x\ A(x) \to C.$$

$$\exists x\ (A(x) \to C) \equiv \forall x\ A(x) \to C.$$

Some Conditionals That Are Not Equivalences

1. $\forall x\ A(x) \to \exists x\ A(x).$

2. $\exists x\ (A(x) \wedge B(x)) \to \exists x\ A(x) \wedge \exists x\ B(x).$

3. $\forall x\ A(x) \vee \forall x\ B(x) \to \forall x\ (A(x) \vee B(x)).$

4. $\forall x\ (A(x) \to B(x)) \to (\forall x\ A(x) \to \forall x\ B(x)).$

5. $\exists y\ \forall x\ W(x, y) \to \forall x\ \exists y\ W(x, y).$

Exercises

1. Prove each of the following equivalences with validity arguments (i.e., use interpretations and models).

   a. $\forall x\, (A(x) \wedge B(x)) \equiv \forall x\, A(x) \wedge \forall x\, B(x)$.

   b. $\exists x\, (A(x) \vee B(x)) \equiv \exists x\, A(x) \vee \exists x\, B(x)$.

   c. $\exists x\, (A(x) \to B(x)) \equiv \forall x\, A(x) \to \exists x\, B(x)$.

   d. $\forall x\, \forall y\, W(x, y) \equiv \forall y\, \forall x\, W(x, y)$.

   e. $\exists x\, \exists y\, W(x, y) \equiv \exists y\, \exists x\, W(x, y)$.

2. Assume that x does not occur in the wff C. Prove each of the following equivalences with validity arguments (i.e., use interpretations and models).

   a. $\forall x\, (C \wedge A(x)) \equiv C \wedge \forall x\, A(x)$.

   b. $\exists x\, (C \wedge A(x)) \equiv C \wedge \exists x\, A(x)$.

   c. $\forall x\, (C \vee A(x)) \equiv C \vee \forall x\, A(x)$.

   d. $\exists x\, (C \vee A(x)) \equiv C \vee \exists x\, A(x)$.

   e. $\forall x\, (C \to A(x)) \equiv C \to \forall x\, A(x)$.

   f. $\exists x\, (C \to A(x)) \equiv C \to \exists x\, A(x)$.

   g. $\forall x\, (A(x) \to C) \equiv \exists x\, A(x) \to C$.

   h. $\exists x\, (A(x) \to C) \equiv \forall x\, A(x) \to C$.

3. Construct a prenex conjunctive normal form for each of the following wffs.

   a. $\forall x\, (p(x) \vee q(x)) \to \forall x\, p(x) \vee \forall x\, q(x)$.

   b. $\exists x\, p(x) \wedge \exists x\, q(x) \to \exists x\, (p(x) \wedge q(x))$.

   c. $\forall x\, \exists y\, p(x, y) \to \exists y\, \forall x\, p(x, y)$.

   d. $\forall x\, (p(x, f(x)) \to p(x, y))$.

4. Construct a prenex disjunctive normal form for each of the following wffs.

   a. $\forall x\, (p(x) \vee q(x)) \to \forall x\, p(x) \vee \forall x\, q(x)$.

   b. $\exists x\, p(x) \wedge \exists x\, q(x) \to \exists x\, (p(x) \wedge q(x))$.

   c. $\forall x\, \exists y\, p(x, y) \to \exists y\, \forall x\, p(x, y)$.

   d. $\forall x\, (p(x, f(x)) \to p(x, y))$.

5. Recall that an equivalence $A \equiv B$ stands for the wff $(A \to B) \wedge (B \to A)$. Let C be a wff that does not contain the variable x.

   a. Find a countermodel to show that the following statement is invalid:
   $(\forall x\, W(x) \equiv C) \equiv \forall x\, (W(x) \equiv C)$.

   b. Find a prenex normal form for the statement $(\forall x\, W(x) \equiv C)$.

6. Formalize each of the following English sentences, where the domain of discourse is the set of all people.

   a. Every committee member is rich and famous.

   b. Some committee members are old.

   c. All college graduates are smart.

   d. No college graduate is dumb.

   e. Not all college graduates are smart.

7.3 Formal Proofs in Predicate Calculus

To reason formally about wffs in the predicate calculus, we need some inference rules. It's nice to know that all the inference rules of the propositional calculus can still be used for the predicate calculus. We just need to replace "tautology" with "valid." In other words, if R is an inference rule for the propositional calculus that maps tautologies to a tautology, then R also maps valid wffs to a valid wff.

For example, let's take the modus ponens inference rule of the propositional calculus and prove that it also works for the predicate calculus. In other words, we'll show that modus ponens maps valid wffs to a valid wff.

Proof: Let A and $A \to B$ be valid wffs. We need to show that B is valid. Suppose we have an interpretation for B with domain D. We can use D to give an interpretation to A by assigning values to all the predicates, functions, free variables, and constants that occur in A but not B. This gives us interpretations for A, B, and $A \to B$ over the domain D. Since we are assuming that A and $A \to B$ are valid, it follows that A and $A \to B$ are true for these interpretations over D. Now we can apply the modus ponens rule for propositions to conclude that B is true with respect to the given interpretation over D. Since the given interpretation of B was arbitrary, it follows that every interpretation of B is a model. Therefore B is valid.   QED.

We can use similar arguments to show that all inference rules of the propositional calculus are also inference rules of the predicate calculus. So we have a built-in collection of rules to do formal reasoning in the predicate calculus. But we need more.

Sometimes it's hard to reason about statements that contain quantifiers. The natural approach is to remove quantifiers from statements, do some reasoning with the unquantified statements, and then restore any needed quantifiers. We might call this the RRR method of reasoning with quantifiers—*remove*, *reason*, and *restore*. But quantifiers cannot be removed and restored at will. There are four rules of inference that govern their use. First we'll discuss the two rules for removing quantifiers, which are called *universal instantiation* and *existential instantiation*. Then we'll discuss the two rules for restoring quantifiers, which are called *universal generalization* and *existential generalization*.

Universal Instantiation (UI)

Let's start by using our intuition and see how far we can get. It seems reasonable to say that if a property holds for everything, then it holds for any particular thing. In other words, we should be able to infer $W(x)$ from $\forall x\, W(x)$. Similarly, we should be able to infer $W(c)$ from $\forall x\, W(x)$ for any constant c.

Can we infer $W(y)$ from $\forall x\, W(x)$? This seems OK too, but there may be a problem if W contains a predicate with two or more arguments. For example, suppose we let

$$W(x) = \exists y\, p(x, y).$$

Then a substitution of y for x in W yields

$$W(y) = \exists y\, p(y, y).$$

For this example, can we infer $W(y)$ from $\forall x\, W(x)$? In other words, can we infer $\exists y\, p(y, y)$ from $\forall x\, \exists y\, p(x, y)$? Let's look at an interpretation to get some insight. Suppose we let $p(x, y) = $ "$x < y$," over the domain of real numbers. Then the statement $\forall x\, \exists y\, x < y$ makes perfectly good sense because for every number x there is a larger number y. But we cannot conclude that $\exists y\, y < y$, which says that some real number is less than itself. So the answer to our question is NO, we can't infer $\exists y\, p(y, y)$ from $\forall x\, \exists y\, p(x, y)$, because we can find an interpretation for which the inference doesn't make sense.

Notice however that we can infer quite a few other things from the statement $\forall x\, \exists y\, x < y$. For example, we can infer statements like the following:

$$\exists y\, 24 < y,$$

$$\exists y\, x < y,$$

$$\exists y\, 39x < y,$$

$$\exists y\, z < y,$$

$$\exists y\, z + 4 < y.$$

But we can't infer statements like the following:

$$\exists y \; y < y,$$

$$\exists y \; |y| < y,$$

$$\exists y \; y + 7 < y.$$

What restriction can we place on our inference rule so that we don't blunder into inferring things that aren't true? Notice in our example that it's OK to infer statements like the following:

$$W(24), \; W(x), \; W(39x), \; W(z), \text{ and } W(z + 4).$$

But problems occur when we try to infer statements like the following:

$$W(y), \; W(|y|), \text{ and } W(y + 7).$$

The problem occurs when we try to infer $W(t)$ from $\forall x \; W(x)$, where t is a term containing y. The reason for the trouble is that y is a quantified variable in $W(x)$. So if a term t contains an occurrence of y, then the substitution of t for x introduces a new bound occurrence of the variable y in the inferred statement $W(t)$. We must restrict our inferences so that this doesn't happen. To make things precise, we'll make the following definition:

A term t is *free to replace* x in $W(x)$ if both $W(t)$ and $W(x)$ have the same bound occurrences of variables.

For example, we always have the following cases:

The variable x is free to replace x in $W(x)$.

Any constant is free to replace x in $W(x)$.

If y does not occur in $W(x)$, then y is free to replace x in $W(x)$.

We need to be careful only when the replacement term contains a variable that is bound in $W(x)$. For example, if $W(x) = \exists y \; p(x, y)$, then y is not free to replace x in $W(x)$ because $W(x)$ has two bound occurrences of y and $W(y) = \exists y \; p(y, y)$ has three bound occurrences of y.

Now we can write down the *universal instantiation rule* as follows:

Universal Instantiation Rule (UI)

$$\frac{\forall x\ W(x)}{\therefore\ W(t)} \qquad \text{if } t \text{ is free to replace } x \text{ in } W(x). \qquad (7.12)$$

The following special cases of (7.12) always satisfy the restriction, so they can be used any time:

$$\frac{\forall x\ W(x)}{\therefore\ W(x)} \quad \text{and} \quad \frac{\forall x\ W(x)}{\therefore\ W(c)}\ , \ \text{where } c \text{ is any constant.} \qquad (7.13)$$

EXAMPLE 1. From the wff $\forall x\ (p(x, y) \wedge q(x))$ we can use UI to infer statements like the following:

$$p(c, y) \wedge q(c),$$
$$p(x, y) \wedge q(x),$$
$$p(y, y) \wedge q(y),$$
$$p(f(x, y, z), y) \wedge q(f(x, y, z)).$$

In fact we can infer $p(t, y) \wedge q(t)$ for any term t we wish, without restriction. This is because $p(x, y) \wedge q(x)$ contains no quantifiers. So a substitution x/t will never create any new bound occurrences of variables.   ♦

EXAMPLE 2. From the wff $\forall x\ (\forall y\ p(x, y) \wedge q(x))$ we can use UI to infer the following wffs:

$$\forall y\ p(x, y) \wedge q(x),$$
$$\forall y\ p(c, y) \wedge q(c),$$
$$\forall y\ p(f(x, z, a), y) \wedge q(f(x, z, a)).$$

But we cannot infer the following wffs:

$$\forall y\ p(y, y) \wedge q(y),$$
$$\forall y\ p(f(x, y, a), y) \wedge q(f(x, y, a)). \quad ♦$$

Existential Instantiation (EI)

It seems reasonable that whenever a property holds for some thing, then the property holds for a particular thing. In other words, from the statement

$$\exists x \; W(x),$$

we should be able to infer $W(c)$ for some constant c. Although this may seem OK, we have to be careful about our choice of the constant c. For example, suppose we let $W(x) = p(x, b)$. Can we infer $W(b)$ from the statement $\exists x \; W(x)$? In other words, does $p(b, b)$ follow from $\exists x \; p(x, b)$? Let's look at an interpretation to get some idea of what's going on. Suppose we let $p(x, b) = $ "x is the mother of b." Then $\exists x \; p(x, b)$ is true, but $p(b, b)$ is false. Therefore the answer is NO.

On the other hand, for our example interpretation we know that $p(c, b)$ is true for some constant $c \ne b$, where c is the mother of b. So we can infer $W(c)$ in this case. The constant c must be considered an arbitrary constant that can represent an element in an arbitrary domain for which the wff $W(c)$ is true. This condition will be satisfied if c is distinct from any constant in the wff $\exists x \; W(x)$.

Are there any other restrictions on the constant chosen? Suppose, for example, that we have the following statement:

$$\exists x \; p(x) \land \exists x \; q(x).$$

What can we conclude from it? Consider the following "attempted" proof:

1.	$\exists x \; p(x) \land \exists x \; q(x)$	P
2.	$\exists x \; p(x)$	1, Simp
3.	$\exists x \; q(x)$	1, Simp
4.	$p(c)$	2, proposed EI rule
5.	$q(c)$	3, proposed EI rule
6.	$p(c) \land q(c)$	4, 5, Conj.

If this proof is correct, then we have the conditional theorem

$$\exists x \; p(x) \land \exists x \; q(x) \rightarrow p(c) \land q(c).$$

But this wff is NOT valid. Consider the following interpretation: Let the domain be the natural numbers, and let $p(x) = $ "x is odd" and $q(x) = $ "x is even." Then $\exists x \; p(x) \land \exists x \; q(x)$ is true, since there is an odd number and there is an even number. But the consequent $p(c) \land q(c)$ is false because no natural number c can be both odd and even. What went wrong? The problem was in line 5, where we used the same constant c that had already been used in the proof. So each time EI is used, we need to make sure a new constant is introduced. The *existential instantiation rule* is stated as follows:

Existential Instantiation Rule (EI)

$$\frac{\exists x\ W(x)}{\therefore\ W(c)} \qquad \text{if } c \text{ is a new constant in the proof.} \qquad (7.14)$$

EXAMPLE 3. We'll give an indirect formal proof of the following statement:

$$\forall x \neg\ W(x) \to \neg\ \exists x\ W(x).$$

Proof:

1.	$\forall x \neg\ W(x)$	P
2.	$\neg\ \neg\ \exists x\ W(x)$	P for IP
3.	$\exists x\ W(x)$	2, T
4.	$W(c)$	3, EI
5.	$\neg\ W(c)$	1, UI
6.	$W(c) \wedge \neg\ W(c)$	4, 5, Conj
7.	false	6, T
	QED	1, 2, 7, IP.

We'll prove the converse of $\forall x \neg\ W(x) \to \neg\ \exists x\ W(x)$ in Example 11.   ♦

Universal Generalization (UG)

We want to consider the possibility of generalizing a wff by attaching a universal quantifier. In other words, we want to consider the circumstances under which we can infer $\forall x\ W(x)$ from $W(x)$. Unfortunately, there are some restrictions on the use of such an inference. We'll introduce the restrictions with some simple examples.

For our first example, suppose we let $p(x)$ mean "x is a prime number." Most of us will agree that we can't infer $\forall x\ p(x)$ from $p(x)$. In other words, it doesn't make sense to conclude that every x is a prime number from the assumption that x is a prime number. So the following attempted proof is wrong:

1.	$p(x)$	P
2.	$\forall x\ p(x)$	1, proposed UG rule. It doesn't work!

We can't do Step 2 because $p(x)$ is a premise in which x occurs free. This leads us to our first restriction. To make things more precise, we'll need to make a definition.

> A variable x in a wff W is a *flagged variable* in W if x is free in W and either W is a premise or W is inferred by a wff containing x as a flagged variable.

For example, all free variables in a premise are flagged variables. If there are no free variables in any premise, then there are no flagged variables in the proof because the chain of flagged variables must start from free variables in premises.

We'll often keep track of flagged variables in the reason column of each line where they appear. Here's an example proof segment that shows which variables are flagged.

1. $p(x)$ P x is flagged
2. $\forall x\, q(x)$ P
3. $q(x)$ 2, UI
4. $p(x) \wedge q(x)$ 1, 3, Conj x is flagged.

Now we can state our first restriction on universal generalization:

> *Do not infer $\forall x\, W(x)$ from $W(x)$ if x is a flagged variable.*

For example, in the preceding proof this restriction forbids us from making the following blunder:

5. $\forall x\, (p(x) \wedge q(x))$ 4, UG. NO because x is flagged on line 4.

If we had allowed such an inference, we would have proved the following invalid wff to be valid:

$$p(x) \wedge \forall x\, q(x) \rightarrow \forall x\, (p(x) \wedge q(x)).$$

Now let's look at an example in which it seems reasonable to generalize with the universal quantifier. We'll attempt a proof of the following statement:

$$\forall x\, (p(x) \rightarrow q(x)) \wedge \forall x\, (q(x) \rightarrow r(x)) \rightarrow \forall x\, (p(x) \rightarrow r(x)).$$

Proof: 1. $\forall x \, (p(x) \to q(x))$ P
           2. $\forall x \, (q(x) \to r(x))$ P
           3. $p(x) \to q(x)$ 1, UI
           4. $q(x) \to r(x)$ 2, UI
           5. $p(x) \to r(x)$ 3, 4, HS
           6. $\forall x \, (p(x) \to r(x))$ 5, proposed UG rule. It works.
              QED   1, 2, 6, CP.

In this proof, Step 6 is OK because x is not flagged on line 5.

There is one more restriction on our ability to infer $\forall x \, W(x)$ from $W(x)$. We'll illustrate it with the following proof sequence about the natural numbers:

    1. $\forall x \, \exists y \, x < y$ P
    2. $\exists y \, x < y$ 1, UI
    3. $x < c$ 2, EI
    4. $\forall x \, x < c$ 3, proposed UG rule. It doesn't work.

The statement $\forall x \, x < c$ on line 4 is certainly false, because it says that every natural number is less than a particular natural number c. What went wrong? Suppose we let $W(x) =$ "$x < c$" from line 3. Then the statement on line 4 can be written $\forall x \, W(x)$. In this case we cannot infer $\forall x \, W(x)$ from $W(x)$, even though x is not free in a premise. To see why things break down, we need to look back at line 2 of the proof. In line 2, x is a free variable, and the variable y depends on x. So in line 3 the constant c is not arbitrary. It depends on x.

When a constant c depends on x, we can think of c as a function of x and write down either $c(x)$ or c_x to denote the fact. This is the kind of situation that gets us into trouble when we try to perform UG with respect to x. Before we state the second restriction, we'll make the following definition:

> A variable x is a *subscripted variable* of W if x is free in W and there is a constant c in W that was created by the EI rule, where c and x occur in the same predicate of W.

For example, if we apply EI to $\exists y \, p(x, y)$ and obtain $p(x, c)$, then x is subscripted in $p(x, c)$. Similarly, if we apply EI to $\exists y \, p(f(x, y))$ and obtain $p(f(x, c))$, then x is subscripted in $p(f(x, c))$. But if we apply EI to $\exists y \, (q(x, z) \vee q(y, z))$ and obtain $q(x, z) \vee q(c, z)$, then z is subscripted and x is not subscripted in $q(x, z) \vee q(c, z)$ because c and x do not occur in the same predicate. In other words, c does not depend on x.

Now we can state our second restriction on universal generalization:

Do not infer $\forall x\, W(x)$ from $W(x)$ if x is a subscripted variable.

We can keep track of a subscripted variable by writing something in the reason column of each line where such a variable occurs. Three possibilities are $c(x)$, c_x, or the statement "x is subscripted." We'll use all three ways to indicate the subscripted variables in the following proof segment:

1. $\forall x\, \exists y\, p(x, y)$ P
2. $\exists y\, p(x, y)$ 1, UI
3. $p(x, c)$ 2, EI $c(x)$, c_x, x is subscripted
4. $p(x, c) \lor q(x, y)$ 3, Add $c(x)$, c_x, x is subscripted.

Since x is subscripted on lines 3 and 4, we cannot use $\forall x$ to generalize the wffs on these lines. In other words, we cannot infer $\forall x\, p(x, c)$ from line 3, and we cannot infer $\forall x\, (p(x, c) \lor q(x, y))$ from line 4. On the other hand, we can use line 4 to infer $\forall y\, (p(x, c) \lor q(x, y))$.

Now we're finally in position to state the *universal generalization rule* with its two restrictions:

Universal Generalization Rule (UG)

$$\frac{W(x)}{\therefore \forall x\, W(x)}$$ if x is not flagged AND x is not subscripted.   (7.15)

Although flagged and subscripted variables are a bit complicated, it's nice to know that the restrictions of the UG rule are almost always satisfied. For example, if the premises in the proof don't contain any free variables, then there can't be any flagged variables. And if the proof doesn't use the EI rule, then there can't be any subscripted variables. So go ahead and use the UG rule with abandon. After you have finished the proof, go back and check to make sure that the two restrictions are satisfied.

The UG inference rule has a natural use whenever we give an informal proof that a statement $W(x)$ is true for all elements x in a domain D. Such a proof goes something like the following:

First we let x be an arbitrary, but fixed, element of the domain D. After we've proved that $W(x)$ is true, we can then say that "Since x was arbitrary, it follows that $W(x)$ is true for all x in D."

Let's do an example that uses the UG rule in this way. We'll give a formal proof of the following statement:

$$\forall x \ (p(x) \rightarrow q(x) \vee p(x)).$$

Proof: 1. $p(x)$ P x is flagged
         2. $q(x) \vee p(x)$ 1, Add x is flagged
         3. $p(x) \rightarrow q(x) \vee p(x)$ 1, 2, CP
         4. $\forall x \ (p(x) \rightarrow q(x) \vee p(x))$ 3, UG
             QED.

We can use the UG rule to generalize the wff on line 3 because the variable x on line 3 is neither flagged nor subscripted. But suppose someone argues against this as follows:

> The variable x is free in the premise on line 1, which is used to indirectly infer line 3. Thus x should be flagged on line 3, and so we can't use UG to generalize the wff on line 3.

This reasoning is wrong because it assumes that CP is an inference rule. But CP is a proof rule, not an inference rule. So x is flagged on lines 1 and 2. But x is not flagged on line 3. Therefore UG can be applied to line 3.

Let's look at a couple more examples.

EXAMPLE 4 (*Lewis Carroll's Logic*). The following argument is from *Symbolic Logic* by Lewis Carroll:

> *Babies are illogical. Nobody is despised who can manage a crocodile.*
> *Illogical persons are despised. Therefore babies cannot manage crocodiles.*

Let's try a formalization over the domain of people. Let $B(x)$ mean "x is a baby," $L(x)$ mean "x is logical," $D(x)$ mean "x is despised," and $C(x)$ mean "x can manage a crocodile." So the four sentences in the argument can be formally represented as follows:

$$\forall x \ (B(x) \rightarrow \neg L(x)).$$
$$\forall x \ (C(x) \rightarrow \neg D(x)).$$
$$\forall x \ (\neg L(x) \rightarrow D(x)).$$
$$\text{Therefore } \forall x \ (B(x) \rightarrow \neg C(x)).$$

The justification of the argument can be given in formal terms as the following proof:

Proof: 1. $\forall x\, (B(x) \to \neg L(x))$ P
       2. $\forall x\, (C(x) \to \neg D(x))$ P
       3. $\forall x\, (\neg L(x) \to D(x))$ P
       4. $B(x) \to \neg L(x)$ 1, UI
       5. $C(x) \to \neg D(x)$ 2, UI
       6. $\neg L(x) \to D(x)$ 3, UI
       7. $B(x)$ P x is flagged
       8. $\neg L(x)$ 4, 7, MP x is flagged
       9. $D(x)$ 6, 8, MP x is flagged
      10. $\neg C(x)$ 5, 9, MT x is flagged
      11. $B(x) \to \neg C(x)$ 7, 10, CP
      12. $\forall x\, (B(x) \to \neg C(x))$ 11, UG
          QED   1, 2, 3, 12, CP.

This argument holds without any particular interpretation. In other words, we've shown that the wff $A \to B$ is valid, where A and B are defined as follows:

$$A = \forall x\, (B(x) \to \neg L(x)) \wedge \forall x\, (C(x) \to \neg D(x)) \wedge \forall x\, (\neg L(x) \to D(x)),$$
$$B = \forall x\, (B(x) \to \neg C(x)). \quad \blacklozenge$$

EXAMPLE 5. We'll prove the following general statement about swapping universal quantifiers: $\forall x\, \forall y\, W \to \forall y\, \forall x\, W$:

Proof: 1. $\forall x\, \forall y\, W$ P
       2. $\forall y\, W$ 1, UI
       3. W 2 , UI
       4. $\forall x\, W$ 3, UG
       5. $\forall y\, \forall x\, W$ 4, UG
          QED   1, 5, CP.

Of course, the converse of the statement can be proved in the same manner. Therefore we have a formal proof of the following equivalence in (7.3):

$$\forall x\, \forall y\, W \equiv \forall y\, \forall x\, W. \quad \blacklozenge$$

Existential Generalization (EG)

It seems to make sense that if a property holds for a particular thing, then the property holds for some thing. For example, we know that 5 is a prime number, and it makes sense to conclude that there is some prime number. If we let $p(x)$ = "x is a prime number," then we can infer $\exists x\, p(x)$ from $p(5)$. So far, so good. If W is a wff, can we infer $\exists x\, W(x)$ from $W(c)$ for a constant c? Can we infer $\exists x\, W(x)$ from $W(x)$? Can we infer $\exists x\, W(x)$ from $W(t)$ for any term t? The answers to these questions depend on whether certain restrictions hold.

Let's look at some examples to see why we need some restrictions. In the following proof, assume that we're talking about natural numbers:

1. $\forall x\, \exists y\, x < y$	P	(7.16)
2. $\exists y\, x < y$	1, UI	
3. $x < c$	2, EI	
4. $\exists x\, x < x$	3, proposed EG rule. It doesn't work.	
5. $\exists z\, x < z$	3, proposed EG rule. It works.	
6. $\exists x\, x < c$	3, proposed EG rule. It works.	

The statement $\exists x\, x < x$ on line 4 is clearly false, since there is no natural number less than itself. How did we get it in the first place? Let the statement on line 3 be $W(c)$ = "$x < c$." Then the statement on line 4 takes the form $\exists x\, W(x)$. So we can't always infer $\exists x\, W(x)$ from $W(c)$ for a constant c.

On the other hand, the statement $\exists z\, x < z$ on line 5 is clearly true. Let the statement on line 3 be $W(c)$ = "$x < c$," as before. Then the statement on line 5 has the form $\exists z\, W(z)$. So we inferred $\exists z\, W(z)$ from $W(c)$. The difference between lines 4 and 5 is the choice of variable for quantification. The variable z worked, and x didn't.

Notice that the statement $\exists x\, x < c$ on line 6 is also clearly true. In this case, let the statement on line 3 be $W(x)$ = "$x < c$." Then the statement on line 6 has the form $\exists x\, W(x)$. So we inferred $\exists x\, W(x)$ from $W(x)$.

The same type of problem exhibited in (7.16) can occur with terms that are not constants. For example, consider the following proof, where you can think of $f(x)$ as the successor function on natural numbers:

1. $\forall x\, x < f(x)$	P	(7.17)
2. $x < f(x)$	1, UI	
3. $\exists x\, x < x$	2, proposed EG rule. It doesn't work.	
4. $\exists z\, x < z$	2, proposed EG rule. It works.	
5. $\exists x\, x < f(x)$	2, proposed EG rule. It works.	

Line 3 is clearly false. We got it by letting $t = f(x)$ and $W(t) =$ "$x < t$" in line 2. Then line 3 becomes $\exists x\ W(x)$. So we can't always infer $\exists x\ W(x)$ from $W(t)$. Notice that line 4 is OK. Here we let $W(t) =$ "$x < t$" on line 2 and inferred the statement $\exists z\ W(z)$ on line 4. Line 5 has the form $\exists x\ W(x)$, where we let $W(x) =$ "$x < f(x)$" on line 3.

We need a restriction that will force us to make the right choices when introducing the existential quantifier. We'll state the restriction as follows and then discuss how it works:

> *To infer $\exists x\ W(x)$ from $W(t)$ for a term t, the following relationship must hold: $W(t) = W(x)\ (x/t)$.*

In other words, $W(t)$ must equal the wff obtained from $W(x)$ by replacing all occurrences of x by t. When we check to see whether $W(t) = W(x)(x/t)$, we'll call this the *backwards check*, since $W(t)$ appears earlier in the proof.

For example, let's see why we can't use EG on line 4 of (7.16). We can write the wff on line 3 of (7.16) as $W(c) =$ "$x < c$." Then line 4 becomes $\exists x\ W(x)$, where $W(x) =$ "$x < x$." The backwards check fails because

$$W(x)(x/c) = \text{"}c < c\text{"} \text{ and } W(c) = \text{"}x < c\text{."}$$

Since $W(x)(x/c) \neq W(c)$, we can't use EG to infer line 4. On the other hand, it's OK to use EG on lines 5 and 6 of (7.16). Check it out. Also do the backwards check on lines 3, 4, and 5 of (7.17).

There is one more restriction on the use of EG. It occurs if we try to infer $\exists x\ W(x)$ from $W(t)$ when t is not free to replace x during the backwards check. In other words, we have the following restriction:

> *To infer $\exists x\ W(x)$ from $W(t)$, the term t must be free to replace x in $W(x)$.*

For example, consider the following proof segment, where we can think of $f(y)$ as the successor function on the natural numbers:

1. $\forall y\ y < f(y)$ P
2. $\exists x\ \forall y\ y < x$ 1, proposed EG rule. It doesn't work.

The statement on line 2 is clearly false. To see what's happening, let the statement in line 1 of the proof be $W(t) =$ "$\forall y\ y < t$," where $t = f(y)$. Then the statement in line 2 has the form $\exists x\ W(x)$. When we apply the backwards check, it works! We get $W(x)\ (x/t) = W(t)$. But t is NOT free to replace x. In other words, $W(t)$ contains two bound occurrences of the variable y, while $W(x)$ has only one bound occurrence of y.

Now we're in a position to state the *existential generalization rule* in its full generality.

Existential Generalization Rule (EG)

$$\frac{W(t)}{\therefore\ \exists x\ W(x)}$$ if the following two restrictions hold: (7.18)

a) $W(t) = W(x)(x/t)$.
b) t is free to replace x in $W(x)$.

The following special case of (7.18) always satisfies the two restrictions, so it can be used any time:

$$\frac{W(x)}{\therefore\ \exists x\ W(x)}.$$ (7.19)

Examples of Formal Proofs

The following examples show the usefulness of the four quantifier rules. Notice in most cases that we can use the less restrictive forms of the rules.

EXAMPLE 6 (*Renaming Variables*). Let's give formal proofs of the equivalences that rename variables (7.6): Let $W(x)$ be a wff, and let y be a variable that does not occur in $W(x)$. Then the following renaming equivalences hold:

$$\exists x\ W(x) \equiv \exists y\ W(y),$$
$$\forall x\ W(x) \equiv \forall y\ W(y).$$

We'll start by proving the equivalence $\exists x\ W(x) \equiv \exists y\ W(y)$, which we'll do by proving the following two statements:

$$\exists x\ W(x) \rightarrow \exists y\ W(y) \quad \text{and} \quad \exists y\ W(y) \rightarrow \exists x\ W(x).$$

Proof of $\exists x\ W(x) \rightarrow \exists y\ W(y)$:

1.	$\exists x\ W(x)$	P
2.	$W(c)$	1, EI
3.	$\exists y\ W(y)$	2, EG, since both EG conditions hold
	QED	1, 3, CP.

Proof of $\exists y\ W(y) \to \exists x\ W(x)$:

   1. $\exists y\ W(y)$ P
   2. $W(c)$ 1, EI
   3. $\exists x\ W(x)$ 2, EG, since both EG conditions hold
      QED   1, 3, CP.

Next, we'll prove the equivalence $\forall x\ W(x) \equiv \forall y\ W(y)$ by proving the two statements $\forall x\ W(x) \to \forall y\ W(y)$ and $\forall y\ W(y) \to \forall x\ W(x)$. We'll combine the two proofs into one proof as follows:

Proof: 1. $\forall x\ W(x)$ P Start first proof
        2. $W(y)$ 1, UI y is free to replace x
        3. $\forall y\ W(y)$ 2, UG
        4. $\forall x\ W(x) \to \forall y\ W(y)$ 1, 3, CP   Finish first proof
        5. $\forall y\ W(y)$ P Start second proof
        6. $W(x)$ 5, UI x is free to replace y
        7. $\forall x\ W(x)$ 6, UG
        8. $\forall y\ W(y) \to \forall x\ W(x)$ 5, 7, CP   Finish second proof
           QED   4, 8, T.   ◆

EXAMPLE 7. We'll prove the statement

$$\forall x\ p(x) \wedge \exists x\ q(x) \to \exists x\ (p(x) \wedge q(x)).$$

Proof: 1. $\forall x\ p(x)$ P
        2. $\exists x\ q(x)$ P
        3. $q(c)$ 2, EI
        4. $p(c)$ 1, UI
        5. $p(c) \wedge q(c)$ 3, 4, Conj
        6. $\exists x\ (p(x) \wedge q(x))$ 5, EG.
           QED   1, 2, 6, CP.   ◆

EXAMPLE 8. Consider the following three statements:

> Every computer science major is a logical thinker.
> John is a computer science major.
> Therefore there is some logical thinker.

We'll formalize these statements as follows: Let $C(x)$ mean "x is a computer science major," let $L(x)$ mean "x is a logical thinker," and let the constant b mean "John." Then the three statements can be written more concisely as follows, over the domain of people:

$$\forall x\, (C(x) \rightarrow L(x))$$
$$C(b)$$
$$\therefore\ \exists x\, L(x).$$

These statements can be written in the form of a conditional wff with two hypotheses:

$$\forall x\, (C(x) \rightarrow L(x)) \wedge C(b) \rightarrow \exists x\, L(x).$$

Although we started with a specific set of English sentences, we now have a wff of the first-order predicate calculus. We'll prove that this conditional wff is valid as follows:

Proof:
1.	$\forall x\, (C(x) \rightarrow L(x)) \wedge C(b)$	P
2.	$\forall x\, (C(x) \rightarrow L(x))$	1, Simp
3.	$C(b)$	1, Simp
4.	$C(b) \rightarrow L(b)$	2, UI
5.	$L(b)$	3, 4, MP
6.	$\exists x\, L(x)$	5, EG
	QED	1, 6, CP. ◆

EXAMPLE 9. Let's consider the following argument:

> All computer science majors are people.
> Some computer science majors are logical thinkers.
> Therefore, some people are logical thinkers.

We'll give a formalization of this argument. Let $C(x)$ mean "x is a computer science major," $P(x)$ mean "x is a person," and $L(x)$ mean "x is a logical thinker." Now the statements can be represented by the following wff:

$$\forall x\, (C(x) \rightarrow P(x)) \wedge \exists x\, (C(x) \wedge L(x)) \rightarrow \exists x\, (P(x) \wedge L(x)).$$

We'll prove that this wff is valid as follows:

Proof: 1. $\forall x\,(C(x) \rightarrow P(x))$ P
      2. $\exists x\,(C(x) \wedge L(x))$ P
      3. $C(c) \wedge L(c)$ 2, EI
      4. $C(c) \rightarrow P(c)$ 1, UI
      5. $C(c)$ 3, Simp
      6. $P(c)$ 4, 5, MP
      7. $L(c)$ 3, Simp
      8. $P(c) \wedge L(c)$ 6, 7, Conj
      9. $\exists x\,(P(x) \wedge L(x))$ 8, EG
         QED   1, 2, 9, CP.   ♦

EXAMPLE 10. We'll give a correct proof of the validity of the following wff:

$$\exists x\,(P(x) \wedge Q(x)) \rightarrow \exists x\, P(x) \wedge \exists x\, Q(x).$$

Proof: 1. $\exists x\,(P(x) \wedge Q(x))$ P
      2. $P(a) \wedge Q(a)$ 1, EI
      3. $P(a)$ 2, Simp
      4. $\exists x\, P(x)$ 3, EG
      5. $Q(a)$ 2, Simp
      6. $\exists x\, Q(x)$ 5, EG
      7. $\exists x\, P(x) \wedge \exists x\, Q(x)$ 4, 6, Conj
         QED   1, 7, CP.   ♦

EXAMPLE 11. In Example 3 we gave a formal proof of the statement

$$\forall x \neg W(x) \rightarrow \neg\,\exists x\, W(x).$$

Now we're in a position to give a formal proof of its converse. Thus we'll have a formal proof of the following equivalence (7.2):

$$\forall x \neg W(x) \equiv \neg\,\exists x\, W(x).$$

The converse that we want to prove is the wff $\neg\,\exists x\, W(x) \rightarrow \forall x \neg W(x)$. To prove this statement, we'll divide the proof into two parts. First, we'll prove the statement $\neg\,\exists x\, W(x) \rightarrow \neg\, W(x)$. Our proof will be indirect.

Proof: 1. $\neg\,\exists x\ W(x)$ P
   2. $W(x)$ P for IP x is flagged
   3. $\exists x\ W(x)$ 2, EG
   4. $\neg\,\exists x\ W(x) \wedge \exists x\ W(x)$ 1, 3, Conj
   5. false   4, T
    QED   1, 2, 5, IP.

Now we can easily prove the statement $\neg\,\exists x\ W(x) \to \forall x\,\neg\,W(x)$.

Proof: 1. $\neg\,\exists x\ W(x)$ P
   2. $\neg\,\exists x\ W(x) \to \neg\,W(x)$ T, proved above
   3. $\neg\,W(x)$ 1, 2, MP
   4. $\forall x\,\neg\,W(x)$ 3, UG
    QED   1, 4, CP.   ◆

EXAMPLE 12 (*An Incorrect Proof*). The converse of the wff in Example 10 is the following wff:

$$\exists x\ P(x) \wedge \exists x\ Q(x) \to \exists x\ (P(x) \wedge Q(x)).$$

This wff is not valid. For example, let $D = \{0, 1\}$, and set $P(0) = Q(1) = \text{true}$ and $P(1) = Q(0) = \text{false}$. Then $\exists x\ P(x) \wedge \exists x\ Q(x)$ is true but $\exists x\ (P(x) \wedge Q(x))$ is false. We'll give an incorrect proof sequence that claims to show that the wff is valid.

   1. $\exists x\ P(x) \wedge \exists x\ Q(x)$ P
   2. $\exists x\ P(x)$ 1, Simp
   3. $P(c)$ 2, EI
   4. $\exists x\ Q(x)$ 1, Simp
   5. $Q(c)$ 4, EI   NO: c already occurs in line 3
   6. $P(c) \wedge Q(c)$ 3, 5, Conj
   7. $\exists x\ (P(x) \wedge Q(x))$ 6, EG.
    Not QED   1, 7, CP.   ◆

EXAMPLE 13 (*An Incorrect Proof*). In Exercise 10e of Section 7.1 we asked for a countermodel to prove that the following conditional wff is not valid:

$$\forall x\ \exists y\ p(x, y) \to \exists y\ \forall x\ p(x, y).$$

We'll give an incorrect proof sequence that claims to show that the wff is valid:

1. $\forall x \, \exists y \, p(x, y)$ P
2. $\exists y \, p(x, y)$ 1, UI
3. $p(x, c)$ 2, EI x is subscripted, $c(x)$
4. $\forall x \, p(x, c)$ 3, UG   NO: x is subscripted on line 3
5. $\exists y \, \forall x \, p(x, y)$ 4, EG
     Not QED   1, 5, CP.

Another way to see that the proof is incorrect is to give an interpretation. Let the domain be $\mathbb{N}$, and let $p(x, y)$ mean "x has successor y." Then the five steps in the proof become:

1. Every natural number x has a successor y.

2. There is a natural number y such that x has successor y.

3. The successor of x is c.

4. Every natural number x has successor c.

5. There is a natural number y that is the successor of every x.

It becomes clear that lines 4 and 5 don't make sense.   ◆

Summary

Let's begin the summary by mentioning, as we did in Chapter 6, the following important usage note for inference rules:

Don't apply inference rules to subexpressions of wffs.

In other words, you can't apply an inference rule to just part of a wff. You have to apply it to the whole wff and nothing but the whole wff. So when applying the quantifier rules, remember to make sure that the wff in the numerator of the rule matches the wff on the line that you intend to use.

Now let's summarize the four inference rules. First we'll list the rules that can be used anytime without thinking because they don't have any restrictions. Then we'll list the rules that can be used only when certain restrictions are satisfied.

Inference Rules Without Restrictions (Use Them Any Time)

(UI): $\dfrac{\forall x\; W(x)}{\therefore\; W(x)}$ and $\dfrac{\forall x\; W(x)}{\therefore\; W(c)}$ where c is any constant. (7.13)

(EG): $\dfrac{W(x)}{\therefore\; \exists x\; W(x)}$. (7.19)

Inference Rules with Restrictions (Be Careful)

(UI): $\dfrac{\forall x\; W(x)}{\therefore\; W(t)}$ if t is free to replace x in $W(x)$. (7.12)

(EI): $\dfrac{\exists x\; W(x)}{\therefore\; W(c)}$ if c is a new constant in the proof. (7.14)

(UG): $\dfrac{W(x)}{\therefore \forall x\; W(x)}$ if x is not flagged and not subscripted. (7.15)

(EG): $\dfrac{W(t)}{\therefore\; \exists x\; W(x)}$ if the following two restrictions hold: (7.18)

      a) $W(t) = W(x)\,(x/t)$.

      b) t is free to replace x in $W(x)$.

Exercises

1. Let W be the wff $\forall x\,(p(x) \vee q(x)) \rightarrow \forall x\, p(x) \vee \forall x\, q(x)$. It's easy to see that W is not valid. For example, let $p(x)$ mean "x is odd" and $q(x)$ mean "x is even" over the domain of integers. Then the antecedent is true, and the consequent is false. Suppose someone claims that the following sequence of statements is a "proof" of W:

   1. $\forall x\,(p(x) \vee q(x))$ P
   2. $p(x) \vee q(x)$ 1, UI
   3. $\forall x\, p(x) \vee q(x)$ 2, UG
   4. $\forall x\, p(x) \vee \forall x\, q(x)$ 3, UG
       QED   1, 4, CP.

   What is wrong with this "proof" of W?

2. a. Find a countermodel to show that the following wff is not valid:

$$\exists x \, P(x) \wedge \exists x \, (P(x) \rightarrow Q(x)) \rightarrow \exists x \, Q(x).$$

b. The following argument attempts to prove that the wff in part (a) is valid. Find an error in the argument.

1. $\exists x \, P(x)$ P
2. $P(d)$ 1, EI
3. $\exists x \, (P(x) \rightarrow Q(x))$ P
4. $P(d) \rightarrow Q(d)$ 3, EI
5. $Q(d)$ 2, 4, MP
6. $\exists x \, Q(x)$ 5, EG.

3. Use the CP rule to prove that each of the following wffs is valid.

a. $\forall x \, p(x) \rightarrow \exists x \, p(x)$.

b. $\forall x \, (p(x) \rightarrow q(x)) \wedge \exists x \, p(x) \rightarrow \exists x \, q(x)$.

c. $\exists x \, (p(x) \wedge q(x)) \rightarrow \exists x \, p(x) \wedge \exists x \, q(x)$.

d. $\forall x \, (p(x) \rightarrow q(x)) \rightarrow (\exists x \, p(x) \rightarrow \exists x \, q(x))$.

e. $\forall x \, (p(x) \rightarrow q(x)) \rightarrow (\forall x \, p(x) \rightarrow \exists x \, q(x))$.

f. $\forall x \, (p(x) \rightarrow q(x)) \rightarrow (\forall x \, p(x) \rightarrow \forall x \, q(x))$.

g. $\exists y \, \forall x \, p(x, y) \rightarrow \forall x \, \exists y \, p(x, y)$.

h. $\exists x \, \forall y \, p(x, y) \wedge \forall x \, (p(x, x) \rightarrow \exists y \, q(y, x)) \rightarrow \exists y \, \exists x \, q(x, y)$.

4. Use the IP rule to prove each that each of the following wffs is valid.

a. $\forall x \, p(x) \rightarrow \exists x \, p(x)$.

b. $\forall x \, (p(x) \rightarrow q(x)) \wedge \exists x \, p(x) \rightarrow \exists x \, q(x)$.

c. $\exists y \, \forall x \, p(x, y) \rightarrow \forall x \, \exists y \, p(x, y)$.

d. $\exists x \, \forall y \, p(x, y) \wedge \forall x \, (p(x, x) \rightarrow \exists y \, q(y, x)) \rightarrow \exists y \, \exists x \, q(x, y)$.

e. $\forall x \, p(x) \vee \forall x \, q(x) \rightarrow \forall x \, (p(x) \vee q(x))$.

5. Transform each informal argument into a formalized wff. Then give a formal proof of the wff, using either CP or IP.

a. Every dog either likes people or hates cats. Rover is a dog. Rover loves cats. Therefore some dog likes people.

b. Every committee member is rich and famous. Some committee members are old. Therefore some committee members are old and famous.

c. No human beings are quadrupeds. All men are human beings. Therefore no man is a quadruped.

   d. Every rational number is a real number. There is a rational
      number. Therefore there is a real number.

   e. Some freshmen like all sophomores. No freshman likes any junior.
      Therefore no sophomore is a junior.

6. Give a formal proof for each of the following equivalences as follows: To
    prove $W \equiv V$, prove the two statements $W \to V$ and $V \to W$. Use either CP
    or IP.

   a. $\exists x\, \exists y\, W(x, y) \equiv \exists y\, \exists x\, W(x, y)$.

   b. $\forall x\, (A(x) \wedge B(x)) \equiv \forall x\, A(x) \wedge \forall x\, B(x)$.

   c. $\exists x\, (A(x) \vee B(x)) \equiv \exists x\, A(x) \vee \exists x\, B(x)$.

   d. $\exists x\, (A(x) \to B(x)) \equiv \forall x\, A(x) \to \exists x\, B(x)$.

7. Give a formal proof of $A \to B$, where A and B are defined as follows:

$$A = \forall x\, (\exists y\, (q(x, y) \wedge s(y)) \to \exists y\, (p(y) \wedge r(x, y))),$$
$$B = \neg\, \exists x\, p(x) \to \forall x\, \forall y\, (q(x, y) \to \neg\, s(y)).$$

8. Give a formal proof of $A \to B$, where A and B are defined as follows:

$$A = \exists x\, (r(x) \wedge \forall y\, (p(y) \to q(x, y))) \wedge \forall x\, (r(x) \to \forall y\, (s(y) \to \neg\, q(x, y)))$$
$$B = \forall x\, (p(x) \to \neg\, s(x)).$$

9. Each of the following proof segments contains an invalid use of a quan-
    tifier inference rule. In each case, state why the inference rule cannot be
    used.

   a. 1. $x < 4$ P
       2. $\forall x\, (x < 4)$ 1, UG.

   b. 1. $\exists x\, (y < x)$ P
       2. $y < c$ 1, EI
       3. $\forall y\, (y < c)$ 2, UG.

   c. 1. $\forall y\, (y < f(y))$ P
       2. $\exists x\, \forall y\, (y < x)$ 1, EG.

   d. 1. $q(x, c)$ P
       2. $\exists x\, q(x, x)$ 1, EG.

   e. 1. $\exists x\, p(x)$ P
       2. $\exists x\, q(x)$ P
       3. $p(c)$ 1, EI
       4. $q(c)$ 2, EI.

   f. 1. $\forall x\, \exists y\, x < y$ P
       2. $\exists y\, y < y$ 1, UI.

10. Each of the following wffs is INVALID. Nevertheless, for each wff you are to construct a proof sequence that claims to be a proof of the wff but that fails because of the improper use of one or more inference rules. Also indicate which rules you use improperly and why the use is improper.

    a. $\exists x\, A(x) \rightarrow \forall x\, A(x)$.

    b. $\exists x\, A(x) \wedge \exists x\, B(x) \rightarrow \exists x\, (A(x) \wedge B(x))$.

    c. $\forall x\, (A(x) \vee B(x)) \rightarrow \forall x\, A(x) \vee \forall x\, B(x)$.

    d. $(\forall x\, A(x) \rightarrow \forall x\, B(x)) \rightarrow \forall x\, (A(x) \rightarrow B(x))$.

    e. $\forall x\, \exists y\, W(x, y) \rightarrow \exists y\, \forall x\, W(x, y)$.

11. Assume that x does not occur in the wff C. Use either CP or IP to give a formal proof for each of the following equivalences.

    a. $\forall x\, (C \wedge A(x)) \equiv C \wedge \forall x\, A(x)$.

    b. $\exists x\, (C \wedge A(x)) \equiv C \wedge \exists x\, A(x)$.

    c. $\forall x\, (C \vee A(x)) \equiv C \vee \forall x\, A(x)$.

    d. $\exists x\, (C \vee A(x)) \equiv C \vee \exists x\, A(x)$.

    e. $\forall x\, (C \rightarrow A(x)) \equiv C \rightarrow \forall x\, A(x)$.

    f. $\exists x\, (C \rightarrow A(x)) \equiv C \rightarrow \exists x\, A(x)$.

    g. $\forall x\, (A(x) \rightarrow C) \equiv \exists x\, A(x) \rightarrow C$.

    h. $\exists x\, (A(x) \rightarrow C) \equiv \forall x\, A(x) \rightarrow C$.

12. Any inference rule for the propositional calculus can be converted to an inference rule for the predicate calculus. In other words, suppose R is an inference rule for the propositional calculus. If the hypotheses of R are valid wffs, then the conclusion of R is a valid wff. Prove this statement for each of the following inference rules.

    a. Modus tollens.

    b. Hypothetical syllogism.

13. Show that the existential generalization rule (EG) can be derived from the universal instantiation rule (UI), and conversely. *Hint:* Use the fact that each quantifier can be written in terms of the other.

Chapter Summary

The first-order predicate calculus extends propositional calculus by allowing wffs to contain predicates and quantifiers of variables. Meanings for these wffs are defined in terms of interpretations over nonempty sets called domains. A wff is valid if it's true for all possible interpretations. A wff is unsatisfiable if it's false for all possible interpretations.

There are basic equivalences (7.2)–(7.9) that allow us to simplify and transform wffs into other wffs. We can use equivalences to transform any wff into a prenex DNF or prenex CNF. Equivalences can also be used to compare different formalizations of the same English sentence.

To decide whether a wff is valid, we can try to transform it into an equivalent wff that we know to be valid. But in general we must rely on some type of informal or formal reasoning. A formal reasoning system for the first-order predicate calculus can use all the rules and proof techniques of the propositional calculus. But we need four additional inference rules for the quantifiers: universal instantiation, existential instantiation, universal generalization, and existential generalization.

Notes

Now we have the basics of logic—the propositional calculus and the first-order predicate calculus. In Chapter 6 we introduced a formal system for the propositional calculus that we called Hilbert's system. We observed that the system is complete, which means that every tautology can be proven as a theorem within the system.

It's nice to know that there is a similar statement for the predicate calculus, which is due to the logician and mathematician Kurt Gödel (1906–1978). Gödel showed that the first-order predicate calculus is complete. In other words, there are formal systems for the first-order predicate calculus such that every valid wff can be proven as a theorem. The formal system presented by Gödel [1930] used fewer axioms and fewer inference rules than the system that we've been using in this chapter.

8

Applied Logic

Once the people begin to reason, all is lost.
—Voltaire (1694–1778)

When we reason, we usually do it in a particular domain of discourse. For example, we might reason about computer science, politics, mathematics, physics, automobiles, or cooking. But these domains are usually too large to do much reasoning. So we normally narrow our scope of thought and reason in domains such as imperative programming languages, international trade, plane geometry, optics, suspension systems, or pasta recipes.

No matter what the domain of discussion, we usually try to correctly apply inferences while we are reasoning. Since each of us has our own personal reasoning system, we sometimes find it difficult to understand one another. In an attempt to find common ground among the various ways that people reason, we introduced the propositional calculus and first-order predicate calculus. So we've looked at some formalizations of logic.

Can we go a step further and formalize the things that we talk about? Many subjects can be formalized by giving some axioms that define the properties of the objects being discussed. For example, when we reason about geometry, we make assumptions about points and lines. When we reason about automobile engines, we make certain assumptions about how they work. When we combine first-order predicate calculus with the formalization of some subject, we obtain a reasoning system called a *first-order theory*.

Chapter Guide

Section 8.1 shows how the fundamental notion of equality can become part of a first-order theory.

Section 8.2 introduces a first-order theory for proving the correctness of imperative programs.

411

Section 8.3 introduces logics that are beyond the first order. We'll give some examples to show how higher-order logics can be used to formalize much of our natural discourse.

8.1 Equality

Equality is a familiar notion to most of us. For example, we might compare two things to see whether they are equal, or we might replace a thing by an equal thing during some calculation. In fact, equality is so familiar that we might think that it does not need to be discussed further. But we are going to discuss it further because different domains of discourse often use equality in different ways. If we want to formalize some subject that uses the notion of equality, then it should be helpful to know basic properties that are common to all equalities.

A first-order theory is called a *first-order theory with equality* if it contains a two-argument predicate, say e, that captures the properties of equality required by the theory. We usually denote $e(x, y)$ by the familiar

$$x = y.$$

Similarly, we let $x \neq y$ denote $\neg\, e(x, y)$.

Let's examine how we use equality in our daily discourse. We always assume that any term is equal to itself. For example, $x = x$ and $f(c) = f(c)$. We might call this "syntactic equality."

Another familiar use of equality might be called "semantic equality." For example, although the expressions $2 + 3$ and $1 + 4$ are not syntactically equal, we still write $2 + 3 = 1 + 4$ because they both represent the same number.

Another important use of equality is to replace equals for equals in an expression. The following examples should get the point across:

If $x + y = 2z$, then $(x + y) + w = 2z + w$.

If $x = y$, then $f(x) = f(y)$.

If $f(x) = f(y)$, then $g(f(x)) = g(f(y))$.

If $x = y + z$, then $8 < x \equiv 8 < y + z$.

If $x = y$, then $p(x) \vee q(w) \equiv p(y) \vee q(w)$.

Describing Equality

Let's try to describe some fundamental properties that all first-order theories with equality should satisfy. Of course, we want equality to satisfy the basic property that each term is equal to itself. The following axiom will suffice for this purpose:

Equality Axiom (EA)

$$\forall x \, (x = x). \tag{8.1}$$

This axiom tells us that $x = x$ for all variables x. The axiom is sometimes called the *law of identity*. But we also want to say that $t = t$ for any term t. For example, if a theory contains a term such as $f(x)$, we certainly want to say that $f(x) = f(x)$. Do we need another axiom to tell us that each term is equal to itself? No. All we need is a little proof sequence as follows:

1. $\forall x \, (x = x)$ EA
2. $t = t$ 1, UI.

So for any term t we have $t = t$. Because this is such a useful result, we'll also refer to it as EA. In other words, we have

Equality Axiom (EA)

$$t = t \text{ for all terms } t. \tag{8.2}$$

Now let's try to describe that well-known piece of folklore, *equals can replace equals*. Since this idea has such a wide variety of uses, it's hard to tell where to begin. So we'll start with a rule that describes the process of replacing some occurrence of a term in a predicate by an equal term. In this rule, p denotes an arbitrary predicate with one or more arguments. The letters t and u represent arbitrary terms:

Equals for Equals Rule (EE)

$$t = u \wedge p(\ldots t \ldots) \rightarrow p(\ldots u \ldots). \tag{8.3}$$

The notations $\ldots t \ldots$ and $\ldots u \ldots$ indicate that t and u occur in the same argument place of p. In other words, u replaces the indicated occurrence of t. Since (8.3) is an implication, we can use it as an inference rule in the following equivalent form:

Equals for Equals Rule (EE)

$$\frac{t = u, \, p(\ldots t \ldots)}{\therefore \, p(\ldots u \ldots)}. \tag{8.4}$$

The EE rule is sometimes called the *principle of extensionality*. Let's see what we can conclude from EE. Whenever we discuss equality of terms, we usually want the following two properties to hold for all terms:

Symmetric: $t = u \rightarrow u = t.$

Transitive: $t = u \wedge u = v \rightarrow t = v.$

We'll use the EE rule to prove the symmetric property in the next example and leave the transitive property as an exercise.

EXAMPLE 1. We'll prove the symmetric property $t = u \rightarrow u = t.$

Proof: 1. $t = u$ P
         2. $t = t$ EA
         3. $u = t$ 1, 2, EE
         QED   1, 3, CP.

To see why the statement on line 3 follows from the EE rule, we can let $p(x, y)$ mean "$x = y$." Then the proof can be rewritten in terms of the predicate p as follows:

Proof: 1. $t = u$ P
         2. $p(t, t)$ EA
         3. $p(u, t)$ 1, 2, EE
         QED   1, 3, CP.   ◆

Another thing we would like to conclude from EE is that equals can replace equals in a term like $f(\dots t \dots)$. In other words, we would like the following wff to be valid:

$$t = u \rightarrow f(\dots t \dots) = f(\dots u \dots).$$

To prove that this wff is valid, we'll let $p(t, u)$ mean "$f(\dots t \dots) = f(\dots u \dots)$." Then the proof goes as follows:

Proof: 1. $t = u$ P
         2. $p(t, t)$ EA
         3. $p(t, u)$ 1, 2, EE
         QED   1, 3, CP.

When we're dealing with axioms for a theory, we sometimes write down more axioms than we really need. For example, some axiom might be deducible as a theorem from the other axioms. The practical purpose for this is to have a listing of the useful properties all in one place. For example, to describe equality for terms, we might write down the following five statements as axioms.

Equality Axioms for Terms (8.5)

In these axioms the letters t, u, and v denote arbitrary terms, f is an arbitrary function, and p is an arbitrary predicate.

EA: $t = t$.

Symmetric: $t = u \rightarrow u = t$.

Transitive: $t = u \wedge u = v \rightarrow t = v$.

EE (functional form): $t = u \rightarrow f(\ldots t \ldots) = f(\ldots u \ldots)$.

EE (predicate form): $t = u \wedge p(\ldots t \ldots) \rightarrow p(\ldots u \ldots)$.

The EE axioms in (8.5) allow only a single occurrence of t to be replaced by u. We may want to substitute more than one "equals for equals" at the same time. For example, if $x = a$ and $y = b$, we would like to say that $f(x, y) = f(a, b)$. It's nice to know that simultaneous use of equals for equals can be deduced from the axioms. For example, we'll prove the following statement:

$$x = a \wedge y = b \rightarrow f(x, y) = f(a, b).$$

Proof:
1. $x = a$ P
2. $y = b$ P
3. $f(x, y) = f(a, y)$ 1, EE
4. $f(a, y) = f(a, b)$ 2, EE
5. $f(x, y) = f(a, b)$ 3, 4, Transitive
     QED   1, 2, 5, CP.

This proof can be extended to substituting any number of equals for equals simultaneously in a function or in a predicate. In other words, we could have written the two EE axioms of (8.5) in the following form:

Multiple replacement EE (8.6)

EE (function): $t_1 = u_1 \wedge \cdots \wedge t_k = u_k \rightarrow f(t_1, \ldots, t_k) = f(u_1, \ldots, u_k)$.

EE (predicate): $t_1 = u_1 \wedge \cdots \wedge t_k = u_k \wedge p(t_1, \ldots, t_k) \rightarrow p(u_1, \ldots, u_k)$.

So the two axioms (8.1) and (8.3) are sufficient for us to deduce all the axioms in (8.5) together with those of (8.6). Let's look at a couple more examples to get some practice working with equality.

EXAMPLE 2. Let's consider the set of arithmetic expressions over the domain of integers, together with the usual arithmetic operations. The terms in this theory are arithmetic expressions, such as the following:

$$35, x, 2 + 8, x + y, 6x - 5 + y.$$

Equality of terms comes into play when we write statements like the following:

$$3 + 6 = 2 + 7, 4 \neq 2 + 3.$$

We have axioms that tell how the operations work. For example, we know that the + operation is associative, and we know that $x + 0 = x$ and $x - x = 0$. We can do formal reasoning in such a theory by using the predicate calculus with equality. For example, let's prove the following well-known statement:

$$\forall x \, (x + x = x \rightarrow x = 0).$$

First we'll do an informal equational type proof. Let x be any number such that $x + x = x$. Then we have the following equations:

$$
\begin{aligned}
x &= x + 0 & \text{(property of 0)} \\
&= x + (x + - x) & \text{(property of } -) \\
&= (x + x) + - x & \text{(associativity of +)} \\
&= x + - x & \text{(hypothesis } x + x = x) \\
&= 0 & \text{(property of } -)
\end{aligned}
$$

QED.

In this proof we used several instances of equals for equals. Now let's look at a formal proof in all its glory.

Proof:
1. $x + x = x$ P
2. $-x = -x$ EA
3. $(x + x) + - x = x + - x$ 1, 2, EE
4. $x + (x + - x) = (x + x) + - x.$ Associativity
5. $x + (x + - x) = x + - x$ 3, 4, Transitivity
6. $x + - x = 0$ Property of $-$

7. $x + 0 = 0$	5, 6, EE
8. $x = x + 0$	Property of 0
9. $x = 0$	7, 8, Transitivity
10. $x + x = x \rightarrow x = 0$	1, 9, CP
11. $\forall x\, (x + x = x \rightarrow x = 0)$	10, UG
QED.	

Let's explain the two uses of EE. For line 3, let $f(u, v) = u + v$. Then the wff on line 3 results from lines 1 and 2 together with the following instance of EE in functional form:

$$x + x = x \rightarrow f(x + x, -x) = f(x, -x).$$

For line 7, let $p(u, v)$ denote the statement "$u + v = v$." Then the wff on line 7 results from lines 5 and 6 together with the following instance of EE in predicate form:

$$x + -x = 0 \wedge p(x, x + -x) \rightarrow p(x, 0). \quad \blacklozenge$$

EXAMPLE 3 (*A Partial Order Theory*). A *partial order theory* is a first-order theory with equality that also contains an ordering predicate. If the ordering predicate is reflexive, we denote it by $\leq$. Otherwise, we denote it by $<$. The three defining axioms for a reflexive partial order are as follows, for all x, y, and z:

Reflexive:	$x \leq x$.
Antisymmetric:	$x \leq y \wedge y \leq x \rightarrow x = y$.
Transitive:	$x \leq y \wedge y \leq z \rightarrow x \leq z$.

We can do formal reasoning in such a theory using the predicate calculus. For example, recall that $<$ and $\leq$ can be defined in terms of each other by using equality as follows:

$$x < y \text{ means } x \leq y \wedge x \neq y,$$
$$x \leq y \text{ means } x < y \vee x = y.$$

From either one of these statements we can write down a formal proof of the following well-known statement:

$$x < y \rightarrow x \leq y.$$

The two proofs of the statement are given as follows:

Proof: 1. $x < y$ P
         2. $x \le y \wedge x \ne y$ 1, T
         3. $x \le y$ 2, Simp
         4. $x < y \to x \le y$ 1, 3, CP
            QED.

Proof: 1. $x < y$ P
         2. $x < y \vee x = y$ 1, Addition
         3. $x \le y$ 2, T
         4. $x < y \to x \le y$ 1, 3, CP
            QED.

A model for a partial order theory is called a *partially ordered structure*. For example, the set of integers ordered by $\le$ is a partially ordered structure. Try to name some other partially ordered structures.   ◆

Extending Equals for Equals

The EE rule for replacing equals for equals in a predicate can be extended to other wffs. For example, we can use the EE rule to prove the following more general statement about wffs without quantifiers:

If $W(x)$ is a wff without quantifiers, then the following wff is valid:

$$t = u \wedge W(t) \to W(u). \tag{8.7}$$

We assume in this case that $W(t)$ is obtained from $W(x)$ by replacing one or more occurrences of x by t and that $W(u)$ is obtained from $W(t)$ by replacing one or more occurrences of t by u.

For example, if $W(x) = p(x, y) \wedge q(x, x)$, then we might have $W(t) = p(t, y) \wedge q(x, t)$, where only two of the three occurrences of x are replaced by t. In this case we might have $W(u) = p(u, y) \wedge q(x, t)$, where only one occurrence of t is replaced by u. In other words, the following wff is valid:

$$t = u \wedge p(t, y) \wedge q(x, t) \to p(u, y) \wedge q(x, t).$$

What about wffs that contain quantifiers? Even when a wff has quantifiers, we can use the EE rule if we are careful not to introduce new bound occurrences of variables. Here is the full blown version of EE:

If $W(x)$ is a wff and t and u are terms that are free to replace x in $W(x)$, then the following wff is valid:

$$t = u \wedge W(t) \rightarrow W(u). \tag{8.8}$$

Again, we can assume in this case that $W(t)$ is obtained from $W(x)$ by replacing one or more occurrences of x by t and that $W(u)$ is obtained from $W(t)$ by replacing one or more occurrences of t by u.

For example, suppose $W(x) = \exists y\, p(x, y)$. Then for any terms t and u that do not contain occurrences of y, the following wff is valid:

$$t = u \wedge \exists y\, p(t, y) \rightarrow \exists y\, p(u, y).$$

The exercises contain some samples to show how EE for predicates (8.3) can be used to prove some simple extensions of EE to more general wffs.

Exercises

1. Use the EE rule to prove the double replacement rule
$$s = v \wedge t = w \wedge p(s, t) \rightarrow p(v, w).$$

2. Show that the transitive property of equality can be deduced from the other axioms for equality (8.5). In other words, prove that $(t = u) \wedge (u = v) \rightarrow (t = v)$ from the other axioms for equality.

3. Let Ux mean "there exists a unique x." If $A(x)$ is a wff of the first-order predicate calculus, then $Ux\, A(x)$ means "There exists a unique x such that $A(x)$." Find a definition for $Ux\, A(x)$ as a wff from the first-order predicate calculus with equality.

4. Give a formal proof of the following statement about the integers:
$$c = a^i \wedge i \leq b \wedge \neg\,(i < b) \rightarrow c = a^b.$$

5. Use the the equality axioms (8.5) to prove each of the following versions of EE, where p and q are predicates, t and u are terms, and $x, y,$ and z are variables.
    a. $t = u \wedge \neg\, p(...\, t\, ...) \rightarrow \neg\, p(...\, u\, ...)$.
    b. $t = u \wedge p(...\, t\, ...) \wedge q(...\, t\, ...) \rightarrow p(...\, u\, ...) \wedge q(...\, u...)$.
    c. $t = u \wedge (p(...\, t\, ...) \vee q(...\, t\, ...)) \rightarrow p(...\, u\, ...) \vee q(...\, u...)$.
    d. $x = y \wedge \exists z\, p(...\, x\, ...) \rightarrow \exists z\, p(...\, y\, ...)$.
    e. $x = y \wedge \forall z\, p(...\, x\, ...) \rightarrow \forall z\, p(...\, y\, ...)$.

6. Prove the validity of the wff $\forall x \, \exists y \, (x = y)$.

7. Prove each of the following equivalences.
   a. $p(x) \equiv \exists y \, (x = y \land p(y))$.
   b. $p(x) \equiv \forall y \, (x = y \rightarrow p(y))$.

8.2 Program Correctness

An important and difficult problem of computer science can be stated as follows:

$$\text{Prove that a program is correct.} \qquad (8.9)$$

This takes some discussion. One major question to ask before we can prove that a program is correct is "What is the program supposed to do?" If we can state in English what a program is supposed to do, and English is the programming language, then the statement of the problem may itself be a proof of its correctness. Normally, a problem is stated in some language X, and its solution is given in some language Y. For example, the statement of the problem might use English mixed with some symbolic notation, while the solution might be in a programming language. How do we prove correctness in cases like this? Often the answer depends on the programming language. As an example, we'll look at a formal theory for proving the correctness of imperative programs.

Imperative Program Correctness

An imperative program consists of a sequence of statements that represent commands. The most important statement is the assignment statement. Other statements are used for control, such as looping and taking alternate paths. To prove things about such programs, we need a formal theory consisting of wffs, axioms, and inference rules.

Suppose we want to prove that a program does some particular thing. We must represent the thing that we want to prove in terms of a precondition P, which states what is supposed to be true before the program starts, and a postcondition Q, which states what is supposed to be true after the program halts. If S denotes the program, then we will describe this informal situation with the following wff:

$$\{P\} \, S \, \{Q\}.$$

The letters P and Q denote logical statements that describe properties of the variables that occur in S. P is called a *precondition* for S, and Q is called a *postcondition* for S. We assume that P and Q are wffs from a first-order theory with equality that depends on the program S. For example, if the program manipulates numbers, then the first-order theory must include the numerical operations and properties that are required to describe the problem at hand. If the program processes strings, then the first-order theory must include the string operations.

For example, suppose S is the single assignment statement $x := x + 1$. Then the following expression is a wff in our logic:

$$\{x > 4\}\ x := x + 1\ \{x > 5\}.$$

If we're going to have a logic, we need to assign a meaning to any wff of the form $\{P\}\ S\ \{Q\}$. In other words, we want to assign a truth value to the wff $\{P\}\ S\ \{Q\}$.

Meaning of $\{P\}\ S\ \{Q\}$

The meaning of $\{P\}\ S\ \{Q\}$ is the truth value of the following statement:

If P is true before S is executed and the execution of S halts, then Q is true after the execution of S halts.

If $\{P\}\ S\ \{Q\}$ is true, we can say that S is *correct* with respect to precondition P and postcondition Q.

For example, from our knowledge of the assignment statement, most of us will agree that the following wff is true:

$$\{x > 4\}\ x := x + 1\ \{x > 5\}.$$

On the other hand, most of us will also agree that the following wff is false:

$$\{x > 4\}\ x := x + 1\ \{x > 6\}.$$

A formal theory for proving correctness of these wffs needs some axioms and some inference rules. The axioms depend on the types of assignments allowed by the assignment statement. The inference rules depend on the control structures of the language. So we had better agree on a language before we go any further in our discussion. To keep things simple, we'll assume that the assignment statement has the following form, where x is a variable and t is a term:

$$x := t.$$

So the only thing we can do is assign a value to a variable. This effectively restricts the language so that it cannot use other structures, such as arrays and records. In other words, we can't make assignments like $a[i] := t$ or $a.b := t$.

Since our assignment statement is restricted to the form $x := t$, we need only one axiom. It's called the *assignment axiom*, and we'll motivate the discovery of the axiom by an example. Suppose we're told that the following wff is correct:

$$\{P\}\ x := 4\ \{y > x\}.$$

In other words, if P is true before the execution of the assignment statement, then after its execution the statement $y > x$ is true. What should P be? From our knowledge of the assignment statement we might guess that P has the following definition:

$$P = \text{``}y > 4.\text{''}$$

This is about the most general statement we can make. Notice that P can be obtained from the postcondition $y > x$ by replacing x by 4. The assignment axiom generalizes this idea. We'll state it as follows:

Assignment Axiom (AA)

$$\{Q(x/t)\}\ x := t\ \{Q\}. \tag{8.10}$$

The notation $Q(x/t)$ denotes the wff obtained from Q by replacing all free occurrences of x by t. The axiom is often called the "backwards" assignment axiom because the precondition is constructed from the postcondition.

Let's see how the assignment axiom works in a backwards manner. When using AA, always start by writing down the form of (8.10) with an empty precondition as follows:

$$\{\quad\}\ x := t\ \{Q\}.$$

Now the task is to construct the precondition by replacing all free occurrences of x in Q by t.

For example, suppose we know that $x < 5$ is the postcondition for the assignment statement $x := x + 1$. We start by writing down the following partially completed version of AA:

$$\{\quad\}\ x := x + 1\ \{x < 5\}.$$

Then we use AA to construct the precondition. In this case we replace the x by $x + 1$ in the postcondition $x < 5$. This gives us the precondition $x + 1 < 5$, and we can write down the completed instance of the assignment axiom:

$$\{x + 1 < 5\}\ x := x + 1\ \{x < 5\}.$$

It happens quite often that the precondition constructed by AA doesn't quite match what we're looking for. For example, most of us will agree that the following wff is correct:

$$\{x < 3\}\ x := x + 1\ \{x < 5\}.$$

But we've already seen that AA applied to this assignment statement gives

$$\{x + 1 < 5\}\ x := x + 1\ \{x < 5\}.$$

Since the two preconditions don't match, we have some more work to do. In this case we know that for any number x we have $x < 3 \rightarrow x + 1 < 5$.

Let's see why this is enough to prove that $\{x < 3\}\ x := x + 1\ \{x < 5\}$ is correct. If $x < 3$ is true before the execution of $x := x + 1$, then we also know that $x + 1 < 5$ is true before execution of $x := x + 1$. Now AA tells us that $x < 5$ is true after execution of $x := x + 1$. So $\{x < 3\}\ x := x + 1\ \{x < 5\}$ is correct.

This kind of argument happens so often that we have an inference rule to describe the situation for any program S. It's called the *consequence rule*:

Consequence Rule

$$\frac{P \rightarrow R \text{ and } \{R\}\ S\ \{Q\}}{\therefore\ \{P\}\ S\ \{Q\}} \quad \text{and} \quad \frac{\{P\}\ S\ \{T\} \text{ and } T \rightarrow Q}{\therefore\ \{P\}\ S\ \{Q\}} . \tag{8.11}$$

Notice that each consequence rule requires two proofs: a proof of correctness and a proof of an implication. Let's do an example.

EXAMPLE 1. We'll prove the correctness of the following wff:

$$\{x < 5\}\ x := x + 1\ \{x < 7\}.$$

To start things off, we'll apply (8.10) to the assignment statement and the postcondition to obtain the following wff:

$$\{x + 1 < 7\}\ x := x + 1\ \{x < 7\}.$$

This isn't what we want. We got the precondition $x + 1 < 7$, but we need the precondition $x < 5$. Let's see whether we can apply (8.11) to the problem. In other words, let's see whether we can prove the following statement:

$$x < 5 \rightarrow x + 1 < 7.$$

This statement is certainly true, and we'll include its proof in the following formal proof of correctness of the original wff:

Proof: 1. $\{x + 1 < 7\}\ x := x + 1\ \{x < 7\}$ AA
      2. $x < 5$ P
      3. $x + 1 < 6$ $2, T$
      4. $6 < 7$ T
      5. $x + 1 < 7$ $3, 4$, Transitive
      6. $x < 5 \rightarrow x + 1 < 7$ $2, 5$, CP
        QED $1, 6$, Consequence.   ♦

Although assignment statements are the core of imperative programming, they can't do much without control structures. So let's look at a few fundamental control structures together with their corresponding inference rules.

The most basic control structure is the composition of two statements S_1 and S_2, which we denote by $S_1; S_2$. This means execute S_1 and then execute S_2. The *composition rule* can be used to prove the correctness of the composition of two statements.

Composition Rule

$$\frac{\{P\}\ S_1\ \{R\}\ \ \text{and}\ \ \{R\}\ S_2\ \{Q\}}{\therefore\ \{P\}\ S_1;\ S_2\ \{Q\}}.\tag{8.12}$$

The composition rule extends naturally to any number of program statements in a sequence. For example, suppose we prove that the following three wffs are correct:

$$\{P\}\ S_1\ \{R\},\ \{R\}\ S_2\ \{T\},\ \text{and}\ \{T\}\ S_3\ \{Q\}.$$

Then we can infer that $\{P\}\ S_1;\ S_2;\ S_3\ \{Q\}$ is correct.

For (8.12) to work, we need an intermediate condition R to place between the two statements. Intermediate conditions often appear naturally during a proof, as the next example shows.

EXAMPLE 2. We'll show the correctness of the following wff:

$$\{x > 2 \land y > 3\}\ x := x + 1;\ y := y + x\ \{y > 6\}.$$

This wff matches the bottom of the composition inference rule (8.12). Since the program statements are assignments, we can use the AA rule to move backward from the postcondition to find an intermediate condition to place between the two assignments. Then we can use AA again to move backward from the intermediate condition. The proof goes as follows:

Proof: First we'll use AA to work backward from the postcondition through the second assignment statement:

    1. $\{y + x > 6\}\ y := y + x\ \{y > 6\}$ AA

Now we can take the new precondition and use AA to work backward from it through the first assignment statement:

    2. $\{y + x + 1 > 6\}\ x := x + 1\ \{y + x > 6\}$ AA

Now we can use the composition rule (8.12) together with lines 1 and 2 to obtain line 3 as follows:

    3. $\{y + x + 1 > 6\}\ x := x + 1;\ y := y + x\ \{y > 6\}$ 1, 2, Comp

At this point the precondition on line 3 does not match the precondition for the wff that we are trying to prove correct. Let's try to apply the consequence rule (8.11) to the situation:

4.	$x > 2 \land y > 3$	P
5.	$x > 2$	4, Simp
6.	$y > 3$	4, Simp
7.	$x + y > 2 + y$	5, T
8.	$2 + y > 2 + 3$	6, T
9.	$x + y > 2 + 3$	7, 8, Transitive
10.	$x + y + 1 > 6$	9, T
11.	$x > 2 \land y > 3 \to x + y + 1 > 6$	4, 10, CP

Now we're in position to apply the consequence rule to lines 3 and 11:

    12. $\{x > 2 \land y > 3\}\ x := x + 1;\ y := y + x\ \{y > 6\}$ 3, 11, Consequence
        QED.   ◆

Let's discuss a few more control structures. We'll start with the *if-then rule* for if-then statements. We should recall that the statement **if** C **then** S means that S is executed if C is true and S is bypassed if C is false. We obtain the following inference rule:

If-Then Rule

$$\frac{\{P \wedge C\}\, S\, \{Q\} \ \text{ and } \ P \wedge \neg C \to Q}{\therefore \ \{P\} \ \textbf{if } C \textbf{ then } S\, \{Q\}}. \tag{8.13}$$

The two wffs in the hypothesis of (8.13) are of different type. The logical wff $P \wedge \neg C \to Q$ needs a proof from the predicate calculus. This wff is necessary in the hypothesis of (8.13) because if C is false, then S does not execute. But we still need Q to be true after C has been determined to be false during the execution of the if-then statement. Let's do an example.

EXAMPLE 3. We'll show that the following wff is correct:

$$\{\text{true}\} \ \textbf{if } x < 0 \textbf{ then } x := -x \ \{x \geq 0\}.$$

Proof: Since the wff fits the pattern of (8.13), all we need to do is prove the following two statements:

1. $\{\text{true} \wedge x < 0\}\ x := -x\ \{x \geq 0\}$.
2. $\text{true} \wedge \neg(x < 0) \to x \geq 0$.

The proofs are easy. We'll combine them into one formal proof:

Proof:
1. $\{-x \geq 0\}\ x := -x\ \{x \geq 0\}$ AA
2. $\text{true} \wedge x < 0$ P
3. $x < 0$ 2, Simp
4. $-x > 0$ 3, T
5. $-x \geq 0$ 4, Add
6. $\text{true} \wedge x < 0 \to -x \geq 0$ 2, 5, CP
7. $\{\text{true} \wedge x < 0\}\ x := -x\ \{x \geq 0\}$ 1, 6, Consequence
8. $\text{true} \wedge \neg(x < 0)$ P
9. $\neg(x < 0)$ 8, Simp
10. $x \geq 0$ 9, T
11. $\text{true} \wedge \neg(x < 0) \to x \geq 0$ 8, 10, CP
    QED   7, 11, If-then.   ◆

Next comes the *if-then-else rule* for the alternative conditional statement. The statement **if** C **then** S_1 **else** S_2 means that S_1 is executed if C is true and S_2 is executed if C is false. We obtain the following inference rule:

If-Then-Else Rule

$$\frac{\{P \wedge C\}\, S_1\, \{Q\} \ \text{ and } \ \{P \wedge \neg\, C\}\, S_2\, \{Q\}}{\therefore \ \{P\}\ \textbf{if } C \textbf{ then } S_1 \textbf{ else } S_2\ \{Q\}}. \qquad (8.14)$$

EXAMPLE 4. Suppose we're given the following wff, where even(x) means that x is an even integer:

$$\{\text{true}\}\ \textbf{if } \text{even}(x) \textbf{ then } y := x \textbf{ else } y := x + 1\ \{\text{even}(y)\}.$$

We'll give a formal proof that this wff is correct. The wff matches the bottom of rule (8.14). Therefore the wff will be correct by (8.14) if we can show that the following two wffs are correct:

1. $\{\text{true} \wedge \text{even}(x)\}\, y := x\, \{\text{even}(y)\}$.
2. $\{\text{true} \wedge \text{odd}(x)\}\, y := x + 1\, \{\text{even}(y)\}$.

To make the proof formal, we need to give formal descriptions of even(x) and odd(x). This is easy to do over the domain of integers:

$$\text{even}(x) = \exists k\, (x = 2k),$$
$$\text{odd}(x) = \exists k\, (x = 2k + 1).$$

To avoid clutter, we'll use even(x) and odd(x) in place of the formal expressions. If you want to see why a particular line holds, you might make the substitution for even or odd and then see whether the statement makes sense. We'll combine the two proofs into the following formal proof:

Proof:
1. $\{\text{even}(x)\}\, y := x\, \{\text{even}(y)\}$ — AA
2. $\quad\quad\text{true} \wedge \text{even}(x)$ — P
3. $\quad\quad\text{even}(x)$ — 2, Simp
4. $\text{true} \wedge \text{even}(x) \rightarrow \text{even}(x)$ — 2, 3, CP
5. $\{\text{true} \wedge \text{even}(x)\}\, y := x\, \{\text{even}(y)\}$ — 1, 4, Consequence
6. $\{\text{even}(x + 1)\}\, y := x + 1\, \{\text{even}(y)\}$ — AA
7. $\quad\quad\text{true} \wedge \text{odd}(x)$ — P

8.	$\text{odd}(x)$	7, Simp
9.	$\text{even}(x + 1)$	8, T
10.	$\text{true} \wedge \text{odd}(x) \to \text{even}(x + 1)$	7, 9, CP
11.	$\{\text{true} \wedge \text{odd}(x)\}\ y := x + 1\ \{\text{even}(y)\}$	6, 10, Consequence
	QED	5, 11, If-then-else. ◆

The last inference rule that we will consider is the *while rule*. The statement **while** C **do** S means that S is executed if C is true, and if C is still true after S has executed, then the process is started over again. Since the body S may execute more than once, there must be a close connection between the precondition and postcondition for S. This can be seen by the appearance of P in all preconditions and postconditions of the rule:

While Rule

$$\frac{\{P \wedge C\}\ S\ \{P\}}{\therefore\ \{P\}\ \textbf{while}\ C\ \textbf{do}\ S\ \{P \wedge \neg C\}}. \tag{8.15}$$

The wff P is called a *loop invariant* because it must be true before and after each execution of the body S. Loop invariants can be tough to find in programs with no documentation. On the other hand, in writing a program, a loop invariant can be a helpful tool for specifying the actions of while loops.

To illustrate the idea of working with while loops, we'll work our way through an example that will force us to discover a loop invariant in order to prove the correctness of a wff. Suppose we want to prove the correctness of the following program to compute the power a^b of two natural numbers a and b, where $a > 0$ and $b \geq 0$:

$$\{a > 0 \wedge b \geq 0\}$$
$$i := 0;$$
$$p := 1;$$
$$\textbf{while}\ i < b\ \textbf{do}$$
$$\qquad p := p * a;$$
$$\qquad i := i + 1$$
$$\textbf{od}$$
$$\{p = a^b\}$$

The program consists of three statements. So we can represent the program and its precondition and postcondition in the following form:

$$\{a > 0 \wedge b \geq 0\}\ S_1;\ S_2;\ S_3\ \{p = a^b\}.$$

In this form, S_1 and S_2 are the first two assignment statements, and S_3 represents the while statement. The composition rule (8.12) tells us that we can prove that the wff is correct if we can find proofs of the following three statements for some wffs P and Q:

$$\{a > 0 \wedge b \geq 0\}\, S_1\, \{Q\},$$

$$\{Q\}\, S_2\, \{P\},$$

$$\{P\}\, S_3\, \{p = a^b\}.$$

Where do P and Q come from? If we know P, then we can use AA to work backward through S_2 to find Q. But how do we find P? Since S_3 is a while statement, P should be a loop invariant. So we need to do a little work.

From (8.15) we know that a loop invariant P for the while statement S_3 must satisfy the following form:

$$\{P\}\ \textbf{while}\ i < b\ \textbf{do}\ p := p * a;\ i := i + 1\ \textbf{od}\ \{P \wedge \neg\, (i < b)\}.$$

Let's try some possibilities for P. Suppose we set $P \wedge \neg\, (i < b)$ equivalent to the program's postcondition $p = a^b$ and try to solve for P. This won't work because $p = a^b$ does not contain the letter i. So we need to be more flexible in our thinking. Since we have the consequence rule, all we really need is an invariant P such that $P \wedge \neg\, (i < b)$ implies $p = a^b$.

After staring at the program, we might notice that the equation $p = a^i$ holds both before and after the execution of the two assignment statements in the body of the while statement. It's also easy to see that the inequality $i \leq b$ holds before and after the execution of the body. So let's try the following definition for P:

$$(p = a^i) \wedge (i \leq b).$$

This P has more promise. Notice that $P \wedge \neg\, (i < b)$ implies $i = b$, which gives us the desired postcondition $p = a^b$. Next, by working backward from P through the two assignment statements, we wind up with the statement

$$1 = a^0 \wedge 0 \leq b.$$

This statement can certainly be derived from the precondition $a \geq 0 \wedge b > 0$. So P does OK from the start of the program down to the beginning of the while loop. All that remains is to prove the following statement:

$$\{P\}\ \textbf{while}\ i < b\ \textbf{do}\ p := p * a;\ i := i + 1\ \textbf{od}\ \{P \wedge \neg\, (i < b)\}.$$

By (8.15), all we need to prove is the following statement:

$$\{P \wedge i < b\}\, p := p * a;\ i := i + 1\ \{P\}.$$

This can be done easily, working backward from P through the two assignment statements. We'll put everything together in the following example.

EXAMPLE 5. We'll prove the correctness of the following program to compute the power a^b of two natural numbers a and b, where $a > 0$ and $b \geq 0$:

$$\{a > 0 \wedge b \geq 0\}$$
$$i := 0;$$
$$p := 1;$$
$$\textbf{while } i < b \textbf{ do}$$
$$\qquad p := p * a;$$
$$\qquad i := i + 1$$
$$\textbf{od}$$
$$\{p = a^b\}$$

The proof will use the loop invariant $P = (p = a^i) \wedge (i \leq b)$ for the while statement. To keep things straight, we can insert $\{P\}$ as the precondition for the while loop and $\{P \wedge \neg (i < b)\}$ as the postcondition for the while loop as follows:

$$\{a > 0 \wedge b \geq 0\}$$
$$i := 0;$$
$$p := 1;$$
$$\{P\} = \{p = a^i \wedge i \leq b\}$$
$$\textbf{while } i < b \textbf{ do}$$
$$\qquad p := p * a;$$
$$\qquad i := i + 1$$
$$\textbf{od}$$
$$\{P \wedge \neg C\} = \{p = a^i \wedge i \leq b \wedge \neg (i < b)\}$$
$$\{p = a^b\}$$

We'll start by proving that the condition after the while loop implies the postcondition of the program. That is, we prove $P \wedge \neg C \to p = a^b$.

1.	$p = a^i \wedge i \leq b \wedge \neg (i < b)$	P
2.	$p = a^i$	1, Simp
3.	$i \leq b \wedge \neg (i < b)$	1, Simp

4. $i = b$ 3, T
5. $p = a^b$ 2, 4, Conj, EE
6. $p = a^i \wedge i \le b \wedge \neg (i < b) \to p = a^b$ 1, 5, CP

Next, we'll prove the correctness of $\{P\}$ **while** $i < b$ **do** S $\{P \wedge \neg (i < b)\}$. The while inference rule tells us to prove the correctness of $\{P \wedge i < b\}$ S $\{P\}$:

7. $\{p = a^{i+1} \wedge i + 1 \le b\} \, i := i + 1 \, \{p = a^i \wedge i \le b\}$ AA
8. $\{p * a = a^{i+1} \wedge i + 1 \le b\} \, p := p * a \, \{p = a^{i+1} \wedge i + 1 \le b\}$ AA
9. $p = a^i \wedge i \le b \wedge i < b$ P
10. $p = a^i$ 9, Simp
11. $i < b$ 9, Simp
12. $b < i + 1$ P for IP
13. $i < b \wedge b < i + 1$ 11, 12, Conj
14.   false   13, T (for integers i and b)
15. $i + 1 \le b$ 11, 14, IP
16. $a = a$ EA
17. $p * a = a^{i+1}$ 10, 16, EE
18. $p * a = a^{i+1} \wedge i + 1 \le b$ 15, 17, Conj
19. $P \wedge i < b \to p * a = a^{i+1} \wedge i + 1 \le b$ 9, 18, CP
20. $\{P \wedge i < b\} \, p := p * a; \, i := i + 1 \, \{P\}$ 7, 8, 19, Conseq, Comp
21. $\{P\}$ **while** $i < b$ **do** $p := p * a; \, i := i + 1$ **od** $\{P \wedge \neg (i < b)\}$ 20, While

Now let's work on the two assignment statements that begin the program. So we'll prove the correctness of $\{a > 0 \wedge b \ge 0\} \, i := 0; \, p := 1 \, \{P\}$:

22. $\{1 = a^i \wedge i \le b\} \, p := 1 \, \{p = a^i \wedge i \le b\}$ AA
23. $\{1 = a^0 \wedge 0 \le b\} \, i := 0 \, \{1 = a^i \wedge i \le b\}$ AA
24. $a > 0 \wedge b \ge 0$ P
25. $a > 0$ 24, Simp
26. $b \ge 0$ 24, Simp
27. $1 = a^0$ 25, T
28. $1 = a^0 \wedge 0 \le b$ 26, 27, Conj
29. $a > 0 \wedge b \ge 0 \to 1 = a^0 \wedge 0 \le b$ 24, 28, CP
30. $\{a > 0 \wedge b \ge 0\} \, i := 0; \, p := 1 \, \{P\}$ 22, 23, 29, Comp, Conseq

The proof is finished by using the Composition rule on the three parts.

QED   30, 21, 6, Comp, Conseq.   ◆

Arrays

Since arrays are fundamental structures in imperative languages, we'll modify our theory so that we can handle assignment statements like $a[i] := t$. In other words, we want to be able to construct a precondition for the following partial wff:

$$\{ \quad \} \; a[i] := t \; \{Q\}.$$

What do we do? We might try to work backward, as with AA, and replace all occurrences of $a[i]$ in Q by t. Let's try it and see what happens. Let $Q(a[i]/t)$ denote the wff obtained from Q by replacing all occurrences of $a[i]$ by t. We'll call the following statement the "attempted" array assignment axiom:

Attempted AAA: $\{Q(a[i]/t)\} \; a[i] := t \; \{Q\}.$ (8.16)

Since we're calling (8.16) the Attempted AAA, let's see whether we can find something wrong with it. For example, suppose we have the following wff, where the letter i is a variable:

$$\{true\} \; a[i] := 4 \; \{a[i] = 4\}.$$

This wff is clearly correct, and we can prove it with (8.16) as follows:

1. $\{4 = 4\} \; a[i] := 4 \; \{a[i] = 4\}$ Attempted AAA
2. true $\rightarrow 4 = 4$ T
     QED   1, 2, Conseq.

At this point, things seem OK. But let's try another example. Suppose we have the following wff, where i and j are variables:

$$\{i = j \land a[i] = 3\} \; a[i] := 4 \; \{a[j] = 4\}.$$

This wff is also clearly correct because $a[i]$ and $a[j]$ both represent the same indexed array variable. Let's try to prove that the wff is correct by using (8.16). The first line of the proof looks like the following:

1. $\{a[j] = 4\} \; a[i] := 4 \; \{a[j] = 4\}$ Attempted AAA

Since the precondition on line 1 is not the precondition of the wff, we need to use the consequence rule, which states that we must prove the following wff:

$$i = j \land a[i] = 3 \rightarrow a[j] = 4.$$

But this wff is invalid because a single array element can't have two distinct values.

So we now have an example of an array assignment statement that we "know" is correct, but we don't yet have the proper tools to prove that it's correct:

$$\{i = j \wedge a[i] = 3\} \; a[i] := 4 \; \{a[j] = 4\}.$$

What went wrong? Well, since the expression $a[i]$ does not appear in the postcondition $\{a[j] = 4\}$, the attempted AAA (8.16) just gives us back the postcondition as the precondition. This stops us in our tracks because we are now forced to prove an invalid conditional wff.

The problem is that (8.16) does not address the possibility that i and j might be equal. So we need a more sophisticated assignment axiom for arrays. Let's start again and try to incorporate the preceding remarks. We want an axiom to fill in the precondition of the following partial wff:

$$\{ \quad \} \; a[i] := t \; \{Q\}.$$

Of course, we need to replace all occurrences of $a[i]$ in Q by t. But we also need to replace all occurrences of $a[j]$ in Q, where j is any arithmetic expression, by an expression that allows the possibility that $j = i$. We can do this by replacing each occurrence of $a[j]$ in Q by the following if-then-else statement:

"if $j = i$ then t else $a[j]$."

For example, if the equation $a[j] = s$ occurs in Q, then the precondition will contain the following equation:

$$(\text{if } j = i \text{ then } t \text{ else } a[j]) = s.$$

When an equation contains an if-then-else statement, we can write it without if-then-else as a conjunction of two wffs. For example, the following two statements are equivalent for terms $s, t,$ and u:

$$(\text{if } C \text{ then } t \text{ else } u) = s,$$

$$(C \rightarrow t = s) \wedge (\neg C \rightarrow u = s).$$

So when we use the if-then-else form in a wff, we are still within the bounds of a first-order theory with equality.

For example, if $a[j] = s$ occurs in the postcondition for the array assignment $a[i] := t$, then the precondition for the assignment should replace $a[j] = s$ with either one of the following two equivalent statements:

$$(\text{if } j = i \text{ then } t \text{ else } a[j]) = s,$$

$$(j = i \rightarrow t = s) \wedge (j \neq i \rightarrow a[j] = s).$$

Now let's put things together and state the correct axiom for array assignment.

Array Assignment Axiom (AAA) (8.17)

$$\{P\}\ a[i] := t\ \{Q\},$$

where P is constructed from Q by applying the following rules:

1. Replace all occurrences of $a[i]$ in Q by t.
2. Replace all occurrences of $a[j]$ in Q by "if $j = i$ then t else $a[j]$."

Note: i and j may be any arithmetic expressions that do not contain a.

It's important that the index expressions i and j don't contain the array name. For example, $a[a[k]]$ is NOT OK, but $a[k + 1]$ is OK. To see why we can't use arrays within arrays when applying AAA, consider the following wff:

$$\{a[1] = 2 \wedge a[2] = 2\}\ a[a[2]] := 1\ \{a[a[2]] = 1\}.$$

This wff is false because the assignment statement sets $a[2] = 1$, which makes the postcondition into the equation $a[1] = 1$, contradicting the fact that $a[1] = 2$. But we can use AAA to improperly "prove" that the wff is correct as follows:

1. $\{1 = 1\}\ a[a[2]] := 1\ \{a[a[2]] = 1\}$ AAA attempt with $a[a[...]]$
2. $a[1] = 2 \wedge a[2] = 2 \rightarrow 1 = 1$ T
    Not QED   1, 2, Conseq.

The exclusion of arrays within arrays is no real handicap because an assignment statement like $a[a[i]] := t$ can be rewritten as a sequence of two assignment statements as follows:

$$j := a[i];\ a[j] := t.$$

Similarly, a logical statement like $a[a[i]] = t$ appearing in a precondition or postcondition can be rewritten as the following wff:

$$\exists x\ (x = a[i] \wedge a[x] = t).$$

Now let's see whether we can use (8.17) to prove the correctness of the wff that we could not prove before.

EXAMPLE 6. We want to prove the correctness of the following wff:

$$\{i = j \wedge a[i] = 3\}\ a[i] := 4\ \{a[j] = 4\}.$$

This wff represents a simple reassignment of an array element, where the index of the array element is represented by two variable names. We'll include all the details of the consequence part of the proof, which uses the conjunction form of an if-then-else equation.

Proof:
1. $\{(\text{if } j = i \text{ then } 4 \text{ else } a[j]) = 4\}\ a[i] := 4\ \{a[j] = 4\}$ AAA
2. $i = j \wedge a[i] = 3$ P
3. $i = j$ 2, Simp
4. $j = i$ 3, Symmetry
5. $4 = 4$ EA
6. $j = i \rightarrow 4 = 4$ 5, T (trivial)
7. $j \neq i \rightarrow a[j] = 4$ 4, T (vacuous)
8. $(j = i \rightarrow 4 = 4) \wedge (j \neq i \rightarrow a[j] = 4)$ 6, 7, Conj
9. $(\text{if } j = i \text{ then } 4 \text{ else } a[j]) = 4$ 8, T
10. $i = j \wedge a[i] = 3 \rightarrow (\text{if } j = i \text{ then } 4 \text{ else } a[j]) = 4$ 2, 9, CP
  QED   1, 10, Conseq.   ◆

Termination

Program correctness as we have been discussing it does not consider whether loops terminate. In other words, the correctness of the wff $\{P\}\ S\ \{Q\}$ includes the assumption that S halts. This kind of correctness is often called *partial correctness*. For *total correctness* we can't assume that loops terminate. We must prove that they terminate.

For example, suppose we're presented with the following while loop, and the only information we know is that the variables take integer values:

$$
\begin{aligned}
&x := a;\\
&y := b;\\
&\textbf{while } x \neq y \textbf{ do}\\
&\quad x := x - 1;\\
&\quad y := y + 1;\\
&\quad c := c + 1\\
&\textbf{od}
\end{aligned}
$$
(8.18)

We don't have enough information to be able to tell for certain whether the loop terminates. For example, if we initialize $a = 4$ and $b = 5$, then the loop will run forever. In fact the loop will run forever if $a < b$. If $a = 6$ and $b = 3$, the loop will run forever. After a little study and thought, we can see that the loop will terminate if $a \geq b$ and $a - b$ is an even number. The value of c doesn't affect the termination, but it probably has something to do with counting the number of times through the loop. In fact, if we initialize $c = 0$, then the value of c at the termination of the loop satisfies the equation $a - b = 2c$.

This example shows that the precondition for a loop must contain enough information to decide whether the loop terminates. We're going to present a formal condition that, if satisfied, will ensure the termination of a loop. But first we need to discuss a few preliminary ideas.

A *state* of a computation is a tuple representing the values of the variables at some point in the computation. For example, the tuple $\langle x, y, c \rangle$ denotes an arbitrary state of program (8.18), where we'll assume that a and b are constants. For our purposes the only time a state will change is when an assignment statement is executed. For example, suppose the initial state of a computation for (8.18) is $\langle 10, 6, 0 \rangle$. For this state the loop condition is true because $10 \neq 6$. After the execution of the assignment statement $x := x - 1$ in the body of the loop, the state becomes $\langle 9, 6, 0 \rangle$. After one iteration of the loop the state will be $\langle 9, 7, 1 \rangle$. For this state the loop condition is true because $9 \neq 7$. After a second iteration of the entire loop the state becomes $\langle 8, 8, 2 \rangle$. For this state the loop condition is false, which causes the loop to terminate. We'll use "States" to represent the set of all possible states for a program's variables. For program (8.18) we have States $= \mathbb{Z} \times \mathbb{Z} \times \mathbb{Z}$.

Program (8.18) terminates for the initial state $\langle 10, 6, 0 \rangle$ because the value $x - y$ gets smaller with each iteration of the loop, eventually equaling zero. In other words, $x - y$ takes on the sequence of values 4, 2, 0. This is the key point in showing loop termination. There must be some decreasing sequence of numbers that stops at some point. In more general terms, the numbers must form a decreasing sequence in some well-founded set. For example, in (8.18) the well-founded set is the natural numbers $\mathbb{N}$.

To show loop termination, we need to find a well-founded set $\langle W, \prec \rangle$ that we can use to associate an element $w_i \in W$ with the ith iteration of the loop such that the elements form a decreasing sequence as follows:

$$w_1 \succ w_2 \succ w_3 \cdots.$$

Since W is well-founded, the sequence must stop. Thus the loop must halt.

Let's put things together and describe in more detail the general process required to prove termination of the following while loop with respect to a precondition P:

$$\textbf{while } C \textbf{ do } S.$$

We'll assume that we already know, or we have already proven, that S halts. This reflects the normal process of working our way from the inside out when doing termination proofs.

We need to introduce a little terminology to describe the termination condition. Let P be a precondition for the loop **while** C **do** S. For any state s, let $P(s)$ denote the wff obtained from P by replacing its variables by their corresponding values in s. Similarly, let $C(s)$ denote the wff obtained from C by replacing its variables by their corresponding values in s. Let s represent an arbitrary state prior to the execution of S, and let t be the state which follows that same execution of S.

Suppose we have a well-founded set $\langle W, \prec \rangle$ together with a function

$$f : \text{States} \to W,$$

where f may be a partial function. Now we can state the loop termination condition for the loop **while** C **do** S. The loop *terminates with respect to precondition P* if the following wff is valid:

Termination Condition

$$P(s) \wedge C(s) \wedge f(s) \in W \to f(t) \in W \wedge f(s) \succ f(t). \tag{8.19}$$

Notice that (8.19) contains the two statements $f(s) \in W$ and $f(t) \in W$. Since f could be a partial function, these statements ensure that $f(s)$ and $f(t)$ are both defined and members of W in good standing. Therefore the statement $f(s) \succ f(t)$ will ensure that the loop terminates. For example, if s_i represents the state prior to the ith execution of S, then we can apply (8.19) to obtain the following decreasing sequence of elements in W:

$$f(s_1) \succ f(s_2) \succ f(s_3) \succ \cdots .$$

Since W is well-founded, the sequence must stop. Therefore the loop halts. Let's do an example to cement the idea.

EXAMPLE 7 (*A Termination Proof*). Let's see how the formal definition of termination relates to program (8.18). From our discussion it seems likely that P should be

$$x \geq y \wedge \text{even}(x - y).$$

We'll leave the proof that P is a loop invariant as an exercise. For a well-founded set W we'll choose $\mathbb{N}$. Then $f :$ States $\to \mathbb{N}$ can be defined by

$$f(\langle x, y, c \rangle) = x - y.$$

If $s = \langle x, y, c \rangle$ represents the state prior to the execution of the loop body and t represents the state after execution of the loop body, then

$$t = \langle x - 1, y + 1, c + 1 \rangle.$$

In this case, $f(s)$ and $f(t)$ have the following values:

$$f(s) = f(x, y, c) = x - y,$$
$$f(t) = f(x - 1, y + 1, c + 1) = (x - 1) - (y + 1) = x - y - 2.$$

With these interpretations, (8.19) can be written as follows:

$$x \geq y \land \text{even}(x - y) \land x \neq y \land x - y \in \mathbb{N} \to x - y - 2 \in \mathbb{N} \land x - y > x - y - 2.$$

Now let's give a proof of this statement.

Proof: 1. $x \geq y \land \text{even}(x - y) \land x \neq y \land x - y \in \mathbb{N}$ P
2. $x \geq y \land x \neq y$ 1, Simp
3. $x > y$ 2, T
4. $\text{even}(x - y)$ 1, Simp
5. $x - y \in \mathbb{N}$ 1, Simp
6. $x - y \geq 2$ 3, 4, 5, T
7. $x - y - 2 \in \mathbb{N}$ 5, 6, T
8. $x - y > x - y - 2$ T
9. $x - y - 2 \in \mathbb{N} \land x - y > x - y - 2$ 7, 8, Conj
QED   1, 9, CP.   ◆

As a final remark to this short discussion, we should remember the fundamental requirement that programs with loops need preconditions that contain enough restrictions to ensure that the loops terminate.

Note

Hopefully, this introduction has given you the flavor of proving properties of programs. There are many mechanical aspects to the process. For example, the backwards application of the AA and AAA rules is a simple substitution problem that can be automated. We've omitted many important results. For example, if the programming language has other control structures, such as for-loops and repeat-loops, then new inference rules must be constructed. The original papers in these areas are by Hoare [1969] and Floyd [1967]. A good place to start reading more about this subject is the survey paper by Apt [1981].

Different languages usually require different formal theories to handle the program correctness problem. For example, declarative languages, in which programs can consist of recursive definitions, require methods of inductive proof in their formal theories for proving program correctness.

Exercises

1. Prove that the following wff is correct over the domain of integers:

    $$\{\text{true} \wedge \text{even}(x)\} \ y := x + 1 \ \{\text{odd}(y)\}.$$

2. Prove that each of the following wffs is correct. Assume that the domain is the set of integers.

    a. $\{a > 0 \wedge b > 0\} \ x := a; \ y := b \ \{x + y > 0\}$.

    b. $\{a > b\} \ x := -a; \ y := -b \ \{x < y\}$.

3. Both of the following wffs claim to correctly perform the swapping process. The first one uses a temporary variable. The second does not. Prove that each wff is correct. Assume that the domain is the real numbers.

    a. $\{x < y\} \ temp := x; \ x := y; \ y := temp \ \{y < x\}$.

    b. $\{x < y\} \ y := y + x; \ x := y - x; \ y := y - x \ \{y < x\}$.

4. Prove that each of the following wffs is correct. Assume that the domain is the set of integers.

    a. $\{x < 10\}$ **if** $x \geq 5$ **then** $x := 4 \ \{x < 5\}$.

    b. $\{\text{true}\}$ **if** $x \neq y$ **then** $x := y \ \{x = y\}$.

    c. $\{\text{true}\}$ **if** $x < y$ **then** $x := y \ \{x \geq y\}$.

    d. $\{\text{true}\}$ **if** $x > y$ **then** $x := y + 1; \ y := x + 1$ **fi** $\{x \leq y\}$.

5. Prove that each of the following wffs is correct. Assume that the domain is the set of integers.

    a. $\{\text{true}\}$ **if** $x < y$ **then** max $:= y$ **else** max $:= x \ \{\text{max} \geq x \wedge \text{max} \geq y\}$.

    b. $\{\text{true}\}$ **if** $x < y$ **then** $y := y - 1$ **else** $x := -x; \ y := -y$ **fi** $\{x \leq y\}$.

6. Show that each of the following wffs is NOT correct over the domain of integers.

    a. $\{x < 5\}$ **if** $x \geq 2$ **then** $x := 5 \ \{x = 5\}$.

    b. $\{\text{true}\}$ **if** $x < y$ **then** $y := y - x \ \{y > 0\}$.

7. Prove that the following wff is correct, where x and y are integers:

$$\{x \geq y \wedge \text{even}(x - y)\}$$
while $x \neq y$ **do**
$\quad x := x - 1;$
$\quad y := y + 1$
od
$$\{x \geq y \wedge \text{even}(x - y) \wedge x = y\}.$$

8. Prove that each of the following wffs is correct.

a. The program computes the floor of a nonnegative real number x. *Hint:* For a loop invariant, use the inequality $i \leq x$.

$$\{x \geq 0\}$$
$i := 0;$
while $i \leq x - 1$ **do** $i := i + 1$ **od**
$$\{i = \text{floor}(x)\}.$$

b. The program computes the floor of a negative real number x. *Hint:* For a loop invariant, use the inequality $x < i + 1$.

$$\{x < 0\}$$
$i := -1;$
while $x < i$ **do** $i := i - 1$ **od**
$$\{i = \text{floor}(x)\}.$$

c. The program computes the floor of an arbitrary real number x, where the statements S_1 and S_2 are the two programs from parts (a) and (b).

$$\{\text{true}\} \ \textbf{if} \ x \geq 0 \ \textbf{then} \ S_1 \ \textbf{else} \ S_2 \ \{i = \text{floor}(x)\}.$$

9. Given a natural number n, the following program computes the sum of the first n natural numbers. Prove that the wff is correct. *Hint:* For a loop invariant, try $s = \frac{i(i + 1)}{2} \wedge i \leq n$.

$$\{n \geq 0\}$$
$i := 0;$
$s := 0;$
while $i < n$ **do**
$\quad i := i + 1;$
$\quad s := s + i$
od
$$\left\{s = \frac{n(n + 1)}{2}\right\}.$$

10. The following program implements the division algorithm for natural numbers. It computes the quotient and the remainder of the division of a natural number by a positive natural number. Prove that the wff is correct. *Hint:* For a loop invariant, try $(a = yb + x) \wedge (0 \le x)$.

$$\{a \ge 0 \wedge b > 0\}$$
$$x := a;$$
$$y := 0;$$
while $b \le x$ **do**
$$\quad x := x - b;$$
$$\quad y := y + 1$$
od;
$$r := x;$$
$$q := y$$
$$\{a = qb + r \wedge 0 \le r < b\}.$$

11. (Greatest Common Divisor) The following program claims to find the greatest common divisor (a, b) of two positive integers a and b. Prove that the wff is correct.

$$\{a > 0 \wedge b > 0\}$$
$$x := a;$$
$$y := b;$$
while $x \ne y$ **do**
$$\quad \textbf{if } x > y \textbf{ then } x := x - y \textbf{ else } y := y - x$$
od;
$$great := x$$
$$\{(a, b) = great\}.$$

Hints: Use $(a, b) = (x, y)$ as the loop invariant. You may use the following useful fact derived from (2.1) for any integers w and z: $(w, z) = (w - z, z)$.

12. Write a program to compute the ceiling of an arbitrary real number. Give the program a precondition and a postcondition, and prove that the resulting wff is correct. *Hint:* Look at Exercise 8.

13. For each of the following partial wffs, fill in the precondition that results by applying the array assignment axiom (8.17).

    a. $\{ \quad \} a[i - 1] := 24 \; \{a[j] = 24\}$.
    b. $\{ \quad \} a[i] := 16 \; \{a[i] = 16 \wedge a[j + 1] = 33\}$.
    c. $\{ \quad \} a[i + 1] := 25; a[j - 1] := 12 \; \{a[i] = 12 \wedge a[j] = 25\}$.

14. Prove that each of the following wffs is correct.

    a. $\{i = j + 1 \wedge a[j] = 39\} a[i - 1] := 24 \; \{a[j] = 24\}$.

b. $\{even(a[i]) \wedge i = j + 1\} \ a[j] := a[i] + 1 \ \{odd(a[i - 1])\}.$

c. $\{i = j - 1 \wedge a[i] = 25 \wedge a[j] = 12\}$
$a[i + 1] := 25; \ a[j - 1] := 12$
$\{a[i] = 12 \wedge a[j] = 25\}.$

15. The following wffs are not correct. For each wff, apply the array assignment axiom to the postcondition and assignment statements to obtain a condition Q. Show that the precondition does not imply Q.

a. $\{even(a[i])\} \ a[i + 1] := a[i] + 1 \ \{even(a[i + 1])\}.$

b. $\{a[2] = 2\} \ i := a[2]; \ a[i] := 1 \ \{a[i] = 1 \wedge i = a[2]\}.$

c. $\{\forall j \ (1 \leq j \leq 5 \rightarrow a[j] = 23)\} \ i := 3; \ a[i] := 355 \ \{\forall j \ (1 \leq j \leq 5 \rightarrow a[j] = 23)\}.$

d. $\{a[1] = 2 \wedge a[2] = 2\} \ a[a[2]] := 1 \ \{\exists x \ (x = a[2] \wedge a[x] = 1\}.$

16. Prove that each of the following loops terminates for the given precondition P. Assume that all variables take integer values. *Hint:* Find an appropriate well-founded set W and function $f :$ States $\rightarrow W$.

a. **while** $i < x$ **do** $i := i + 1$ **od** and $P = i \leq x.$

b. **while** $i < x$ **do** $x := x - 1$ **od** and $P = i \leq x.$

17. Prove that the following loop terminates with respect to the precondition $P =$ true. Assume that all variables take values in the positive integers.

$$\textbf{while } x \neq y \textbf{ do if } x < y \textbf{ then } y := y - x \textbf{ else } x := x - y \textbf{ od}.$$

8.3 Higher-Order Logics

In first-order predicate calculus the only things that can be quantified are individual variables, and the only things that can be arguments for predicates are terms (i.e., constants, variables, or functional expressions with terms as arguments). If we loosen up a little and allow our wffs to quantify other things like predicates or functions, or if we allow our predicates to take arguments that are predicates or functions, then we move to a *higher-order logic*. Is higher-order logic necessary? The purpose of this section is to convince you that the answer is yes. After some examples we'll give a general definition that will allow us to discuss nth-order logic for any natural number n.

We often need higher-order logic to express simple statements about the things that interest us. For example, let's try to formalize the statement

"There is a function that is larger than the log function."

This statement asserts the existence of a function. So if we want to formalize the statement, we'll need to use higher-order logic to quantify a function. We might formalize the statement as follows:

$$\exists f \ \forall x \ (f(x) > \log x).$$

This wff is an instance of the following more general wff, where > is an instance of p and log is an instance of g:

$$\exists f \ \forall x \ p(f(x), g(x)).$$

For another example, let's see whether we can formalize the notion of equality. Suppose we agree to say that x and y are identical if all their properties are the same. We'll signify this by writing $x = y$. Can we express this thought in formal logic? Sure. If P is some property, then we can think of P as a predicate, and we'll agree that $P(x)$ means that x has property P. Then we can define $x = y$ as the following higher-order wff:

$$\forall P \ ((P(x) \rightarrow P(y)) \wedge (P(y) \rightarrow P(x))).$$

This wff is higher-order because the predicate P is quantified.

Now that we have some examples, let's get down to business and discuss higher-order logic in a general setting that allows us to classify the different orders of logic.

Classifying Higher-Order Logics

To classify higher-order logics, we need to make an assumption about the relationship between predicates and sets. We'll assume that predicates are sets and that sets are predicates. Let's see why we can think of predicates and sets as the same thing. For example, if P is a predicate with one argument, we can think of P as a set in which $x \in P$ if and only if $P(x)$ is true. Similarly, if S is a set of 3-tuples, we can think of S as a predicate in which $S(x, y, z)$ is true if and only if $\langle x, y, z \rangle \in S$.

The relationship between sets and predicates allows us to look at some wffs in a new light. For example, consider the following wff:

$$\forall x \ (A(x) \rightarrow B(x)).$$

In addition to the usual reading of this wff as "For every x, if $A(x)$ is true, then $B(x)$ is true," we can now read it in terms of sets by saying, "For every x, if $x \in A$, then $x \in B$." In other words, we have a wff that represents the statement "A is a subset of B."

The identification of predicates and sets puts us in position to define higher-order logics. A logic is called *higher-order* if it allows sets to be quantified or if it allows sets to be elements of other sets. A wff that quantifies a set or has a set as an argument to a predicate is called a *higher-order wff*. For example, the following two wffs are higher-order wffs:

$\exists S\ S(x)$ The set S is quantified.

$S(x) \wedge T(S)$ The set S is an element of the set T.

Let's see how functions fit into the picture. Recall that a function can be thought of as a set of 2-tuples. For example, if $f(x) = 3x$ for all $x \in \mathbb{N}$, then we can think of f as the set

$$f = \{\langle x, 3x \rangle \mid x \in \mathbb{N}\}.$$

So whenever a wff contains a quantified function name, the wff is actually quantifying a set and thus is a higher-order wff by our definition. Similarly, if a wff contains a function name as an argument to a predicate, then the wff is higher-order. For example, the following two wffs are higher-order wffs:

$\exists f\ \forall x\ p(f(x), g(x))$ The function f is a set and is quantified.

$p(f(x)) \wedge q(f)$ The function f is a set and is an element of the set q.

Since we can think of a function as a set and we are identifying sets with predicates, we can also think of a function as a predicate. For example, let f be the function

$$f = \{\langle x, 3x \rangle \mid x \in \mathbb{N}\}.$$

We can think of f as a predicate with two arguments. In other words, we can write the wff $f(x, 3x)$ and let it mean "x is mapped by f to $3x$," which of course we usually write as $f(x) = 3x$.

Now let's see whether we can classify the different orders of logic. We'll start with the two logics that we know best. A propositional calculus is called a *zero-order logic* and a first-order predicate calculus is called a *first-order logic*. We want to continue the process by classifying the higher-order logics as second-order, third-order, and so on. To do this, we need to attach an order to each predicate and each quantifier that occurs in a wff. We'll define the *order of a predicate* as follows:

A predicate has *order* 1 if all its arguments are terms (i.e., constants, individual variables, or function values). Otherwise, the predicate has *order* $n + 1$, where n is the highest order among its arguments that are not terms.

For example, for each of the following wffs we've given the order of its predicates (i.e., sets):

$S(x) \wedge T(S)$ S has order 1, and T has order 2.

$p(f(x)) \wedge q(f)$ p has order 1, f has order 1, and q has order 2.

The reason that the function f has order 1 is that any function when thought of as a predicate takes only terms for arguments. Thus any function name has order 1. Remember to distinguish between $f(x)$ and f; $f(x)$ is a term, and f is a function (i.e., a set or a predicate).

We can also relate the order of a predicate to the level of nesting of its arguments, where we think of a predicate as a set. For example, if a wff contains the three statements $S(x)$, $T(S)$, and $P(T)$, then we have $x \in S$, $S \in T$, and $T \in P$. The orders of S, T, and P are 1, 2, and 3. So the order of a predicate (or set) is the maximum number of times the symbol $\in$ is used to get from the set down to its most basic elements.

Now we'll define the *order of a quantifier* as follows:

A quantifer has *order* 1 if it quantifies an individual variable. Otherwise, the quantifier has *order* $n + 1$, where n is the order of the predicate being quantified.

For example, let's find the orders of the quantifiers in the wff that follows. Try your luck before you read the answers.

$$\forall x \; \exists S \; \exists T \; \exists f \; (S(x, f(x)) \wedge T(S)).$$

The quantifier $\forall x$ has order 1 because x is an individual variable. $\exists S$ has order 2 because S has order 1. $\exists T$ has order 3 because T has order 2. $\exists f$ has order 2 because f is a function name, and all function names have order 1.

Now we can make a simple definition for the order of a wff. The *order of a wff* is the highest of the orders of its predicates and quantifiers. Here are a few examples:

> *First-order wffs*
>
> $S(x)$ S has order 1.
>
> $\forall x \; S(x)$ Both S and $\forall x$ have order 1.
>
> *Second-order wffs*
>
> $S(x) \wedge T(S)$ S has order 1, and T has order 2.
>
> $\exists S \; S(x)$ S has order 1, and $\exists S$ has order 2.
>
> $\exists S \; (S(x) \wedge T(S))$ S has order 1, and $\exists S$ and T have order 2.
>
> $P(x, f, f(x))$ P has order 2 because f has order 1.

Third-order wffs

$S(x) \wedge T(S) \wedge P(T)$ S, T, and P have orders 1, 2, and 3.

$\forall T\, (S(x) \wedge T(S))$ S, T, $\forall T$ have orders 1, 2, and 3.

$\exists T\, (S(x) \wedge T(S) \wedge P(T))$ S, T, P, and $\exists T$ have orders 1, 2, 3, and 3.

Now we can make the definition of a *n*th-order logic. A *nth-order logic* is a logic whose wffs have order *n* or less. Let's do some examples that transform sentences into higher-order wffs.

EXAMPLE 1 (*Subsets*). Suppose we want to represent the following statement in formal logic:

"There is a set of natural numbers that doesn't contain 4."

Since the statement asserts the existence of a set, we'll need an existential quantifier. The set must be a subset of the natural numbers, and it must not contain the number 4. Putting these ideas together, we can write a mixed version (informal and formal) as follows:

$$\exists S\, (S \text{ is a subset of } \mathbb{N} \text{ and } \neg\, S(4)).$$

Let's see whether we can finish the formalization. We've seen that the general statement "*A* is a subset of *B*" can be formalized as follows:

$$\forall x\, (A(x) \rightarrow B(x)).$$

Therefore we can write the following formal version of our statement:

$$\exists S\, (\forall x\, (S(x) \rightarrow \mathbb{N}(x)) \wedge \neg\, S(4)).$$

This wff is second-order because S has order 1, so $\exists S$ has order 2.   ♦

EXAMPLE 2 (*Cities, Streets, and Addresses*). Suppose we think of a city as a set of streets and a street as a set of house addresses. We'll try to formalize the following statement:

"There is a city with a street named Main, and
there is an address 1140 on Main Street."

Suppose C is a variable representing a city and S is a variable representing a street. If x is a name, then we'll let $N(S, x)$ mean that the name of S is x. A third-order logic formalization of the sentence can be written as follows:

$$\exists C \; \exists S \; (C(S) \wedge N(S, \text{Main}) \wedge S(1140)).$$

This wff is third-order because S has order 1, so C has order 2 and $\exists C$ has order 3. ◆

Semantics

How do we attach a meaning to a higher-order wff? The answer is that we construct an interpretation for the wff. We start out by specifying a domain D of individuals that we use to give meaning to the constants, the free variables, and the functions and predicates that are not quantified. The quantified individual variables, functions, and predicates are allowed to vary over all possible meanings in terms of D.

Let's try to make the idea of an interpretation clear with an example. We'll give an interpretation for the following second-order wff:

$$\exists S \; \exists T \; \forall x \; (S(x) \to \neg \; T(x)).$$

Suppose we let the domain be $D = \{a, b\}$. We observe that S and T are predicates of order 1, and they are both quantified. So S and T can vary over all possible single-argument predicates over D. For example, the following list shows the four possible predicate definitions for S together with the corresponding set definitions for S:

Predicate definitions for S	*Set definitions for S*
$S(a)$ and $S(b)$ are both true.	$S = \{a, b\}$.
$S(a)$ is true and $S(b)$ is false.	$S = \{a\}$.
$S(a)$ is false and $S(b)$ is true.	$S = \{b\}$.
$S(a)$ and $S(b)$ are both false.	$S = \varnothing$.

We can see from this list that there are as many possibilities for S as there are subsets of D. A similar statement holds for T. Now it's easy to see that our example wff is true for our interpretation. For example, if we choose $S = \{a, b\}$ and $T = \varnothing$, then S is always true and T is always false. Thus $S(a) \to \neg T(a)$ and $S(b) \to \neg T(b)$ are both true. Therefore $\exists S \; \exists T \; \forall x \; (S(x) \to \neg \; T(x))$ is true for the interpretation.

For a second example we'll give an interpretation for the following second-order wff:

$$\exists S \; \forall x \; \exists y \; S(x, y).$$

Again we'll let $D = \{a, b\}$. Since S takes two arguments, it has 16 possible definitions, one corresponding to each subset of 2-tuples over D. For example, if $S = \{\langle a, a \rangle, \langle b, a \rangle\}$, then $S(a, a)$ and $S(b, a)$ are both true, and $S(a, b)$ and $S(b, b)$ are both false. Thus the wff $\exists S \ \forall x \ \exists y \ S(x, y)$ is true for our interpretation.

For a final example we'll give an interpretation for the following third-order wff:

$$\exists T \ \forall x \ (T(S) \rightarrow S(x)).$$

We'll let $D = \{a, b\}$. Since S is not quantified, it is a normal predicate and we must give it a meaning. Suppose we let $S(a)$ be true and $S(b)$ be false. This is the same thing as setting $S = \{a\}$. Now T is an order 2 predicate because it takes an order 1 predicate as its argument. T is also quantified, so it is allowed to vary over all possible predicates that take arguments like S. From the standpoint of sets the arguments to T can be any of the four subsets of D. Therefore T can vary over any of the 16 subsets of $\{\varnothing, \{a\}, \{b\}, \{a, b\}\}$. For example, one possible value for T is $T = \{\varnothing, \{a\}\}$. If we think of T as a predicate, this means that $T(\varnothing)$ and $T(\{a\})$ are both true, while $T(\{b\})$ and $T(\{a, b\})$ are both false. This value of T makes the wff $\forall x \ (T(S) \rightarrow S(x))$ true. Thus the wff $\exists T \ \forall x \ (T(S) \rightarrow S(x))$ is true for our interpretation.

So we can give interpretations to higher-order wffs. This means that we can also use the following familiar terms in our discussions about higher-order wffs:

model, countermodel, valid, invalid, satisfiable, and unsatisfiable.

What about formal reasoning with higher-order wffs? That's next.

Higher-Order Reasoning

Gödel proved a remarkable result in 1931. He proved that if a formal system is powerful enough to describe all the arithmetic formulas of the natural numbers and the system is consistent, then it is not complete. In other words, there is a valid wff that can't be proven as a theorem in the system. Even if additional axioms were added to make the wff provable, then there would exist a new valid wff that is not provable in the larger system. A very readable account of Gödel's proof is given by Nagel and Newman [1958].

The formulas of arithmetic can be described in a first-order theory with equality, so it follows from Gödel's result that first-order theories with equality are not complete. Similarly, we can represent the idea of equality with second-order predicate calculus. So it follows that second-order predicate calculus is not complete.

What does it really mean when we have a logic that is not complete? It means that we might have to leave the formalism of the logic to prove that some wffs are valid. In other words, we may need to argue informally—using only our wits and imaginations—to prove some logical statements. In some sense this is nice because it justifies our existence as reasoning beings. Since most theories cannot be captured by using only first-order logic, there will always be enough creative work for us to do—perhaps aided by computers.

Even though higher-order logic does not give us completeness, we can still do formal reasoning to prove the validity of many higher-order wffs. Let's look at a familiar example to see how higher-order logic comes into play when we discuss elementary geometry.

EXAMPLE 3 (*Euclidean Geometry*). From an informal standpoint the wffs of Euclidean geometry are English sentences. For example, the following four statements describe part of Hilbert's axioms for Euclidean plane geometry:

1. On any two distinct points there is always a line.
2. On any two distinct points there is not more than one line.
3. Every line has at least two distinct points.
4. There are at least three points not on the same line.

Can we formalize these axioms? Let's assume that a line is a set of points. So two lines are equal if they have the same set of points. We'll also assume that points are denoted by the variables x, y, and z and by the constants a, b, and c. We'll denote lines by the variables L, M, and N and by the constants l, m, and n. We'll let the predicate $L(x)$ denote the fact that x is a point on line L or, equivalently, L is a line on the point x. Now we can write the four axioms as second-order wffs as follows:

1. $\forall x \, \forall y \, (x \neq y \rightarrow \exists L \, (L(x) \wedge L(y)))$.
2. $\forall x \, \forall y \, (x \neq y \rightarrow \forall L \, \forall M \, (L(x) \wedge L(y) \wedge M(x) \wedge M(y) \rightarrow L = M))$.
3. $\forall L \, \exists x \, \exists y \, (x \neq y \wedge L(x) \wedge L(y))$.
4. $\exists x \, \exists y \, \exists z \, (x \neq y \wedge x \neq z \wedge y \neq z \wedge \forall L \, (L(x) \wedge L(y) \rightarrow \neg \, L(z)))$.

Let's prove the following theorem: *There are at least two distinct lines.*

Informal proof: Axiom 4 tells us that there are three distinct points a, b, and c not on the same line. By axiom 1 there is a line l on a and b, and again by axiom 1 there is a line m on a and c. By Axiom 4, c is not on line l. Therefore we have $l \neq m$.   QED.

Now we'll formalize the theorem and give a formal proof. A formalized version of the theorem can be written as follows:

$$\exists L \; \exists M \; \exists x \, (\neg L(x) \wedge M(x)).$$

Before we give a formal proof, we need to say something about inference rules for quantifiers of second-order logic, since there are now two kinds of quantified variables: lines and points. We'll use the same rules that we used for first-order logic but will apply them to second-order logic when the need arises. For example, from the statement $\exists L \; L(x)$, we will use EI to infer the existence of a particular line l such that $l(x)$. Basically, our rules are a reflection of our natural discourse. Now we're ready to give a formalized proof of the theorem:

Proof:

1.	$\exists x \; \exists y \; \exists z \, (x \neq y \wedge x \neq z \wedge y \neq z \wedge \forall L \, (L(x) \wedge L(y) \rightarrow \neg L(z)))$	Axiom 4
2.	$a \neq b \wedge a \neq c \wedge b \neq c \wedge \forall L \, (L(a) \wedge L(b) \rightarrow \neg L(c))$	1,EI, EI, EI
3.	$\forall x \; \forall y \, (x \neq y \rightarrow \exists L \, (L(x) \wedge L(y)))$	Axiom 1
4.	$a \neq b \rightarrow \exists L \, (L(a) \wedge L(b))$	3, UI, UI
5.	$a \neq b$	2, Simp
6.	$\exists L \, (L(a) \wedge L(b))$	4, 5, MP
7.	$l(a) \wedge l(b)$	6, EI
8.	$a \neq c \rightarrow \exists L \, (L(a) \wedge L(c))$	3, UI, UI
9.	$a \neq c$	2, Simp
10.	$\exists L \, (L(a) \wedge L(c))$	8, 9, MP
11.	$m(a) \wedge m(c)$	10, EI
12.	$\forall L \, (L(a) \wedge L(b) \rightarrow \neg L(c))$	2, Simp
13.	$l(a) \wedge l(b) \rightarrow \neg l(c)$	12, UI
14.	$\neg l(c)$	7, 13, MP
15.	$m(c)$	11, Simp
16.	$\neg l(c) \wedge m(c)$	14, 15, Conj
17.	$\exists x \, (\neg l(x) \wedge m(x))$	16, EG
18.	$\exists M \; \exists x \, (\neg l(x) \wedge M(x))$	17, EG
19.	$\exists L \; \exists M \; \exists x \, (\neg L(x) \wedge M(x)).$	18, EG

QED.   ♦

The exercises contain some more proof practice (informal and formal) for the geometry of this example.

Exercises

1. State the minimal order of logic needed to describe each of the following wffs.

   a. $\forall x \ (Q(x) \rightarrow P(Q))$.

   b. $\exists x \ \forall g \ \exists p \ (q(c, g(x)) \wedge p(g(x)))$.

   c. $A(B) \wedge B(C) \wedge C(D) \wedge D(E) \wedge E(F)$.

   d. $\exists P \ (A(B) \wedge B(C) \wedge C(D) \wedge P(A))$.

   e. $S(x) \wedge T(S, x) \rightarrow U(T, S, x)$.

   f. $\forall x \ (S(x) \wedge T(S, x) \rightarrow U(T, S, x))$.

   g. $\forall x \ \exists S \ (S(x) \wedge T(S, x) \rightarrow U(T, S, x))$.

   h. $\forall x \ \exists S \ \exists T \ (S(x) \wedge T(S, x) \rightarrow U(T, S, x))$.

   i. $\forall x \ \exists S \ \exists T \ \exists U \ (S(x) \wedge T(S, x) \rightarrow U(T, S, x))$.

2. Formalize each of the following sentences as a wff in second-order logic.

   a. There are sets A and B such that $A \cap B = \varnothing$.

   b. There is a set S with two subsets A and B such that $S = A \cup B$.

3. Formalize each of the following sentences as a wff in an appropriate higher-order logic. Also figure out the order of the logic that you use in each case.

   a. Every state has a city named Springfield.

   b. There is a nation with a state that has a county named Washington.

   c. A house has a room with a bookshelf containing a book by Thoreau.

   d. There is a continent with a nation containing a state with a county named Lincoln, which contains a city named Central City that has a street named Broadway.

   e. Some set has a partition consisting of two subsets.

4. Find a formalization of the following statement upon which mathematical induction is based: If S is a subset of $\mathbb{N}$ and $0 \in S$ and whenever $x \in S$ then $\text{succ}(x) \in S$, then $S = \mathbb{N}$.

5. Show that each of the following wffs is valid by giving an informal validity argument.

   a. $\forall S \ \exists x \ S(x) \rightarrow \exists x \ \forall S \ S(x)$.

   b. $\forall x \ \exists S \ S(x) \rightarrow \exists S \ \forall x \ S(x)$.

   c. $\exists S \ \forall x \ S(x) \rightarrow \forall x \ \exists S \ S(x)$.

   d. $\exists x \ \forall S \ S(x) \rightarrow \forall S \ \exists x \ S(x)$.

6. Use the facts about geometry given in Example 3 to give an informal proof for each of the following statements. You may use any of these statements to prove a subsequent statement.

   a. For each line there is a point not on the line.

   b. Two lines cannot intersect in more than one point.

   c. Through each point there exist at least two lines.

   d. Not all lines pass through the same point.

7. Formalize each of the following statements as a wff in second-order logic, using the variable names from Example 3. Then provide a formal proof for each wff.

   a. Not all points lie on the same line.

   b. Two lines cannot intersect in more than one point.

   c. Through each point there exist at least two lines.

   d. Not all lines pass through the same point.

Chapter Summary

A first-order theory is a formal treatment of some subject that uses first-order predicate calculus. We often need the idea of equality when applying logic in a formal manner to a particular subject. Equality can be added to first-order logic in such a way that the following familiar notion is included: Equals can replace—be substituted for—equals.

We can prove elementary statements about imperative programs within a first-order theory where each program is bounded by two conditions—a precondition and a postcondition. The theory uses only one axiom—the assignment axiom. Some useful inference rules are the consequence rule and the rules for composition, if-then, if-then-else, and while statements. The theory can be extended by adding axioms and inference rules for items that are normally found in imperative languages, such as arrays and other loop forms. The termination of while loops can also be proven in a formal manner.

When formalizing a subject, we often need higher forms of logic to express statements. Higher-order logic extends first-order logic by allowing objects other than variables—such as predicates and function names—to be quantified and to be arguments in predicates. We can classify the order of a logic if we make the association that a predicate is a set. Even though higher-order logics are not complete, we can still reason formally within these logics just as we do in propositional logic and first-order logic.

9

Computational Logic

Let us not dream that reason can ever be popular.
Passions, emotions, may be made popular, but
reason remains ever the property of the few.
—Johann Wolfgang von Goethe (1749–1832)

Can reasoning be automated? The answer is yes for some logics. In this chapter we'll discuss how to automate the reasoning process for first-order logic. We might start by automating the "natural deduction" proof techniques that we introduced in Chapters 6, 7, and 8. A problem with this approach is that there are many inference rules that can be applied in many different ways. In this chapter we'll look at a more mechanical way to perform deduction.

We'll see that there is a single inference rule, called resolution, that can be applied automatically by a computer. We'll also see how the resolution rule is adapted to the execution of logic programs.

Chapter Guide

Section 9.1 introduces the resolution inference rule. To understand the rule, we'll need to discuss clauses, clausal forms, substitution, and unification. We'll see how the rule can be applied in a mechanical fashion to prove theorems.

Section 9.2 introduces logic programming and shows how resolution is applied to perform the computation of a logic program. We'll also give some elementary techniques for constructing logic programs.

9.1 Automatic Reasoning

Let's look at the mechanical side of logic. We're going to introduce an inference rule that can be applied automatically. As fate would have it, the rule must be applied while trying to prove that a wff is unsatisfiable. This is not really a problem, because we know that a wff is valid if and only if its negation is unsatisfiable. In other words, if we want to prove that the wff W is valid, then we can do so by trying to prove that $\neg W$ is unsatisfiable. For example, if we want to prove the validity of the conditional $A \to B$, then we can try to prove the unsatisfiability of its negation $A \wedge \neg B$.

The new inference rule, which is called the *resolution rule*, can be applied over and over again in an attempt to show unsatisfiability. We can't present the resolution rule yet because it can be applied only to wffs that are written in a special form, called *clausal form*. So let's get to it.

Clauses and Clausal Forms

We need to introduce a little terminology before we can describe a clausal form. Recall that a *literal* is either an atom or the negation of an atom. For example, $p(x)$ and $\neg q(x, b)$ are literals. To distinguish whether a literal has a negation sign, we may use the terms *positive literal* and *negative literal*. $p(x)$ is a positive literal, and $\neg q(x, b)$ is a negative literal.

A *clause* is a disjunction of zero or more literals. For example, the following wffs are clauses:

$$p(x),$$
$$\neg q(x, b),$$
$$\neg p(a) \vee p(b),$$
$$p(x) \vee \neg q(a, y) \vee p(a).$$

The clause that is a disjunction of zero literals is called the *empty clause*, and it's denoted by the following special box symbol:

$$\square.$$

The empty clause is assigned the value false. We'll soon see why this makes sense when we discuss resolution.

A *clausal form* is the universal closure of a conjunction of clauses. In other words, a clausal form is a prenex conjunctive normal form, in which all quantifiers are universal and there are no free variables. For ease of notation we'll often represent a clausal form by the set consisting of its clauses. For example, if $S = \{C_1, ..., C_n\}$, where each C_i is a clause, and if $x_1, ..., x_m$ are the free variables in the clauses of S, then S denotes the following clausal form:

$$\forall x_1 \cdots \forall x_m (C_1 \wedge \cdots \wedge C_n).$$

For example, the following list shows five wffs in clausal form together with their corresponding sets of clauses:

Wffs in Clausal Form	*Sets of Clauses*
$\forall x \, p(x)$	$\{p(x)\}$
$\forall x \, \neg \, q(x, b)$	$\{\neg \, q(x, b)\}$
$\forall x \, \forall y \, (p(x) \wedge \neg \, q(y, b))$	$\{p(x), \neg \, q(y, b)\}$
$\forall x \, \forall y \, (p(y, f(x)) \wedge (q(y) \vee \neg \, q(a)))$	$\{p(y, f(x)), q(y) \vee \neg \, q(a)\}$
$(p(a) \vee p(b)) \wedge q(a, b)$	$\{p(a) \vee p(b), q(a, b)\}$

Notice that the last clausal form does not need quantifiers because it doesn't have any variables. In other words, it's a proposition. In fact, for propositions a clausal form is just a conjunctive normal form (CNF).

When we talk about an interpretation for a set S of clauses, we mean an interpretation for the clausal form that S denotes. Thus we can use the words "valid," "invalid," "satisfiable," and "unsatisfiable" to describe S because these words have meaning for the clausal form that S denotes.

It's easy to see that some wffs are not equivalent to any clausal form. For example, let's consider the following wff:

$$\forall x \, \exists y \, p(x, y).$$

This wff is not a clausal form, and it isn't equivalent to any clausal form because it has an existential quantifier. Since clausal forms are the things that resolution needs to work on, it's nice to know that we can associate a clausal form with each wff in such a way that the clausal form is unsatisfiable if and only if the wff is unsatisfiable. Let's see how to find such a clausal form for each wff.

To construct a clausal form for a wff, we can start by constructing a prenex conjunctive normal form for the wff. If there are no free variables and all the quantifiers are universal, then we have a clausal form. Otherwise, we need to get rid of the free variables and the existential quantifiers and still retain enough information to be able to detect whether the original wff is unsatisfiable. Luckily, there's a way to do this. The technique is due to the mathematician Thoralf Skolem (1887–1963), and it appears in his paper [1928].

Let's introduce Skolem's idea by considering the following example wff:

$$\forall x \, \exists y \, p(x, y).$$

In this case the quantifier $\exists y$ is inside the scope of the quantifier $\forall x$. So it may be that y depends on x. For example, if we let $p(x, y)$ mean "x has a successor y," then y certainly depends on x. If we're going to remove the quantifier $\exists y$ from $\forall x\, \exists y\, p(x, y)$, then we'd better leave some information about the fact that y may depend on x. Skolem's idea was to use a new function symbol, say f, and replace each occurrence of y within the scope of $\exists y$ by the term $f(x)$. After performing this operation, we obtain the following wff, which is now in clausal form:

$$\forall x\, p(x, f(x)).$$

We can describe the general method for eliminating existential quantifiers as follows:

Skolem's Rule   (9.1)

Let $\exists x\, W(x)$ be a wff or part of a larger wff. If $\exists x$ is not inside the scope of a universal quantifier, then pick a new constant c, and

$$\text{replace } \exists x\, W(x) \text{ by } W(c).$$

If $\exists x$ is inside the scope of universal quantifiers $\forall x_1, ..., \forall x_n$, then pick a new function symbol f, and

$$\text{replace } \exists x\, W(x) \text{ by } W(f(x_1, ..., x_n)).$$

The constants and functions introduced by the rule are called *Skolem functions*.

EXAMPLE 1. Let's apply Skolem's rule to the following wff:

$$\exists x\, \forall y\, \forall z\, \exists u\, \forall v\, \exists w\, p(x, y, z, u, v, w).$$

Since the wff contains three existential quantifiers, we'll use (9.1) to create three Skolem functions to replace the existentially quantified variables as follows:

replace x by b because $\exists x$ is not in the scope of a universal quantifier;
replace u by $f(y, z)$ because $\exists u$ is in the scope of $\forall y$ and $\forall z$;
replace w by $g(y, z, v)$ because $\exists w$ is in the scope of $\forall y$, $\forall z$, and $\forall v$.

Now we can apply (9.1) to eliminate the existential quantifiers by making the above replacements to obtain the following clausal form:

$$\forall y\, \forall z\, \forall v\, p(b, y, z, f(y, z), v, g(y, z, v)). \quad \blacklozenge$$

Now we have the ingredients necessary to construct clausal forms with the property that a wff and its clausal form are either both unsatisfiable or both satisfiable.

Skolem's Algorithm   (9.2)

Given a wff W, there exists a clausal form such that W and the clausal form are either both unsatisfiable or both satisfiable. The clausal form can be constructed from W as follows:

1. Construct the prenex conjunctive normal form of W.
2. Replace all occurrences of each free variable by a new constant.
3. Use Skolem's rule (9.1) to eliminate the existential quantifiers.

End Algorithm

Before we do some examples, let's make a couple of remarks about the steps of the algorithm. Step 2 could be replaced by the statement "Take the existential closure." But then Step 3 would remove these same quantifiers by replacing each of the newly quantified variables with a new constant name. So we saved time and did it all in one step. Step 2 can be done at any time during the process. We need Step 2 because we know that a wff and its existential closure are either both unsatisfiable or both satisfiable.

Step 3 can be applied during Step 1 after all implications have been eliminated and after all negations have been pushed to the right but before all quantifiers have been pushed to the left. Often this will reduce the number of variables in the Skolem function. Another way to simplify the Skolem function is to push all quantifiers to the right as far as possible before applying Skolem's rule. For example, suppose W is the following wff:

$$W = \forall x \, \neg \, p(x) \wedge \forall y \, \exists z \, q(y, z).$$

First, we'll apply (9.2) as stated. In other words, we calculate the prenex form of W by moving the quantifiers to the left to obtain

$$\forall x \, \forall y \, \exists z \, (\neg \, p(x) \wedge q(y, z)).$$

Next, we apply Skolem's rule (9.1), which says that we replace z by $f(x, y)$ to obtain the following clausal form for W:

$$\forall x \, \forall y \, (\neg p(x) \wedge q(y, f(x, y))).$$

Now let's start over with W and apply (9.1) during Step 1 before we move the quantifiers to the left. In this case the quantifier $\exists z$ is only within the scope of $\forall y$, so we replace z by $f(y)$ to obtain

$$\forall x \, \neg p(x) \wedge \forall y \, q(y, f(y)).$$

Now finish constructing the prenex form by moving the universal quantifiers to the left to obtain the following clausal form for W:

$$\forall x \, \forall y \, (\neg p(x) \wedge q(y, f(y))).$$

So we get a simpler clausal form for W in this case.

Let's look at a few examples that construct clausal forms with Skolem's algorithm (9.2).

EXAMPLE 2. Suppose we have a wff with no variables (i.e., a propositional wff). For example, let W be the wff

$$(p(a) \rightarrow q) \wedge ((q \wedge s(b)) \rightarrow r).$$

To find the clausal form for W, we need only apply equivalences from propositional calculus to find a CNF as follows:

$$
\begin{aligned}
(p(a) \rightarrow q) \wedge ((q \wedge s(b)) \rightarrow r) \;&\equiv\; (\neg p(a) \vee q) \wedge (\neg (q \wedge s(b)) \vee r) \\
&\equiv\; (\neg p(a) \vee q) \wedge ((\neg q \vee \neg s(b)) \vee r) \\
&\equiv\; (\neg p(a) \vee q) \wedge (\neg q \vee \neg s(b) \vee r). \quad \blacklozenge
\end{aligned}
$$

EXAMPLE 3. We'll use (9.2) to find a clausal form for the following wff:

$$\exists y \, \forall x \, (p(x) \rightarrow q(x, y)) \wedge \forall x \, \exists y \, ((q(x, x) \wedge s(y)) \rightarrow r(x)).$$

The first step is to find the prenex conjunctive normal form. Since there are two quantifiers with the same name, we'll do some renaming to obtain the following wff:

$$\exists y \, \forall x \, (p(x) \rightarrow q(x, y)) \wedge \forall w \, \exists z \, ((q(w, w) \wedge s(z)) \rightarrow r(w)).$$

Next, we eliminate the conditionals to obtain the following wff:

$$\exists y \, \forall x \, (\neg p(x) \vee q(x, y)) \wedge \forall w \, \exists z \, (\neg (q(w, w) \wedge s(z)) \vee r(w)).$$

Now, push negation to the right to obtain the following wff:

$$\exists y \; \forall x \; (\neg \, p(x) \lor q(x, y)) \land \forall w \; \exists z \; (\neg \, q(w, w) \lor \neg \, s(z) \lor r(w)).$$

Next, we'll apply Skolem's rule (9.1) to eliminate the existential quantifiers and obtain the following wff:

$$\forall x \; (\neg \, p(x) \lor q(x, a)) \land \forall w \; (\neg \, q(w, w) \lor \neg \, s(f(w)) \lor r(w)).$$

Lastly, we push the universal quantifiers to the left, obtaining the desired clausal form:

$$\forall x \; \forall w \; ((\neg \, p(x) \lor q(x, a)) \land (\neg \, q(w, w) \lor \neg \, s(f(w)) \lor r(w))). \quad \blacklozenge$$

EXAMPLE 4. We'll construct a clausal form for the following wff:

$$\forall x \; (p(x) \to \exists y \; \forall z \; ((p(w) \lor q(x, y)) \to \forall w \; r(x, w))).$$

The free variable w is also used in the quantifier $\forall w$, and the quantifier $\forall z$ is redundant. So we'll do some renaming, and we'll remove $\forall z$ to obtain the following wff:

$$\forall x \; (p(x) \to \exists y \; ((p(w) \lor q(x, y)) \to \forall z \; r(x, z))).$$

We remove the conditionals in the usual way to obtain the following wff:

$$\forall x \; (\neg \, p(x) \lor \exists y \; (\neg \, (p(w) \lor q(x, y)) \lor \forall z \; r(x, z))).$$

Next, we move negation inward to obtain the following wff:

$$\forall x \; (\neg \, p(x) \lor \exists y \; ((\neg \, p(w) \land \neg \, q(x, y)) \lor \forall z \; r(x, z))).$$

Now we can apply Skolem's rule (9.1) to eliminate $\exists y$ and replace the free variable w by b to get the following wff:

$$\forall x \; (\neg \, p(x) \lor ((\neg \, p(b) \land \neg \, q(x, f(x))) \lor \forall z \; r(x, z))).$$

Next, we push the universal quantifier $\forall z$ to the left, obtaining the following wff:

$$\forall x \; \forall z \; (\neg \, p(x) \lor ((\neg \, p(b) \land \neg \, q(x, f(x))) \lor r(x, z))).$$

Lastly, we distribute ∨ over ∧ to obtain the following clausal form:

$$\forall x \ \forall z \ ((\neg p(x) \vee \neg p(b) \vee r(x, z)) \wedge (\neg p(x) \vee \neg q(x, f(x)) \vee r(x, z))). \quad \blacklozenge$$

So we can transform any wff into a wff in clausal form in which the two wffs are either both unsatisfiable or both satisfiable. Since the resolution rule tests clausal forms for unsatisfiability, we're a step closer to describing the idea of resolution. Before we introduce the general idea of resolution, we're going to pause and discuss resolution for the simple case of propositions.

A Primer of Resolution for Propositions

It's easy to see how resolution works for propositional clauses (i.e., clauses with no variables). The resolution inference rule works something like a cancellation process. It takes two clauses and constructs a new clause from them by deleting all occurrences of a positive literal p from one clause and all occurrences of $\neg p$ from the other clause. For example, suppose we are given the following two propositional clauses:

$$p \vee q,$$
$$\neg p \vee r \vee \neg p.$$

We obtain a new clause by first eliminating p from the first clause and eliminating the two occurrences of $\neg p$ from the second clause. Then we take the disjunction of the leftover clauses to form the new clause:

$$q \vee r.$$

Let's write down the resolution rule in a more general way. Suppose we have two propositional clauses of the following forms:

$$p \vee A,$$
$$\neg p \vee B.$$

Let $A - p$ denote the disjunction obtained from A by deleting all occurrences of p. Similarly, let $B - \neg p$ denote the disjunction obtained from B by deleting all occurrences of $\neg p$. The resolution rule allows us to infer the propositional clause

$$A - p \vee B - \neg p.$$

We'll write the rule as follows:

Resolution Rule for Propositions: (9.3)

$$\frac{p \vee A, \; \neg p \vee B}{\therefore \;\; A - p \vee B - \neg p}.$$

Although the rule may look strange, it's a good rule. That is, it maps tautologies to a tautology. To see this, we can suppose that $(p \vee A) \wedge (\neg p \vee B) =$ true. If p is true, then the equation reduces to $B =$ true. Since $\neg p$ is false, we can remove all occurrences of $\neg p$ from B and still have $B - \neg p =$ true. Therefore $A - p \vee B - \neg p =$ true. We obtain the same result if p is false. So the inference rule does its job.

A proof by resolution is a refutation that uses only the resolution rule. So we can define a *resolution proof* as a sequence of clauses, ending with the empty clause, in which each clause in the sequence either is a premise or is inferred by the resolution rule from two preceding clauses in the sequence. Notice that the empty clause is obtained from (9.3) when A either is empty or contains only copies of p and when B either is empty or contains only copies of $\neg p$. For example, the simplest version of (9.3) can be stated as follows:

$$\frac{p, \; \neg p}{\therefore \; \Box}.$$

In other words, we obtain the well known tautology $p \wedge \neg p \rightarrow$ false.

For example, let's prove that the following clausal form is unsatisfiable:

$$(\neg p \vee q) \wedge (p \vee q) \wedge (\neg q \vee p) \wedge (\neg p \vee \neg q).$$

In other words, we'll prove that the following set of clauses is unsatisfiable:

$$\{\neg p \vee q, p \vee q, \neg q \vee p, \neg p \vee \neg q\}.$$

The following resolution proof does the job:

Proof:
1. $\neg p \vee q$ P
2. $p \vee q$ P
3. $\neg q \vee p$ P
4. $\neg p \vee \neg q$ P
5. $q \vee q$ 1, 2, Resolution
6. p 3, 5, Resolution
7. $\neg p$ 4, 5, Resolution
8. $\Box$ 6, 7, Resolution

QED.

Now let's get back on our original track, which is to describe the resolution rule for clauses of the first-order predicate calculus.

Substitution and Unification

When we discuss the resolution inference rule for clauses that contain variables, we'll see that a certain kind of matching is required. For example, suppose we are given the following two clauses:

$$p(x, y) \vee q(y),$$
$$r(z) \vee \neg q(b).$$

The matching that we will discuss allows us to replace all occurrences of the variable y by the constant b, thus obtaining the following two clauses:

$$p(x, b) \vee q(b),$$
$$r(z) \vee \neg q(b).$$

Notice that one clause contains $q(b)$ and the other contains its negation $\neg q(b)$. Resolution will allow us to cancel them and construct the disjunction of the remaining parts, which is the clause $p(x, b) \vee r(z)$.

We need to spend a little time to discuss the process of replacing variables by terms. If x is a variable and t is a term, then the expression x/t is called a *binding* of x to t and can be read as "x gets t" or "x is bound to t" or "x has value t" or "x is replaced by t." For example, three typical bindings are written as follows:

$$x/a, \quad y/z, \quad w/f(b, v).$$

A *substitution* is a finite set of bindings $\{x_1/t_1, \ldots, x_n/t_n\}$, where variables $x_1, \ldots, x_n$ are all distinct and $x_i \neq t_i$ for each i. We use lowercase Greek letters to denote substitutions. The *empty substitution*, which is just the empty set, is denoted by the Greek letter ε.

What do we do with substitutions? We apply them to expressions, an *expression* being a finite string of symbols. Let E be an expression, and let θ be the following substitution:

$$\theta = \{x_1/t_1, \ldots, x_n/t_n\}.$$

Then the *instance* of E by θ, denoted $E\theta$, is the expression obtained from E by simultaneously replacing all occurrences of the variables $x_1, \ldots, x_n$ in E by the terms $t_1, \ldots, t_n$, respectively. We say that $E\theta$ is obtained from E by *applying* the substitution θ to the expression E. For example, if $E = p(x, y, f(x))$ and $\theta = \{x/a, y/f(b)\}$, then $E\theta$ has the following form:

$$E\theta = p(x, y, f(x))\{x/a, y/f(b)\} = p(a, f(b), f(a)).$$

If S is a set of expressions, then the *instance* of S by θ, denoted $S\theta$, is the set of all instances of expressions in S by θ. For example, if $S = \{p(x, y), q(a, y)\}$ and $\theta = \{x/a, y/f(b)\}$, then $S\theta$ has the following form:

$$S\theta = \{p(x, y), q(a, y)\}\{x/a, y/f(b)\} = \{p(a, f(b)), q(a, f(b)\}.$$

Now let's see how we can combine two substitutions θ and σ into a single substitution that has the same effect as applying θ and then applying σ to any expression. The *composition* of θ and σ, denoted by $\theta\sigma$, is a substitution that satisfies the following property: $E(\theta\sigma) = (E\theta)\sigma$ for any expression E. Although we have described the composition in terms of how it acts on all expressions, we can compute $\theta\sigma$ without any reference to an expression as follows:

Composition of Substitutions (9.4)

Given the two substitutions

$$\theta = \{x_1/t_1, ..., x_n/t_n\} \quad \text{and} \quad \sigma = \{y_1/s_1, ..., y_m/s_m\}.$$

The composition $\theta\sigma$ is constructed as follows:

1. Apply σ to the denominators of θ to form $\{x_1/t_1\sigma, ..., x_n/t_n\sigma\}$.
2. Delete all bindings x_i/x_i from line 1.
3. Delete from σ any binding y_i/s_i, where y_i is a variable in $\{x_1, ..., x_n\}$.
4. $\theta\sigma$ is the union of the two sets constructed on lines 2 and 3.

The process looks complicated, but it's really quite simple. It's just a formalization of the following construction: For each distinct variable v occurring in the numerators of θ and σ, apply θ and then σ to v, obtaining the expression $(v\theta)\sigma$. The composition $\theta\sigma$ consists of all bindings $v/(v\theta)\sigma$ such that $v \neq (v\theta)\sigma$.

It's also nice to know that we can always check whether we constructed a composition correctly. Just make up an example atom containing the distinct variables in the numerators of θ and σ, say $p(v_1, ..., v_k)$, and then check to make sure the following equation holds:

$$((p(v_1, ..., v_k)\theta)\sigma) = p(v_1, ..., v_k)(\theta\sigma).$$

EXAMPLE 5. Let $\theta = \{x/f(y), y/z\}$ and $\sigma = \{x/a, y/b, z/y\}$. To find the composition $\theta\sigma$, we first apply σ to the denominators of θ to form the following set:

$$\{x/f(y)\sigma, y/z\sigma\} = \{x/f(b), y/y\}.$$

Now remove the binding y/y to obtain $\{x/f(b)\}$. Next, delete the bindings x/a and y/b from σ to obtain $\{z/y\}$. Finally, compute $\theta\sigma$ as the union of these two sets $\theta\sigma = \{x/f(b), z/y\}$.

Let's check to see whether the answer is correct. For our example atom we'll pick

$$p(x, y, z)$$

because x, y, and z are the distinct variables occurring in the numerators of θ and σ. We'll make the following two calculations to see whether we get the same answer:

$$((p(x, y, z)\theta)\sigma) = p(f(y), z, z)\sigma = p(f(b), y, y),$$

$$p(x, y, z)(\theta\sigma) = p(f(b), y, y). \quad \blacklozenge$$

Three simple, but useful, properties of composition are listed next. The proofs are left as exercises.

Properties of Composition (9.5)

For any substitutions θ and σ and any expression E the following statements hold:

1. $E(\theta\sigma) = (E\theta)\sigma$.

2. $E\varepsilon = E$.

3. $\theta\varepsilon = \varepsilon\theta = \theta$.

A substitution θ is called a *unifier* of a finite set S of literals if $S\theta$ is a singleton set. For example, if we let $S = \{p(x, b), p(a, y)\}$, then the substitution $\theta = \{x/a, y/b\}$ is a unifier of S because

$$S\theta = \{p(a, b)\}, \text{ which is a singleton set.}$$

Some sets of literals don't have a unifier, while other sets have infinitely many unifiers. The range of possibilities can be shown by the following four simple examples:

$\{p(x), q(y)\}$ doesn't have a unifier.

$\{p(x), \neg p(x)\}$ doesn't have a unifier.

$\{p(x), p(a)\}$ has exactly one unifier $\{x/a\}$.

$\{p(x), p(y)\}$ has infinitely many unifiers: $\{x/y\}$, $\{y/x\}$, and
                                                   $\{x/t, y/t\}$ for any term t.

Among the unifiers of a set there is always at least one unifier that can be used to construct every other unifier. To be specific, a unifier θ for S is called a *most general unifier* (mgu) for S if for every unifier α of S there exists a substitution σ such that $\alpha = \theta\sigma$. In other words, an mgu for S is a factor of every other unifier of S. Let's look at an example.

EXAMPLE 6. The set $S = \{p(x), p(y)\}$ has infinitely many unifiers:

$$\{x/y\}, \{y/x\}, \text{ and } \{x/t, y/t\} \text{ for any term } t.$$

The unifier $\{x/y\}$ is an mgu for S because we can write the other unifiers in terms of $\{x/y\}$ as follows: $\{y/x\} = \{x/y\}\{y/x\}$, and $\{x/t, y/t\} = \{x/y\}\{y/t\}$ for any term t. Similarly, $\{y/x\}$ is an mgu for S.   ♦

We want to find a way to construct an mgu for any set of literals. Before we do this, we need a little terminology. The *disagreement set* of S is a set of terms constructed from the literals of S in the following way:

Find the longest common substring that starts at the left end of each literal of S. The disagreement set of S is the set of all the terms that occur in the literals of S that are immediately to the right of the longest common substring.

For example, let's construct the disagreement set for the following set of literals:

$$S = \{p(x, f(x), y), p(x, y, z), p(x, f(a), b)\}.$$

The longest common substring for the literals in S is the string

$$\text{``}p(x,\text{''}$$

of length four. The terms in the literals of S that occur immediately to the right of this string are $f(x)$, y, and $f(a)$. Thus the disagreement set of S is

$$\{f(x), y, f(a)\}.$$

Now we have the tools to describe a very important algorithm of Robinson [1965]. The algorithm computes, for a set of atoms, a most general unifier, if one exists.

The Unification Algorithm   (9.6)

Input: A finite set S of atoms.

Output: Either a most general unifier for S or a statement that S is not unifiable.

1. Set $k = 0$ and $\theta_0 = \varepsilon$, and go to Step 2.

2. Calculate $S\theta_k$. If it's a singleton set, then stop (θ_k is the mgu for S). Otherwise, let D_k be the disagreement set of $S\theta_k$, and go to Step 3.

3. If D_k contains a variable v and a term t, such that v does not occur in t, then calculate the composition $\theta_{k+1} = \theta_k\{v/t\}$, set $k := k + 1$, and go to Step 2. Otherwise, stop (S is not unifiable).

End of Algorithm

The composition $\theta_k\{v/t\}$ in Step 3 is easy to compute for two reasons. The variable v doesn't occur in t, and v will never occur in the numerator of θ_k. Therefore the middle two steps of the composition construction (9.4) don't change anything. In other words, the composition $\theta_k\{v/t\}$ is constructed by applying $\{v/t\}$ to each denominator of θ_k and then adding the binding v/t to the result.

EXAMPLE 7. Let's try the algorithm on the set $S = \{p(x, f(y)), p(g(y), z)\}$. We'll list each step of the algorithm as we go:

1. Set $\theta_0 = \varepsilon$.

2. $S\theta_0 = S\varepsilon = S$ is not a singleton. $D_0 = \{x, g(y)\}$.

3. Variable x doesn't occur in term $g(y)$ of D_0.
    Put $\theta_1 = \theta_0 \{x/g(y)\} = \{x/g(y)\}$.

2. $S\theta_1 = \{p(g(y), f(y)), p(g(y), z)\}$ is not a singleton. $D_1 = \{f(y), z\}$.

3. Variable z does not occur in term $f(y)$ of D_1.
   Put $\theta_2 = \theta_1 \{z/f(y)\} = \{x/g(y), z/f(y)\}$.

2. $S\theta_2 = \{p(g(y), f(y))\}$ is a singleton. Therefore the algorithm terminates with the mgu $\{x/g(y), z/f(y)\}$ for the set S. ◆

EXAMPLE 8. Let's trace the algorithm on the set $S = \{p(x), p(g(x))\}$. We'll list each step of the algorithm as we go:

1. Set $\theta_0 = \varepsilon$.

2. $S\theta_0 = S\varepsilon = S$, which is not a singleton. $D_0 = \{x, g(x)\}$.

3. The only choices for a variable and a term in D_0 are x and $g(x)$. But the variable x occurs in $g(x)$. So the algorithm stops, and S is not unifiable.

This makes sense too. For example, if we were to apply the substitution $\{x/g(x)\}$ to S, we would obtain the set $\{p(g(x)), p(g(g(x)))\}$, which in turn gives us the same disagreement set $\{x, g(x)\}$. So the process would go on forever. Notice that a change of variables makes a big difference. For example, if we change the second atom in S to $p(g(y))$, then the algorithm unifies the set $\{p(x), p(g(y))\}$, obtaining the mgu $\{x/g(y)\}$. ◆

Resolution: The General Case

Now we've got the tools to discuss resolution of clauses that contain variables. Let's look at a simple example to help us see how unification comes into play. Suppose we're given the following two clauses:

$$p(x, a) \vee \neg\, q(x),$$

$$\neg\, p(b, y) \vee \neg\, q(a).$$

We want to cancel $p(x, a)$ from the first clause and $\neg\, p(b, y)$ from the second clause. But they won't cancel until we unify the two atoms $p(x, a)$ and $p(b, y)$. An mgu for these two atoms is $\{x/b, y/a\}$. If we apply this unifier to the original two clauses, we obtain the following two clauses:

$$p(b, a) \vee \neg\, q(b),$$

$$\neg\, p(b, a) \vee \neg\, q(a).$$

Now we can cancel $p(b, a)$ from the first clause and $\neg\, p(b, a)$ from the second clause and take the disjunction of what's left to obtain the following clause:

$$\neg\, q(b) \vee \neg\, q(a).$$

That's the way the resolution inference rule works when variables are present. Now let's give a detailed description of the rule.

The Resolution Inference Rule

The resolution inference rule takes two clauses and constructs a new clause. But *the rule can be applied only to clauses that possess the following two properties:*

1. The two clauses have no variables in common.

2. There are one or more atoms, $L_1, \ldots, L_k$, in one of the clauses and one or more literals, $\neg\, M_1, \ldots, \neg\, M_n$, in the other clause such that $\{L_1, \ldots, L_k, M_1, \ldots, M_n\}$ is unifiable.

The first property can always be satisfied by renaming variables. For example, the variable x is used in both of the following clauses:

$$q(b, x) \vee p(x), \quad \neg\, q(x, a) \vee p(y).$$

We can replace x in the second clause with a new variable z to obtain the following two clauses that satisfy the first property:

$$q(b, x) \vee p(x), \quad \neg\, q(z, a) \vee p(y).$$

Suppose we have two clauses that satisfy properties 1 and 2. Then they can be written in the following form, where C and D represent the other parts of each clause:

$$L_1 \vee \ldots \vee L_k \vee C \quad \text{and} \quad \neg\, M_1 \vee \ldots \vee \neg\, M_n \vee D.$$

Since the clauses satisfy the second property, we know that there is an mgu θ that unifies the set of atoms $\{L_1, \ldots, L_k, M_1, \ldots, M_n\}$. In other words, there is a unique atom N such that $N = L_i\theta = M_j\theta$ for any i and j. To be specific, we'll set

$$N = L_1\theta.$$

Now we're ready to do our cancelling. Let $C\theta - N$ denote the clause obtained from $C\theta$ by deleting all occurrences of the atom N. Similarly, let $D\theta - \neg N$ denote the clause obtained from $D\theta$ by deleting all occurrences of the atom $\neg N$. The clause that we construct is the disjunction of any literals that are left after the cancellation:

$$(C\theta - N) \vee (D\theta - \neg N).$$

Summing all this up, we can state the resolution inference rule as follows:

Resolution Rule (R)   (9.7)

$$\frac{L_1 \vee \ldots \vee L_k \vee C}{\neg M_1 \vee \ldots \vee \neg M_n \vee D}$$
$$\therefore (C\theta - N) \vee (D\theta - \neg N).$$

The clause constructed in the denominator of (9.7) is called a *resolvant* of the two clauses in the numerator. Let's describe how to use (9.7) to find a resolvant of the two clauses.

1. Check the two clauses for distinct variables (rename if necessary).
2. Find an mgu θ for the set of atoms $\{L_1, \ldots, L_k, M_1, \ldots, M_n\}$.
3. Apply θ to both clauses C and D.
4. Set $N = L_1\theta$.
5. Remove all occurrences of N from $C\theta$.
6. Remove all occurrences of $\neg N$ from $D\theta$.
7. Form the disjunction of the clauses in Steps 5 and 6. This is the resolvant.

Let's do some examples to get the look and feel of resolution before we forget everything.

EXAMPLE 9. We'll try to find a resolvant of the following two clauses:

$$q(b, x) \vee p(x) \vee q(b, a),$$
$$\neg q(y, a) \vee p(y).$$

We'll cancel the atom $q(b, x)$ in the first clause with the literal $\neg q(y, a)$ in the second clause. So we'll write the first clause in the form $L \vee C$, where L and C have the following values:

$$L = q(b, x) \quad \text{and} \quad C = p(x) \vee q(b, a).$$

The second clause can be written in the form $\neg M \vee D$, where M and D have the following values:

$$M = q(y, a) \quad \text{and} \quad D = p(y).$$

Now L and M, namely $q(b, x)$ and $q(y, a)$, can be unified by the mgu $\theta = \{y/b, x/a\}$. We can apply θ to either atom to obtain the common value $N = L\theta = M\theta = q(b, a)$. Now we can apply (9.7) to find the resolvant of the two clauses. First, compute the clauses $C\theta$ and $D\theta$:

$$C\theta = (p(x) \vee q(b, a))\{y/b, x/a\} = p(a) \vee q(b, a),$$

$$D\theta = p(y)\ \{y/b, x/a\} = p(b).$$

Next we'll remove all occurrences of $N = q(b, a)$ from $C\theta$ and remove all occurrences of $\neg N = \neg q(b, a)$ from $D\theta$:

$$C\theta - N = p(a) \vee q(b, a) - q(b, a) = p(a),$$

$$D\theta - \neg N = p(b) - \neg q(b, a) = p(b).$$

Lastly, we'll take the disjunction of the remaining clauses to obtain the desired resolvant: $p(a) \vee p(b)$. ◆

EXAMPLE 10. In this example we'll consider cancelling two literals from one of the clauses. Suppose we have the following two clauses:

$$p(f(x)) \vee p(y) \vee \neg q(x),$$

$$\neg p(z) \vee q(w).$$

We'll pick the disjunction $p(f(x)) \vee p(y)$ from the first clause to cancel with the literal $\neg p(z)$ in the second clause. So we need to unify the set of atoms $\{p(f(x)), p(y), p(z)\}$. An mgu for this set is $\theta = \{y/f(x), z/f(x)\}$. The common value N obtained by applying θ to any of the atoms in the set is $N = p(f(x))$. To see how the cancellation takes place, we'll apply θ to both of the original clauses to obtain the clauses

$$p(f(x)) \lor p(f(x)) \lor \neg q(x),$$

$$\neg p(f(x)) \lor q(w).$$

We'll cancel $p(f(x)) \lor p(f(x))$ from the first clause and $\neg p(f(x))$ from the second clause, with no other deletions possible. Thus the resolvent of the original two clauses is the disjunction of the remaining parts of the preceding two clauses: $\neg q(x) \lor q(w)$. ◆

What's so great about finding resolvents? Two things are great. One great thing is that the process is mechanical—it can be programmed. The other great thing is that the process preserves unsatisfiability. In other words, we have the following result:

Let G be a resolvent of the clauses E and F. Then $\{E, F\}$ is   (9.8)
unsatisfiable if and only if $\{E, F, G\}$ is unsatisfiable.

Now we're almost in position to describe how to prove that a set of clauses is unsatisfiable. Let S be a set of clauses where—after possibly renaming some variables—distinct clauses of S have disjoint sets of variables. We define the *resolution* of S, denoted by $R(S)$, to be the set

$$R(S) = S \cup \{G \mid G \text{ is a resolvent of a pair of clauses in } S\}.$$

We can conclude from (9.8) that S is unsatisfiable if and only if $R(S)$ is unsatisfiable. Similarly, $R(S)$ is unsatisfiable if and only if $R(R(S))$ is unsatisfiable. We can continue on in this way. To simplify the notation, we'll define $R^0(S) = S$ and $R^{n+1}(S) = R(R^n(S))$ for $n > 0$. So for any n we can say that

$$S \text{ is unsatisfiable if and only if } R^n(S) \text{ is unsatisfiable.}$$

Let's look at some examples to demonstrate the calculation of the sequence of sets $S, R(S), R^2(S), \dots$.

EXAMPLE 11. Suppose we start with the following set of clauses:

$$S = \{p(x), \neg p(a)\}.$$

To compute $R(S)$, we must add to S all possible resolvents of pairs of clauses. There is only one pair of clauses in S, and the resolvent of $p(x)$ and $\neg p(a)$, is the empty clause. Thus $R(S)$ is the following set:

$$R(S) = \{p(x), \neg\, p(a), \square\}.$$

Now let's compute $R(R(S))$. The only two clauses in $R(S)$ that can be resolved are $p(x)$ and $\neg\, p(a)$. Since their resolvant is already in $R(S)$, there's nothing new to add. So the process stops, and we have $R(R(S)) = R(S)$.   ◆

EXAMPLE 12. Consider the set of three clauses

$$S = \{p(x), q(y) \vee \neg\, p(y), \neg\, q(a)\}.$$

Let's compute $R(S)$. There are two pairs of clauses in S that have resolvants. The two clauses $p(x)$ and $q(y) \vee \neg\, p(y)$ resolve to $q(y)$. The clauses $q(y) \vee \neg\, p(y)$ and $\neg\, q(a)$ resolve to $\neg\, p(a)$. Thus $R(S)$ is the following set:

$$R(S) = \{p(x), q(y) \vee \neg\, p(y), \neg\, q(a), q(y), \neg\, p(a)\}.$$

Now let's compute $R(R(S))$. The two clauses $p(x)$ and $\neg\, p(a)$ resolve to the empty clause, and nothing new is added by resolving any other pairs from $R(S)$. Thus $R(R(S))$ is the following set:

$$R(R(S)) = \{p(x), q(y) \vee \neg\, p(y), \neg\, q(a), q(y), \neg\, p(a), \square\}.$$

It's easy to see that we can't get anything new by resolving pairs of clauses in $R(R(S))$. Thus we have $R^3(S) = R^2(S)$.   ◆

These two examples have something very important in common. In each case the set S is unsatisfiable, and the empty clause occurs in $R^n(S)$ for some n. This is no coincidence. The following result of Robinson [1965] allows us to test for the unsatisfiability of a set of clauses by looking for the empty clause in the sequence S, $R(S)$, $R^2(S)$, ...:

Resolution Theorem   (9.9)

A finite set S of clauses is unsatisfiable if and only if $\square \in R^n(S)$
for some $n \geq 0$.

The theorem provides us with an algorithm to prove that a wff is unsatisfiable. Let S be the set of clauses that make up the clausal form of the wff.

Start by calculating all the resolvents of pairs of clauses from S. The new resolvents are added to S to form the larger set of clauses $R(S)$. If the empty clause has been calculated, then we are done. Otherwise, calculate resolvents of pairs of clauses in the set $R(S)$. Continue the process until we find a pair of clauses whose resolvent is the empty clause. If we get to a point at which no new clauses are being created and we have not found the empty clause, then the process stops, and we conclude that the wff that we started with is satisfiable.

Theorem Proving with Resolution

Recall that a resolution proof is a sequence of clauses that ends with the empty clause, in which each clause either is a premise or can be inferred from two preceding clauses by the resolution rule. Recall also that a resolution proof is a proof of unsatisfiability. Since we normally want to prove that some wff is valid, we must first take the negation of the wff, then find a clausal form, and then attempt to do a resolution proof. We'll summarize the steps as follows:

> *Steps to Prove That* W *Is Valid*
>
> 1. Form the negation $\neg\, W$. For example, if W is a conditional of the form $A \wedge B \wedge C \to D$, then $\neg\, W$ has the form $A \wedge B \wedge C \wedge \neg\, D$.
>
> 2. Use Skolem's algorithm (9.2) to convert line 1 into clausal form.
>
> 3. Take the clauses from line 2 as premises in the proof.
>
> 4. Apply the resolution rule (9.7) to derive the empty clause.

Let's look at a few examples to see how the process works.

EXAMPLE 13 (*The Family Tree Problem*). Suppose we let p stand for the isParentOf relation and let g stand for the isGrandParentOf relation. We can define g in terms of p as follows:

$$p(x, z) \wedge p(z, y) \to g(x, y).$$

In other words, if x is a parent of z and z is a parent of y, then we conclude that x is a grandparent of y. Suppose we have the following facts about parents, where we use the letters $a, b, c, d,$ and e to denote names:

$$p(a, b) \wedge p(c, b) \wedge p(b, d) \wedge p(a, e).$$

Suppose someone claims that $g(a, d)$ is implied by the given facts. In other words, the claim is that the following wff is valid:

$$p(a, b) \wedge p(c, b) \wedge p(b, d) \wedge p(a, e) \wedge (p(x, z) \wedge p(z, y) \rightarrow g(x, y)) \rightarrow g(a, d).$$

If we're going to use resolution, the first thing we must do is negate the wff to obtain the following wff:

$$p(a, b) \wedge p(c, b) \wedge p(b, d) \wedge p(a, e) \wedge (p(x, z) \wedge p(z, y) \rightarrow g(x, y)) \wedge \neg g(a, d).$$

This wff will be in clausal form if we replace $p(x, z) \wedge p(z, y) \rightarrow g(x, y)$ by the following equivalent clause:

$$\neg p(x, z) \vee \neg p(z, y) \vee g(x, y).$$

Now we'll begin the proof by making each clause of the clausal form a premise, as follows:

Proof:
1. $p(a, b)$ P
2. $p(c, b)$ P
3. $p(b, d)$ P
4. $p(a, e)$ P
5. $\neg p(x, z) \vee \neg p(z, y) \vee g(x, y)$ P
6. $\neg g(a, d)$ P   Negation of conclusion

Now we can construct resolvents with the goal of obtaining the empty clause. In the following proof steps we've listed the mgu used for each application of resolution:

7. $\neg p(a, z) \vee \neg p(z, d)$ 5, 6, R, $\{x/a, y/d\}$
8. $\neg p(b, d)$ 1, 7, R, $\{z/b\}$
9. $\square$ 3, 8, R, $\{\,\}$
       QED.

So we have a refutation. Therefore we can conclude that $g(a, d)$ is implied from the given facts.   ♦

EXAMPLE 14 (*Diagonals of a Trapezoid*). We'll give a resolution proof that the alternate interior angles formed by a diagonal of a trapezoid are equal. This problem is from Chang and Lee [1973]. Let $t(x, y, u, v)$ mean that $x, y, u,$ and v are the four corner points of a trapezoid. Let $p(x, y, u, v)$ mean that edges xy

and uv are parallel lines. Let $e(x, y, z, u, v, w)$ mean that angle xyz is equal to angle uvw. We'll assume the following two axioms about trapezoids:

Axiom 1: $t(x, y, u, v) \rightarrow p(x, y, u, v)$.
Axiom 2: $p(x, y, u, v) \rightarrow e(x, y, v, u, v, y)$.

To prove: $t(a, b, c, d) \rightarrow e(a, b, d, c, d, b)$.

To prepare for a resolution proof, we need to write each axiom in its clausal form. This gives us the following two clauses:

Axiom 1: $\neg\, t(x, y, u, v) \vee p(x, y, u, v)$.

Axiom 2: $\neg\, p(x, y, u, v) \vee e(x, y, v, u, v, y)$.

Next, we need to negate the statement to be proved, which gives us the following two clauses:

$$t(a, b, c, d),$$

$$\neg\, e(a, b, d, c, d, b).$$

The resolution proof process begins by listing as premises the two axioms written as clauses together with the preceding two clauses that represent the negation of the conclusion. A proof follows:

Proof.
1. $\neg\, t(x, y, u, v) \vee p(x, y, u, v)$ P
2. $\neg\, p(x, y, u, v) \vee e(x, y, v, u, v, y)$ P
3. $t(a, b, c, d)$ P Antecedent
4. $\neg\, e(a, b, d, c, d, b)$ P Negation of consequent
5. $\neg\, p(a, b, c, d)$ 2, 4, R, $\{x/a, y/b, v/d, u/c\}$
6. $\neg\, t(a, b, c, d)$ 1, 5, R, $\{x/a, y/b, u/c, v/d\}$
7. $\square$ 3, 6, R, $\{\,\}$
     QED.   ◆

Remarks

In the example proofs we didn't follow a specific strategy to help us choose which clauses to resolve. Strategies are important because they may help reduce the searching required to find a proof. Although a general discussion of strategy is beyond our scope, we'll present a strategy in the next section for the special case of logic programming.

The unification algorithm (9.6) is the original version given by Robinson [1965]. Other researchers have found algorithms that can be implemented more efficiently. For example, the paper by Paterson and Wegman [1978] presents a linear algorithm for unification.

There are also other versions of the resolution inference rule. One approach uses two simple rules, called *binary resolution* and *factoring*, which can be used together to do the same job as resolution. Another inference rule, called *paramodulation*, is used when the equality predicate is present to take advantage of substituting equals for equals. An excellent introduction to automatic reasoning is contained in the book by Wos, Overbeek, Lusk, and Boyle [1984].

Another subject that we haven't discussed is automatic reasoning in higher-order logic. In higher-order logic it's undecidable whether a set of atoms can be unified. Still there are many interesting results about higher-order unification and there are automatic reasoning systems for some higher-order logics. For example, in second-order monadic logic (*monadic logic* restricts predicates to at most one argument) there is an algorithm to decide whether two atoms can be unified. For example, if F is a variable that represents a function, then the two atoms $F(a)$ and a can be unified by letting F be the constant function that returns the value a or by letting F be the identity function. The paper by Snyder and Gallier [1989] contains many results on higher-order unification.

Automatic theorem-proving techniques are an important and interesting part of computer science, with applications to almost every area of endeavor. Probably the most successful applications of automatic theorem proving will be interactive in nature, the proof system acting as an assistant to the person using it. Typical tasks involve such things as finding ways to represent problems and information to be processed by an automatic theorem prover, finding algorithms that make proper choices in performing resolution, and finding algorithms to efficiently perform unification. We'll look at the programming side of theorem proving in the next section.

Exercises

1. Use Skolem's algorithm, if necessary, to transform each of the following wffs into a clausal form.

   a. $(A \wedge B) \vee C \vee D$.

   b. $(A \wedge B) \vee (C \wedge D) \vee (E \rightarrow F)$.

   c. $\exists y \, \forall x \, (p(x, y) \rightarrow q(x))$.

   d. $\exists y \, \forall x \, p(x, y) \rightarrow q(x)$.

   e. $\forall x \, \forall y \, (p(x, y) \vee \exists z \, q(x, y, z))$.

   f. $\forall x \, \exists y \, \exists z \, [(\neg p(x, y) \wedge q(x, z)) \vee r(x, y, z)]$.

2. What is the resolvent of the propositional clause $p \vee \neg p$ with itself? What is the resolvent of $p \vee \neg p \vee q$ with itself?

3. Find a resolution proof to show that each of the following sets of propositional clauses is unsatisfiable.
   a. $\{A \vee B, \neg A, \neg B \vee C, \neg C\}$.
   b. $\{p \vee q, \neg p \vee r, \neg r \vee \neg p, \neg q\}$.
   c. $\{A \vee B, A \vee \neg C, \neg A \vee C, \neg A \vee \neg B, C \vee \neg B, \neg C \vee B\}$.

4. Compute the composition $\theta\sigma$ of each of the following pairs of substitutions.
   a. $\theta = \{x/y\}, \sigma = \{y/x\}$.
   b. $\theta = \{x/y\}, \sigma = \{y/x, x/a\}$.
   c. $\theta = \{x/y, y/a\}, \sigma = \{y/x\}$.
   d. $\theta = \{x/f(z), y/a\}, \sigma = \{z/b\}$.
   e. $\theta = \{x/y, y/f(z)\}, \sigma = \{y/f(a), z/b\}$.

5. Use the unification algorithm to find a most general unifier for each of the following sets of atoms.
   a. $\{p(x, f(y, a), y), p(f(a, b), v, z)\}$.
   b. $\{q(x, f(x)), q(f(x), x)\}$.
   c. $\{p(f(x, g(y)), y), p(f(g(a), z), b)\}$.
   d. $\{p(x, f(x), y), p(x, y, z), p(w, f(a), b)\}$.

6. What is the resolvent of the clause $p(x) \vee \neg p(f(a))$ with itself? What is the resolvent of $p(x) \vee \neg p(f(a)) \vee q(x)$ with itself?

7. Use resolution to show that each of the following sets of clauses is unsatisfiable.
   a. $\{p(x), q(y, a) \vee \neg p(a), \neg q(a, a)\}$.
   b. $\{p(u, v), q(w, z), \neg p(y, f(x, y)) \vee \neg p(f(x, y), f(x, y)) \vee \neg q(x, f(x, y))\}$.
   c. $\{p(a) \vee p(x), \neg p(a) \vee \neg p(y)\}$
   d. $\{p(x) \vee p(f(a)), \neg p(y) \vee \neg p(f(z))\}$.
   e. $\{q(x) \vee q(a), \neg p(y) \vee \neg p(g(a)) \vee \neg q(a), p(z) \vee p(g(w)) \vee \neg q(w)\}$.

8. Prove that each of the following propositional statements is a tautology by using resolution to prove that its negation is a contradiction.
   a. $(A \vee B) \wedge \neg A \rightarrow B$.
   b. $(p \rightarrow q) \wedge (q \rightarrow r) \rightarrow (p \rightarrow r)$.
   c. $(p \vee q) \wedge (q \rightarrow r) \wedge (r \rightarrow s) \rightarrow (p \vee s)$.
   d. $[(A \wedge B \rightarrow C) \wedge (A \rightarrow B)] \rightarrow (A \rightarrow C)$.

9. Prove that each of the following statements is valid by using resolution to prove that its negation is unsatisfiable.

   a. $\forall x\, p(x) \to \exists x\, p(x)$.
   b. $\forall x\, (p(x) \to q(x)) \land \exists x\, p(x) \to \exists x\, q(x)$.
   c. $\exists y\, \forall x\, p(x, y) \to \forall x\, \exists y\, p(x, y)$.
   d. $\exists x\, \forall y\, p(x, y) \land \forall x\, (p(x, x) \to \exists y\, q(y, x)) \to \exists y\, \exists x\, q(x, y)$.
   e. $\forall x\, p(x) \lor \forall x\, q(x) \to \forall x\, (p(x) \lor q(x))$.

10. Translate each of the following arguments into first-order predicate calculus. Then use resolution to prove that the resulting wffs are valid by proving that the negations are unsatisfiable.

    a. All computer science majors are people. Some computer science majors are logical thinkers. Therefore some people are logical thinkers.

    b. Babies are illogical. Nobody is despised who can manage a crocodile. Illogical persons are despised. Therefore babies cannot manage crocodiles.

11. Translate each of the following arguments into first-order predicate calculus. Then use resolution to prove that the resulting wffs are valid by proving the negations are unsatisfiable.

    a. Every dog either likes people or hates cats. Rover is a dog. Rover loves cats. Therefore some dog likes people.

    b. Every committee member is rich and famous. Some committee members are old. Therefore some committee members are old and famous.

    c. No human beings are quadrupeds. All men are human beings. Therefore no man is a quadruped.

    d. Every rational number is a real number. There is a rational number. Therefore there is a real number.

    e. Some freshmen like all sophomores. No freshman likes any junior. Therefore no sophomore is a junior.

12. Let E be any expression, A and B two sets of expressions, and θ, σ, α any substitutions. Show that each of the following statements is true.

    a. $E(\theta\sigma) = (E\theta)\sigma$.
    b. $E\varepsilon = E$.
    c. $\theta\varepsilon = \varepsilon\theta = \theta$.
    d. $(\theta\sigma)\alpha = \theta(\sigma\alpha)$.
    e. $(A \cup B)\theta = A\theta \cup B\theta$.

9.2 Logic Programming

We'll introduce the idea of logic programming by considering some family tree problems.

Family Tree Problem

Given a set of parent-child relationships, find answers to family questions such as "Is x a second cousin of y?"

First of all, we need to decide what is meant by second cousin and the like. In other words, we need to define relationships such as the following:

> isGrandparentOf,
>
> isGrandchildOf,
>
> isNth-CousinOf,
>
> isNth-Cousin-Mth-RemovedOf,
>
> isNth-AncesterOf,
>
> isSiblingOf.

Let's do an example. We'll look at the isGrandParentOf relation. Let $p(a, b)$ mean "a is a parent of b," and let $g(x, y)$ mean "x is a grandparent of y." We often refer to a relational fact like $p(a, b)$ or $g(x, y)$ as an *atom* because it is an atomic formula in the first-order predicate calculus. To see what's going on, we'll do an example as we go. Our example will be a portion of the British royal family consisting of the following five facts:

> p(Edward VII, George V),
>
> p(Victoria, Edward VII),
>
> p(Alexandra, George V),
>
> p(George VI, Elizabeth II),
>
> p(George V, George VI).

We want to find all the grandparent relations. From our family knowledge we know that x is a grandparent of y if there is some z such that x is a parent of z and z is a parent of y. Therefore it seems reasonable to define the isGrandParentOf relation g as follows:

> $g(x, y)$ if $p(x, z)$ and $p(z, y)$.

For the five facts listed, it's easy to compute the isGrandParentOf relation by hand. It consists of the following four facts:

$$g(\text{Victoria, George V}),$$

$$g(\text{Edward VII, George VI}),$$

$$g(\text{Alexandra, George VI}),$$

$$g(\text{George V, Elizabeth II}).$$

But suppose the isParentOf relation contained 1000 facts and we wanted to list all possible grandparent relations. It would be a time-consuming process if we did it by hand. We could program a solution in almost any language. In fact, we can write a logic program to solve the problem by simply listing the parent facts and then giving the simple definition of the isGrandParentOf relation.

Can we discover how such a program does its computation? The answer is a big maybe. First, let's look at how we human beings get the job done. Somehow we notice the following two facts:

$$p(\text{Victoria, Edward VII}),$$

$$p(\text{Edward VII, George V}).$$

We conclude from these facts that Victoria is a grandparent of George V. A computation by a computer will have to do the same thing. Let's suppose we have an interactive system, in which we give commands to be carried out. To compute all possible grandparent relations, we give a command such as the following:

Find all pairs x, y such that x is a grandparent of y.

In a logic program this is usually represented by a statement such as the following, which is called a *goal*:

$$\leftarrow g(x, y).$$

The program executes by trying to carry out the goal. Upon termination a yes or no answer is output to indicate whether the goal was accomplished. Thus we can think of a goal as a question. For our example goal the system should eventually output something like the following:

"Yes, there are pairs x, y such that $g(x, y)$ is true, and here they are:"

$$x = \text{Victoria}, \qquad y = \text{George V};$$
$$x = \text{Edward VII}, \qquad y = \text{George VI};$$
$$x = \text{Alexandra}, \qquad y = \text{George VI};$$
$$x = \text{George V}, \qquad y = \text{Elizabeth II}.$$

As another example, suppose we want to know whether Victoria is a grandparent of Elizabeth II. In this case we write the following goal:

$$\leftarrow g(\text{Victoria, Elizabeth II}).$$

If the program does its job, it will output something like the following:

"No, Victoria is not a grandparent of Elizabeth II."

To get some insight into how the process works, let's introduce a little more terminology from logic programming. If $r(a, b, c)$ is a fact, then we denote it by writing a backwards arrow on its right side as follows:

$$r(a, b, c) \leftarrow.$$

For example, the fact $p(\text{Victoria, Edward VII})$ is written as follows:

$$p(\text{Victoria, Edward VII}) \leftarrow.$$

A conditional statement of the form "if A then B," where A and B are atoms, is written in logic programming as follows:

$$B \leftarrow A.$$

We read this statement as "B is true if A is true." A conditional statement of the form "if A and B then C" is written in logic programming as follows:

$$C \leftarrow A, B.$$

We read this statement as "C is true if A and B are true." For example, if g and p are the isGrandParentOf and isParentOf relations, respectively, then we have the conditional "if $p(x, z)$ and $p(z, y)$ then $g(x, y)$." We write this statement as follows:

$$g(x, y) \leftarrow p(x, z), p(z, y).$$

Now we're in a position to write down a logic program that solves our example problem. It consists of the five isParentOf facts together with the definition for the isGrandParentOf relation:

$$p(\text{Edward VII, George V}) \leftarrow \qquad\qquad (9.10)$$
$$p(\text{Victoria, Edward VII}) \leftarrow$$
$$p(\text{Alexandra, George V}) \leftarrow$$
$$p(\text{George VI, Elizabeth II}) \leftarrow$$
$$p(\text{George V, George VI}) \leftarrow$$
$$g(x, y) \leftarrow p(x, z), p(z, y)$$

How does this program's computation take place? Well, as we agreed, the computation starts with a command in the form of a goal. For example, suppose we want to find all grandparent-grandchild pairs. We'll do this by giving the following goal, where v and w are variables:

$$\leftarrow g(v, w).$$

The computation proceeds by trying to unify the atom $g(v, w)$ with some program fact on the left side of an arrow. Notice that $g(v, w)$ unifies with $g(x, y)$ in the program by the substitution $\{v/x, w/y\}$. But $g(x, y)$ has an antecedent consisting of the two atoms $p(x, z)$ and $p(z, y)$. These atoms have to be unified with some facts before we can return the answer yes. If we start searching from the top of the list, we see that the atom $p(x, z)$ unifies with the first fact, $p(\text{Edward VII, George V})$, by the substitution $\{x/\text{Edward VII}, z/\text{George V}\}$.

Before we try to process the second atom, $p(z, y)$, we need to apply the binding $z/\text{George V}$. Thus we try to find a fact to unify with the atom

$$p(\text{George V}, y).$$

This atom unifies with $p(\text{George V, George VI})$ by $\{y/\text{George VI}\}$. Therefore the computation returns the following answer:

$$\text{Yes, } x = \text{Edward VII and } y = \text{George VI.}$$

If necessary the computation can continue by trying to find other "yes" answers.

In the next few paragraphs we'll define what a logic program is, and we'll show how logic program computations are performed. After the general discussion about logic programming we'll finish with some examples of logic programming techniques.

Definition of a Logic Program

Let's start by introducing the notation for logic programs. A logic program consists of clauses that have a special form—they contain exactly one positive literal. So a *logic program clause* takes one of the following two forms, where $A, B_1, ..., B_n$ are atoms:

$$A \vee \neg B_1 \vee \cdots \vee \neg B_n \quad \text{(one positive and some negative literals)},$$

$$A \qquad\qquad\qquad\qquad \text{(one positive and no negative literals)}.$$

The computation of a logic program begins after it has been given a *goal*, which is a clause containing only negative literals. So a goal clause takes the following form, where $B_1, ..., B_n$ are atoms:

$$\neg B_1 \vee \cdots \vee \neg B_n \qquad\qquad \text{(no positive and all negative literals)}.$$

The program clauses and goal clauses of a logic program are often called Horn clauses because a *Horn clause* is a clause containing at most one positive literal. So a program clause is a Horn clause with one positive literal, and a goal clause is a Horn clause with no positive literals. There's a simple notation for Horn clauses that is used in logic programming. To see where the notation comes from, notice how we can use equivalences to write a program clause as an implication:

$$A \vee \neg B_1 \vee \cdots \vee \neg B_n \equiv A \vee \neg (B_1 \wedge \cdots \wedge B_n) \equiv B_1 \wedge \cdots \wedge B_n \to A.$$

In logic programming the implication $B_1 \wedge \cdots \wedge B_n \to A$ is denoted by writing it backwards and replacing the conjunction symbols by commas, as follows:

$$A \leftarrow B_1, ..., B_n.$$

We can read this program clause as "A is true if $B_1, ..., B_n$ are all true." In a similar manner we denote the clause consisting of a single atom A as follows:

$$A \leftarrow .$$

We read this program clause as "A is true."

Now let's consider goal clauses. A goal clause like $\neg B_1 \vee \cdots \vee \neg B_n$ is denoted as follows:

$$\leftarrow B_1, ..., B_n.$$

We can interpret this goal clause as the question "Are B_1, ..., B_n all true?" However, since a goal is supposed to relate to a program, a more complete interpretation of the goal clause is "Are B_1, ..., B_n inferred by the program?" We'll make more sense out of this shortly.

Let's summarize. A *logic program* is a finite set of program clauses of the following forms:

$$A \leftarrow B_1, ..., B_n$$
$$A \leftarrow .$$

A *goal* for a logic program has the following form:

$$\leftarrow B_1, ..., B_n.$$

EXAMPLE 1. Let P be the logic program consisting of the following three clauses:

$$q(a) \leftarrow$$
$$r(a) \leftarrow$$
$$p(x) \leftarrow q(x), r(x).$$

Suppose we give P the following goal:

$$\leftarrow p(a).$$

We can read this goal as "Is $p(a)$ true?" or "Is $p(a)$ inferred from P?" The answer to these goal questions is yes. We can argue informally. The three program clauses tell us that $q(a)$ and $r(a)$ are both true and the implication $p(a) \leftarrow q(a), r(a)$ is also true. Therefore we infer that $p(a)$ is true by modus ponens. In the next paragraph we'll see how the answer follows from resolution.   ◆

Resolution and Logic Programming

Let's make a closer examination of goals to see why things are set up to use resolution. For this little discussion we'll suppose that P is a logic program and

$$G \text{ is the goal } \leftarrow B_1, ..., B_n.$$

We can read this goal clause as the following question:

$$\text{"Does } P \text{ imply } B_1 \wedge \cdots \wedge B_n?\text{"}$$

Now remember that the goal $\leftarrow B_1, ..., B_n$ is just shorthand for the following equivalent expressions:

$$\neg B_1 \vee \cdots \vee \neg B_n \equiv \neg (B_1 \wedge \cdots \wedge B_n).$$

So the goal is actually represented as the negation of the thing we want to infer from the program. This is exactly what we want because we will be performing resolution. In other words, we will prove the validity of a statement by showing that its negation is unsatisfiable. For example, to prove that $P \rightarrow B$ is valid, we negate the conditional to obtain $P \wedge \neg B$. Then we apply resolution to obtain a contradiction—the empty clause. In other words, we prove the validity of the statement $P \wedge \neg B \rightarrow$ false.

Let's continue our detailed examination of the goal G. First, remember that clauses are written under the assumption that all variables are universally quantified. For this discussion we'll use the notation

$$\forall (\neg B_1 \vee \cdots \vee \neg B_n)$$

to emphasize this fact. Thus we can write the following equivalences for the goal G:

$$\forall (\neg B_1 \vee \cdots \vee \neg B_n) \equiv \forall \neg (B_1 \wedge \cdots \wedge B_n) \equiv \neg \exists (B_1 \wedge \cdots \wedge B_n).$$

This allows us to give the following more detailed version of the question we need answered for G:

"Does P imply $\exists (B_1 \wedge \cdots \wedge B_n)$?"

To prove "P implies $\exists (B_1 \wedge \cdots \wedge B_n)$" by resolution, we need to negate the statement and then use resolution to show that the negation is unsatisfiable. In other words, we want to prove that the following set of clauses is unsatisfiable (we're using sets because P is a set of clauses):

$$P \cup \{\neg \exists (B_1 \wedge \cdots \wedge B_n)\}.$$

Of course, we have the following equivalences and equalities:

$$
\begin{aligned}
\neg \exists (B_1 \wedge \cdots \wedge B_n) \ &\equiv\ \forall \neg (B_1 \wedge \cdots \wedge B_n) \\
&\equiv\ \forall (\neg B_1 \vee \cdots \vee \neg B_n) \\
&=\ \leftarrow B_1, ..., B_n \\
&=\ G.
\end{aligned}
$$

So to answer the question "Does P imply $\exists\,(B_1 \wedge \cdots \wedge B_n)$?" we must show that the following set of clauses is unsatisfiable:

$$P \cup \{G\}.$$

What's the point of all this? The point is that if P is a logic program and G is a goal, then the goal question can be answered in the affirmative if there is a proof that the set of clauses $P \cup \{G\}$ is unsatisfiable.

When we give a goal to a logic program, we usually want more than just the answer yes or no. If the answer is yes, we might want to know the values of any of the variables appearing in the goal. So a more technically accurate reading of the goal statement "Does P imply $\exists\,(B_1 \wedge \cdots \wedge B_n)$?" is the following:

"Does there exist a substitution θ such that P implies $(B_1 \wedge \cdots \wedge B_n)\theta$?"

Let's look at an example to see how the notation for logic program clauses makes it easy to find answers to goal questions.

EXAMPLE 2. Suppose we have a logic program P consisting of the following two clauses:

$$q(a) \leftarrow$$
$$p(f(x)) \leftarrow q(x).$$

Let G be the goal $\leftarrow p(y)$. This means that we want an answer to the question "Does P imply $\exists y\ p(y)$?" In other words, "Is there is a substitution θ such that $p(y)\theta$ is inferred from P?" Let's give the answer first and then see how we got it. The answer is yes. Letting $\theta = \{y/f(a)\}$, we can evaluate $p(y)\theta$ as follows:

$$p(y)\theta = p(y)\{y/f(a)\} = p(f(a)).$$

We claim that $p(f(a))$ is inferred from P. This is easy to see from an informal standpoint. Just apply θ to the second clause. This transforms the two clauses of P into the following:

$$q(a) \leftarrow$$
$$p(f(a)) \leftarrow q(a).$$

Now, since $q(a)$ is a fact, we can apply modus ponens to conclude that $p(f(a))$ is true.

So much for the informal discussion. Now let's give a resolution proof showing that $P \cup \{G\}$ is unsatisfiable. We'll convert the logic program notation for the clauses and the goal into normal clausal notation, and we'll keep track of the most general unifiers as we go.

Proof:	1.	$q(a)$	P	program clause: $q(a) \leftarrow$
	2.	$p(f(x)) \lor \neg q(x)$	P	program clause: $p(f(x)) \leftarrow q(x)$
	3.	$\neg p(y)$	P	goal clause: $\leftarrow p(y)$
	4.	$\neg q(x)$	2, 3, R, $\{y/f(x)\}$	
	5.	$\square$	1, 4, R, $\{x/a\}$	
		QED.		

Therefore by the resolution theorem, $P \cup \{G\}$ is unsatisfiable. So the answer to the goal question is yes. What value of y does the job? The y that does the job can be obtained by composing the mgu's obtained during the resolution process and then applying the result to y, as follows:

$$y \{y/f(x)\} \{x/a\} = y \{y/f(a)\} = f(a).$$

Therefore $p(f(a))$ is a logical consequence of program P.   ◆

There are three important advantages to the notation that we are using for logic programs:

1. The notation is easy to write down because we don't have to use the symbols $\neg$, $\land$, and $\lor$.

2. The notation allows us to interpret a program in two different ways. For example, suppose we have the clause $A \leftarrow B_1, ..., B_n$. This clause has the usual logical interpretation "A is true if $B_1, ..., B_n$ are all true." The clause also has the procedural interpretation "A is a procedure that is executed by executing the procedures $B_1, ..., B_n$ in the order they are written." Most logic programming systems allow this procedural interpretation.

3. The notation makes it easy to apply the resolution rule. We'll discuss this next.

Whenever we apply the resolution rule, we have to do a lot of choosing. We have to choose two clauses to resolve, and we have to choose literals to "cancel" from each clause. Since there are many choices, it's easy to understand why we can come up with many different proof sequences. When resolution is used with logic program clauses, we can specialize the rule.

The specialized rule always picks one clause to be the most recent line of the proof, which is always a goal clause. Start the proof by picking the initial goal. Select the leftmost atom in the goal clause as the literal to "cancel." For the second clause, pick a program clause whose head unifies with the atom selected from the goal clause. The resolvant of these two clauses is created by first replacing the leftmost atom in the goal clause by the body atoms of the program clause and then applying the unifier to the resulting goal. Here is a formal description of the rule, which is called the *SLD-resolution* rule:[†]

SLD-Resolution Rule   (9.11)

Resolve the goal $\leftarrow B_1, ..., B_k$ with the program clause $A \leftarrow A_1, ..., A_n$ by first unifying B_1 with A via mgu θ. Then replace B_1 in the goal by the body $A_1, ..., A_n$ and apply θ to the resulting goal to obtain the resolvant

$$\leftarrow (A_1, ..., A_n, B_2, ..., B_k)\theta.$$

To construct a logic program proof, we start by listing each program clause as a premise. Then we write the goal clause as a premise. Now we use (9.11) repeatedly to add new resolvants to the proof, each new resolvant being constructed from the goal on the previous line together with some program clause. We can summarize the application of (9.11) with the following four-step procedure:

1. Pick the goal clause on the last line of the partial proof, and select its leftmost atom, say B_1.

2. Find a program clause whose head unifies with B_1, say by θ. Be sure the two clauses have distinct sets of variables (rename if necessary).

3. Replace B_1 in the goal clause with the body of the program clause.

4. Apply θ to the goal constructed on line 3 to get the resolvant, which is placed on a new line of the proof.

We'll introduce the use of the SLD-resolution rule with an example. Suppose we are given the following logic program, where p means isParentOf and g means isGrandparentOf:

$$p(a, b) \leftarrow$$
$$p(d, b) \leftarrow$$
$$p(b, c) \leftarrow$$
$$g(x, y) \leftarrow p(x, z), p(z, y).$$

[†] SLD-resolution means Selective Linear resolution of Definite clauses. In our case we always "select" the leftmost atom of the goal clause.

We'll execute the program by giving it the following goal:

$$\leftarrow g(w, c).$$

Since there is a variable w in this goal, we can read the goal as the question

"Is there a grandparent for c?"

The resolution proof starts by letting the program clauses and the goal clause be premises. For this example we have the following five lines:

Proof: 1. $p(a, b) \leftarrow$ P
        2. $p(d, b) \leftarrow$ P
        3. $p(b, c) \leftarrow$ P
        4. $g(x, y) \leftarrow p(x, z), p(z, y).$ P
        5. $\leftarrow g(w, c)$ P Initial goal

The proof starts by resolving the initial goal on line 5 with some program clause. The atom $g(w, c)$ from the initial goal unifies with $g(x, y)$, the head of the program clause on line 4, by the mgu

$$\theta_1 = \{w/x, y/c\}.$$

Therefore we can use (9.11) to resolve the two clauses on lines 4 and 5. So we replace the goal atom $g(w, c)$ on line 5 with the body of the clause on line 4 and then apply the mgu θ_1 to the result to obtain the following resolvant goal clause:

$$\leftarrow p(x, z), p(z, c).$$

Let's compare what we've just done for logic program clauses using (9.11) to the case for regular clauses using (9.7). The following two lines are copies of lines 4 and 5 in which we've included the clausal notation for each logic program clause:

Logic Program Notation	*Clausal Notation*
4. $g(x, y) \leftarrow p(x, z), p(z, y)$	$g(x, y) \vee \neg p(x, z) \vee \neg p(z, y)$
5. $\leftarrow g(w, c)$	$\neg g(w, c)$

We apply (9.11) to the logic program notation clauses, and we apply (9.7) to the clauses in clausal notation. This gives the following pair of resolvents:

Logic Program Notation	Clausal Notation
$\leftarrow p(x, z), p(z, c)$	$\neg p(x, z) \vee \neg p(z, c)$

So we get the same answer with either method.

Now let's continue the proof. We'll write down the new resolvant on line 6 of our proof, in which we've added the mgu to the reason column:

6. $\leftarrow p(x, z), p(z, c)$ 4, 5, R, $\theta_1 = \{w/x, y/c\}$

To continue the proof according to (9.11), we must choose this new goal on line 6 for one of the clauses, and we must choose its leftmost atom $p(x, z)$ for "cancellation." For the second clause we'll choose the clause on line 1 because its head $p(a, b)$ unifies with our chosen atom by the mgu

$$\theta_2 = \{x/a, z/b\}.$$

To apply (9.11), we must replace $p(x, z)$ on line 6 by the body of the clause on line 1 and then apply θ_2 to the result. Since the clause on line 1 does not have a body, we simply delete $p(x, z)$ from line 6 and apply θ_2 to the result, obtaining the resolvant

$$\leftarrow p(b, c).$$

Let's compute this result in terms of both (9.11) and (9.7). The clauses on lines 1 and 6 take the following forms, in which we've added the regular clausal notation for each clause:

Logic Program Notation	Clausal Notation
1. $p(a, b) \leftarrow$	$p(a, b)$
6. $\leftarrow p(x, z), p(z, c)$	$\neg p(x, z) \vee \neg p(z, c)$

After applying (9.11) and (9.7) to the respective notations on lines 1 and 6, we obtain the following pair of resolvants:

Logic Program Notation	Clausal Notation
$\leftarrow p(b, c)$	$\neg p(b, c)$

So we can continue the proof by writing down the new resolvant on line 7 as follows:

7. $\leftarrow p(b, c)$ 1, 6, R, $\theta_2 = \{x/a, z/b\}$

To continue the proof using (9.11), we must choose the goal clause on line 7 together with its only atom $p(b, c)$. It unifies with the head $p(b, c)$ of the clause on line 3 by the empty unifier

$$\theta_3 = \{\ \}.$$

Since there is only one atom in the goal clause of line 7 and there is no body in the clause on line 3, it follows that the resolvent of the clauses on these two lines is just the empty clause. Thus our proof is completed by writing this information on line 8 as follows:

8. $\square$ $3, 7, R, \theta_3 = \{\ \}$
    QED.

To finish things off, we'll collect the eight steps of the proof and rewrite them as a single unit:

Proof: 1. $p(a, b) \leftarrow$ P
           2. $p(d, b) \leftarrow$ P
           3. $p(b, c) \leftarrow$ P
           4. $g(x, y) \leftarrow p(x, z), p(z, y)$ P
           5. $\leftarrow g(w, c)$ P Initial goal
           6. $\leftarrow p(x, z), p(z, c)$ $4, 5, R, \theta_1 = \{w/x, y/c\}.$
           7. $\leftarrow p(b, c)$ $1, 6, R, \theta_2 = \{x/a, z/b\}$
           8. $\square$ $3, 7, R, \theta_3 = \{\ \}$
              QED.

Since $\square$ was obtained, the answer to the question for the goal

$$\leftarrow g(w, c)$$

is yes. Now, what about the variable w in the goal statement? The only reason we got the answer yes is because w was bound to some term. We can recover the value of that binding by composing the three unifiers θ_1, θ_2, and θ_3 and then applying the result to w:

$$w\theta_1\theta_2\theta_3 = a.$$

So the goal question "Is there a grandparent for c?" is answered as follows:

Yes

$$w = a.$$

We should notice for this example that there is another possible yes answer to the goal $\leftarrow g(w, c)$. Namely,

$$\text{Yes}$$
$$w = d.$$

Can this answer be computed? Sure. Keep the first six lines of the proof as they are. Then resolve the goal on line 6 with the clause on line 2 instead of the clause on line 1. The goal atom $p(x, z)$ on line 6 unifies with the head $p(d, b)$ from line 1 by mgu

$$\theta_2 = \{x/d, z/b\}.$$

This θ_2 is different from the previous θ_2. So we get a new line 7 and, in this case, the same line 8 as follows:

7. $\leftarrow p(b, c)$ $2, 6, R, \theta_2 = \{x/d, z/b\}$
8. $\square$ $3, 7, R, \theta_3 = \{\ \}$
    QED.

With this proof we obtain the answer yes, and we calculate a new value of w as follows:

$$w\theta_1\theta_2\theta_3 = d.$$

Computation Trees

Now that we have an example under our belts, let's look again at the general picture. The preceding proof had two possible yes answers. We would like to find a way to represent all possible answers (i.e., proof sequences) for a goal. For our purposes a tree will do the job.

A *computation tree* for a goal is an ordered tree whose root is the goal. The children of any parent node are all the possible goals (i.e., resolvants) that can be obtained by resolving the parent goal with a program clause. We agree to order the children of each node from left to right in terms of the top-to-bottom ordering of the program clauses that are used with the parent to create the children. Each parent-child branch is labeled with the mgu obtained to create the child. A leaf may be the empty clause or a goal. If the empty clause occurs as a leaf, we write "yes" together with the values of any variables that occur in the original goal at the root of the tree. If a goal occurs as a leaf, this means that it can't be resolved with any program clause, so we write "failure." The computation tree will always show all possible answers for the given goal at its root.

For example, the computation tree for the goal $\leftarrow g(w, c)$ with respect to our example program can be pictured as shown in Figure 9.1.

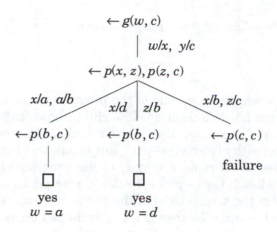

Figure 9.1

Notice that the tree contains all possible answers to the goal question.

A logic programming system needs a strategy to search the computation tree for a leaf with a yes answer. The strategy used by most Prolog systems is the *depth-first* search strategy, which starts by traversing the tree down to the leftmost leaf. If the leaf is the empty clause, then the yes answer is reported. If the leaf is a failure leaf, then the search returns to the parent of the leaf. At this point a depth-first search is started at the next child to the right. If there is no next child, then the search returns to the parent of the parent, and a depth-first search starts with its next child to the right, and so on. If this process eventually returns to the root of the tree and there are no more paths to search, then failure is reported.

It might be desirable for a logic programming system to attempt to find all possible answers to a goal question. One strategy for attempting to find all possible answers is called *backtracking*. For example, with depth-first search we perform backtracking by continuing the depth-first search process from the point at which the last yes answer was found. In other words, when a yes answer is found, the system reports the answer and then continues just as though a failure leaf was encountered.

In the next few examples we'll construct some computation trees and discuss the problems that can arise in trying to find all possible answers to a goal question.

EXAMPLE 3. Let's consider the following two-clause program:

$$p(a) \leftarrow$$
$$p(\text{succ}(x)) \leftarrow p(x).$$

Suppose we give the following goal to the program:

$$\leftarrow p(x).$$

This goal will resolve with either one of the program clauses. So the root of the computation tree has two children. One child, the empty clause, results from the resolution of $\leftarrow p(x)$ with $p(a) \leftarrow$. The other child results from the resolution of $\leftarrow p(x)$ with $p(\text{succ}(x)) \leftarrow p(x)$. But before this happens, we need to change variables. We'll replace x by x_1 in the program clause to obtain $p(\text{succ}(x_1)) \leftarrow p(x_1)$. Resolving $\leftarrow p(x)$ with this clause produces the clause $\leftarrow p(x_1)$, which becomes the second child of the root. The process starts all over again with the goal $\leftarrow p(x_1)$. To keep track of variable names, we'll replace x by x_2 in the second program clause. Then resolve $\leftarrow p(x_1)$ with $p(\text{succ}(x_2)) \leftarrow p(x_2)$ to obtain the clause $\leftarrow p(x_2)$. This process continues forever.

The computation tree for this example is shown in Figure 9.2. It is an infinite tree, which continues the indicated pattern forever.

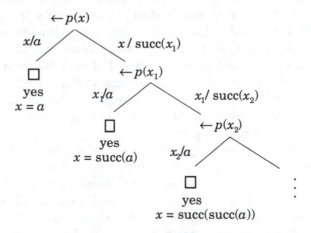

Figure 9.2

If we use the depth-first search rule, the first answer is "yes, $x = a$." If we force backtracking, the next answer we'll get is "yes, $x = \text{succ}(a)$." If we force backtracking again, we'll get the answer "yes, $x = \text{succ}(\text{succ}(a))$." Continuing in this way, we can generate the following infinite sequence of possible values for x:

$$a, \text{succ}(a), \text{succ}(\text{succ}(a)), ..., \text{succ}^k(a), ... \quad \blacklozenge$$

EXAMPLE 4. Consider the following three-clause program, in which the third clause has more than one atom in its body:

$$q(a) \leftarrow$$
$$p(a) \leftarrow$$
$$p(f(x)) \leftarrow p(x), q(x).$$

Figure 9.3 shows a few levels of the computation tree for the goal $\leftarrow p(x)$. Notice that as we travel down the rightmost path from the root, the number of goal atoms at each node is increased by one for each new level.

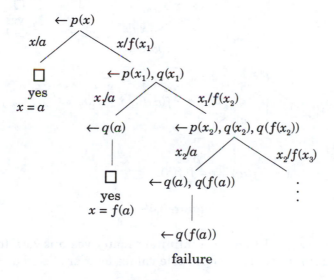

Figure 9.3

Using the depth-first search rule, we obtain the answer "yes, $x = a$." Backtracking works one time to give the answer "yes, $x = f(a)$." If we force backtracking again, then the computation takes an infinite walk down the tree, failing at each leaf. $\blacklozenge$

EXAMPLE 5. Suppose we're given the following three-clause program:

$$p(f(x)) \leftarrow p(x)$$
$$p(a) \leftarrow$$
$$p(b) \leftarrow .$$

If we start with the goal

$$\leftarrow p(x),$$

then the computation tree will be a ternary tree because there are three "*p*" clauses that match each goal. The first few levels of the tree are given in Figure 9.4.

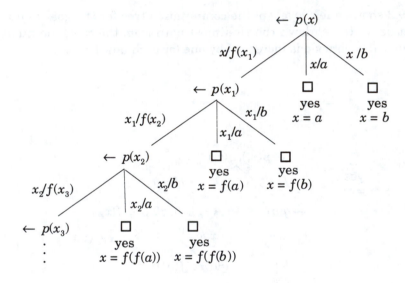

Figure 9.4

The tree is infinite, and there are infinitely many yes answers to the goal question. The infinite sequence of possible values for x are listed as follows:

$$a, b, f(a), f(b), f(f(a)), f(f(b)), ..., f^k(a), f^k(b), ...$$

Notice that if we used the depth-first search strategy, then the computation would take an infinite walk down the left branch of the tree. So although there are infinitely many answers, the depth-first search strategy won't find even one of them.   ♦

In the preceding example, depth-first search did not find any answers to the goal $\leftarrow p(x)$. Suppose we reordered the three program clauses as follows:

$$p(a) \leftarrow$$
$$p(b) \leftarrow$$
$$p(f(x)) \leftarrow p(x).$$

The computation tree corresponding to these three clauses can be searched in a depth-first fashion with backtracking to generate all the answers to the goal $\leftarrow p(x)$. Suppose we write the three clauses in the following order:

$$p(a) \leftarrow$$
$$p(f(x)) \leftarrow p(x)$$
$$p(b) \leftarrow.$$

The computation tree for these three clauses, when searched with depth-first and backtracking, will yield some, but not all, of the possible answers.

So when a logic programming language uses depth-first search, two problems can occur when the computation tree for a goal is infinite:

1. The yes answers found may depend on the order of the clauses.

2. Backtracking might not find all possible yes answers to a goal.

Many logic programming systems use the depth-first search strategy because it's efficient to implement and because it reflects the procedural interpretation of a clause. For example, the clause $A \leftarrow B, C$ represents a procedure named A that is executed by first calling procedure B and then calling procedure C.

Another search strategy is called *breadth-first search*. It looks for a yes answer by examining all the children of a node before it looks at the next level of the tree. This strategy will find all possible answers to a goal question. But it will carry on forever looking for more answers if the computation tree is infinite.

Some implementation strategies for searching the computation tree use breadth-first search with a twist. All children of a node are searched in parallel. A search at a particular node is started only when the goal atom has not already occurred at a higher level in the tree. If the goal atom matches a goal at a higher level in the tree, then the process waits for the answer to the other goal. When it receives the answer, then it continues with its search. This technique requires a table containing previous goal atoms and answers.

It has proved useful in detecting certain kinds of loops that give rise to infinite computation trees. In some cases the search process won't take an infinite walk. An introduction to these ideas is given in Warren [1992].

Logic Programming Techniques

Let's spend some time discussing a few elementary techniques to construct logic programs. First we'll see how to construct logic programs that process relations. Then we'll discuss logic programs that process functions. The clauses in our examples are ordered to take advantage of the depth-first search strategy. This strategy is used by most Prolog systems.

Techniques for Relations

Logic programming allows us to easily process many relations because relations are just predicates. For example, we've already seen an example of how to find the isGrandparentOf relation if we're given the isParentOf relation. The technique is to write down the isParentOf relation as a set of facts of the form

$$p(a, b) \leftarrow .$$

We read this clause as "a is a parent of b." Then we define the isGrandparentOf relation as the following clause:

$$g(x, y) \leftarrow p(x, z), p(z, y).$$

This clause is read as "x is a grandparent of y if x is a parent of z and z is a parent of y."

Suppose we want to write the isAncestorOf relation in terms of the isParentOf relation, where an ancestor is either a parent, or a grandparent, or a great-grandparent, and so on. The next example discusses this problem in general terms.

EXAMPLE 6 (*Transitive Closure*). The isAncestorOf relation is the transitive closure of the isParentOf relation. In general terms, suppose we're given a binary relation r and we need to compute the transitive closure of r. If we let tc denote the transitive closure of r, the following two-clause program does the job:

$$tc(x, y) \leftarrow r(x, y)$$
$$tc(x, y) \leftarrow r(x, z), tc(z, y).$$

For example, suppose r is the isParentOf relation. Then tc is the isAncestorOf relation. The first clause can be read as "x is an ancestor of y if x is a parent of y," and the second clause can be read as "x is an ancestor of y if x is a parent of z and z is an ancestor of y."   ◆

Techniques for Functions

Now let's see whether we can find a technique to construct logic programs to compute functions. Actually, it's pretty easy. The major thing to remember in translating a function definition to a logic definition is the following:

The functional equation $f(x) = y$ can be represented by a predicate expression as follows:

$$pf(x, y).$$

The predicate name "pf" can remind us that we have a "predicate for f." The predicate expression pf(x, y) can still be read as "f of x is y."

Now let's discuss a technique to construct a logic program for a recursively defined function. If f is defined recursively, then there is at least one part of the definition that defines $f(x)$ in terms of some $f(y)$. In other words, some part of the definition of f has the following form, where $E(f(y))$ denotes an expression containing $f(y)$:

$$f(x) = E(f(y)).$$

Using our technique to create a predicate for this functional equation, we get the following expression:

$$pf(x, E(f(y))).$$

But we aren't done yet because the recursive definition of f causes $f(y)$ to occur as an argument in the predicate. Since we're trying to compute f by the predicate pf, we need to get rid of $f(y)$. The solution is to replace $f(y)$ by a new variable z. We can represent this replacement by writing down the following version of the expression:

$$pf(x, E(z)) \text{ where } z = f(y).$$

Now we have a functional equation $z = f(y)$, which we can replace by pf(y, z).

So we obtain the following expression:

$$\text{pf}(x, E(z)) \text{ where } \text{pf}(y, z).$$

The transformation to a logic program is now simple: Replace the word "where" by the symbol $\leftarrow$ to obtain a logic program clause as follows:

$$\text{pf}(x, E(z)) \leftarrow \text{pf}(y, z).$$

Thus we have a general technique to transform a functional equation into a logic program. Here are the steps, all in one place:

$f(x) = E(f(y))$	The given functional equation.
$\text{pf}(x, E(f(y)))$	Create a predicate expression.
$\text{pf}(x, E(z))$ where $z = f(y)$	Let $z = f(y)$.
$\text{pf}(x, E(z))$ where $\text{pf}(y, z)$	Create a predicate expression.
$\text{pf}(x, E(z)) \leftarrow \text{pf}(y, z)$	Create a clause.

Of course, there may be more work to do, depending on the complexity of the expression $E(z)$. Let's do some examples to help get the look and feel of this process.

EXAMPLE 7. Suppose we want to write a logic program to compute the factorial function. Letting $f(x) = x!$, we have the following recursive definition of f:

$$f(0) = 1$$
$$f(x) = x * f(x - 1).$$

To implement f as a logic program, we'll let "fact" be the predicate to compute f. Then the two equations of the recursive definition become the following two predicate expressions:

$$\text{fact}(0, 1)$$
$$\text{fact}(x, x * f(x - 1)).$$

The second statement contains the argument $f(x - 1)$, which we'll replace by a new variable y to obtain the following version of the two expressions:

$$\text{fact}(0, 1)$$
$$\text{fact}(x, x * y) \text{ where } y = f(x - 1).$$

Now we can change the functional equation $y = f(x - 1)$ into a predicate expression to obtain the following version:

$$\text{fact}(0, 1)$$
$$\text{fact}(x, x * y) \text{ where } \text{fact}(x - 1, y).$$

Therefore the desired logic program has the following two clauses:

$$\text{fact}(0, 1) \leftarrow$$
$$\text{fact}(x, x * y) \leftarrow \text{fact}(x - 1, y). \quad \blacklozenge$$

EXAMPLE 8. Suppose we want to write a logic program to compute the length of a list. Let's start with the following recursively defined function L that does the job:

$$L(\langle\,\rangle) = 0.$$
$$L(x :: y) = L(y) + 1.$$

We can start by writing down two predicate expressions to represent these two functional equations. We'll use the predicate name "length" as follows:

$$\text{length}(\langle\,\rangle, 0)$$
$$\text{length}(x :: y, L(y) + 1).$$

The second expression contains an occurrence of the function L, which we're trying to define. So we'll replace $L(y)$ by a new variable z to obtain the following version:

$$\text{length}(\langle\,\rangle, 0)$$
$$\text{length}(x :: y, z + 1) \text{ where } z = L(y).$$

Now replace the functional equation $z = L(y)$ by the predicate expression $\text{length}(y, z)$ to obtain the following version:

$$\text{length}(\langle\,\rangle, 0)$$
$$\text{length}(x :: y, z + 1) \text{ where } \text{length}(y, z).$$

Lastly, convert the expressions to the following logic program clauses:

$$\text{length}(\langle\,\rangle, 0) \leftarrow$$
$$\text{length}(x :: y, z + 1) \leftarrow \text{length}(y, z). \quad \blacklozenge$$

EXAMPLE 9. Suppose we want to delete the first occurrence of an element from a list. A recursively defined function to do the job can be written as follows:

$$\text{delete}(x, L) \ = \ \begin{array}{l} \text{if } L = \langle\,\rangle \text{ then } \langle\,\rangle \\ \text{else if head}(L) = x \text{ then tail}(L) \\ \text{else head}(L) :: \text{delete}(x, \text{tail}(L)). \end{array}$$

Let's construct a logic program to compute this function. It's much easier to write a logic program for a function described as a set of equations. So we'll rewrite the functional definition as three functional equations in the following way:

$$\text{delete}(x, \langle\,\rangle) = \langle\,\rangle$$
$$\text{delete}(x, x :: T) = T$$
$$\text{delete}(x, y :: T) = y :: \text{delete}(x, T).$$

First we'll convert each equation to a predicate expression using the predicate named "remove" as follows:

$$\text{remove}(x, \langle\,\rangle, \langle\,\rangle)$$
$$\text{remove}(x, x :: T, T)$$
$$\text{remove}(x, y :: T, y :: \text{delete}(x, T)).$$

Since the functional value delete(x, T) occurs in the third expression, we'll replace it by a new variable U to obtain the following version:

$$\text{remove}(x, \langle\,\rangle, \langle\,\rangle)$$
$$\text{remove}(x, x :: T, T)$$
$$\text{remove}(x, y :: T, y :: U) \text{ where } U = \text{delete}(x, T).$$

Now replace the functional equation $U = \text{delete}(x, T)$ by the predicate expression remove(x, T, U) as follows:

$$\text{remove}(x, \langle\,\rangle, \langle\,\rangle)$$
$$\text{remove}(x, x :: T, T)$$
$$\text{remove}(x, y :: T, y :: U) \text{ where remove}(x, T, U).$$

Finally, transform these three expressions into the following three-clause logic program:

$$remove(x, \langle \rangle, \langle \rangle) \leftarrow$$
$$remove(x, x :: T, T) \leftarrow$$
$$remove(x, y :: T, y :: U) \leftarrow remove(x, T, U).$$

Exercises

1. Suppose you are given an isParentOf relation. Find a definition for each of the following relations.

   a. isChildOf.

   b. isGrandchildOf.

   c. isGreatGrandparentOf.

2. Suppose you are given an isParentOf relation. Try to find a definition for each of the following relations. *Hint:* You might want to consider some kind of test for equality.

   a. isSiblingOf.

   b. isCousinOf.

   c. isSecondCousinOf.

   d. isFirstCousinOnceRemovedOf.

3. Suppose we're given the following logic program:

   $$p(a, b) \leftarrow$$
   $$p(a, c) \leftarrow$$
   $$p(b, d) \leftarrow$$
   $$p(c, e) \leftarrow$$
   $$g(x, y) \leftarrow p(x, z), p(z, y).$$

   a. Find a resolution proof for the goal $\leftarrow g(a, w)$.

   b. Draw a picture of the computation tree for the goal $\leftarrow g(a, w)$.

4. Suppose we're given the following logic program:

   $$p(a) \leftarrow$$
   $$p(g(x)) \leftarrow p(x)$$
   $$p(b) \leftarrow .$$

   a. Draw at least three levels of the computation tree for the goal $\leftarrow p(x)$.

   b. What are the possible yes answers for the goal $\leftarrow p(x)$?

   c. Describe the values of x that are generated by backtracking with the depth-first search strategy for the goal $\leftarrow p(x)$.

5. The following logic program claims to test an integer to see whether it is a natural number, where pred(x, y) means that the predecessor of x is y:

$$\text{isNat}(0) \leftarrow$$
$$\text{isNat}(x) \leftarrow \text{isNat}(y), \text{pred}(x, y).$$

   a. What happens when the goal is $\leftarrow$ isNat(2)?

   b. What happens when the goal is $\leftarrow$ isNat(−1)?

6. Let r denote a binary relation. Write logic programs to compute each of the following relations.

   a. The symmetric closure of r.

   b. The reflexive closure of r.

7. Translate each of the following functional definitions into a logic program. *Hint:* First, translate the if-then-else definitions into equational definitions.

   a. The function f computes the nth Fibonacci number:

$$f(n) = \text{if } n = 0 \text{ then } 1 \text{ else if } n = 1 \text{ then } 1 \text{ else } f(n-1) + f(n-2).$$

   b. The function "cat" computes the concatenation of two lists:

$$\text{cat}(x, y) = \text{if } x = \langle \, \rangle \text{ then } y \text{ else head}(x) :: \text{cat}(\text{tail}(x), y).$$

   c. The function "nodes" computes the number of nodes in a binary tree:

$$\text{nodes}(t) = \text{if } t = \langle \, \rangle \text{ then } 0 \text{ else } 1 + \text{nodes}(\text{left}(t)) + \text{nodes}(\text{right}(t)).$$

8. Find a logic program to implement each of the following functions, where the variables represent elements or lists.

   a. equalLists(x, y) tests whether the lists x and y are equal.

   b. member(x, y) tests whether x is an element of the list y.

   c. all(x, y) is the list obtained from y by removing all occurrences of x.

   d. makeSet(x) is the list obtained from x by deleting redundant elements.

   e. subset(x, y) tests whether x, considered as a set, is a subset of y.

   f. equalSets(x, y) tests whether x and y, considered as sets, are equal.

   g. subBag(x, y) tests whether x, considered as a bag, is a subbag of y.

   h. equalBags(x, y) tests whether the bags x and y are equal.

9. Suppose we have a schedule of classes with each entry having the form class(i, s, t, p), which means that class i section s meets at time t in place p. Find a logic program to compute the possible schedules available for a given list of classes.

10. Write a logic program to test whether a propositional wff is a tautology. Assume that the wffs use the four operators in the set $\{\neg, \wedge, \vee, \rightarrow\}$. *Hint:* Use the method of Quine together with the fact that if A is a wff containing a letter p, then A is a tautology iff $A(p/\text{true})$ and $A(p/\text{false})$ are both tautologies. To assist in finding the propositional letters, assume that the predicate atom(x) means that x is a propositional letter.

Chapter Summary

The major component of automatic reasoning for the first-order predicate calculus is the resolution inference rule. Resolution proofs work by showing that a wff is unsatisfiable. So to prove that a wff is valid, we can use resolution to show that its negation is unsatisfiable. Resolution requires wffs to be represented as sets of clauses, which can be constructed by Skolem's algorithm. Before each step of a resolution proof involving predicates, the unification algorithm must calculate a substitution—a most general unifier—that will unify a set of atoms. The process of applying the resolution rule can be programmed to perform automatic reasoning.

Logic programs consist of clauses that have one positive literal and zero or more negative literals. A logic program goal is a clause consisting of one or more negative literals. Logic program goals are computed by a modification of resolution called SLD-resolution. Each goal of a logic program has an associated computation tree that can be searched in a variety of ways. The depth-first search strategy is used by most logic programming languages. Elementary techniques for logic programming include the implementation of relations and recursively defined functions.

10

Algebraic Structures and Techniques

*Algebraic rules of procedure were proclaimed as
if they were divine revelations....*
—From *The History of Mathematics*
by David M. Burton

The word "algebra" comes from the word "al-jabr" in the title of the textbook *Hisâb al-jabr w'al-muqâbala*, which was written around 820 by the mathematician and astronomer al-Khowârizmî. The title translates roughly to "calculations by restoration and reduction," where restoration—al-jabr— refers to adding or subtracting a number on both sides of an equation, and reduction refers to simplification. We should also note that the word "algorithm" has been traced back to al-Khowârizmî because people used his name—mispronounced of course—when referring to a method of calculating with Hindu numerals that was contained in another of his books.

Having studied high school algebra, most of us probably agree that algebra has something to do with equations and simplification. In high school algebra we simplified a lot. In fact, we were often given the one word command "simplify" in the exercises. So we tried to somehow manipulate a given expression into one that was simpler than the given one, although this was a bit vague, and there always seemed to be a question about what "simplify" meant. We also tried to describe word problems in terms of algebraic equations and then to apply our simplification methods to extract solutions. Everything we did dealt with numbers and expressions for numbers.

In this chapter we'll clarify and broaden the idea of an algebra. The chapter introduces the notions and notations of algebra with special emphasis on the techniques and applications of algebra in computer science.

Chapter Guide

Section 10.1 introduces the idea of an algebra. We'll see that high school algebra is just one kind of algebra.

Section 10.2 introduces Boolean algebra. We'll discuss some techniques to simplify Boolean expressions, and we'll see how to construct digital circuits.

Section 10.3 introduces the idea of an abstract data type as an algebra. As examples, we'll discuss some properties of the natural numbers, lists, strings, stacks, queues, binary trees, and priority queues.

Section 10.4 introduces some properties of three kinds of algebras that are useful as computationl tools: relational algebras, process algebras, and functional algebras.

Section 10.5 introduces a collection of algebraic ideas that are also useful for computational problems: congruences, subalgebras, quotient algebras, and morphisms.

10.1 What Is an Algebra?

Before we say just what an algebra is, let's see how an algebra is used in the problem-solving process. An important part of problem solving is the process of transforming informal word problems into formal things like equations, expressions, or algorithms. Another important part of problem solving is the process of transforming these formal things into solutions by solving equations, simplifying expressions, or implementing algorithms. For example, in high school algebra we tried to describe certain word problems in terms of algebraic equations, and then we tried to solve the equations. An algebra should provide tools and techniques to help us describe informal problems in formal terms and to help us solve the resulting formal problems.

The Description Problem

How can we describe something to another person in such a way that the person understands exactly what we mean? One way is to use examples. But sometimes examples may not be enough for a proper understanding. It is often useful at some point to try to describe an object by describing some properties that it possesses. So we state the following general problem:

> *The Description Problem*
> Describe an object.

Whatever form a description takes, it should be communicated in a clear and concise manner so that examples or instances of the object can be easily checked for correctness. Try to describe one of the following things to a friend:

> A car.
> The left side of a person.
> The number zero.
> The concept of area.

Most likely, you'll notice that the description of an object often depends on the knowledge level of the audience.

We need some tools to help us describe properties of the things we are talking about, so we can check not only the correctness of examples, but also the correctness of the descriptions. Algebras provide us with natural notations that can help us give precise descriptions for many things, particularly those structures and ideas that are used in computer science.

High School Algebra

A natural example of an algebra that we all know and love is the algebra of numbers. We learned about it in school, and we probably had different ideas about what it was. First we learned about arithmetic of the natural numbers $\mathbb{N}$, using the operation of addition. We came eventually to believe things like

$$7 + 12 = 19, \quad 3 + 5 = 5 + 3, \quad \text{and} \quad 4 + (6 + 2) = (4 + 6) + 2.$$

Soon we learned about multiplication, negative numbers, and the integers $\mathbb{Z}$. It seemed that certain numbers like 0 and 1 had special properties such as

$$14 + 0 = 14, \quad 1 * 47 = 47, \quad \text{and} \quad 0 = 9 + (-9).$$

Somewhere along the line, we learned about division, the rational numbers $\mathbb{Q}$, and the fact that we could not divide by zero.

Then came the big leap. We learned to denote numbers by symbols like the letters x and y and by expressions like $x^2 + y$. We spent much time transforming one expression into another, such as $x^2 + 4x + 4 = (x + 2)(x + 2)$. All this had something to do with algebra, perhaps because that was the name of the class.

There are two main ingredients to the algebra that we studied in high school. The first is a set of numbers to work with, such as the real numbers $\mathbb{R}$. The second is a set of operations on the numbers, such as $-$ and $+$. We learned about the general properties of the operations, such as $x + y = y + x$

and $x + 0 = x$. And we learned to use these properties to simplify expressions and solve equations.

Now we are in position to discuss algebra from a more general point of view. We will see that high school algebra is just one of many different kinds of algebras.

Definition of an Algebra

An *algebra* is a structure consisting of one or more sets together with one or more operations on the sets. The sets are often called *carriers* of the algebra. This is a very general definition. If this is the definition of an algebra, how can it help us solve problems? As we will see, the utility of an algebra comes from knowing how to use the operations.

For example, high school algebra is an algebra with the single carrier $\mathbb{R}$, or maybe $\mathbb{Q}$. The operators of the algebra are $+$, $-$, $\cdot$, and $\div$. The constants 0 and 1 are also important to consider because they have special properties. Recall that a constant can be thought of as a nullary operation (having arity zero). Many familiar properties hold among the operations, such as the fact that multiplication distributes over addition: $a \cdot (b + c) = a \cdot b + a \cdot c$; and the fact that we can cancel: If $a \neq 0$, then $a \cdot b = a \cdot c$ implies $b = c$.

There are many algebras in computer science. For example, the Pascal data type INTEGER is an algebra. The carrier is a finite set of integers that changes from machine to machine. Some of the operators in this algebra are

maxint, $+$, $-$, $*$, div, mod, succ, pred.

Maxint is a constant that represents the largest integer in the carrier. Of course, there are many relationships that hold among the operations. For example, we know that for any $x \neq$ maxint the two operators pred and succ satisfy the equation $\mathrm{pred}(\mathrm{succ}(x)) = x$. The operators don't need to be total functions. For example, succ(maxint) is not defined.

An *algebraic expression* is a string of symbols used to represent an element in a carrier of an algebra. For example, in high school algebra the strings 3, $8 \div x$, and $x^2 + y$ are algebraic expressions. But the string $x + y +$ is not an algebraic expression. The set of algebraic expressions is a language. The symbols in the alphabet are the operators and constants from the algebra, together with variable names and grouping symbols, like parentheses and commas. The language of algebraic expressions over an algebra can be defined inductively as follows:

Basis: Constants and variables are algebraic expressions.

Induction: An operator applied to its arguments is an algebraic expression if the arguments are algebraic expressions.

For example, suppose x and y are variables and c is a constant. If g is a ternary operator, then the following five strings are algebraic expressions:

$$x, \quad y, \quad c, \quad g(x, y, c), \quad g(x, g(c, y, x), x).$$

Different algebraic expressions often mean the same thing. For example, the equation $2x = x + x$ makes sense to us because we look beyond the two strings $2x$ and $x + x$, which are not equal strings. Instead, we look at the possible values of the two expressions and conclude that they always have the same value, no matter what value x has. Two algebraic expressions are *equivalent* if they always evaluate to the same element in a carrier of the algebra. So the expressions $2x$ and $x + x$ are equivalent in high school algebra. We can make the idea of equivalence precise by giving an inductive definition. Assume that C is a carrier of an algebra.

Basis: Any element in C is equivalent to itself.

Induction: Suppose E and E' are two algebraic expressions and x is a variable such that $E(x/b)$ and $E'(x/b)$ are equivalent for all elements b in C. Then E is equivalent to E'.

For example, the two expressions $(x + 2)^2$ and $x^2 + 4x + 4$ are equivalent in high school algebra. But $x + y$ is not equivalent to $5x$ because we can let $x = 1$ and $y = 2$, which makes $x + y = 3$ and $5x = 5$.

The set of operators in an algebra is called the *signature* of the algebra. When describing an algebra, we need to decide which operators to put in the signature. For example, we may wish to list only the primitive operators (the constructors) that are used to build all other operators. On the other hand, we might want to list all the operators that we know about.

Let's look at a convenient way to denote an algebra. We'll list the carrier or carriers first, followed by a semicolon. The operators in the signature are listed next. Then enclose the two listings with tuple markers. For example, this notation is used to denote the following algebras:

$$\langle \mathbb{N}; \text{succ}, 0 \rangle \qquad\qquad \langle \mathbb{N}; +, \cdot, 0, 1 \rangle$$
$$\langle \mathbb{N}; +, 0 \rangle \qquad\qquad \langle \mathbb{Z}; +, \cdot, -, 0, 1 \rangle$$
$$\langle \mathbb{N}; \cdot, 1 \rangle \qquad\qquad \langle \mathbb{Q}; +, \cdot, -, \div, 0, 1 \rangle$$
$$\langle \mathbb{N}; \text{succ}, +, 0 \rangle \qquad\qquad \langle \mathbb{R}; +, \cdot, -, \div, 0, 1 \rangle$$

The constants 0 and 1 are listed as operations to emphasize the fact that they have special properties, such as $x + 0 = x$ and $x \cdot 1 = x$.

It may also be convenient to use a picture to describe an algebra. The diagram in Figure 10.1 represents the algebra $\langle \mathbb{N}; +, 0 \rangle$.

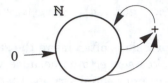

Figure 10.1

The circle represents the carrier $\mathbb{N}$, of natural numbers. The two arrows coming out of $\mathbb{N}$ represent two arguments to the + operator. The arrow from + to $\mathbb{N}$ indicates that the result of + is an element of $\mathbb{N}$. The fact that there are no arrows pointing at 0 means that 0 is a constant (an operator with no arguments), and the arrow from 0 to $\mathbb{N}$ means that 0 is an element of $\mathbb{N}$.

Let's look at some fundamental properties that may be associated with a binary operation. If $\circ$ is a binary operator on a set C, then an element $z \in C$ is called a *zero* for $\circ$ if the following condition holds:

$$z \circ x = x \circ z = z \quad \text{for all } x \in C.$$

For example, the number 0 is a zero for the multiply operation over the real numbers because $0 \cdot x = x \cdot 0 = 0$ for all real numbers x.

Continuing with the same binary operator $\circ$ and carrier C, we call an element $u \in C$ an *identity*, or *unit*, for $\circ$ if the following condition holds:

$$u \circ x = x \circ u = x \quad \text{for all } x \in C.$$

For example, the number 1 is a unit for the multiply operation over the real numbers because $1 \cdot x = x \cdot 1 = x$ for all numbers x. Similarly, the number 0 is a unit for the addition operation over real numbers because $0 + x = x + 0 = x$ for all numbers x.

Suppose u is an identity element for $\circ$, and $x \in C$. An element y in C is called an *inverse* of x if the following equation holds:

$$x \circ y = y \circ x = u.$$

For example, in the algebra $\langle \mathbb{Q}; \cdot, 1 \rangle$ the number 1 is an identity element. We also know that if $x \neq 0$, then

$$x \cdot \frac{1}{x} = \frac{1}{x} \cdot x = 1.$$

In other words, all nonzero rational numbers have inverses.

Each of the following examples presents an algebra together with some observations about its operators.

EXAMPLE 1. Let S be a set. Then the power set of S is the carrier for an algebra described as follows:

$$\langle \text{power}(S); \cup, \cap, \varnothing, S \rangle.$$

Notice that if $A \in \text{power}(S)$, then $A \cup \varnothing = A$, and $A \cap S = A$. So $\varnothing$ is an identity for $\cup$, and S is an identity for $\cap$. Similarly, $A \cap \varnothing = \varnothing$, and $A \cup S = S$. Thus $\varnothing$ is a zero for $\cap$, and S is a zero for $\cup$. This algebra has many well-known properties. For example, $A \cup A = A$ and $A \cap A = A$ for any $A \in \text{power}(S)$. We also know that $\cap$ and $\cup$ are commutative and associative and that they distribute over each other. ◆

EXAMPLE 2. Let $\mathbb{N}_n$ denote the set $\{0, 1, ..., n-1\}$, and let "max" be the function that returns the maximum of its two arguments. Consider the following algebra with carrier $\mathbb{N}_n$:

$$\langle \mathbb{N}_n; \text{max}, 0, n-1 \rangle.$$

Notice that max is commutative and associative. Notice also that for any $x \in \mathbb{N}_n$ it follows that $\text{max}(x, 0) = \text{max}(0, x) = x$. So 0 is an identity for max. It's also easy to see that for any $x \in \mathbb{N}_n$,

$$\text{max}(x, n-1) = \text{max}(n-1, x) = n-1.$$

So $n-1$ is a zero for the operator max. ◆

EXAMPLE 3. Let S be a set, and let F be the set of all functions of type $S \to S$. If we let $\circ$ denote the operation of composition of functions, then F is the carrier of an algebra $\langle F; \circ, \text{id} \rangle$. The function id denotes the identity function. In other words, we have the equation $\text{id} \circ f = f \circ \text{id} = f$ for all functions f in F. Therefore id is an identity for $\circ$. ◆

Notice that we used the equality symbol "=" in the above examples without explicitly defining it as a relation. The first example uses equality of sets, the second uses equality of numbers, and the third uses equality of functions. In our discussions we will usually assume an implicit equality theory on each carrier of an algebra. But, as we have said before, equality relations are operations that may need to be implemented when needed as part of a programming activity.

An algebra is called *concrete* if its carriers are specific sets of elements so that its operators are defined by rules applied to the carrier elements. High school algebra is a concrete algebra. In fact, all the examples that we have seen so far are concrete algebras.

An algebra is called *abstract* if it is not concrete. In other words, its carriers don't have any specific set interpretation. Thus its operators cannot be defined in terms of rules applied to the carrier elements because we don't have a description of them. Therefore the general properties of the operators in an abstract algebra must be given by axioms. An abstract algebra is a powerful description tool because it represents all the concrete algebras that satisfy its axioms. Thus when we talk about an abstract algebra, we are really talking about all possible examples of the algebra. Is this a useful activity? Sure. Many times we are overwhelmed with important concepts, but we aren't given any tools to make sense of them. Abstraction can help to classify things and thus make sense of things that act in similar ways.

If an algebra is abstract, then we must be more explicit when trying to describe it. For example, suppose we write down the following algebra:

$$\langle S; s, a \rangle.$$

All we know at this point is that S is a carrier and there are two operators s and a. We don't even know the arity of s or a. Suppose we're told that a is a constant of S and s is a unary operator on S. Now we know something, but not very much, about the algebra. We can use the operators s and a to construct the following algebraic expressions for elements of S:

$$a, s(a), s(s(a)), ..., s^n(a), ...$$

This is the most we can say about the elements of S. There might be other elements in S, but we have no way of knowing it. The elements of S that we know about can be represented by all possible algebraic expressions made up from the operator symbols in the signature together with left and right parentheses. A concrete example of such an algebra is $\langle \mathbb{N}; \text{succ}, 0 \rangle$. We can write down all possible algebraic expressions for this algebra as

$$0, \text{succ}(0), \text{succ}(\text{succ}(0)), ..., \text{succ}^n(0), ...$$

Of course, we normally abbreviate $\text{succ}(0) = 1$, $\text{succ}(\text{succ}(0)) = 2$, and so on.

Another concrete example of such an algebra is $\langle \mathbb{N}_4; \text{succ}_4, 0 \rangle$, where we define the operator succ_4 as

$$\text{succ}_4(x) = \text{succ}(x) \bmod 4.$$

The algebraic expressions for the carrier elements are

$$0, \text{succ}_4(0), \text{succ}_4(\text{succ}_4(0)), ..., \text{succ}_4^n(0), ...$$

Notice in this case that we have infinitely many expressions that represent the four distinct numbers 0, 1, 2, 3 because $\text{succ}_4^4(x) = x$ for the numbers x in $\mathbb{N}_4$.

Interesting things can happen when we add axioms to an algebra. For example, the algebra $\langle S; s, a \rangle$ changes its character when we add the single axiom $s^6(x) = x$ for all $x \in S$. All we can say about this algebra is that the algebraic expressions define a finite set of elements, which can be represented by the following six expressions:

$$a, s(a), s^2(a), s^3(a), s^4(a), s^5(a).$$

A complete definition of an abstract algebra can be given by listing the carriers, operations, and axioms. For example, the abstract algebra that we've just been discussing can be defined as follows:

Carrier: S

Operations: $a \in S$

              $s : S \to S$

Axiom: $s^6(x) = x.$

We'll always assume that the variable x is universally quantified over S.

The algebra $\langle \mathbb{N}_6; \text{succ}_6, 0 \rangle$ is a concrete example of the preceding algebra. But the algebra $\langle \mathbb{N}; \text{succ}, 0 \rangle$ is not a concrete example because it doesn't satisfy the axiom.

EXAMPLE 4 (*Induction Algebra*). An algebra $\langle S; s, a \rangle$ is called an *induction algebra* if s is a unary operator on S and a is a constant of S such that

$$S = \{a, s(a), s(s(a)), ..., s^n(a), ...\}.$$

The word "induction" is used because of the natural ordering on the carrier that can be used in inductive proofs. For example, the algebra $\langle \mathbb{N}; \text{succ}, 0 \rangle$ is a concrete example of an induction algebra. Similarly, suppose we define the set

$$A = \{2, 1, 0, -1, -2, -3, ...\}.$$

Then the algebra $\langle A; \text{pred}, 2 \rangle$ is an induction algebra, and its elements can be represented by the expressions

$$2, \text{pred}(2), \text{pred}(\text{pred}(2)), ... \quad \blacklozenge$$

Working in Algebras

The goal of the following paragraphs is to get familiar with some elementary algebraic techniques that are used to solve problems.

Operation Tables

Any binary operation on a finite set can be represented by a table, called an *operation table*. For example, if $\circ$ is a binary operation on the set $\{a, b, c, d\}$, then the operation table for $\circ$ might look like Table 10.1, where the elements of the set are used as row labels and column labels.

$\circ$	a	b	c	d
a	a	b	c	d
b	b	c	d	a
c	c	d	a	b
d	d	a	b	c

Table 10.1

If x is a row label and y is a column label, then the element in the table at row x and column y represents the element $x \circ y$. For example, we have $c \circ d = b$.

We can often find out many things about a binary operation by observing its operation table. For example, notice in Table 10.1 that the row labeled a and the column labeled a are copies of the row label and column label sequence $a \ b \ c \ d$. This tells us that a is an identity for $\circ$. It's also easy to see that $\circ$ is commutative and that each element has an inverse. Does $\circ$ have a zero? It's easy to see that the answer is no. Is $\circ$ associative? The answer is yes, but it's not very easy to check. We'll leave these problems as exercises.

It's also easy to see that there cannot be more than one identity for a binary operation. Can you see why from Table 10.1? We'll prove the following general fact about identities for any binary operation:

Any binary operation has at most one identity.   (10.1)

Proof: Let ∘ be a binary operation on a set S. To show that ∘ has at most one identity, we'll assume that u and e are identities for ∘. Then we'll show that $u = e$. Remember, since u and e are identities, we know that $u \circ x = x \circ u = x$ and $e \circ x = x \circ e = x$ for all x in S. Thus we have the following equality:

$$e = e \circ u \qquad u \text{ is an identity for } \circ$$
$$= u \qquad e \text{ is an identity for } \circ \quad \text{QED.}$$

Algebras with One Binary Operation

Some algebras are used so frequently that they have been given names. For example, any algebra of the form $\langle A; \circ \rangle$, where ∘ is a binary operation, is called a *groupoid*. If we know that the binary operation is associative, then the algebra is called a *semigroup*. If we know that the binary operation is associative and also has an identity, then the algebra is called a *monoid*. If we know that the binary operation is associative, it has an identity, and each element has an inverse, then the algebra is called a *group*. So these words are used to denote certain properties of the binary operation. We can display the names and properties as follows:

Groupoid: ∘ is a binary operation.

Semigroup: ∘ is an associative binary operation.

Monoid: ∘ is an associative binary operation with an identity.

Group: ∘ is an associative binary operation with an identity and every element has an inverse.

We can classify these algebras as follows, where each name denotes the set of all algebras of that kind:

$$\text{Groups} \subset \text{Monoids} \subset \text{Semigroups} \subset \text{Groupoids.}$$

These containments are proper. For example, every group is a monoid. But there are monoids that are not groups. For example, the algebra in Example 3 is a monoid but not a group, since not every function has an inverse.

We can have some fun with these names. For example, we can describe a group as a monoid with inverses, and we can describe a monoid as a semigroup with identity. When an algebra contains an operation that satisfies some special property beyond the axioms of the algebra, we often modify the name of the algebra with the name of the property. For example, the algebra

$\langle \mathbb{N}; +, 0 \rangle$ is a group. But we know that the operation $+$ is commutative. Therefore we can call the algebra a "commutative" group.

Now let's discuss a few elementary results. To get our feet wet, we'll prove the following simple property that holds in any monoid:

> If an element in a monoid has an inverse, then that inverse is unique. (10.2)

Proof: Let $\langle M; \circ, u \rangle$ be a monoid. We will show that if an element x in M has an inverse, then the inverse is unique. In other words, if y and z are both inverses of x, then $y = z$. We can prove this result as follows:

$$
\begin{aligned}
y &= y \circ u && (u \text{ is the identity for } \circ) \\
  &= y \circ (x \circ z) && (z \text{ is an inverse of } x) \\
  &= (y \circ x) \circ z && (\circ \text{ is associative}) \\
  &= u \circ z && (y \text{ is an inverse of } x) \\
  &= z && (u \text{ is the identity for } \circ). \quad \text{QED.}
\end{aligned}
$$

If we have a group, then we know that every element has an inverse. Thus we can conclude from (10.2) that every element x in a group has a unique inverse, which is usually denoted by writing the symbol

$$ x^{-1}. $$

EXAMPLE 5 (*Working with Groups*). We can use the elementary properties of a group to obtain other properties. For example, let $\langle G; \circ, e \rangle$ be a group. This means that we know that $\circ$ is associative, e is an identity, and every element of G has an inverse. The first property that we want to prove is:

> Whenever $x \circ x = x$ holds for some element $x \in G$, then $x = e$. (10.3)

Proof: To prove this statement, we need all the properties of a group:

$$
\begin{aligned}
x &= x \circ e && (e \text{ is an identity for } \circ) \\
  &= x \circ (x \circ x^{-1}) && (x^{-1} \text{ is the inverse of } x) \\
  &= (x \circ x) \circ x^{-1} && (\circ \text{ is associative}) \\
  &= x \circ x^{-1} && (x \circ x = x \text{ is the hypothesis}) \\
  &= e && (x^{-1} \text{ is the inverse of } x). \quad \text{QED.}
\end{aligned}
$$

Another property of groups is cancellation on the left. This property can be stated as follows:

$$\text{if } x \circ y = x \circ z \text{ then } y = z. \qquad (10.4)$$

Proof: We can prove this statement by equational reasoning as follows:

$$
\begin{aligned}
y \quad &= e \circ y & &(e \text{ is an identity}) \\
&= (x^{-1} \circ x) \circ y & &(x^{-1} \text{ is the inverse of } x) \\
&= x^{-1} \circ (x \circ y) & &(\circ \text{ is associative}) \\
&= x^{-1} \circ (x \circ z) & &(\text{hypothesis}) \\
&= (x^{-1} \circ x) \circ z & &(\circ \text{ is associative}) \\
&= e \circ z & &(x^{-1} \text{ is the inverse of } x) \\
&= z & &(e \text{ is an identity}). \quad \text{QED.}
\end{aligned}
$$

The properties of groups are too numerous to mention. We'll discuss a few more simple properties in the exercises. ◆

Algebras with Several Operations

A natural example of an algebra with two binary operations is the integers together with the usual operations of addition and multiplication. We can denote this algebra by the structure $\langle \mathbb{Z}; +, \cdot, 0, 1 \rangle$. This algebra is a concrete example of an algebra called a ring, which we'll now define. A *ring* is an algebra with the structure

$$\langle A; +, \cdot, 0, 1 \rangle,$$

where $\langle A; +, 0 \rangle$ is a commutative group, $\langle A; \cdot, 1 \rangle$ is a monoid, and the operation $\cdot$ distributes over $+$ from the left and the right. This means that

$$a \cdot (b + c) = a \cdot b + a \cdot c \quad \text{and} \quad (b + c) \cdot a = b \cdot a + c \cdot a.$$

Check to see that $\langle \mathbb{Z}; +, \cdot, 0, 1 \rangle$ is indeed a ring.

If $\langle A; +, \cdot, 0, 1 \rangle$ is a ring with the additional property that $\langle A - \{0\}; \cdot, 1 \rangle$ is a commutative group, then it's called a *field*. The ring $\langle \mathbb{Z}; +, \cdot, 0, 1 \rangle$ is not a field because, for example, 3 does not have an inverse for multiplication. On the other hand, if we replace $\mathbb{Z}$ by $\mathbb{Q}$, the rational numbers, then $\langle \mathbb{Q}; +, \cdot, 0, 1 \rangle$ is a field. For example, 3 has inverse $\frac{1}{3}$ in $\mathbb{Q} - \{0\}$.

For another example of a field, let $\mathbb{N}_5 = \{0, 1, 2, 3, 4\}$ and let $+_5$ and $\cdot_5$ be addition mod 5 and multiplication mod 5, respectively. Then $\langle \mathbb{N}_5; +_5, \cdot_5, 0, 1 \rangle$ is a field. Tables 10.2 and 10.3 are the operation tables for $+_5$ and $\cdot_5$. We'll leave the verification of the field properties as an exercise.

$+_5$	0	1	2	3	4
0	0	1	2	3	4
1	1	2	3	4	0
2	2	3	4	0	1
3	3	4	0	1	2
4	4	0	1	2	3

$\cdot_5$	0	1	2	3	4
0	0	0	0	0	0
1	0	1	2	3	4
2	0	2	4	1	3
3	0	3	1	4	2
4	0	4	3	2	1

Table 10.2   **Table 10.3**

The next examples show some algebras that might be familiar to you.

EXAMPLE 6 (*Polynomial Algebras*). Let $\mathbb{R}[x]$ denote the set of all polynomials over x with real numbers as coefficients. It's a natural process to add and multiply two polynomials. So we have an algebra $\langle \mathbb{R}[x]; +, \cdot, 0, 1 \rangle$, where $+$ and $\cdot$ represent addition and multiplication of polynomials and 0 and 1 represent themselves. This algebra is a ring. Why isn't it a field?   ◆

EXAMPLE 7 (Matrix Algebras). Suppose we let $M_n(\mathbb{R})$ denote the set of all n by n matrices with elements in $\mathbb{R}$. We can add two matrices A and B by letting $A_{ij} + B_{ij}$ be the element in the ith row and jth column of the sum. We can multiply A and B by letting $\sum_{k=1}^{n} A_{ik} B_{kj}$ be the element in the ith row and jth column of the product. Thus we have an algebra $\langle M_n(\mathbb{R}); +, \cdot, 0, 1 \rangle$, where $+$ and $\cdot$ represent matrix addition and multiplication, 0 represents the matrix with all entries zero, and 1 represents the matrix with 1's along the main diagonal and 0's elsewhere. This algebra is a ring. Why isn't it a field?   ◆

EXAMPLE 8 (*Vector Algebras*). The algebra of n-dimensional vectors, with real numbers as components, can be described by listing two carriers $\mathbb{R}$ and $\mathbb{R}^n$. We can multiply a vector $\langle x_1, \ldots, x_n \rangle \in \mathbb{R}^n$ by number $b \in \mathbb{R}$ to obtain a new vector by multiplying each component of the vector by b, obtaining

$$\langle bx_1, \ldots, bx_n \rangle.$$

If we let $\cdot$ denote this operation, then we have

$$b \cdot \langle x_1, \ldots, x_n \rangle = \langle bx_1, \ldots, bx_n \rangle.$$

We can add vectors by adding corresponding components. For example,

$$\langle x_1, ..., x_n \rangle + \langle y_1, ..., y_n \rangle = \langle x_1 + y_1, ..., x_n + y_n \rangle.$$

Thus we have an algebra of n-dimensional vectors, which we can write in tuple form as $\langle \mathbb{R}, \mathbb{R}^n; \cdot, + \rangle$. Notice that the algebra has two carriers, $\mathbb{R}$ and $\mathbb{R}^n$. This is because they are both necessary to define the $\cdot$ operation, which has type $\mathbb{R} \times \mathbb{R}^n \to \mathbb{R}^n$.   ♦

EXAMPLE 9 (*Power Series Algebras*). If we extend polynomials over x to allow infinitely many terms, then we obtain what are called *power series* (we also know them as generating functions). Letting $\mathbb{R}[[x]]$ denote the set of power series with real numbers as coefficients, we obtain the algebra $\langle \mathbb{R}[[x]]; +, \cdot, 0, 1 \rangle$, where $+$ and $\cdot$ represent addition and multiplication of power series and 0 and 1 represent themselves. This algebra is a ring. Why isn't it a field?   ♦

Exercises

1. Let m and n be two integers with $m < n$. Let $A = \{m, m + 1, ..., n\}$, and let "min" be the function that returns the smaller of its two arguments. Does min have a zero? Identity? Inverses? If so, describe them.

2. Let $A = \{\text{true, false}\}$. For each of the following binary operations on A, answer the three questions: Does the operation have a zero? Does the operation have an identity? What about inverses?
    a. Conditional, $\to$.
    b. Conjunction, $\wedge$.
    c. Disjunction, $\vee$.

3. Given the algebra $\langle S; f, a \rangle$, where f is a unary operation and a is a constant of S, suppose that all elements of S are described by algebraic expressions involving f and a. Suppose also that the axiom $f^5(x) = f^3(x)$ holds. Find a finite set of algebraic expressions that will represent the distinct elements of S.

4. Given a binary operation on a finite set in table form, for each of the following parts, describe an easy way to detect whether the binary operation has the listed property.
    a. There is a zero.
    b. The operation is commutative.
    c. Inverses exist for each element of the set (assume that there is an identity).

5. Let $A = \{a, b, c, d\}$, and let $\circ$ be a binary operation on A. For each of the following problems, write down a table for $\circ$ that satisfies the given properties.

   a. a is an identity for $\circ$, but no other element of A has an inverse.

   b. a is an identity for $\circ$, and every element of A has an inverse.

   c. a is a zero for $\circ$, and $\circ$ is not associative.

   d. a is an identity, and exactly two elements have inverses.

   e. a is an identity for $\circ$, and $\circ$ is commutative but not associative.

6. Let $A = \{a, b\}$. For each of the following problems, find an operation table satisfying the given condition for a binary operation $\circ$ on A.

   a. $\langle A; \circ \rangle$ is a group.

   b. $\langle A; \circ \rangle$ is a monoid but not a group.

   c. $\langle A; \circ \rangle$ is a semigroup but not a monoid.

   d. $\langle A; \circ \rangle$ is a groupoid but not a semigroup.

7. Write an algorithm to check a binary operation table for associativity.

8. Prove each of the following facts about a group $\langle G; \circ, e \rangle$.

   a. Cancellation on the right: If $y \circ x = z \circ x$, then $y = z$.

   b. The inverse of $x \circ y$ is $y^{-1} \circ x^{-1}$. In other words, $(x \circ y)^{-1} = y^{-1} \circ x^{-1}$.

9. Let $\mathbb{N}_5 = \{0, 1, 2, 3, 4\}$, and let $+_5$ and $\cdot_5$ be the two operations of addition mod 5 and multiplication mod 5, respectively. Show that $\langle \mathbb{N}_5; +_5, \cdot_5, 0, 1 \rangle$ is a field.

10.2 Boolean Algebra

Do the techniques of set theory and the techniques of logic have anything in common? Let's do an example to see that the answer is yes. When working with sets, we know that the following equation holds for all sets A, B, and C:

$$A \cup (B \cap C) = (A \cup B) \cap (A \cup C).$$

When working with propositions, we know that the following equivalence holds for all propositions A, B, and C:

$$A \vee (B \wedge C) \equiv (A \vee B) \wedge (A \vee C).$$

Certainly these two examples have a similar pattern. As we'll see shortly, sets and logic have a lot in common. They can both be described as

concrete examples of a Boolean algebra. The name "Boolean" comes from the
mathematician George Boole (1815–1864), who studied relationships between
set theory and logic. Let's get to the definition.

An algebra is a *Boolean algebra* if it has the structure

$$\langle B; +, \cdot, ^-, 0, 1 \rangle,$$

where the following properties hold:

1. $\langle B; +, 0 \rangle$ and $\langle B; \cdot, 1 \rangle$ are commutative monoids. In other words, the
   following properties hold for all $x, y, z \in B$:

$$(x + y) + z = x + (y + z), \qquad (x \cdot y) \cdot z = x \cdot (y \cdot z),$$

$$x + y = y + x, \qquad\qquad x \cdot y = y \cdot x,$$

$$x + 0 = x, \qquad\qquad x \cdot 1 = x.$$

2. $+$ and $\cdot$ distribute over each other. In other words, the following properties hold for all $x, y, z \in B$:

$$x \cdot (y + z) = (x \cdot y) + (x \cdot z) \quad \text{and} \quad x + (y \cdot z) = (x + y) \cdot (x + z).$$

3. $x + \bar{x} = 1$ and $x \cdot \bar{x} = 0$ for all elements $x \in B$. The element $\bar{x}$ is called
   the *complement* of x or the *negation* of x.

We often drop the dot and write xy in place of $x \cdot y$. We'll also reduce the
need for parentheses by agreeing to the following precedence hierarchy:

$$^- \qquad \text{highest (do it first)},$$

$$\cdot$$

$$+ \qquad \text{lowest (do it last)}.$$

For example, the expression $a + b\,\bar{c}$ means the same thing as $(a + (b(\bar{c})))$.

Let's look at a few examples.

EXAMPLE 1 (*Sets*). Suppose $B = \text{power}(S)$ for some set S. Then B is the carrier
of a Boolean algebra if we let union and intersection act as the operations $+$
and $\cdot$, let X' be the complement of X, let $\varnothing$ act as 0, and let S act as 1. For example, the two properties in part 3 of the definition are represented by the
following equations, where X is any subset of S:

$$X \cup X' = S \quad \text{and} \quad X \cap X' = \varnothing. \quad \blacklozenge$$

EXAMPLE 2 (*Logic*). Suppose we let B be the set of all propositional wffs of the propositional calculus. Then B is the carrier of a Boolean algebra if we let disjunction and conjunction act as the operations $+$ and $\cdot$, let $\neg X$ be the complement of X, let false act as 0, let true act as 1, and let logical equivalence act as equality. For example, the two properties in part 3 of the definition are represented by the following equivalences, where X is any proposition:

$$X \vee \neg X \equiv \text{true} \quad \text{and} \quad X \wedge \neg X \equiv \text{false}.$$

We can also obtain a very simple Boolean algebra by using just the carrier {false, true} together with the operations $\vee$, $\wedge$, and $\neg$.   ◆

EXAMPLE 3. Suppose we let B_n be the set of positive divisors of n, where n is a product of distinct prime numbers. For example, n could be 10 because it is the product of primes 2 and 5. But n cannot be 12 because it is the product of primes 2, 2, and 3, which are not distinct. Then B_n is the carrier of a Boolean algebra if we let "least common multiple" and "greatest common divisor" act as the operations $+$ and $\cdot$, let n/x be the complement of x, let 1 act as the zero, and let n act as the one. For example, the two properties in part 3 of the definition are represented by the following equations, where $x \in B_n$:

$$\text{lcm}(x, n/x) = n \quad \text{and} \quad \gcd(x, n/x) = 1.$$

For example, if $n = 10$, then $B_{10} = \{1, 2, 5, 10\}$, 1 is the zero, 10 is the one, the complement of 2 is 5, $\text{lcm}(2, 5) = 10$ (the one), and $\gcd(2, 5) = 1$ (the zero).

Notice what happens if we let $n = 12$. We get $B_{12} = \{1, 2, 3, 4, 6, 12\}$. The reason B_{12} does not yield a Boolean algebra under our definitions is because 2 and its complement 6 don't satisfy the properties in part 3 of the definition. Notice that $\text{lcm}(2, 6) = 6$, which is not the one, and $\gcd(2, 6) = 2$, which is not the zero.   ◆

Simplifying Boolean Expressions

A fundamental problem of Boolean algebra, with applications to such areas as logic design and theorem-proving systems, is to represent a Boolean expression in a simple way. Here we define *simple* to mean a small number of operations. We can use the axioms of Boolean algebra to obtain some useful properties that can help us simplify Boolean expressions.

For example, in the Boolean algebra of propositions we have $P \wedge P = P$ for any proposition P. Similarly, in the Boolean algebra of sets we have $S \cap S = S$ for any set S. Can we generalize these properties to all Boolean algebras?

In other words, can we say $b \cdot b = b$ for every element b in the carrier of a Boolean algebra? The answer is yes. Let's prove it with equational reasoning. Be sure you can provide a reason for each step of the following proof:

$$b = b \cdot 1 = b \cdot (b + \bar{b}) = b \cdot b + b \cdot \bar{b} = b \cdot b + 0 = b \cdot b.$$

A related statement is $b + b = b$ for all elements b. Can you provide the proof? We'll state these two properties for the record.

Idempotent Properties

$$b \cdot b = b \quad \text{and} \quad b + b = b. \tag{10.5}$$

A nice property of Boolean algebras is that results come in pairs. This is because the axioms come in pairs. In other words, $\langle B; +, 0 \rangle$ and $\langle B; \cdot, 1 \rangle$ are both commutative monoids; $+$ and $\cdot$ distribute over each other; and $b + \bar{b} = 1$ and $b \cdot \bar{b} = 0$ for all elements $b \in B$. The *duality principle* states that whenever a result A is true for a Boolean algebra, then a dual result A' is also true, where A' is obtained from the A by simultaneously replacing all occurrences of $\cdot$ by $+$, all occurrences of $+$ by $\cdot$, all occurrences of 1 by 0, and all occurrences of 0 by 1. A proof for the result A' can be obtained by making these same changes in the proof of A.

There are lots of properties that we can discover. For example, if S is a set, then $\varnothing \cap A = \varnothing$ for any subset A of S. This is an instance of a general property that holds for any Boolean algebra: $0 \cdot b = 0$ for every element b. This follows readily from (10.5) as follows:

$$0 \cdot b = (\bar{b} \cdot b) \cdot b = \bar{b} \cdot (b \cdot b) = \bar{b} \cdot b = 0.$$

Again, there is a dual result: $1 + b = 1$ for every b. Can you prove this result? We'll also state these two properties for the record:

$$0 \cdot b = 0 \quad \text{and} \quad 1 + b = 1. \tag{10.6}$$

Let's do an example to see how we can put our new knowledge to use in simplifying a Boolean expression. Suppose the function f is defined over a Boolean algebra by

$$f(x, y, z) = x + yz + z\,\bar{x}y + \bar{y}xz.$$

To evaluate f, we need to perform three $+$ operations, five $\cdot$ operations, and two $^-$ operations. Can we do any better? Sure we can. We can simplify the expression for $f(x, y, z)$ as follows—make sure you can state a reason for each line:

$$
\begin{aligned}
f(x, y, z) &= x + yz + z\bar{x}y + \bar{y}xz \\
&= x + yz\,(1 + \bar{x}) + \bar{y}xz \\
&= x + yz\,1 + \bar{y}xz \\
&= x\,(1 + \bar{y}z) + yz \\
&= x\,1 + yz \\
&= x + yz.
\end{aligned}
$$

So f can be evaluated with only one $+$ operation and one $\cdot$ operation.

To simplify Boolean expressions, it's important to have a good knowledge of Boolean algebra together with some luck and ingenuity. We'll give a few more general properties that are very useful simplification tools. The following properties can be used to simplify an expression by reducing the number of operations by two. We'll leave the proofs as exercises.

Absorption Laws

$$
a + ab = a \quad \text{and} \quad a\,(a + b) = a. \tag{10.7}
$$
$$
a + \bar{a}b = a + b \quad \text{and} \quad a\,(\bar{a} + b) = ab.
$$

In a Boolean algebra, complements are unique in the following sense: If an element acts like a complement of some element, then it is in fact the only complement of the element. Using symbols, we can state the result as follows:

$$
\text{if } a + b = 1 \text{ and } a\,b = 0, \text{ then } b = \bar{a}. \tag{10.8}
$$

Proof: To prove this statement, we write the following equations:

$$
\begin{aligned}
b &= b\,1 \\
&= b\,(a + \bar{a}) \\
&= ba + b\bar{a} \\
&= 0 + b\bar{a} \qquad \text{(since } ab = 0\text{)} \\
&= a\bar{a} + b\bar{a} \\
&= (a + b)\bar{a} \\
&= 1\,\bar{a} \qquad \text{(since } a + b = 1\text{)} \\
&= \bar{a} \qquad \text{QED.}
\end{aligned}
$$

Complements are quite useful in Boolean algebra. As a consequence of the uniqueness of complements (10.8), we have the following property:

Involution law

$$\bar{\bar{a}} = a. \tag{10.9}$$

Proof: Notice that $\bar{a} + a = 1$ and $\bar{a}a = 0$. Therefore a acts like the complement of $\bar{a}$. Thus a is indeed equal to the complement of $\bar{a}$. That is, $a = \bar{\bar{a}}$.   QED.

Recall from the propositional calculus that we have the following logical equivalence:

$$\neg\,(p \wedge q) \equiv \neg\,p \vee \neg\,q.$$

This is an example of one of De Morgan's laws:

De Morgan's laws

$$\overline{a+b} = \bar{a}\,\bar{b} \quad \text{and} \quad \overline{ab} = \bar{a} + \bar{b}. \tag{10.10}$$

Proof: We'll prove the first of the two laws and leave the second as an exercise. We'll use (10.8) to show that $\overline{a+b} = \bar{a}\,\bar{b}$. In other words, we'll show that $\bar{a}\,\bar{b}$ acts like the complement of $a + b$. Then we'll use (10.8) to conclude the result. First we'll show that $(a + b) + \bar{a}\,\bar{b} = 1$ as follows:

$$
\begin{aligned}
(a + b) + \bar{a}\,\bar{b} \ &= \ (a + b + \bar{a})(a + b + \bar{b}) \\
&= \ (a + \bar{a} + b)(a + b + \bar{b}) \\
&= \ (1 + b)(a + 1) \\
&= \ 1 \cdot 1 \\
&= \ 1.
\end{aligned}
$$

Next we'll show that $(a + b) \cdot \bar{a}\,\bar{b} = 0$ as follows:

$$
\begin{aligned}
(a + b) \cdot \bar{a}\,\bar{b} \ &= \ a\bar{a}\,\bar{b} + b\bar{a}\,\bar{b} \\
&= \ a\bar{a}\,\bar{b} + \bar{a}b\bar{b} \\
&= \ 0\,\bar{b} + \bar{a}\,0 \\
&= \ 0 + 0 \\
&= \ 0.
\end{aligned}
$$

Thus $\bar{a}\,\bar{b}$ acts like a complement of $a + b$. So we can apply (10.8) to conclude that $\bar{a}\,\bar{b}$ is the complement of $a + b$. In other words, $\overline{a+b} = \bar{a}\,\bar{b}$.   QED.

Recall that for any wff in the propositional calculus we can find a disjunctive normal form (DNF) and a conjunctive normal form (CNF). These ideas carry over to any Boolean algebra, where + corresponds to disjunction and · corresponds to conjunction. So we can make the following statement:

Any Boolean expression has a DNF and a CNF.

For example, a DNF for the expression

$$(\bar{a} + \bar{b} + c)a$$

can be found as follows:

$$(\bar{a} + \bar{b} + c)a = (\bar{a} + \bar{b})a + ca = \bar{a}a + \bar{b}a + ca.$$

Digital Circuits

Now let's see what Boolean algebra has to do with digital circuits. A *digital circuit* (also called a *logic circuit*) is an electronic representation of a function whose input values are either high or low voltages and whose output value is either a high or low voltage. Digital circuits are used to represent and process information in digital computers. The high- and low-voltage values are normally represented by the two digits 1 and 0. The basic electronic components used to build digital circuits are called *gates*. The three basic "logic" gates are the AND gate, the OR gate, and the NOT gate. These gates work just like the corresponding logical operations, where 1 means true and 0 means false. So we can represent digital circuits as Boolean expressions with values in the Boolean algebra whose carrier is {0, 1}, where 0 means false, 1 means true, and the operations +, ·, and $^-$ stand for $\lor$, $\land$, and $\neg$, respectively.

The three logic gates are represented graphically as shown in Figure 10.2, where the inputs are on the left and the outputs are on the right.

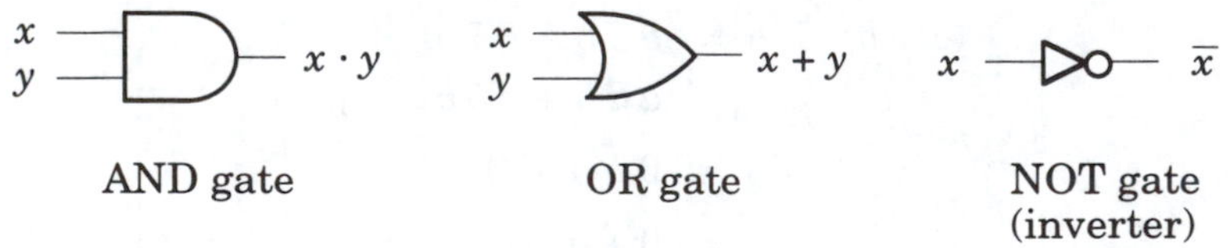

AND gate   OR gate   NOT gate
                                   (inverter)

Figure 10.2

These gates can be combined in various ways to form digital circuits. For example, suppose we want to add two binary digits x and y. The first thing to notice is that the result has a summand digit and a carry digit. We'll consider two functions, "carry" and "summand." Let's look at the carry function first. Notice that carry$(x, y) = 1$ if and only if $x = 1$ and $y = 1$. Thus we can define the carry function as follows:

$$\mathrm{carry}(x, y) = x \cdot y.$$

A circuit to implement the carry function consists of the simple AND gate shown in Figure 10.2.

Now let's look at the summand. It's clear that summand$(x, y) = 1$ if and only if either $x = 0$ and $y = 1$ or $x = 1$ and $y = 0$. Thus we can define the summand function as follows:

$$\mathrm{summand}(x, y) = \bar{x}y + x\bar{y}.$$

A circuit to implement the summand function is shown in Figure 10.5.

Figure 10.5

We can combine the two circuits for the carry and the summand into one circuit that gives both outputs. The circuit is shown in Figure 10.4.

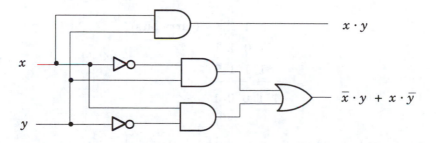

Figure 10.4   Half-adder circuit.

Such a circuit is called a *half-adder*, and it's a fundamental building block in all arithmetic circuits. Let's look at some more examples.

EXAMPLE 4 (*A Simpler Half-Adder*). Let's see whether we can simplify the circuit for a half-adder. The preceding circuit for a half-adder has six gates. Can we do better? The answer is yes. First, notice that the expression for the summand, $\bar{x}y + x\bar{y}$ has five operations: two negations, two conjunctions, and one disjunction. Let's rewrite it as follows (be sure to fill in the reasons for each step):

$$\begin{aligned}
\bar{x}y + x\bar{y} &= (\bar{x}y + x)(\bar{x}y + \bar{y}) \\
&= (x + y)(\bar{x} + \bar{y}) \\
&= (x + y)\,\overline{xy}.
\end{aligned}$$

This latter expression has four operations: two conjunctions, one disjunction, and one negation. Also, note that the expression $x \cdot y$ is computed before the negation is applied. Therefore we can also use this expression for the carry. So we have a simpler version of the half-adder, as shown in Figure 10.5.   ◆

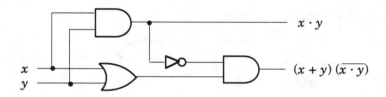

Figure 10.5   Simpler half-adder circuit.

EXAMPLE 5 (*A Full Adder*). Suppose we want to add the two binary numbers 1 0 1 1 and 1 1 1 0. The school method can be pictured as follows:

$$\begin{array}{cccc}
1 & 1 & & \text{(carry bits)} \\
1 & 0 & 1 & 1 \\
1 & 1 & 1 & 0 \\
\hline
1 & 1 & 0 & 0 & 1
\end{array}$$

So if we want to add two binary numbers, then we can start by using a half-adder on the two rightmost digits of each number. After that, we must be able to handle the addition of three binary digits: two binary digits and a carry from the preceding addition. A digital circuit to accomplish this latter feat is called a *full adder*. We can build a full adder by using half-adders as components. Let's see how it goes. First, to get the big picture, we will denote a half-adder by a box with two input lines and two output lines, as shown in Figure 10.6.

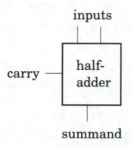

Figure 10.6

To get an idea about the kind of circuit we need, let's look at a table of values for the outputs sum and carry. Table 10.4 shows the values of sum and carry that are obtained by adding three binary digits.

x	y	z	Sum	Carry
0	0	0	0	0
0	0	1	1	0
0	1	0	1	0
0	1	1	0	1
1	0	0	1	0
1	0	1	0	1
1	1	0	0	1
1	1	1	1	1

Table 10.4

Let's use Table 10.4 to find DNFs for the sum and carry functions in terms of x, y and z. Notice that the value of the sum is 1 in four places, on lines 2, 3, 5, and 8 of the table. So the DNF for the sum function will consist of the disjunction of four terms, each a conjunction constructed from the values of x, y, and z on the four lines. Similarly, the value of the carry is 1 in four places, on lines 4, 6, 7, and 8. So the DNF for the carry function depends on conjunctions that depend on these latter four lines. We obtain the following forms for sum and carry:

$$\text{sum}(x, y, z) = \bar{x}\bar{y}z + \bar{x}y\bar{z} + x\,\bar{y}\,\bar{z} + xyz$$
$$= \bar{x}\,(\bar{y}z + y\bar{z}\,) + x\,(\bar{y}\,\bar{z} + yz).$$

$$\text{carry}(x, y, z) = \bar{x}yz + x\bar{y}z + xy\bar{z} + xyz$$
$$= yz + x(\bar{y}z + y\bar{z}\,).$$

At this point we could build a circuit for sum and carry. But let's look at the expressions that we obtained. Notice first that the expression

$$\bar{y}z + y\bar{z}$$

occurs in both the sum formula and the carry formula. Recall also that this is the expression for the summand output of the half-adder. It can be shown that the expression $\bar{y}\,\bar{z} + yz$, in the sum function, is equal to the negation of the expression $\bar{y}z + y\bar{z}$ (the proof is left as an exercise). In other words, if we let $e = \bar{y}z + y\bar{z}$, then we can write the sum in the following form:

$$\text{sum}(x, y, z) = \bar{x}e + x\bar{e}.$$

This shows us that $\text{sum}(x, y, z)$ is just the summand output of a half-adder. So we can let y and z be inputs to a half-adder and then feed the summand output along with x into another half-adder to obtain the desired $\text{sum}(x, y, z)$. Before we draw the diagram, we need to look at the carry function. We have written carry in the following form:

$$\text{carry}(x, y, z) = yz + x(\bar{y}z + y\bar{z}\,).$$

Notice that the term yz is the carry output of a half-adder with input values y and z. Further, the term $x(\bar{y}z + y\bar{z})$ is the carry output of a half-adder with inputs x and $\bar{y}z + y\bar{z}$, where $\bar{y}z + y\bar{z}$ is the output of the half-adder with inputs y and z. So we can draw a picture of the circuit for a full adder as shown in Figure 10.7. ◆

As we've seen in the previous two examples, we can get simpler digital circuits if we spend some time simplifying the corresponding Boolean expressions. Often a digital circuit must be built with the minimum number of components, where the components correspond to DNFs or CNFs. This brings up the question of finding a *minimal* DNF for a Boolean expression. Here the

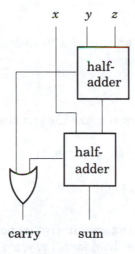

Figure 10.7

word "minimal" is usually defined to mean the fewest number of fundamental conjunctions in a DNF, and if two DNFs have the same number of fundamental conjunctions, then the one with the fewest literals is minimal. The term *minimal* CNF is defined analogously.

It's not always easy to find a minimal DNF for a Boolean expression. For example, $yz + yx$ is a minimal DNF for the expression

$$\overline{x}yz + xyz + xy\overline{z}.$$

We can show that these two expressions are equivalent as follows:

$$
\begin{aligned}
\overline{x}yz + xyz + xy\overline{z} &= (\overline{x} + x)yz + xy\overline{z} \\
&= yz + xy\overline{z} \\
&= y(z + x\overline{z}) \\
&= y(z + x) \\
&= yz + yx.
\end{aligned}
$$

But it takes some work to see that $yz + xy$ is a minimal DNF. First we need to argue that there is no equivalent DNF with just a single fundamental conjunction. Then we need to argue that there is no equivalent DNF with two fundamental conjunctions with fewer than four literals.

There are formal methods that can be applied to the problem of finding minimal DNFs and minimal CNFs. We'll leave them to more specialized texts.

Summary of Properties

Let's collect in one place the axioms of a Boolean algebra together with the properties that we have developed.

Boolean Algebra Axioms

An algebra is a *Boolean algebra* if it has the structure

$$\langle B; +, \cdot, {}^-, 0, 1 \rangle,$$

where the following properties hold:

1. $\langle B; +, 0 \rangle$ and $\langle B; \cdot, 1 \rangle$ are commutative monoids. In other words, the following properties hold for all $x, y, z \in B$:

   $$(x + y) + z = x + (y + z), \qquad (x \cdot y) \cdot z = x \cdot (y \cdot z),$$

   $$x + y = y + x, \qquad\qquad x \cdot y = y \cdot x,$$

   $$x + 0 = x, \qquad\qquad x \cdot 1 = x.$$

2. $+$ and $\cdot$ distribute over each other. In other words, the following properties hold for all $x, y, z \in B$:

   $$x \cdot (y + z) = (x \cdot y) + (x \cdot z) \quad \text{and} \quad x + (y \cdot z) = (x + y) \cdot (x + z).$$

3. $x + \bar{x} = 1$ and $x \cdot \bar{x} = 0$ for all elements $x \in B$. The element $\bar{x}$ is called the *complement* of x or the *negation* of x.

Boolean Algebra Properties

$$b \cdot b = b \quad \text{and} \quad b + b = b \qquad \text{(idempotent);} \qquad (10.5)$$

$$0 \cdot b = 0 \quad \text{and} \quad 1 + b = 1; \qquad\qquad\qquad (10.6)$$

$$a + ab = a \quad \text{and} \quad a(a + b) = a \qquad \text{(absorption);} \qquad (10.7)$$

$$a + \bar{a}b = a + b \quad \text{and} \quad a(\bar{a} + b) = ab \quad \text{(absorption);} \qquad (10.7)$$

$$\text{if } a + b = 1 \text{ and } ab = 0, \text{ then } b = \bar{a}; \qquad\qquad (10.8)$$

$$\bar{\bar{a}} = a \qquad\qquad\qquad \text{(involution);} \qquad (10.9)$$

$$\overline{a + b} = \bar{a}\,\bar{b} \quad \text{and} \quad \overline{ab} = \bar{a} + \bar{b} \qquad \text{(De Morgan).} \qquad (10.10)$$

Exercises

1. Let S be a set and $B = \text{power}(S)$. Suppose someone makes the following definitions for the operators of a Boolean algebra with carrier B. Is the result a Boolean algebra? Why or why not?

$$+ \quad \text{is union,}$$
$$\cdot \quad \text{is difference,}$$
$$- \quad \text{is complement with respect to } S,$$
$$0 \quad \text{is } S,$$
$$1 \quad \text{is } \varnothing.$$

2. Prove each of the following four absorption laws (10.7).

   a. $x + xy = x$.
   b. $x(x + y) = x$.
   c. $x + \bar{x}y = x + y$.
   d. $x(\bar{x} + y) = xy$.

3. Let $e = \bar{y}z + y\bar{z}$. Prove that $\bar{e} = \bar{y}\bar{z} + yz$.

4. Use Boolean algebra properties to prove each of the following equalities.

   a. $\bar{x} + \bar{y} + xyz = \bar{x} + \bar{y} + z$.
   b. $\bar{x} + \bar{y} + xy\bar{z} = \bar{x} + \bar{y} + \overline{xyz}$.

5. Simplify each of the following Boolean expressions.

   a. $x + \bar{x}y$.
   b. $x\bar{y}\bar{x} + xy\bar{x}$.
   c. $\bar{x}\bar{y}z + x\,\bar{y}\bar{z} + x\bar{y}z$.
   d. $xy + x\bar{y} + \bar{x}y$.
   e. $x(y + \bar{y}z) + \bar{y}z + yz$.
   f. $x + yz + \bar{x}y + \bar{y}xz$.

6. For each part of Exercise 2, draw two logic circuits. One circuit should implement the expression on the left side of the equality. The other circuit should implement the expression on the right side of the equality. Each circuit should use the same number of gates as there are operations in the expression.

7. Write down the dual of each of the following Boolean expressions.

   a. $x + 1$.
   b. $x(y + z)$.
   c. $xy + xz$.
   d. $xy + z$.
   e. $y + \bar{x}z$.
   f. $z\bar{y}\bar{x} + xz\bar{x}$.

8. Show that, in a Boolean algebra, $1 + b = 1$ for every element b.

9. Show that, in a Boolean algebra, $\overline{ab} = \bar{a} + \bar{b}$ for all elements a and b.

10. Let B be the carrier of a Boolean algebra. Suppose B is a finite set, and suppose $0 \neq 1$. Show that the cardinality of B is an even number.

11. A Boolean algebra can be made into a partially ordered set by letting $a \preceq b$ mean $a = ab$.

    a. Show that $\preceq$ is reflexive, antisymmetric, and transitive.

    b. Show that $a \preceq b$ if and only if $b = a + b$.

12. A Boolean algebra, when considered as a poset—as in Exercise 11—is also a lattice. Prove that $\mathrm{glb}(a, b) = ab$ and $\mathrm{lub}(a, b) = a + b$.

13. In Example 3 we considered the set B_n of positive divisors of n together with the operations of lcm, gcd, n/x, where n is one and 1 is zero. Prove that this algebra is not Boolean if a prime p occurs more than once as a factor of n. *Hint:* It has something to do with the complement of p.

10.3 Abstract Data Types as Algebras

Programming problems involve data objects that are processed by a computer. To process data objects, we need operations to act on them. So algebra enters the programming picture. In computer science, an *abstract data type* consists of one or more sets of data objects together with one or more operations on the sets and some axioms to describe the operations. In other words, an abstract data type is an algebra. There is, however, a restriction on the carriers of abstract data types. A carrier must be able to be constructed in some way that will allow the data objects and the operations to be implemented on a computer.

    Programming languages normally contain some built-in abstract data types. But it's not possible for a programming language to contain all possible ways to represent and operate on data objects. Therefore programmers must often design and implement new abstract data types. The axioms of an abstract data type can be used by a programmer to check whether an implementation is correct. In other words, the implemented operations can be checked to see whether they satisfy the axioms.

    An abstract data type allows us to program with its data objects and operations without having to worry about implementation details. For example, suppose we need to create an abstract data type for processing polynomials. We might agree to use the expression add(p, q) to represent the sum of two polynomials p and q. To implement the abstract data type, we might represent a polynomial as an array of coefficients and then implement the add operation by adding corresponding array components. Of course, there are other interesting and useful ways to represent polynomials and their addition. But no matter what implementation is used, the statement add(p, q) always means the same thing. So we've abstracted away the implementation details.

    In this section we'll introduce some of the basic abstract data types of computer science.

Natural Numbers

In Chapter 3 we discussed the problem of trying to describe the natural numbers to a robot. Let's revisit the problem by trying to describe the natural numbers to ourselves from an algebraic point of view. We can start by trying out the following inductive definition:

1. $0 \in \mathbb{N}$.

2. There is a function $s : \mathbb{N} \to \mathbb{N}$ called "successor" with the following property: If $x \in \mathbb{N}$, then $s(x) \in \mathbb{N}$.

Does this inductive definition adequately describe the natural numbers? It depends on what we mean by "successor." For example, if $s(0) = 0$, then the set {0} satisfies the definition. So the property $s(0) = 0$ must be ruled out. Let's try the additional axiom:

3. $s(x) \neq 0$ for all $x \in \mathbb{N}$.

Now {0} doesn't satisfy the three axioms. But if we assume that $s(s(0)) = s(0)$, then the set {0, s(0)} satisfies them. The problem here is that s sends two elements to the same place. We can eliminate this problem if we require s to be injective (one to one):

4. If $s(x) = s(y)$, then $x = y$.

This gives us the set of natural numbers in the form 0, $s(0)$, $s(s(0))$, ..., where we set $s(0) = 1$, $s(s(0)) = 2$, and so on.

Historically, the first description of the natural numbers using axioms 1–4 was given by Peano. He also included a fifth axiom to describe the principle of mathematical induction:

5. If $q(x)$ is a predicate for every $x \in \mathbb{N}$, and $q(0)$ is true, and $\forall x \in \mathbb{N} \ (q(x) \to q(s(x)))$ is true, then $\forall x \in \mathbb{N} \ q(x)$ is true.

Let's stop for a minute to write down an algebraic description of the natural numbers in terms of the first four rules:

$$\begin{aligned}
\text{Carrier:} \quad & \mathbb{N}. \\
\text{Operations:} \quad & 0 \in \mathbb{N}, \\
& s : \mathbb{N} \to \mathbb{N}. \\
\text{Axioms:} \quad & s(x) \neq 0, \\
& \text{If } s(x) = s(y), \text{ then } x = y.
\end{aligned}$$

An algebra is useful as an abstract data type if we can define useful operations on the type in terms of its primitive operations. For example, can we define addition of natural numbers in this algebra? Sure. We can define the "plus" operation using only the successor operation as follows:

$$\text{plus}(0, y) = y,$$
$$\text{plus}(s(x), y) = s(\text{plus}(x, y)).$$

For example, plus(2, 1) is computed by first writing $2 = s(s(0))$ and $1 = s(0)$. Then we can apply the definition recursively as follows:

$$
\begin{aligned}
\text{plus}(2, 1) &= \text{plus}(s(s(0)), s(0)) \\
&= s(\text{plus}(s(0), s(0))) \\
&= s(s(\text{plus}(0, s(0)))) \\
&= s(s(s(0))) \\
&= 3.
\end{aligned}
$$

An alternative definition for the plus operation can be given as

$$\text{plus}(0, y) = y,$$
$$\text{plus}(s(x), y) = \text{plus}(x, s(y)).$$

For example, using this definition, we can evaluate plus(2, 2) as follows:

$$\text{plus}(2, 2) = \text{plus}(1, 3) = \text{plus}(0, 4) = 4.$$

Now that we have the plus operation, we can use it to define the multiplication operation as follows:

$$\text{mult}(0, y) = 0,$$
$$\text{mult}(s(x), y) = \text{plus}(\text{mult}(x, y), y).$$

For example, we'll evaluate mult(3, 4) as follows—assuming that plus does its job properly:

$$
\begin{aligned}
\text{mult}(3, 4) &= \text{plus}(\text{mult}(2, 4), 4) \\
&= \text{plus}(\text{plus}(\text{mult}(1, 4), 4), 4) \\
&= \text{plus}(\text{plus}(\text{plus}(\text{mult}(0, 4), 4), 4), 4) \\
&= \text{plus}(\text{plus}(\text{plus}(0, 4), 4), 4) \\
&= \text{plus}(\text{plus}(4, 4), 4) \\
&= \text{plus}(8, 4) = 12.
\end{aligned}
$$

Let's see whether we can write the definitions for plus and mult in if-then-else form. To do so, we need the idea of a predecessor. Letting $p(x)$ denote the "predecessor" of x, we can write the definition of plus in either of the following ways:

$$\text{plus}(x, y) = \text{if } x = 0 \text{ then } y \text{ else } s(\text{plus}(p(x), y))$$

or

$$\text{plus}(x, y) = \text{if } x = 0 \text{ then } y \text{ else plus}(p(x), s(y)).$$

We'll leave it as an exercise to prove that these two definitions are equivalent. We can write the definition of mult as follows:

$$\text{mult}(x, y) = \text{if } x = 0 \text{ then } 0 \text{ else plus}(\text{mult}(p(x), y), y).$$

We can define the predecessor operation in terms of successor using the equation $p(s(x)) = x$. Since we're dealing only with natural numbers, we should either make $p(0)$ undefined or else define it so that it won't cause trouble. The usual definition is to say that $p(0) = 0$. It's interesting to note that we can't write an if-then-else definition for predecessor using only the successor operation. So in some sense, predecessor is a primitive operation too. So we'll add the definition of p to our algebra. We also need a test for zero to handle the test "$x = 0$" that occurs in if-then-else definitions.

To describe the algebra that includes these notions, we'll need another carrier to contain the true and false results that are returned by the test for zero. Letting Boolean = {true, false} and replacing s and p by the more descriptive names "succ" and "pred," we obtain the following algebra to represent the *abstract data type of natural numbers*:

Carriers:	$\mathbb{N}$, Boolean.	(10.11)
Operations:	$0 \in \mathbb{N}$,	
	isZero : $\mathbb{N} \to$ Boolean,	
	succ : $\mathbb{N} \to \mathbb{N}$,	
	pred : $\mathbb{N} \to \mathbb{N}$.	
Axioms:	isZero(0) = true,	
	isZero(succ(x)) = false,	
	pred(0) = 0,	
	pred(succ(x)) = x.	

Notice that we've made some replacements. The old axiom succ(x) $\neq 0$ has been replaced by the new axiom, isZero(succ(x)) = false, which expresses the same idea. Also, the old axiom "If succ(x) = succ(y), then $x = y$" has been replaced by the new axiom pred(succ(x)) = x. To see this, notice that succ(x) = succ(y) implies that pred(succ(x)) = pred(succ(y)). Therefore we can conclude that $x = y$ because x = pred(succ(x)) = pred(succ(y)) = y.

For example, we can rewrite the plus function in terms of the primitives of this algebra as

$$\text{plus}(x, y) = \text{if isZero}(x) \text{ then } y \text{ else succ(plus(pred}(x), y)).$$

We can also write the mult function in terms of the primitives of (10.11) together with the plus function as follows:

$$\text{mult}(x, y) = \text{if isZero}(x) \text{ then } 0 \text{ else plus(mult(pred}(x), y), y).$$

EXAMPLE 1. Let's define the "less" relation on natural numbers using only the primitives of the algebra (10.11). To get an idea of how we might proceed, consider the following evaluation of the expression less(2, 4):

$$\text{less}(2, 4) = \text{less}(1, 3) = \text{less}(0, 2) = \text{true}.$$

We simply replace each argument by its predecessor until one of the arguments is zero. Therefore less can be computed from a recursive definition such as the following:

$$\text{less}(0, 0) = \text{false},$$
$$\text{less(succ}(x), 0) = \text{false},$$
$$\text{less}(0, \text{succ}(y)) = \text{true},$$
$$\text{less(succ}(x), \text{succ}(y)) = \text{less}(x, y).$$

Using the if-then-else form, we obtain the following definition:

$$\begin{aligned}
\text{less}(x, y) \quad = \quad &\text{if isZero}(y) \text{ then false} \\
&\text{else if isZero}(x) \text{ then true} \\
&\text{else less(pred}(x), \text{pred}(y)). \quad \blacklozenge
\end{aligned}$$

The following paragraphs describe several fundamental algebras of computer science. As we have said, they are also called abstract data types. The need for abstraction can be seen by considering questions like the following: What do lists and stacks have in common? How can we be sure that a queue is implemented correctly? How can we be sure that any data structure is implemented correctly? The answers to these questions depend on how we define the structures that we are talking about, without regard to any particular implementation.

Lists and Strings

Lists

Recall that the set of lists over a set A can be defined inductively by using the empty list, $\langle\,\rangle$, and the cons operation (with infix form ::), as constructors. If we denote the set of all lists over A by Lists$[A]$, we have the following inductive definition:

Basis: $\langle\,\rangle \in$ Lists$[A]$.

Induction: If $x \in A$ and $L \in$ Lists$[A]$, then cons$(x, L) \in$ Lists$[A]$.

The algebra of lists can be defined by the constructors $\langle\,\rangle$ and cons together with the primitive operations isEmptyL, head, and tail. With these operations we can describe the *list abstract data type* as the following algebra of lists over A:

Carriers:	Lists$[A]$, A, Boolean.
Operations:	$\langle\,\rangle \in$ Lists$[A]$,
	isEmptyL : Lists$[A] \to$ Boolean,
	cons : $A \times$ Lists$[A] \to$ Lists$[A]$,
	head : Lists$[A] \to A$,
	tail : Lists$[A] \to$ Lists$[A]$.
Axioms:	isEmptyL$(\langle\,\rangle)$ = true,
	isEmptyL$(\text{cons}(x, L))$ = false,
	head$(\text{cons}(x, L))$ = x,
	tail$(\text{cons}(x, L))$ = L.

Can all desired list functions be written in terms of the "primitive" operations of this algebra? The answer probably depends on the definition of "desired." For example, we saw in Chapter 3 that the following functions can be written in terms of the operations of the list algebra:

length:	Lists$[A] \to \mathbb{N}$	Finds length of a list
member:	$A \times$ Lists$[A] \to$ Boolean	Tests membership in a list
last:	Lists$[A] \to A$	Finds last element of a list
concatenate:	Lists$[A] \times$ Lists$[A] \to$ Lists$[A]$	
putLast:	$A \times$ Lists$[A] \to$ Lists$[A]$	Puts element at right end

Let's look at a couple of these functions to see whether we can implement them. Assume that all the operations in the signature of the list algebra are implemented. Then a definition for "length" can be written as follows:

$$\text{length}(L) \;=\; \text{if isEmptyL}(L) \text{ then } 0$$
$$\text{else } 1 + \text{length}(\text{tail}(L)).$$

In this case the algebra $\langle \mathbb{N}; +, 0 \rangle$ must also be implemented for the length function to work properly.

Similarly, suppose we define "member" as follows:

$$\text{member}(a, L) \;=\; \text{if isEmptyL}(L) \text{ then false}$$
$$\text{else if } a = \text{head}(L) \text{ then true}$$
$$\text{else member}(a, \text{tail}(L)).$$

In this case the predicate "$a = \text{head}(L)$" must be computed. Thus an equality relation must be implemented for the carrier A.

As these two examples have shown, although we can define list functions in terms of the algebra of lists, we often need other algebras, such as $\langle \mathbb{N}; +, 0 \rangle$, or other relations, such as equality on A.

Strings

Strings may look different than lists, but these structures have a lot in common. For example, they both have length, and their constructions are similar. For example, the set of all strings over an alphabet A can be defined inductively from the empty string and the append operation. Letting A^* denote the set of all strings over A, we have the following inductive definition:

Basis: $\Lambda \in A^*$.

Induction: If $x \in A$ and $s \in A^*$, then $x \cdot s \in A^*$.

We can describe the *string abstract data type* as the following algebra of strings over A:

Carriers:	$A, A^*,$ Boolean.
Operations:	$\Lambda \in A^*,$
	$\text{isEmptyS} : A^* \to \text{Boolean},$
	$\cdot : A \times A^* \to A^*,$
	$\text{headS} : A^* \to A,$
	$\text{tailS} : A^* \to A^*.$
Axioms:	$\text{isEmptyS}(\Lambda) = \text{true},$
	$\text{isEmptyS}(a \cdot s) = \text{false},$
	$\text{headS}(a \cdot s) = a,$
	$\text{tailS}(a \cdot s) = s.$

When working with strings, we want to be able to combine strings, compare strings, and so on. We can define functions to accomplish these things using the string algebra. For example, let's write a definition for the "cat" function to combine two strings. For example, cat(*cb*, *aba*) = *cbaba*. Cat has type $A^* \times A^* \to A^*$ and can be defined as follows:

$$\text{cat}(s, t) \quad = \quad \text{if isEmptyS}(s) \text{ then } t$$
$$\text{else headS}(s) \cdot \text{cat(tailS}(s), t).$$

Stacks and Queues

Stacks

A *stack* is a structure satisfying the LIFO property of last in, first out. In other words, the last element input is the first element output. The main stack operations are *push*, which pushes a new element onto a stack; *pop*, which removes the top element from a stack; and *top*, which examines the top element of a stack. We also need an indication of when a stack is empty.

Let's describe the *stack abstract data type* as an algebra. For any set *A*, let Stks[*A*] denote the set of stacks whose elements are from *A*. We'll include error messages in our description for those cases in which the operators are not defined. Here's the algebra:

Carriers: *A*, Stks[*A*], Boolean, Errors.

Operations: emptyStk $\in$ Stks[*A*],

isEmptyStk : Stks[*A*] $\to$ Boolean,

push : $A \times$ Stks[*A*] $\to$ Stks[*A*],

pop : Stks[*A*] $\to$ Stks[*A*] $\cup$ Errors,

top : Stks[*A*] $\to A \cup$ Errors.

Axioms: isEmptyStk(emptyStk) = true,

isEmptyStk(push(*a*, *s*)) = false,

pop(push(*a*, *s*)) = *s*,

pop(emptyStk) = stackError,

top(push(*a*, *s*)) = *a*,

top(emptyStk) = valueError.

Notice the similarity between the stack algebra and the list algebra. In fact, we can implement the stack algebra as a list algebra by assigning the following meanings to the stack symbols:

$$\begin{aligned}
\text{Stks}[A] &= \text{Lists}[A], \\
\text{emptyStk} &= \langle\,\rangle, \\
\text{isEmptyStk} &= \text{isEmptyL}, \\
\text{push} &= \text{cons}, \\
\text{pop} &= \text{tail}, \\
\text{top} &= \text{head}.
\end{aligned}$$

To prove that this implementation is correct, we need to show that the axioms of a stack are true for the above assignment. They are all trivial. For example, the proof of the third axiom is a one-liner:

$$\text{pop}(\text{push}(a, s)) = \text{tail}(\text{cons}(a, s)) = s. \quad \text{QED.}$$

EXAMPLE 2 (*Evaluating a Postfix Expression*). Let's look at the general approach to evaluate an arithmetic expression represented in postfix notation. For example, the postfix expression $abc+-$ can be evaluated by pushing a, b, and c onto a stack. Then b and c are popped, and the value $b + c$ is pushed onto the stack. Finally, a and $b + c$ are popped, and the value $a - (b + c)$ is pushed. We'll assume that there is a function "val," which takes an operator and two operands and returns the value of the operator applied to the two operands.

The general algorithm for evaluating a postfix expression can be given as follows, where the initial call has the form $\text{post}(L, \langle\,\rangle)$ and L is the list representation of the postfix expression:

$$\begin{aligned}
\text{post}(\langle\,\rangle, \text{stk}) &= \text{top}(\text{stk}) \\
\text{post}(x :: t, \text{stk}) &= \text{if } x \text{ is an argument then} \\
&\qquad\quad \text{post}(t, \text{push}(x, \text{stk})) \\
&\qquad \text{else } \{x \text{ is an operator}\} \\
&\qquad\quad \text{post}(t, \text{eval}(x, \text{stk})),
\end{aligned}$$

where

$$\text{eval}(\text{op}, \text{push}(a, \text{push}(b, \text{stk}))) = \text{push}(\text{val}(b, \text{op}, a), \text{stk}).$$

For example, let's evaluate the expression $\text{post}(\langle 2, 5, +\rangle, \langle\,\rangle)$.

$$\begin{aligned}
\text{post}(\langle 2, 5, +\rangle, \langle\,\rangle) &= \text{post}(\langle 5, +\rangle, \langle 2\rangle) \\
&= \text{post}(\langle +\rangle, \langle 5, 2\rangle) \\
&= \text{post}(\langle\,\rangle, \text{eval}(+, \langle 5, 2\rangle)) \\
&= \text{top}(\text{eval}(+, \langle 5, 2\rangle))
\end{aligned}$$

$$= \text{top(push(val}(2, +, 5), \langle \rangle))$$
$$= \text{val}(2, +, 5)$$
$$= 7. \quad \blacklozenge$$

Queues

A *queue* is a structure satisfying the FIFO property of first in, first out. In other words, the first element input is the first element output. So a queue is a fair waiting line. The main operations on a queue involve adding a new element, examining the front element, and deleting the front element.

To describe the *queue abstract data type* as an algebra, we'll let A be a set and $Q[A]$ be the set of queues over A. The algebra can be described as follows:

Carriers:	A, $Q[A]$, Boolean.
Operations:	$\text{emptyQ} \in Q[A]$,
	$\text{isEmptyQ} : Q[A] \to \text{Boolean}$,
	$\text{addQ} : A \times Q[A] \to Q[A]$,
	$\text{frontQ} : Q[A] \to A$,
	$\text{delQ} : Q[A] \to Q[A]$.
Axioms:	$\text{isEmptyQ(emptyQ)} = \text{true}$,
	$\text{isEmptyQ(addQ}(a, q)) = \text{false}$,
	$\text{frontQ(addQ}(a, q)) = \text{ if isEmptyQ}(q) \text{ then } a$
	$\text{else frontQ}(q)$,
	$\text{delQ(addQ}(a, q)) = \text{ if isEmptyQ}(q) \text{ then } q$
	$\text{else addQ}(a, \text{delQ}(q))$.

Although we haven't stated it in the axioms, an error will occur if either frontQ or delQ is applied to an empty queue.

Suppose we represent a queue as a list. For example, the list $\langle a, b \rangle$ represents a queue with a at the front and b at the rear. If we add a new item c to this queue, we obtain the queue $\langle a, b, c \rangle$. So $\text{addQ}(c, \langle a, b \rangle) = \langle a, b, c \rangle$. Thus addQ can be implemented as the putLast function. The implementation of a queue algebra as a list algebra can be given as follows:

$Q[A]$	=	$\text{Lists}[A]$,
emptyQ	=	$\langle \rangle$,
isEmptyQ	=	isEmptyL,
frontQ	=	head,
delQ	=	tail,
addQ	=	putLast.

The proof of correctness of this implementation is more interesting (not trivial) because two queue axioms include conditionals, and putLast is written in terms of the list primitives. For example, we'll prove the correctness of the third axiom for the algebra of queues, leaving the proof of the fourth axiom as an exercise. Since the third axiom is an if-then-else statement, we'll consider two cases:

Case 1: Assume that q = emptyQ. In this case the axiom becomes

$$
\begin{aligned}
\text{frontQ(addQ}(a, \text{emptyQ)}) &= \text{head(putLast}(a, \text{emptyQ})) \\
&= \text{head}(a :: \text{emptyQ}) \\
&= a.
\end{aligned}
$$

Case 2: Assume that $q \neq$ emptyQ. In this case the axiom becomes

$$
\begin{aligned}
\text{frontQ(addQ}(a, q)) &= \text{head(putLast}(a, q)) \\
&= \text{head(head}(q) :: \text{putLast}(a, \text{tail}(q))) \\
&= \text{head}(q) \\
&= \text{frontQ}(q).
\end{aligned}
$$

EXAMPLE 3. Let's use the queue algebra to define the append function, apQ, that joins two queues together. It can be written in terms of the primitive operations of a queue algebra as follows:

$$
\begin{aligned}
\text{apQ}(x, y) = \ &\text{if isEmptyQ}(y) \text{ then } x \\
&\text{else apQ(addQ(frontQ}(y), x), \text{delQ}(y)).
\end{aligned}
$$

For example, suppose $x = \langle a, b \rangle$ and $y = \langle c, d \rangle$ are two queues, where a is the front of x and c is the front of y. We can evaluate the expression apQ(x, y) as follows:

$$
\begin{aligned}
\text{apQ}(x, y) &= \text{apQ}(\langle a, b \rangle, \langle c, d \rangle) \\
&= \text{apQ}(\langle a, b, c \rangle, \langle d \rangle) \\
&= \text{apQ}(\langle a, b, c, d \rangle, \langle \ \rangle) \\
&= \langle a, b, c, d \rangle. \quad \blacklozenge
\end{aligned}
$$

EXAMPLE 4 (*Decimal to Binary*). Let's convert a natural number to a binary number and represent the output as a queue of binary digits. Let bin(n) represent the queue of binary digits representing n. For example, we should have bin(4) = $\langle 1, 0, 0 \rangle$, assuming that the front of the queue is the head of the list. Let's get to the definition. If $n = 0$ or $n = 1$, we should return the queue $\langle n \rangle$, which is constructed by addQ(n, emptyQ). If n is not 0 or 1, then we should return the queue addQ(n mod 2, bin(floor($n/2$))). In other words, we can define bin as follows:

$$\text{bin}(n) \quad = \quad \text{if } n = 0 \text{ or } n = 1 \text{ then}$$
$$\text{addQ}(n, \text{emptyQ})$$
$$\text{else}$$
$$\text{addQ}(n \text{ mod } 2, \text{bin}(\text{floor}(n/2))).$$

We leave it as an exercise to check that bin works. For example, try to evaluate the expression bin(4) to see whether you get the tuple $\langle 1, 0, 0 \rangle$. ◆

Binary Trees and Priority Queues

Binary Trees

Let $B[A]$ denote the set of binary trees over a set A. The main operations on binary trees involve constructing a tree, picking the root, and picking the left and right subtrees. If $a \in A$ and $l, r \in B[A]$, let tree(l, a, r) denote the tree whose root is a, whose left subtree is l, and whose right subtree is r. We can describe the *binary tree abstract data type* as the following algebra of binary trees:

Carriers: $A, B[A]$, Boolean.

Operations: emptyTree $\in B[A]$,
isEmptyTree : $B[A] \rightarrow$ Boolean,
root : $B[A] \rightarrow A$,
tree : $B[A] \times A \times B[A] \rightarrow B[A]$,
left : $B[A] \rightarrow B[A]$,
right : $B[A] \rightarrow B[A]$.

Axioms: isEmptyTree(emptyTree) = true,
isEmptyTree(tree(l, a, r)) = false,
left(tree(l, a, r)) = l,
right(tree(l, a, r)) = r,
root(tree(l, a, r)) = a.

Although we haven't stated it in the axioms, an error will occur if the functions left, right, and root are applied to the empty tree. Next, we'll give a few examples to show how useful functions can be constructed from the basic tree operations.

EXAMPLE 5. We'll look at two typical functions, "count" and "depth." Count returns the number of nodes in a binary tree. Its type is $B[A] \to \mathbb{N}$, and its definition follows:

$$\text{count}(t) = \text{if isEmptyTree}(t) \text{ then } 0$$
$$\text{else } 1 + \text{count}(\text{left}(t)) + \text{count}(\text{right}(t)).$$

Depth returns the length of the longest path from the root to the leaves of a binary tree. Assume that an empty binary tree has depth -1. Its type is $B[A] \to \mathbb{Z}$, and its definition follows:

$$\text{depth}(t) = \text{if isEmptyTree}(t) \text{ then } -1$$
$$\text{else } 1 + \max(\text{depth}(\text{left}(t)), \text{depth}(\text{right}(t))). \quad \blacklozenge$$

EXAMPLE 6. Suppose we want to write a function "inorder" to perform an inorder traversal of a binary tree and place the nodes in a queue. So we want to define a function of type $B[A] \to Q[A]$. For example, we might use the following definition:

$$\text{inorder}(t) = \text{if isEmptyTree}(t) \text{ then emptyQ}$$
$$\text{else}$$
$$\text{apQ}(\text{addQ}(\text{root}(t), \text{inorder}(\text{left}(t))), \text{inorder}(\text{right}(t))).$$

We'll leave the preorder and postorder traversals as exercises. $\blacklozenge$

Priority Queues

A *priority queue* is a structure satisfying the BIFO property: best in, first out. For example, a stack is a priority queue if we let Best = Last. Similarly, a queue is a priority queue if we let Best = First. The main operations of a priority queue involve adding a new element, accessing the best element, and deleting the best element.

Let $P[A]$ denote the set of priority queues over A. If $a \in A$ and $p \in P[A]$, then insert(a, p) denotes the priority queue obtained by adding a to p. We can describe the *priority queue abstract data type* as the following algebra:

> Carriers: A, $P[A]$, Boolean.
>
> Operations: emptyP $\in P[A]$,
> isEmptyP : $P[A] \to$ Boolean,
> better : $A \times A \to$ Boolean,
> best : $P[A] \to A$,
> insert : $A \times P[A] \to P[A]$,
> delBest : $P[A] \to P[A]$.

We'll note here that we are assuming that the function "better" is a binary relation on A. Now for the axioms:

Axioms: isEmptyP(emptyP) = true,

isEmptyP(insert(a, p)) = false,

best(insert(a, p))   =   if isEmptyP(p) then a

else if better(a, best(p)) then a

else best(p),

delBest(insert(a, p)) =   if isEmptyP(p) then emptyP

else if better(a, best(p)) then p

else insert(a, delBest(p)).

We should note that the operations best and delBest are defined only on nonempty priority queues. Priority queues can be implemented in many different ways, depending on the definitions of "better" and "best" for the set A.

To show the power of priority queues, we'll write a sorting function that sorts the elements of a priority queue into a sorted list. The initial call to sort the priority queue p is sort($p, \langle \rangle$). The definition can be written as follows:

sort(p, L)   =   if isEmptyP(p) then L

else sort(delBest(p), best(p) :: L).

Exercises

1. The *monus* operation on natural numbers is like subtraction, except that it always gives a natural number as a result. An informal definition of monus can be written as follows:

$$\text{monus}(x, y) = \text{if } x \geq y \text{ then } x - y \text{ else } 0.$$

Write down a recursive definition of monus that uses only the primitive operations isZero, succ, and pred.

2. The exponentiation function is defined by $\exp(a, b) = a^b$. Write down a recursive definition of exp that uses primitive operations or functions that are defined in terms of the primitive operations on the natural numbers. *Note:* Assume that $\exp(0, 0) = 0$.

3. Use the algebra of lists to write a definition of the function "reverse" to reverse the elements of a list. For example, reverse($\langle x, y, z \rangle$) = $\langle z, y, x \rangle$.

4. Use lists to describe an implementation of the algebra of strings over some alphabet.

5. Write an algebraic specification for general lists over a set A (where the elements of a list may also be lists). Then define the functions length, member, and so on for this specification.

6. Use the algebra of lists to write a definition for the "flatten" function that takes a general list over a set A and returns the list of its elements from A. For example, flatten($\langle\langle a, b\rangle, c, d\rangle$) = $\langle a, b, c, d\rangle$. *Hint:* Assume that there is a function isAtom to check whether its argument is an atom (not a list). Also assume that the other list operations work on general lists.

7. Evaluate the expression post($\langle 4, 5, -, 2, +\rangle, \langle\,\rangle$) by unfolding the definition in Example 2.

8. Evaluate the expression bin(4) by unfolding the definition in Example 4.

9. Write down a definition for the function "preorder," which performs a preorder traversal of a binary tree and places the node values in a queue.

10. Write down a definition for the function "postorder," which performs a postorder traversal of a binary tree and places the node values in a queue.

11. Find a descriptive name for the "mystery" function f, which has the type $A \times \text{Stks}[A] \to \text{Stks}[A]$ and is defined by the following equations:

$$f(a, \text{emptyStk}) = \text{emptyStk},$$
$$f(a, \text{push}(a, s)) = f(a, s),$$
$$f(a, \text{push}(b, s)) = \text{push}(b, f(a, s)) \quad \text{if } a \neq b.$$

12. Find a descriptive name for the "mystery" function f, which has type $Q[A] \to Q[A]$ and is defined as follows:

$$f(q) = \begin{array}{l} \text{if isEmptyQ}(q) \text{ then } q \\ \text{else addQ}(\text{frontQ}(q), f(\text{delQ}(q))). \end{array}$$

13. A *deque*, pronounced "deck," is a double-ended queue in which insertions and deletions can be made at either end of the deque. Write down an algebraic specification for deques over a set A.

14. For the list implementation of a queue, prove the correctness of the following axiom:

$$\text{delQ}(\text{addQ}(a, q)) = \begin{array}{l} \text{if isEmptyQ}(q) \text{ then } q \\ \text{else addQ}(a, \text{delQ}(q)). \end{array}$$

15. Implement a queue by using the operations of a deque. Prove the correctness of your implementation.

16. Suppose the "better" function used in a priority queue has the following type definition:

$$\text{better} : A \times A \to A.$$

How would the axioms change? Do we need any new operations?

17. Consider the following two definitions for adding natural numbers, where p and s denote the predecessor and successor operations:

$$\text{plus}(x, y) = \text{if } x = 0 \text{ then } y \text{ else } s(\text{plus}(p(x), y)),$$
$$\text{add}(x, y) = \text{if } x = 0 \text{ then } y \text{ else } \text{add}(p(x), s(y)).$$

   a. Use induction to prove that $\text{plus}(x, s(y)) = s(\text{plus}(x, y))$ for all $x, y \in \mathbb{N}$.
   b. Use induction to prove that $\text{plus}(x, y) = \text{add}(x, y)$ for all $x, y \in \mathbb{N}$. *Hint:* Part (a) can be useful.

18. Use induction to prove the following property over a queue algebra, where apQ is the append function defined in Example 3:

$$\text{apQ}(x, \text{addQ}(a, y)) = \text{addQ}(a, \text{apQ}(x, y)).$$

Hint: To simplify notation, let $x{:}a$ denote $\text{addQ}(a, x)$. Then the equation becomes $\text{apQ}(x, y{:}a) = \text{apQ}(x, y){:}a$.

19. Use induction to prove the following property over a queue algebra, where apQ is the append function defined in Example 3:

$$\text{apQ}(x, \text{apQ}(y, z)) = \text{apQ}(\text{apQ}(x, y), z).$$

Hint: Exercise 17 may be helpful.

10.4 Computational Algebras

In this section we present some important examples of algebras that are useful in the computation process. First we look at relational algebra as a tool for representing relational data. Then we look at processes from an algebraic point of view. Lastly, we discuss functional algebra as a means of programming and as a useful tool for reasoning about programs.

Relational Algebras

Relations can be combined in various ways to build new relations that solve problems. An algebra is called a *relational algebra* if its carrier is a set of relations. We begin with three useful operations on relations: selection, projection, and joining. Each of these operations builds a new relation by selecting certain tuples, by eliminating certain attributes, or by combining attributes of two relations. We'll give examples to motivate the definitions. Let's consider the Parts relation, which represents the collection of parts in an auto parts store. Each part has the following five attributes:

PartNumber, Price, Cost, DateBought, NumberOnHand.

So Parts is a collection of 5-tuples.

Suppose we're interested in finding all parts that cost $5.00. In other words, we want to form a new relation, which can be described by,

$$\{t \mid t \in \text{Parts and } t(\text{Cost}) = \$5.00\}.$$

The operation to construct such a relation is called the *select* operation, and it can be defined as follows: Let R be a relation, let A be one of the indices (or attributes) of R, and let a denote a possible value of A. The relation consisting of all tuples in R whose Ath element is a, which is denoted by sel(R, A, a), is defined as follows:

$$\text{sel}(R, A, a) = \{t \mid t \in R \text{ and } t(A) = a\}.$$

If A and a are fixed, then sel(R, A, a) is sometimes denoted by sel$_{A=a}(R)$. For example, sel(Parts, Cost, $5.00) is the set of all parts costing five dollars.

Suppose we want to know the age of the parts inventory. In other words, we're interested in two attributes PartNumber and DateBought. Thus we need the relation defined as follows:

$$\{\langle t(\text{PartNumber}), t(\text{DateBought})\rangle \mid t \in \text{Parts}\}.$$

The operation to construct such a relation is called the *project* operation, and it can be defined as follows: Let R be a relation and X a set of indices (or attributes) of R. The relation proj(R, X) consists of all tuples indexed by X constructed from the tuples of R. In formal terms we can write

$$\text{proj}(R, X) = \{s \mid \text{there exists } t \in R \text{ such that } s(A) = t(A) \text{ for each } A \in X\}.$$

If X is fixed, then proj(R, X) is sometimes written as proj$_X(R)$. For example, proj(Parts, {PartNumber, DateBought}) denotes the desired "age-of-inventory" relation.

A familiar use of the projection operation is with three dimensional geometric figures when we need to observe two-dimensional features. For example, a sphere of radius 1 can be described in an *xyz*-coordinate system by the relation

$$S = \{\langle x, y, z\rangle \mid 0 \le x^2 + y^2 + z^2 \le 1\}.$$

The projection of S in the xy-plane is the relation

$$\text{proj}(S, \{x, y\}) = \{\langle x, y \rangle \mid 0 \leq x^2 + y^2 \leq 1\}.$$

Suppose we have two relations and we wish to combine their tuples into a new relation. This is called *joining*. Let R and S be two relations with index sets I and J, respectively. The *join* of R and S is the set of all tuples over the index set $I \cup J$ that are constructed from R and S by "joining" those tuples that are equal on the common index set $I \cap J$. We denote the join of R and S by join(R, S). In formal terms we can write

$$\text{join}(R, S) = \{t \mid \text{there are tuples } r \in R \text{ and } s \in S \text{ such that}$$
$$t(a) = r(a) \text{ for all } a \in I \text{ and } t(b) = s(b) \text{ for all } b \in J\}.$$

For example, suppose we have another auto parts relation, History, with attributes PartNumber and MonthlySales. Then we can form the new relation join(Parts, History), which is a set of 6-tuples with the following attributes:

PartNumber, Price, Cost, DateBought, NumberOnHand, MonthlySales.

Let's look at two special cases. Suppose R and S are relations with index sets I and J. If $I \cap J = \varnothing$, then join(R, S) is obtained by concatenating all pairs of tuples in $R \times S$. For example, if we have tuples $\langle a, b \rangle \in R$ and $\langle c, d, e \rangle \in S$, then $\langle a, b, c, d, e \rangle \in$ join(R, S). If $I = J$, then join$(R, S) = R \cap S$.

Now we have the ingredients of a relational algebra. The carrier is the set of all possible relations, and the three operations are sel, proj, and join. We should remark that join(R, S) is often denoted by $R \bowtie S$. The properties of this relational algebra are too numerous to mention here. But we'll list a few properties that can be readily verified from the definitions:

$$\text{sel}_{A=a}(\text{sel}_{B=b}(R)) = \text{sel}_{B=b}(\text{sel}_{A=a}(R)),$$

$$R \bowtie R = R,$$

$$R \bowtie S = S \bowtie R,$$

$$(R \bowtie S) \bowtie T = R \bowtie (S \bowtie T),$$

$$\text{proj}_X(\text{sel}_{A=a}(R)) = \text{sel}_{A=a}(\text{proj}_X(R)) \qquad \text{where if } I \text{ is the index set for } R,$$
$$\text{then } A \in I \text{ and } X \subset I.$$

If R and S both have the same index set, then the select operation has some nice properties when combined with the set operations: $\cup$, $\cap$, and $-$.

For example, we have the following properties:

$$\text{sel}_{A=a}(R \cup S) = \text{sel}_{A=a}(R) \cup \text{sel}_{A=a}(S),$$

$$\text{sel}_{A=a}(R \cap S) = \text{sel}_{A=a}(R) \cap \text{sel}_{A=a}(S),$$

$$\text{sel}_{A=a}(R - S) = \text{sel}_{A=a}(R) - \text{sel}_{A=a}(S).$$

There are many other useful operations on relations, some of which can be defined in terms of the ones we have discussed. Relational algebra provides a set of tools for constructing, maintaining, and accessing databases.

Process Algebras

A *process* is a sequence of actions. Although there are many kinds of processes, we will concentrate on processes performed by machines. For example, we can think of a process as a sequence of actions performed by a computer. In particular, the execution of a computer program is a process. When we talk about a process, we'll think of it as an entity that performs its sequence of actions.

Suppose we are interested in answering questions about the processes—sequences of actions—that a machine can perform. Then we need to describe the sequences in some way. The basic building blocks of processes are actions. Let Act denote the set of all *actions*. The simplest process is no action, which we denote by Nil. If a is an action and t is a process, let the expression

$$a \cdot t$$

denote the process that first performs action a and then acts like process t. This is called *prefixing*, since t is prefixed by action a. If t_1 and t_2 are processes, then the expression

$$t_1 + t_2$$

denotes a process that can act like t_1 or t_2, but we don't know which. This represents the idea of *nondeterminism*, which means that there is no definite thing to do next. Now we can represent processes algebraically. To simplify things, we agree to give $\cdot$ higher precedence than $+$, and we will use parentheses for grouping when necessary. For example, the expression

$$a\,(b\,c + a)$$

is shorthand for the expression $a \cdot (b \cdot c + a)$. It represents a process that performs a and then either performs b followed by c or does a again. The expression

$$a\,b + a\,c$$

represents the process that either does a followed by b or does a followed by c. The string $a\,(b + \text{Nil})$ represents the process that does a first and then either does b or does nothing.

We've described a *process algebra* for nondeterministic processes. Such an algebra has the structure $\langle \text{Act}, P; \text{Nil}, \cdot, + \rangle$, where P is the set of processes. Of course, the description so far has been abstract. Hopefully, there are concrete examples of process algebras that can help answer questions about processes. The following example presents a concrete process algebra in which certain sets of strings represent processes.

EXAMPLE 1 (*A String Oriented Process Algebra*). Suppose we are interested in the sequences of actions that a process performs. It makes sense, in this case, to represent a process as a set of strings of actions in Act*. We say that a subset S of Act* is *prefix-closed* if, whenever $s \in S$ and r is a prefix of s, then $r \in S$.

For example, the set of strings $\{\Lambda, a, b, ab, abb\}$ is prefix-closed. The set $\{a\}$ is not prefix-closed because Λ is a prefix of $a = \Lambda a$, but $\Lambda \notin \{a\}$. So Λ is a member of every nonempty prefix-closed set. The set $\{\Lambda, a, ba\}$ is not prefix-closed. Let P denote the collection of all finite prefix-closed subsets of Act*:

$$P = \{S \mid S \text{ is a finite prefix closed subset of Act*}\}.$$

Then P is the carrier of a process algebra, where each set in P denotes a process. For example, the set $\{\Lambda\}$ denotes the Nil process, and $\{\Lambda, a\}$ denotes the action a. If a is an action and t is a process, then how will we denote the process $a \cdot t$? Well, after some thought (maybe a lot of thought) it seems reasonable to denote $a \cdot t$ as the following prefix-closed set:

$$a \cdot t = \{\Lambda\} \cup \{as \mid s \in t\}.$$

For example, if $t = \{\Lambda, b, bb\}$, then $a \cdot t = \{\Lambda, a, ab, abb\}$. Now let's find a prefix-closed set to denote the idea of nondeterminism. If t_1 and t_2 are processes, then the expression $t_1 + t_2$ is defined as follows:

$$t_1 + t_2 = t_1 \cup t_2.$$

For example, if $t_1 = \{\Lambda, a, ab\}$ and $t_2 = \{\Lambda, b, ba\}$, then $t_1 + t_2 = \{\Lambda, a, b, ab, ba\}$.

Thus we have defined a concrete example of a process algebra. The carrier is the collection P of all prefix closed subsets of Act*, and the operations Nil, $\cdot$, and +, are all defined in terms of elements of P.   ◆

More information on the ideas of process algebras can be found in the book by Hennessy [1988].

Functional Algebras

Suppose that for some set O we let Fun[O] denote the set of all functions that take arguments from O and return results in O. We can combine functions to obtain new functions by composition and tupling. These two combining operations satisfy the following property:

$$\langle f_1, ..., f_n \rangle \circ h = \langle f_1 \circ h, ..., f_n \circ h \rangle.$$

From a programming standpoint we want to be able to define programs as functions, but we also want to represent data using objects in O. So how should we define O to satisfy programmer needs?

For example, if we assume that $O = A \cup$ GenLists[A] for some set A, then there are more interesting kinds of operations and axioms. For example, we'll need primitive operations like cons, head, tail, $\langle \rangle$, and isEmptyL. Does the if-then-else operation make sense in this context? The answer is yes if there is a Boolean component to A. Of course, we also want numbers and strings in our data set. So we might define A as follows:

$$A = \text{Boolean} \cup \text{Numbers} \cup C^*,$$

where C is the set of characters on a computer terminal keyboard. Of course, we now need operations that act on Booleans, numbers, and strings.

What we're building here is an algebra for programming with functions. A *functional algebra* consists of a set of objects O and a set of functions Fun[O] as carriers. The algebra has operations to combine functions and operations to process data objects. If we allow functions to take functions as arguments and as results, then there is a single carrier that contains both functions and data.

Let's look at a particular functional algebra that is both a programming language and an algebra for reasoning about programs.

FP: A Functional Algebra

The correctness problem for programs can sometimes be solved by showing that the program under consideration is equivalent to another program that we "know" is correct. Methods for showing equivalence depend very much on the programming language. The FP language was introduced by Backus [1978]. FP stands for *functional programming*, and it is a fundamental example of a programming language that allows us to reason about programs in the programming language itself. To do this, we need a set of rules that allow us to do some reasoning. In this case the rules are axioms in the algebra of FP programs.

FP functions are defined on a set of objects that include atoms (numbers, strings of characters), and lists. A list can itself have lists as elements. For example, $\langle 1, a, \langle 4, x \rangle \rangle$ is a list. To apply an FP function f to an object x, we write down $f : x$ instead of the familiar $f(x)$. To compose two FP functions f and g, we write $f @ g$ instead of the familiar $f \circ g$.

We'll introduce FP with some programming examples. For example, suppose we want to write an FP program "sumSeq" that will add up a sequence of numbers like the following:

$$\text{sumSeq} : \langle x_1, x_2, ..., x_n \rangle = x_1 + x_2 + \cdots + x_n.$$

The right-hand side of this equation is not an FP expression, so we must transform it. FP has a + function that works only with a two-argument list. The expression $+ : \langle 1, 2, 3 \rangle$ has the value "?" which means "undefined." FP has an "insert" operator for binary functions, which is denoted by the symbol !. It is applied by placing it to the left of its argument, in juxtaposition. For example $!+ : \langle 1, 2, 3 \rangle$ yields the value 6 as follows:

$$!+ : \langle 1, 2, 3 \rangle = + : \langle 1, !+ : \langle 2, 3 \rangle \rangle = + : \langle 1, + : \langle 2, 3 \rangle \rangle = + : \langle 1, 5 \rangle = 6.$$

Therefore we can make the definition: sumSeq = !+.

As another example, suppose "sumFun" totals up the values obtained by applying f to the individual elements of the input sequence $\langle x_1, x_2, ..., x_n \rangle$. In other words, we want

$$\text{sumFun} : \langle x_1, x_2, ..., x_n \rangle = f(x_1) + f(x_2) + \cdots + f(x_n).$$

Again, the right-hand side of this equation is not in the form of a known FP function applied to the input sequence. But we can rewrite the expression on the right-hand side of the equation by using the insert operator:

$$f(x_1) + f(x_2) + \cdots + f(x_n) = !+ : \langle f(x_1), f(x_2), ..., f(x_n) \rangle.$$

Now we must apply f to each element of the sequence $\langle x_1, x_2, ..., x_n \rangle$. FP has an "apply to all" operator, denoted by &, which does exactly that, as follows:

$$\&f : \langle x_1, x_2, ..., x_n \rangle = \langle f(x_1), f(x_2), ..., f(x_n) \rangle.$$

So we have the following series of equations, where the last expression is an FP expression:

$$
\begin{aligned}
\text{sumFun} : \langle x_1, x_2, ..., x_n \rangle \quad &= \quad f(x_1) + f(x_2) + \cdots + f(x_n) \\
&= \quad !+ : \langle f(x_1), f(x_2), ..., f(x_n) \rangle \\
&= \quad !+ @ \&f : \langle x_1, x_2, ..., x_n \rangle.
\end{aligned}
$$

Now we can cancel the input sequence from the first and last expression to obtain the FP definition: sumFun = !+ @ &f.

Suppose we have a function f defined by

$$f(x) = \text{if } a(x) \text{ then } b(x) \text{ else } c(x).$$

In FP we define f by writing

$$f = a \rightarrow b; c.$$

In this case the expression $f : x$ is evaluated by first computing the condition $a : x$. If $a : x$ is true, then the result is the value of the expression $b : x$. Otherwise, the result is the value of the expression $c : x$.

A function f is a tupled function if it has a form like

$$f = [g, h, k].$$

In this case the value of the expression $f : x$ is the 3-tuple $\langle g : x, h : x, k : x \rangle$. A function is a constant function if it has the form $f = \sim c$, where c is any object. In this case the expression $f : x$ has the value c for all objects x.

Let's look at a few examples.

EXAMPLE 2. Let eq0 denote the function that tests its argument for zero. An FP definition for eq0 can be written as follows, where eq is the FP function that tests two atoms for equality:

$$\text{eq0} = \text{eq} @ [\text{id}, \sim 0].$$

For example, we'll evaluate the expression eq0 : 3 as follows:

$$\text{eq0} : 3 = \text{eq} @ [\text{id}, \sim 0] : 3 = \text{eq} : \langle \text{id} : 3, \sim 0 : 3 \rangle = \text{eq} : \langle 3, 0 \rangle = \text{false}. \quad \blacklozenge$$

EXAMPLE 3. Let sub1 be the function that subtracts 1 from its argument. An FP definition for sub1 can be written as follows, where – is the subtract function:

$$\text{sub1} = -\ @\ [\text{id},\ {\sim}1].$$

For example, we'll evaluate the expression sub1 : 4 as follows:

$$\text{sub1} : 4 = -\ @\ [\text{id},\ {\sim}1] : 4 = -\ :\ \langle \text{id} : 4,\ {\sim}1 : 4 \rangle = -\ :\ \langle 4,\ 1 \rangle = 3. \quad \blacklozenge$$

EXAMPLE 4. Let "length" be the function that returns the length of its list argument. Using variables, we have the usual definition:

$$\text{length}(x) = \text{if } x = \langle\,\rangle \text{ then } 0 \text{ else } 1 + \text{length}(\text{tail}(x)).$$

We can transform this definition into FP as follows, where "null" is a test for the empty list and tl computes the tail of a list:

$$\text{length} = \text{null} \rightarrow {\sim}0;\ +\ @\ [{\sim}1,\ \text{length}\ @\ \text{tl}].$$

For example, we'll evaluate the expression length : $\langle a \rangle$.

$$
\begin{aligned}
\text{length} : \langle a \rangle\ &=\ (\text{null} \rightarrow {\sim}0;\ +\ @\ [{\sim}1,\ \text{length}\ @\ \text{tl}]) : \langle a \rangle \\
&=\ \text{null} : \langle a \rangle \rightarrow {\sim}0 : \langle a \rangle;\ +\ @\ [{\sim}1,\ \text{length}\ @\ \text{tl}] : \langle a \rangle \\
&=\ \text{false} \rightarrow 0;\ +\ :\ \langle 1,\ \text{length} : \langle\,\rangle \rangle \\
&=\ +\ :\ \langle 1,\ \text{length} : \langle\,\rangle \rangle \\
&=\ +\ :\ \langle 1,\ (\text{null} \rightarrow {\sim}0;\ +\ @\ [{\sim}1,\ \text{length}\ @\ \text{tl}]) : \langle\,\rangle \rangle \\
&=\ +\ :\ \langle 1,\ \text{true} \rightarrow 0;\ +\ @\ [{\sim}1,\ \text{length}\ @\ \text{tl}] : \langle\,\rangle \rangle \\
&=\ +\ :\ \langle 1,\ 0 \rangle \\
&=\ 1. \quad \blacklozenge
\end{aligned}
$$

Our objective is to get the flavor of the language to see its algebraic nature. So let's describe an algebra of FP programs. We'll limit the operations to those that will be useful in the examples and exercises. We'll also include a few axioms to show some useful relationships between the three constructors composition, tupling, and if-then-else. Let O be the set of objects and let F be the set of all FP functions over O. Then the *FP algebra* can be described as follows:

Carriers: O, F.

Operations to construct new functions:

$$@ \qquad \text{composition (e.g., } f @ g),$$

$$\rightarrow \qquad \text{if-then-else (e.g., } p \rightarrow a \; ; b),$$

$$[\ldots] \qquad \text{tuple of functions. (e.g., } [f, g, h]),$$

$$\& \qquad \text{apply to all (e.g., } \& f),$$

$$! \qquad \text{insert (e.g., } !+),$$

$$\sim \qquad \text{constant (e.g., } \sim 2).$$

Primitive operations:

id	the identity function,
hd, tl	head and tail,
apndl, apndr	cons and consR,
1, 2, ...	selectors,
and, or, not	Boolean operations,
null	test for empty list,
atom	test for an atom,
$\langle \rangle$	empty list,
?	undefined symbol,
eq	test for equality of two atoms,
$<, >, +, -, *, /$	arithmetic relations and operations.

Axioms:

$$f @ (a \rightarrow b \; ; c) = a \rightarrow f @ b \; ; f @ c,$$

$$(a \rightarrow b \; ; c) @ d = a @ d \rightarrow b @ d \; ; c @ d,$$

$$[f_1, \ldots, f_n] @ g = [f_1 @ g, \ldots, f_n @ g],$$

$$f @ [\ldots, (a \rightarrow b \; ; c), \ldots] = a \rightarrow f @ [\ldots, b, \ldots] \; ; f @ [\ldots, c, \ldots].$$

Now let's do an example that uses FP algebra to prove the equivalence of two FP programs.

EXAMPLE 5. An FP program to compute $n!$ can be constructed directly from the following recursive definition:

$$\text{fact}(x) = \text{if } x = 0 \text{ then } 1 \text{ else } x * \text{fact}(x - 1).$$

The FP version of fact is

$$\text{fact} = \text{eq0} \to {\sim}1; * @ \, [\text{id, fact} @ \text{sub1}].$$

Another FP program to compute $n!$ can be defined as follows:

$$\text{Ifact} = g @ [{\sim}1, \text{id}],$$

where g is the FP program defined by

$$g = \text{eq0} @ 2 \to 1; g @ [*, \text{sub1} @ 2].$$

Notice that g is iterative because it has a tail-recursive form (i.e., it has the form $g = a \to b; g @ d$), which can be replaced by a loop. Therefore Ifact is also iterative. Thus Ifact may be more efficient than fact. Let's show that the two programs are equivalent: Ifact = fact. We'll need the following relation involving g:

$$* @ \, [a, g @ [b, c]] = g @ [* @ [a, b], c] \, , \tag{10.12}$$

where a, b , and c are any functions that return natural numbers. We'll leave the proof of (10.12) as an exercise. Now we can prove that Ifact = fact.

Proof: If the input to either function is 0 then eq0 is true, which gives us the base case fact : 0 = Ifact : 0. Now we'll make the induction assumption that fact @ sub1 = Ifact @ sub1 and show that fact = Ifact. Starting with Ifact, we have the following sequence of algebraic equations:

$$
\begin{aligned}
\text{Ifact} \;&=\; g @ [{\sim}1, \text{id}] &&\text{(definition)} \\
&=\; (\text{eq0} @ 2 \to 1; g @ [*, \text{sub1} @ 2]) @ [{\sim}1, \text{id}] &&\text{(definition)} \\
&=\; \text{eq0} @ \text{id} \to {\sim}1; g @ [* @ [{\sim}1, \text{id}], \text{sub1} @ \text{id}] &&\text{(FP algebra)} \\
&=\; \text{eq0} \to {\sim}1; g @ [* @ [{\sim}1, \text{id}], \text{sub1}] &&\text{(FP algebra)} \\
&=\; \text{eq0} \to {\sim}1; * @ [\text{id}, g @ [{\sim}1, \text{sub1}]] &&\text{(10.12)} \\
&=\; \text{eq0} \to {\sim}1; * @ [\text{id}, \text{Ifact} @ \text{sub1}] &&\text{(FP algebra)} \\
&=\; \text{eq0} \to {\sim}1; * @ [\text{id}, \text{fact} @ \text{sub1}] &&\text{(induction)} \\
&=\; \text{fact} &&\text{(definition). QED.}
\end{aligned}
$$

Therefore Ifact is correct if we assume that fact is correct. This is a plausible assumption because fact is just a translation of the recursive definition of the factorial function.   ◆

Exercises

1. Given the following relations, where the letters in parentheses represent attributes and the tuples are indicated as rows below each relation:

$$R(A,\ B,\ C,\ D)\qquad S(B,\ C,\ D,\ E)$$

1	a	#	M
2	a	*	N
1	b	#	M
3	a	%	N

a	#	M	x
b	*	N	y
a	#	M	z
b	%	M	w

Compute each of the following relations.

a. sel(R, B, a).

b. proj($R, \{B, D\}$).

c. join(R, S).

d. proj($R, \{B, C, D\}$) $\cup$ proj($S, \{B, C, D\}$).

e. proj($R, \{B, D\}$) $\cap$ proj($S, \{B, D\}$).

2. Use the definitions for the operators select, project, and join to prove each of the following listed properties.

a. $\mathrm{sel}_{A=a}(\mathrm{sel}_{B=b}(R)) = \mathrm{sel}_{B=b}(\mathrm{sel}_{A=a}(R))$.

b. $R \bowtie R = R$.

c. $R \bowtie S = S \bowtie R$.

d. $(R \bowtie S) \bowtie T = R \bowtie (S \bowtie T)$.

3. Suppose R is a relation with index I, A is an attribute in I, and X is a subset of I. Prove the following relationship between project and select:

$$\mathrm{proj}_X(\mathrm{sel}_{A=a}(R)) = \mathrm{sel}_{A=a}(\mathrm{proj}_X(R)).$$

4. Suppose we are interested in representing processes as trees. We could start by letting the Nil process be represented by a single node tree with root labeled Nil. If a is an action and t is a process, let $a \cdot t$ be represented by the tree whose root is a and whose single subtree is the tree representing t.

a. Draw a picture of the tree for the action a.

b. Draw a picture of the tree for process $a \cdot b$, where a and b are actions.

c. Give a definition for the tree to represent the process $t_1 + t_2$.

d. Using your definition for + in part (c), draw pictures of the trees to represent the processes $a(b + c)$ and $ab + ac$.

5. Write an FP function to implement the seqPairs function, which takes a natural number n as input and produces as output the list of pairs

$$\langle\langle 0, 0\rangle, \langle 1, 1\rangle, ..., \langle n, n\rangle\rangle.$$

6. Write an FP function to implement the pairs function, which takes two n-tuples, $\langle x_1, ..., x_n\rangle$ and $\langle y_1, ..., y_n\rangle$, and produces the result

$$\langle\langle x_1, y_1\rangle, ..., \langle x_n, y_n\rangle\rangle.$$

7. Write an FP function to implement the dotProduct function, which takes two n-tuples of numbers, $\langle x_1, ..., x_n\rangle$ and $\langle y_1, ..., y_n\rangle$, and produces the value $x_1 * y_1 + \cdots + x_n * y_n$. *Hint:* Use the pairs function from the previous exercise.

8. Prove each of the following FP equations.
   a. $+ @ [1, 2] = + @ [2, 1] = +.$
   b. $1 @ \sim\langle a, b\rangle = \sim a$ and $2 @ \sim\langle a, b\rangle = \sim b.$

9. Prove the FP equation (10.12): $* @ [a, g @ [b, c]] = g @ [* @ [a, b], c],$ where a, b, and c are any functions that return natural numbers and

$$g = eq0 @ 2 \rightarrow 1; g @ [*, sub1 @ 2].$$

10. Prove that "slow" and "fast" are equivalent FP programs to compute the nth Fibonacci number. *Note:* sub2 is the FP function that subtracts 2.

$$slow = eq0 \rightarrow \sim 0; eq1 \rightarrow \sim 1; + @ [slow @ sub1, slow @ sub2],$$

$$fast = 1 @ g, \text{ where } g = eq0 \rightarrow \sim\langle 0, 1\rangle; [2, +] @ g @ sub1.$$

10.5 Other Algebraic Ideas

In this section we introduce some algebraic tools that can be used to solve some computational problems.

Congruences

We're going to examine a useful property that sometimes occurs when an equivalence relation interacts with an algebraic operation in a certain way. We'll introduce the idea by using the familiar mod function. Recall that if x is an integer and n is a positive integer, then $x \bmod n$ denotes the remainder upon division of x by n. We can define an equivalence relation on the integers by relating two numbers x and y if the following equation holds:

$$x \bmod n = y \bmod n.$$

When x and y are related by this equation, we'll indicate the fact by one of the following notations:

$$x \equiv y \pmod{n} \quad \text{or} \quad x \equiv_n y.$$

In other words, $x \equiv y \pmod{n}$ and $x \equiv_n y$ both mean $x \bmod n = y \bmod n$. For example, we have $12 \equiv_5 7$ because $12 \bmod 5 = 7 \bmod 5$. Similarly, we have $3 \equiv_5 13$. If the modulus n is fixed throughout the discussion, then we write $x \equiv y$.

Notice the following interesting property of the relation $\equiv_n$, where n is a fixed positive integer:

$$\text{If } a \equiv_n b \text{ and } c \equiv_n d, \text{ then } a + c \equiv_n b + d \text{ and } ac \equiv_n bd. \tag{10.13}$$

Proof: Since $a \equiv_n b$ and $c \equiv_n d$, it follows that there are integers k and k' such that

$$a = b + kn \quad \text{and} \quad c = d + k'n.$$

Adding the two equations, we get

$$a + c = b + d + (k + k')n,$$

which tells us that $a + c \equiv_n b + d$. Multiplying the two equations yields

$$ac = bd + (bk' + kd + kk')n,$$

which tells us that $ac \equiv_n bd$.   QED.

The two properties (10.13) are a special case of operations that interact with an equivalence relation in a special way, which we'll now describe. Suppose $\sim$ is an equivalence relation on a set A. An n-ary operation f on A is said to *preserve* $\sim$ if it satisfies the following property:

$$\text{If } a_1 \sim b_1, ..., a_n \sim b_n, \text{ then } f(a_1, ..., a_n) \sim f(b_1, ..., b_n).$$

For example, (10.13) says that $\equiv_n$ is preserved by plus and times.

When an equivalence relation on the carrier of an algebra is preserved by each operation of the algebra, then the relation is called a *congruence relation* on the algebra, and the expression $x \sim y$ is called a *congruence*. For example, the relation $\equiv_n$ is a congruence relation on the algebra $\langle \mathbb{N}; +, \cdot \rangle$.

Are congruence relations good for anything? Sure. Let's look at some examples that involve encoding and decoding.

Encoding and Decoding Numbers

Suppose we wish to represent a large natural number x by a pair of smaller numbers $\langle r, s \rangle$. Of course there are many ways to do this. But we want to consider a special way that uses congruences. We also want an algorithm to recover x from the pair $\langle r, s \rangle$. The idea is to find a pair of numbers m and n such that $x < m \cdot n$. Then define the pair $\langle r, s \rangle$ by the following two congruences:

$$r \equiv x \pmod{m} \quad \text{and} \quad s \equiv x \pmod{n}.$$

For example, we can represent the number 48 by the pair of numbers $\langle 6, 3 \rangle$, where $6 = 48 \bmod 7$ and $3 = 48 \bmod 9$. We chose the numbers 7 and 9 because their product is greater than 48. Another reason for the choice is that we will be able to reverse the process and recover the original number from the pair of numbers. For example, suppose we're given the pair of numbers $\langle 6, 3 \rangle$ and the following additional facts:

$$x \equiv_7 6 \quad \text{and} \quad x \equiv_9 3 \quad \text{and} \quad 0 \le x < 63 = 7 \cdot 9.$$

Is this enough information to find x? The answer is yes. Let's start by looking at the congruence $x \equiv_7 6$. This congruence implies that x must have the form

$$x = 6 + 7k,$$

for some k. This says that x must be in the set of numbers

$$\{\dots, -1, 6, 13, 20, 27, 34, 41, 48, 55, 62, 69, \dots\}.$$

The second congruence $x \equiv_9 3$ tells us that x must have the following form for some integer j:

$$x = 3 + 9j.$$

This says that x must be in the set of numbers

$$\{\dots, -6, 3, 12, 21, 30, 39, 48, 57, 66, 75, \dots\}.$$

Scanning the intersection of the two sets we find that there is exactly one solution in the range $0 \le x < 63$, namely, $x = 48$.

If we drop the restriction that $0 \le x < 63$, then we can find other values for x that satisfy the two congruences $x \equiv_7 6$ and $x \equiv_9 3$. For example, $x = 111$ works. The interesting part is that all solutions differ from each other by a multiple of $63 = 7 \cdot 9$. So the unrestricted solutions form the infinite set

$$\{\dots, -15, 48, 111, 174, \dots\}.$$

Suppose we started with the two congruences $x \equiv_6 2$ and $x \equiv_3 1$. The solution sets for these congruences are $\{\dots, -4, 2, 8, 14, \dots\}$ and $\{\dots, -3, 1, 5, 9, \dots\}$. So there is no common solution. To ensure a common solution for two congruences, we need m and n to be relatively prime. For example, $(7, 9) = 1$, and we found common solutions to the congruences $x \equiv_7 6$ and $x \equiv_9 3$.

Let's see why things work out nicely when the moduli are relatively prime. Suppose we have the following two congruences, where $(m, n) = 1$:

$$x \equiv_m r \quad \text{and} \quad x \equiv_n s.$$

To solve for x, we'll start with the congruence $x \equiv_m r$. It follows that x must have the following form for any integer k:

$$x = r + k \cdot m.$$

Since m and n are relatively prime, there must exist integers c and d such that $1 = m \cdot c + n \cdot d$ (by 2.1c). Now we come to the major step: Let $k = (s - r) \cdot c$ mod n. Then we can write $k \equiv_n (s - r) \cdot c$. For this value of k, let's see whether x satisfies the second congruence:

$$
\begin{aligned}
x &= r + k \cdot m \\
&\equiv_n r + (s - r) \cdot c \cdot m \\
&\equiv_n r + (s - r) \cdot 1 \qquad (\text{since } c \cdot m \equiv_n 1) \\
&\equiv_n r + s - r \\
&\equiv_n s.
\end{aligned}
$$

So x satisfies the two congruences $x \equiv_m r$ and $x \equiv_n s$. Do we know whether x satisfies the inequality $0 \le x < m \cdot n$? We can make sure that this inequality holds if we choose r so that $0 \le r < m$. Let's see why this is the case. We chose $k = (s - r) \cdot c$ mod n. Therefore $0 \le k < n$. Thus we have the two inequalities

$$r \le m - 1 \text{ and } k \le n - 1.$$

It follows that

$$x = r + k \cdot m \le (m - 1) + (n - 1) \cdot m = n \cdot m - 1.$$

Therefore we have $x < m \cdot n$. Since r, k, and m are all nonnegative, it follows that $0 \le x < m \cdot n$.

If r is not smaller than m, then we can still use the above procedure. But first we need to replace r by the value r mod m when doing the calculations. There is no restriction on s. The algorithm follows.

Solving Two Congruences (10.14)

Given two congruences $x \equiv r$ (mod m) and $x \equiv s$ (mod n), where r and s are integers and m and n are positive integers that are relatively prime, the following four-step process solves the congruences for the unique x such that $0 \leq x < m \cdot n$:

1. Make sure $0 \leq r < m$. If not, replace r by the value r mod m.
2. Find integers c and d such that $1 = m \cdot c + n \cdot d$.
3. Set $k = (s - r) \cdot c$ mod n.
4. Set $x = r + k \cdot m$.

EXAMPLE 1. Suppose we want to find the smallest natural number satisfying the two congruences

$$x \equiv_6 17 \quad \text{and} \quad x \equiv_{11} 15.$$

In terms of the algorithm, think of the first congruence as $x \equiv_m r$ and the second congruence as $x \equiv_m s$. So $m = 6$ and $r = 17$. Since 17 is not less than 6, we must apply Step 1 of the algorithm, which says to replace r by the number 5, where $5 = 17$ mod 6. Now perform the second step of the algorithm. We must find some integers c and d such that $1 = 6c + 11d$. Trial and error give us

$$1 = (6)(2) + (11)(-1).$$

So $c = 2$. Step 3 is next. We set $k = (15 - 5)(2)$ mod 11. Thus $k = 9$. Therefore we apply Step 4 to obtain the solution $x = 5 + 9 \cdot 6 = 59$. ◆

The technique for solving two congruences can be extended to solving three or more congruences. The general technique for solving n congruences follows from a famous theorem called the Chinese Remainder Theorem.

New Algebras from Old Algebras

We want to give a short description of some tools and techniques for constructing new algebras from existing algebras. The ideas that we consider are directly related to congruences and to defining new abstract data types from existing ones.

Quotient Algebras

When we worked with congruences in the previous paragraphs, we were working in a new kind of algebra. This algebra can be described in terms of an existing algebra and an equivalence relation. Let's describe the general method to construct such algebras.

Suppose the carrier A of an algebra can be partitioned by an equivalence relation ~ that is also a congruence relation (i.e., ~ is preserved by each operation of the algebra). Then we can construct a new algebra whose carrier is the partition $A/\!\sim$. This new algebra is called the *quotient* algebra of the original algebra by ~. For example, if $c \in A$, then the class $[c]$ is an element of $A/\!\sim$. The original operations are used to define corresponding operations on $A/\!\sim$. For example, if $\circ$ is a binary operation on A, then we use the same symbol to define the corresponding operation on $A/\!\sim$ as follows:

$$[a] \circ [b] = [a \circ b].$$

If f is an n-ary operation on A, then we define the operation f on $A/\!\sim$ as follows:

$$f([a_1], ..., [a_n]) = [f(a_1, ..., a_n)].$$

The classic example of a quotient algebra starts with the algebra of integers $\langle \mathbb{Z}; +, \cdot, 0, 1 \rangle$. For an equivalence relation on $\mathbb{Z}$ we pick the relation $\equiv_n$ for some fixed positive integer n. We showed in (10.13) that $\equiv_n$ is preserved by plus and times. Therefore the quotient algebra induced by $\equiv_n$ has carrier $\mathbb{Z}/\!\equiv_n$, which consists of the n classes $[0], [1], ..., [n-1]$.

For example, if $n = 4$, then $\mathbb{Z}/\!\equiv_4$ has four elements: the classes $[0], [1], [2]$, and $[3]$. The following examples show the evaluation of a few expressions in the algebra:

$$[2] + [3] = [2 + 3] = [5] = [1],$$
$$[2] \cdot [3] = [2 \cdot 3] = [6] = [2],$$
$$[0] + [3] = [0 + 3] = [3],$$
$$[1] \cdot [2] = [1 \cdot 2] = [2].$$

Subalgebras

Programmers often need to create new data types to represent information. Sometimes a new data type can use the same operations of an existing type. For example, suppose we have an integer type available to us but we need to detect an error condition whenever a negative integer is encountered. One way to solve the problem is to define a new type for the natural numbers that uses some of the operations of the integer type. For example, we can still use +, *, mod, and div (integer divide) because $\mathbb{N}$ is "closed" with respect to these operations. In other words, these operations return values in $\mathbb{N}$ if their arguments are in $\mathbb{N}$. On the other hand, we can't use the subtraction operation because $\mathbb{N}$ isn't closed with respect to it (e.g., $3 - 4 \notin \mathbb{N}$). We say that our new type "inherits" the operations +, *, mod, and div from the existing integer type. In algebraic terms we've created a new algebra $\langle \mathbb{N}; +, *, \text{mod}, \text{div} \rangle$, which is a "subalgebra" of $\langle \mathbb{Z}; +, *, \text{mod}, \text{div} \rangle$.

Let's describe the general idea of a subalgebra. Let A be the carrier of an algebra, and let B be a subset of A. We say that B is *closed* with respect to an operation if the operation returns a value in B whenever its arguments are from B. The diagram in Figure 10.8 gives a graphical picture showing B is closed with respect to the binary operation ∘.

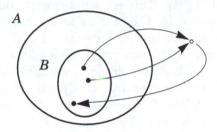

Figure 10.8

If A is the carrier of an algebra and B is a subset of A that is closed with respect to all the operations of A, then B is the carrier of an algebra called a *subalgebra* of the algebra of A. In other words, if $\langle A; \Omega \rangle$ is an algebra, where Ω is the set of operations, and if B is a subset of A that is closed with respect to all operations in Ω, then $\langle B; \Omega \rangle$ is an algebra, called a *subalgebra* of $\langle A; \Omega \rangle$. We'll denote this fact by writing $\langle B; \Omega \rangle \le \langle A; \Omega \rangle$, or simply $B \le A$ if the operations in Ω are known. We'll list a few examples.

1. $\langle \mathbb{N}; +, *, \text{mod}, \text{div} \rangle$ is a subalgebra of $\langle \mathbb{Z}; +, *, \text{mod}, \text{div} \rangle$.

2. If A is a set, then Lists[A] $\le$ GenLists[A], where the four operations are cons, head, tail, and $\langle\,\rangle$ (a nullary operation).

3. We have a sequence of subalgebras $\mathbb{Z} \le \mathbb{Q} \le \mathbb{R}$ with the arithmetic operations $+, -, \cdot, \div, 0$ and 1.

4. Consider the algebra $\langle \mathbb{N}_8; +_8, 0 \rangle$, where $+_8$ means addition modulo 8. The set {0, 2, 4, 6} forms the carrier of a subalgebra. But {0, 3, 6} is not the carrier of a subalgebra because $3 +_8 6 = 1$ and $1 \notin \{0, 3, 6\}$.

We can combine subalgebras by forming the intersection of the carriers. For example, consider the algebra $\langle N_{12}; +_{12}, 0 \rangle$. Two subalgebras of this algebra have carriers $\{0, 2, 4, 6, 8, 10\}$ and $\{0, 3, 6, 9\}$. The intersection of these two carriers is the set $\{0, 6\}$, which forms the carrier of another subalgebra of $\langle N_{12}; +_{12}, 0 \rangle$. This is no fluke because the carrier of each subalgebra is closed with respect to the operations. Thus the intersection of carriers is also closed.

One way to generate a new subalgebra is to take any subset you like, say S, from the carrier of an algebra. If the operations of the algebra are closed with respect to S, then we have a new subalgebra. If not, then keep applying the operations of the algebra to elements of S. If an operation gives a result x and $x \notin S$, then enlarge S by adding x to form the bigger set $S \cup \{x\}$. Each time a bigger set is constructed, the process must start over again until the set is closed under the operations of the algebra. The resulting subalgebra has the smallest carrier that contains S.

For example, suppose we start with the algebra $\langle N_{12}; +_{12}, 0 \rangle$, and we choose the subset $\{4, 10\}$ of N_{12}. This set is not closed under the operation $+_{12}$ because $4 +_{12} 4 = 8$, and $8 \notin \{4, 10\}$. So we add the number 8 to get the new subset $\{4, 8, 10\}$. Still there are problems because $4 +_{12} 8 = 0$, and $8 +_{12} 10 = 6$. So we add 0 and 6 to our set to obtain the subset $\{0, 4, 6, 8, 10\}$. We aren't done yet, because $6 +_{12} 8 = 2$. After adding 2, we obtain the set $\{0, 2, 4, 6, 8, 10\}$, which is closed under the operation of $+_{12}$ and contains the constant 0. Therefore the algebra $\langle \{0, 2, 4, 6, 8, 10\}; +_{12}, 0 \rangle$ is the smallest subalgebra of $\langle N_{12}; +_{12}, 0 \rangle$ that contains the set $\{4, 10\}$.

Morphisms

This little discussion is about some tools and techniques that can be used to compare two different entities for common properties. For example, if A is an alphabet, then we know that a string over A is different from a list over A. In other words, we know that A^* and Lists[A] contain different kinds of objects. But we also know that A^* and Lists[A] have a lot in common. For example, we know that the operations on A^* are similar to the operations on Lists[A]. We know that the algebra of strings and the algebra of lists both have an empty object and that they construct new objects in a similar way. In fact, we know that strings can be represented by lists.

On the other hand, we know that A^* is quite different from the set of binary trees over A. For example, the construction of a string is not at all like the construction of a binary tree.

We would like to be able to decide whether two different entities are alike in some way. When two things are alike, we are often more familiar with one of the things. So we can apply our knowledge about the familiar one and learn something about the unfamiliar one. This is a bit vague. So let's start off with a general problem of computer science:

The Transformation Problem

Transform an object into another object with some particular property.

This is a very general statement. So let's look at a few interpretations. For example, we may want the transformed object to be "simpler" than the original object. This usually means that the new object has the same meaning as the given object but uses fewer symbols. For example, the expression $x + 1$ might be a simplification of $(x^2 + x)/x$, and the FP program $f \, @ \, (\text{true} \to c; d)$ can be simplified to $f \, @ \, c$.

We may want the transformed object to act as the meaning of the given object. For example, we usually think of the meaning of the expression $3 + 4$ as its value, which is 7. On the other hand, the meaning of the expression $x + 1$ is $x + 1$ if we don't know the value of x.

Whenever a light bulb goes on in our brain and we finally understand the meaning of some idea or object, we usually make statements like "Oh yes, I see it now" or "Yes, I understand." These statements usually mean that we have made a connection between the thing we're trying to understand and some other thing that is already familiar to us. So there is a transformation (i.e., a function) from the new idea to a familiar old idea.

For example, suppose we want to describe the meaning of the base 10 numerals (i.e., nonempty strings of decimal digits) or the base 2 numerals (i.e., nonempty strings of binary digits). Let m_{ten} denote the meaning function for base 10 numerals, and let m_{two} denote the meaning function for base 2 numerals. If we can agree on anything, we most probably will agree that $m_{\text{ten}}(16) = m_{\text{two}}(10000)$ and $m_{\text{ten}}(14) = m_{\text{two}}(1110)$. Further, if we let m_{rom} denote the meaning function for Roman numerals, then we most probably also agree that $m_{\text{rom}}(\text{XII}) = m_{\text{ten}}(12) = m_{\text{two}}(1100)$.

For this example we'll use the set $\mathbb{N}$ of natural numbers to represent the meanings of the numerals. For base 10 and base 2 numerals there may be some confusion because, for example, the string 25 denotes a base 10 numeral and it also represents the natural number that we call twenty-five. Given that this confusion exists, we have

$$m_{\text{ten}}(25) = m_{\text{two}}(11001) = m_{\text{rom}}(\text{XXV}) = 25.$$

So we can write down three functions from the three kinds of numerals (the syntax) to natural numbers (the semantics):

$$m_{\text{ten}} : \text{DecimalNumerals} \to \mathbb{N},$$

$$m_{\text{two}} : \text{BinaryNumerals} \to \mathbb{N},$$

$$m_{\text{rom}} : \text{RomanNumerals} \to \mathbb{N}.$$

Can we give definitions of these functions? Sure. For example, a natural definition for m_{ten} can be given recursively as follows:

Let $m_{\text{ten}}(d) = d$ for each decimal digit d (here the argument is a character, and the value is a natural number). If $d_k d_{k-1} \ldots d_1 d_0$ is a string of base 10 numerals, then

$$m_{\text{ten}}(d_k d_{k-1} \ldots d_1 d_0) = 10^k d_k + \cdots + 10 d_1 + d_0.$$

What properties, if any, should a semantics function possess? Certain operations defined on numerals should be, in some sense, "preserved" by the semantics function. For example, suppose we let $+_{\text{bi}}$ denote the usual binary addition defined on binary numerals. We would like to say that the meaning of the binary sum of two binary numerals is the same as the result obtained by adding the two individual meanings in the algebra $\langle \mathbb{N}; + \rangle$. In other words, for any binary numerals x and y the following equation holds:

$$m_{\text{two}}(x +_{\text{bi}} y) = m_{\text{two}}(x) + m_{\text{two}}(y).$$

The idea of a function preserving an operation can be defined in a general way. Let $f : A \to A'$ be a function between the carriers of two algebras. Suppose ω is an n-ary operation on A. We say that f *preserves* the operation ω if there is a corresponding operation ω' on A' such that for every $x_1, \ldots, x_n \in A$ the following equality holds:

$$f(\omega(x_1, \ldots, x_n)) = \omega'(f(x_1), \ldots, f(x_n)).$$

Of course, if ω is a binary operation, then we can write the above equation in its infix form as follows:

$$f(x \, \omega \, y)) = f(x) \, \omega' \, f(y).$$

For example, the binary numeral meaning function m_{two} preserves $+_{\text{bi}}$. We can write the equation using the prefix form of $+_{\text{bi}}$ as follows:

$$m_{\text{two}}(+_{\text{bi}}(x, y)) = + (m_{\text{two}}(x), m_{\text{two}}(y)).$$

Here's the thing to remember about an operation that is preserved by a function $f : A \to A'$: You can apply the operation to arguments in A and then use f to map the result to A', or you can use f to map each argument from A to A' and then apply the corresponding operation on A' to these arguments. In either case you get the same result.

The following diagram illustrates this property for two binary operators ∘ and ∘':

$$a \; \circ \; b \; = \; c$$

$$\downarrow \qquad \downarrow \quad_{?} \quad \downarrow$$

$$f(a) \; \circ' \; f(b) \; = \; f(c) \quad \text{Yes, if } f \text{ preserves } \circ.$$

We say that $f : A \to A'$ is a *morphism* (also called a *homomorphism*) if every operation in the algebra of A is preserved by f. If a morphism is injective, then it's called a *monomorphism*. If a morphism is surjective, then it's called an *epimorphism*. If a morphism is bijective, then it's called an *isomorphism*. If there is an isomorphism between two algebras, we say that the algebras are *isomorphic*. Two isomorphic algebras are very much alike, and, hopefully, one of them is easier to understand.

For example, m_{two} is a morphism from $\langle \text{BinaryNumerals}; +_{\text{bi}} \rangle$ to $\langle \mathbb{N}; + \rangle$. In fact, we can say that m_{two} is an epimorphism because it's surjective. Notice that distinct binary numerals like 011 and 11 both represent the number 3. Therefore m_{two} is not injective, so it is not a monomorphism, and thus it is not an isomorphism.

EXAMPLE 2. Suppose we define $f : \mathbb{Z} \to \mathbb{Q}$ by $f(n) = 2^n$. Notice that

$$f(n + m) = 2^{n+m} = 2^n \cdot 2^m = f(n) \cdot f(m).$$

Therefore f is a morphism from the algebra $\langle \mathbb{Z}; + \rangle$ to the algebra $\langle \mathbb{Q}; \cdot \rangle$. Notice that $f(0) = 2^0 = 1$. So f is a morphism from the algebra $\langle \mathbb{Z}; +, 0 \rangle$ to the algebra $\langle \mathbb{Q}; \cdot, 1 \rangle$. Notice that $f(-n) = 2^{-n} = (2^n)^{-1} = f(n)^{-1}$. Therefore f is a morphism from the algebra $\langle \mathbb{Z}; +, -, 0 \rangle$ to the algebra $\langle \mathbb{Q}; \cdot, {}^{-1}, 1 \rangle$. It's easy to see that f is injective. It's also easy to see that f is not surjective. Therefore f is a monomorphism, but it is neither an epimorphism nor an isomorphism.   ◆

EXAMPLE 3. Let $m > 1$ be a natural number, and let the function $f : \mathbb{N} \to \mathbb{N}_m$ be defined by $f(x) = x \bmod m$. Let's see whether f is a morphism from the algebra $\langle \mathbb{N}, +, *, 0, 1 \rangle$ to the algebra $\langle \mathbb{N}_m, +_m, *_m, 0, 1 \rangle$. For f to be a morphism we must have $f(0) = 0$, $f(1) = 1$, and for all $x, y \in \mathbb{N}$:

$$f(x + y) = f(x) +_m f(y) \quad \text{and} \quad f(x * y) = f(x) *_m f(y).$$

It's clear that $f(0) = 0$ and $f(1) = 1$. The other equations are just restatements of the congruences (10.13).   ◆

EXAMPLE 4. For any alphabet A we can define a function $f : A^* \to$ Lists[A] by mapping any string to the list consisting of all letters in the string. For example, $f(\Lambda) = \langle\,\rangle$, $f(a) = \langle a \rangle$, and $f(aba) = \langle a, b, a \rangle$. We can give a formal definition of f as follows:

$$f(\Lambda) = \langle\,\rangle,$$

$$f(a \cdot t) = a :: f(t) \text{ for every } a \in A \text{ and } t \in A^*.$$

For example, if $a \in A$, then $f(a) = f(a \cdot \Lambda) = a :: f(\Lambda) = a :: \langle\,\rangle = \langle a \rangle$. It's easy to see that f is bijective because any two distinct strings get mapped to two distinct lists and that any list is the image of some string.

We'll show that f preserves the concatenation of strings. Let "cat" denote both the concatenation of strings and the concatenation of lists. Then we must verify that $f(\text{cat}(s, t)) = \text{cat}(f(s), f(t))$ for any two strings s and t. We'll do it by induction on the length of s. If $s = \Lambda$, then we have

$$f(\text{cat}(\Lambda, t)) = f(t) = \text{cat}(\langle\,\rangle, f(t)) = \text{cat}(f(\Lambda), f(t)).$$

Now assume that s has length $n > 0$ and $f(\text{cat}(u, t)) = \text{cat}(f(u), f(t))$ for all strings u of length less than n. Since the length of s is greater than 0, we can write $s = a \cdot x$ for some $a \in A$ and $x \in A^*$. Then we have

$$
\begin{aligned}
f(\text{cat}(a \cdot x, t)) &= f(a \cdot \text{cat}(x, t)) && \text{(definition of string cat)} \\
&= a :: f(\text{cat}(x, t)) && \text{(definition of } f) \\
&= a :: \text{cat}(f(x), f(t)) && \text{(induction assumption)} \\
&= \text{cat}(a :: f(x), f(t)) && \text{(definition of list cat)} \\
&= \text{cat}(f(a \cdot x), f(t)) && \text{(definition of } f).
\end{aligned}
$$

Therefore f preserves concatenation. Thus f is a morphism from the algebra $\langle A^*; \text{cat}, \Lambda \rangle$ to the algebra $\langle \text{Lists}[A]; \text{cat}, \langle\,\rangle \rangle$. Since f is also a bijection, it follows that the two algebras are isomorphic.   ◆

Constructing Morphisms

Now let's consider the problem of constructing a morphism. We'll demonstrate the ideas with an example function $f : \mathbb{N}_8 \to \mathbb{N}_8$ such that $f(1) = 3$. We want to finish the definition of f so that it becomes a morphism from the algebra $\langle \mathbb{N}_8; +_8, 0 \rangle$ to itself. By $+_8$ we mean the operation of addition modulo 8. For example, $6 +_8 7 = 5$. For f to be a morphism it must preserve $+_8$ and 0. So we must set $f(0) = 0$. What value should we assign to $f(2)$? Notice that we can write $2 = 1 +_8 1$. Since $f(1) = 3$ and f must preserve the operation $+_8$, we can obtain the value $f(2)$ as follows:

$$f(2) = f(1 +_8 1) = f(1) +_8 f(1) = 3 +_8 3 = 6.$$

Now we can compute $f(3) = f(1 +_8 2) = f(1) +_8 f(2) = 3 +_8 6 = 1$. Continuing, we get the following values: $f(4) = 4$, $f(5) = 7$, $f(6) = 2$, and $f(7) = 5$. So the two facts $f(0) = 0$ and $f(1) = 3$ are sufficient to define f.

But is the resulting definition a morphism? We must be sure that $f(x +_8 y) = f(x) +_8 f(y)$ for all $x, y \in \mathbb{N}_8$. For example, is $f(3 +_8 6) = f(3) +_8 f(6)$? We can check it out easily by computing the left- and right-hand sides of the equation:

$$f(3 +_8 6) = f(1) = 3 \quad \text{and} \quad f(3) +_8 f(6) = 1 +_8 2 = 3.$$

Do we have to check the function for all possible pairs $\langle x, y \rangle$? NO. Remember that our rule for defining f was to force the following equation to be true:

$$f(1 +_8 \cdots +_8 1) = f(1) +_8 \cdots +_8 f(1).$$

Since any number in $\mathbb{N}_8$ is a sum of 1's, we are assured that f is a morphism. Let's write this out for an example:

$$
\begin{aligned}
f(3 +_8 4) &= f(1 +_8 1 +_8 1 +_8 1 +_8 1 +_8 1 +_8 1) \\
&= f(1) +_8 f(1) +_8 f(1) +_8 f(1) +_8 f(1) +_8 f(1) +_8 f(1) \\
&= [f(1) +_8 f(1) +_8 f(1)] +_8 [f(1) +_8 f(1) +_8 f(1) +_8 f(1)] \\
&= f(1 +_8 1 +_8 1) +_8 f(1 +_8 1 +_8 1 +_8 1) \\
&= f(3) +_8 f(4).
\end{aligned}
$$

The above discussion might convince you that once we pick $f(1)$, then we know $f(x)$ for all x. But if the codomain is a different carrier, then things can break down. For example, suppose we want to define a morphism f from the algebra $\langle \mathbb{N}_3; +_3, 0 \rangle$ to the algebra $\langle \mathbb{N}_6; +_6, 0 \rangle$. Then we must have $f(0) = 0$. Now, suppose we try to set $f(1) = 3$. Then we must have $f(2) = 0$. To see this, note that $f(2) = f(1 +_3 1) = f(1) +_6 f(1) = 3 +_6 3 = 0$. Is this definition of f a morphism? The answer is NO! Notice that $f(1 +_3 2) \neq f(1) +_6 f(2)$, because $f(1 +_3 2) = f(0) = 0$ and $f(1) +_6 f(2) = 3 +_6 0 = 3$.

So morphisms are not as numerous as one might think. Let's look at a couple more examples.

EXAMPLE 5 (*Language Morphisms*). If A and B are alphabets, we call a function $f : A^* \to B^*$ a *language morphism* if $f(\Lambda) = \Lambda$ and $f(uv) = f(u)f(v)$ for any strings $u, v \in A^*$. In other words, a language morphism from A^* to B^* is a morphism from the algebra $\langle A^*; \text{cat}, \Lambda \rangle$ to the algebra $\langle B^*; \text{cat}, \Lambda \rangle$. Since concatenation must be preserved, a language morphism is completely determined by defining the values $f(a)$ for each $a \in A$. For example, we'll let $A = B = \{a, b\}$ and define a language morphism $f : \{a, b\}^* \to \{a, b\}^*$ by setting $f(a) = b$ and $f(b) = ab$. Then we can make statements like $f(bab) = f(b)f(a)f(b) = abbab$ and $f(b^2) = (ab)^2$.

Language morphisms can be used to transform one language into another language with a similar grammar. For example, the grammar

$$S \to aSb \mid \Lambda$$

defines the language $\{a^n b^n \mid n \in \mathbb{N}\}$. Since $f(a^n b^n) = b^n(ab)^n$ for $n \in \mathbb{N}$, the language $\{a^n b^n \mid n \in \mathbb{N}\}$ is transformed by f into the language $\{b^n(ab)^n \mid n \in \mathbb{N}\}$. This language can be generated by the grammar $S \to f(a)Sf(b) \mid f(\Lambda)$, which becomes $S \to bSab \mid \Lambda$.   ◆

EXAMPLE 6 (*Casting Out by Nines, Threes, etc*). An old technique for finding some answers and checking errors in some arithmetic operations is called "casting out by nines". We want to study the technique and see why it works (so it's not magic). Is 44,820 divisible by 9? Is 43·768 + 9579 = 41593? We can use casting out by nines to answer yes to the first question and no to the second question. How does the idea work? It's a consequence of the following result:

Casting Out by Nines   (10.15)

If K is a natural number and $d_n \ldots d_0$ is the decimal representation of K, then

$$K \bmod 9 = (d_n \bmod 9) +_9 \cdots +_9 (d_0 \bmod 9).$$

Proof: Considering the algebras $\langle \mathbb{N}; +, *, 0, 1 \rangle$ and $\langle \mathbb{N}_9; +_9, *_9, 0, 1 \rangle$, the function $f : \mathbb{N} \to \mathbb{N}_9$ defined by $f(x) = x \bmod 9$ is a morphism. Notice also that $f(10) = 1$, and in general $f(10^n) = 1$ for any natural number n. Since $d_n \ldots d_0$ is the decimal representation of K, we can write

$$K = d_n * 10^n + \cdots + d_1 * 10 + d_0.$$

Now apply f to both sides of the equation to get the desired result:

$$
\begin{aligned}
f(K) &= f(d_n * 10^n + \cdots + d_1 * 10 + d_0) \\
&= f(d_n) *_9 f(10^n) +_9 \cdots +_9 f(d_1) *_9 f(10) +_9 f(d_0) \\
&= f(d_n) *_9 1 +_9 \cdots +_9 f(d_1) *_9 1 +_9 f(d_0) \\
&= f(d_n) +_9 \cdots +_9 f(d_1) +_9 f(d_0). \quad \text{QED.}
\end{aligned}
$$

Casting out by nines works because 10 mod 9 = 1. Therefore casting out by threes also works because 10 mod 3 = 1. More generally, in a radix B number system, casting out by the predecessor of B works if the following equation holds:

$$
B \bmod \mathrm{pred}(B) = 1.
$$

For example, in octal, casting out by sevens works. (Do any other numbers work in octal?) But in binary, casting out by ones does not work because 2 mod 1 = 0. ◆

Exercises

1. For each of the following pairs of congruences, find the smallest natural number x satisfying both congruences in the pair.

   a. $x \equiv 8 \pmod{13}$,
      $x \equiv 3 \pmod 8$.

   b. $x \equiv 34 \pmod 9$,
      $x \equiv 23 \pmod{10}$.

2. Let $x = 398$. Find a pair of numbers $\langle r, s \rangle$ and moduli m and n such that x can be decoded as the unique solution to congruences

   $$x \equiv r \pmod m \quad \text{and} \quad x \equiv s \pmod n, \text{ where } 0 \le x < m \cdot n.$$

3. Given the algebra $\langle \mathbb{N}; +, \cdot, 0, 1 \rangle$ and the relation ~ defined by $x \sim y$ if and only if $x \equiv y \pmod 6$. Calculate the values of each of the following expressions in the quotient algebra $\langle \mathbb{N}/\sim; +, \cdot, [0], [1] \rangle$.

   a. $[4] + [5]$.   b. $[16] \cdot [5]$.   c. $[0] + [1]$.   d. $[0] \cdot [1]$.

4. For each of the following sets, state whether the set is the carrier of a subalgebra of the algebra $\langle \mathbb{N}_9; +_9, 0 \rangle$.

   a. $\{0, 3, 6\}$.   b. $\{1, 4, 5\}$.   c. $\{0, 2, 4, 6, 8\}$.

5. Given the algebra $\langle \mathbb{N}_{12}; +_{12}, 0 \rangle$, find the carriers of the subalgebras generated by each of the following sets.

   a. $\{6\}$.   b. $\{3\}$.   c. $\{5\}$.

6. Find the three morphisms that exist from the algebra $\langle \mathbb{N}_3; +_3, 0 \rangle$ to the algebra $\langle \mathbb{N}_6; +_6, 0 \rangle$.

7. Let A be an alphabet and $f : A^* \to \mathbb{N}$ be defined by $f(x) = \text{length}(x)$. Show that f is a morphism from the algebra $\langle A^*; \text{cat}, \Lambda \rangle$ to $\langle \mathbb{N}; +, 0 \rangle$, where cat denotes the concatenation of strings.

8. Give an example to show that the absolute value function abs : $\mathbb{Z} \to \mathbb{N}$ defined by $\text{abs}(x) = |x|$ is not a morphism from the algebra $\langle \mathbb{Z}; + \rangle$ to the algebra $\langle \mathbb{N}; + \rangle$.

9. Let's assume that we know that the operation $+_n$ is associative over $\mathbb{N}_n$. Let $\circ$ be the binary operation over $\{a, b, c\}$ defined by the following table:

$\circ$	a	b	c
a	c	a	b
b	a	b	c
c	b	c	a

Show that $\circ$ is associative by finding an isomorphism of the two algebras $\langle \{a, b, c\}; \circ \rangle$ and $\langle \mathbb{N}_3; +_3 \rangle$.

10. Given the language morphism $f : \{a, b\}^* \to \{a, b\}^*$ defined by $f(a) = b$ and $f(b) = ab$, compute the value of each of the following expressions.
    a. $f(\{b^n a \mid n \in \mathbb{N}\})$.
    b. $f(\{ba^n \mid n \in \mathbb{N}\})$.
    c. $f^{-1}(\{b^n a \mid n \in \mathbb{N}\})$.
    d. $f^{-1}(\{ba^n \mid n \in \mathbb{N}\})$.
    e. $f^{-1}(\{ab^{n+1} \mid n \in \mathbb{N}\})$.

Chapter Summary

An algebra consists of one or more sets, called carriers, together with operations on the sets. An algebra is useful for solving problems when we have a good knowledge of its operations. We can use the properties of the operations to transform algebraic expressions into equivalent simpler expressions. In high school algebra the carrier is the set of real numbers, and the operations are addition, multiplication, and so on.

An abstract algebra is described by giving a set of axioms to describe the properties of its operations. An abstract algebra is useful when it has lots of concrete examples. Two especially useful concrete examples of Boolean algebra are the algebra of sets and the algebra of propositions. Some important properties of Boolean algebra operations are the idempotent properties, the absorption laws, the involution law, and De Morgan's laws. Digital circuits are modeled by Boolean algebraic expressions. Thus Boolean algebra can be used to simplify a digital circuit by simplifying the corresponding algebraic expression.

The abstract data types of computer science can be described as algebras. When an abstract data type is described as an algebra, its operations can be implemented and then checked for correctness against the axioms. Some fundamental abstract data types are the natural numbers, lists, strings, stacks, queues, binary trees, and priority queues.

Three algebras that are useful as computational tools are relational algebras for databases, process algebras to model computational processes, and functional algebras to describe functional programming.

Many other algebraic ideas are quite useful for computational problems. Congruences are useful for encoding and decoding numbers. Quotient algebras and subalgebras can be used to define new abstract data types. Morphisms allow us to transform one algebra into another—often simpler—algebra and still preserve the meaning of the operations. Language morphisms can be used to generate new languages along with their grammars.

Answers to Selected Exercises

Chapter 1

Section 1.1

1. Consider the four statements "if $1 = 1$ then $2 = 2$," "if $1 = 1$ then $2 = 3$," "if $1 = 0$ then $2 = 2$," and "if $1 = 0$ then $2 = 3$." The second statement is the only one that is false.

3. a. 47 is a prime between 45 and 54. **c.** 9 is odd but not prime. Therefore the statement is false.

4. Let $d \mid m$ and $m \mid n$. Then there are integers k and j such that $m = d{\cdot}k$ and $n = m{\cdot}j$. Therefore we can write $n = m{\cdot}j = (d{\cdot}k){\cdot}j = d{\cdot}(k{\cdot}j)$, which says that $d \mid n$.

5. a. Let x and y be any two even integers. Then they can be written in the form $x = 2m$ and $y = 2n$ for some integers m and n. Therefore the sum $x + y$ can be written as $x + y = 2m + 2n = 2(m + n)$, which is an even integer.

7. Let $x = 3m + 4$, and let $y = 3n + 4$ for some integers m and n. Then the product $x{\cdot}y$ has the form $x{\cdot}y = (3m + 4)(3n + 4) = 9mn + 12m + 12n + 16 = 3(3mn + 4m + 4n + 4) + 4$, which has the same form as x and y.

9. First we'll prove the statement "if x is odd then x^2 is odd." If x is odd, then $x = 2n + 1$ for some integer n. Therefore $x^2 = (2n + 1)(2n + 1) = 4n^2 + 4n + 1 = 2(2n^2 + 2n) + 1$, which is an odd integer. Now we must prove the second statement "if x^2 is odd then x is odd." We'll do it indirectly by proving the contrapositive of the statement, which is "if x is even then x^2 is even." If x is even, then $x = 2n$ for some integer n. Therefore $x^2 = 2n{\cdot}2n = 2(2n^2)$, which is even. Therefore the second statement is also true.

Section 1.2

1. a. $\{x \mid x \text{ is a lion}\}$. **c.** $D = \{x \mid x \in \mathbb{N} \text{ and } 1 \leq x \leq 31\}$. **e.** $\{x \mid x = 2k+1 \text{ and } k \in \mathbb{N} \text{ and } 0 \leq k \leq 7\}$.

2. a. True. **c.** False. **e.** True. **g.** True.

3. $\{a, 4, x, 3, b, c, d\}$.

5. $A = \{x\}$ and $B = \{x, \{x\}\}$.

6. a. $\{\varnothing, \{x\}, \{y\}, \{z\}, \{w\}, \{x, y\}, \{x, z\}, \{x, w\}, \{y, z\}, \{y, w\}, \{z, w\}, \{x, y, z\}, \{x, y, w\}, \{x, z, w\}, \{y, z, w\}, \{x, y, z, w\}$. **c.** $\{\varnothing\}$. **e.** $\{\varnothing, \{\{a\}\}, \{\varnothing\}, \{\{a\}, \varnothing\}\}$.

8. a. $A \cup B = \{1, 5, 8, 9, 11, 13, 14, 17, 20, 21, ...\}$.

9. a. Certainly A is a subset of $A \cup \varnothing$. On the other hand, $A \cup \varnothing$ doesn't contain any elements other than those of A. Therefore $A = A \cup \varnothing$. **c.** An element $x \in A \cup A$ is defined by the property $x \in A$ or $x \in A$, which is the same as the property $x \in A$. Therefore $A \cup A = A$.

10. a. No elements can be in both A and $\varnothing$ at the same time. Therefore $A \cap \varnothing = \varnothing$. **c.** An element $x \in A \cap (B \cap C)$ is defined by the property $x \in A$ and $x \in B \cap C$, which is the same as the property $x \in A$ and $x \in B$ and $x \in C$. This property is the same as $x \in A \cap B$ and $x \in C$, which is the property that describes the fact that $x \in (A \cap B) \cap C$. Therefore $A \cap (B \cap C) = (A \cap B) \cap C$. **e.** First of all, notice that $A \cap B \subset A$ without any assumptions. Now assume that $A \subset B$, and show that $A \cap B = A$. But if $A \subset B$, then any element of A must also be an element of B. Therefore $A \subset A \cap B$, and thus it follows that $A \cap B = A$. Now assume that $A \cap B = A$, and show that $A \subset B$. If $x \in A$, then certainly $x \in A \cap B$ because of our assumption. Therefore $A \subset B$. Thus we have proven that $A \subset B$ if and only if $A \cap B = A$.

11. Let $S \in \text{power}(A \cap B)$. Then $S \subset A \cap B$, which says that $S \subset A$ and $S \subset B$. Therefore $S \in \text{power}(A)$ and $S \in \text{power}(B)$, which says that $S \in \text{power}(A) \cap \text{power}(B)$. This proves that $\text{power}(A \cap B) \subset \text{power}(A) \cap \text{power}(B)$. The other containment is similar.

12. No. A counterexample is $A = \{a\}$ and $B = \{b\}$.

14. a. Let $x \in A \cap (B \cup A)$. Then $x \in A$, so we have $A \cap (B \cup A) \subset A$. For the other containment, let $x \in A$. Then $x \in B \cup A$. Therefore $x \in A \cap (B \cup A)$, which says that $A \subset A \cap (B \cup A)$. This proves the equality by set containment. We can also prove the equality by using a property of intersection (1.11e) applied to the two sets A and $B \cup A$. Thus (1.11e) becomes $A \subset B \cup A$ if and only if $A \cap (B \cup A) = A$. Since we know that $A \subset B \cup A$ is always true, the equality follows.

15. Assume that $(A \cap B) \cup C = A \cap (B \cup C)$, and let $x \in C$. Therefore $x \in (A \cap B) \cup C = A \cap (B \cup C)$, which says that $x \in A$. Thus $C \subset A$. Assume that $C \subset A$, and let $x \in (A \cap B) \cup C$. Then $x \in A \cap B$ or $x \in C$. In either case it follows that $x \in A \cap (B \cup C)$ because $C \subset A$. Thus $(A \cap B) \cup C \subset A \cap (B \cup C)$. The other containment is similar. Thus $(A \cap B) \cup C = A \cap (B \cup C)$.

16. a. Counterexample: $A = \{a\}$, $B = \{b\}$. **c.** Counterexample: $A = \{a\}$, $B = \{b\}$, $C = \{b\}$.

17. a. $A_0 = \mathbb{Z} - \{0\}$, $A_1 = \{0\}$, $A_2 = \mathbb{Z} - \{-2, -1, 0, 1, 2\}$, $A_3 = \{-2, -1, 0, 1, 2\}$, $A_{-2} = \mathbb{Z}$ and $A_{-3} = \varnothing$. **c.** $\mathbb{Z}$. **e.** $\mathbb{Z}$. **g.** $\varnothing$. **i.** $\varnothing$.

18. a. $A_0 = \mathbb{N} - \{0\}$, $A_1 = \{1\}$, $A_2 = \{1, 2\}$, $A_3 = \{1, 3\}$, $A_4 = \{1, 2, 4\}$, $A_5 = \{1, 5\}$, $A_6 = \{1, 2, 3, 6\}$, $A_7 = \{1, 7\}$, and $A_{100} = \{1, 2, 4, 5, 10, 20, 25, 50, 100\}$. **c.** $\{1\}$. **e.** $\{1\}$.

20. a. $A \cap B - C$. **c.** $B \oplus C$.

21. a. For any element x we have $x \in (A')'$ if and only if $x \in U$ and $x \notin A'$ if and only if $x \notin U - A$ if and only if $x \in A$. **c.** Note that an element $x \in A \cap A'$ means that $x \in A$ and $x \in U - A$. But this says that $x \in A$ and $x \notin A$, which can't happen. Therefore $A \cap A' = \varnothing$. To see that $A \cup A' = U$, just note that any element of U must be either in A or not in A.

22. $|A| + |B| + |C| + |D| - |A \cap B| - |A \cap C| - |A \cap D| - |B \cap C| - |B \cap D| - |C \cap D| + |A \cap B \cap C| + |A \cap B \cap D| + |A \cap C \cap D| + |B \cap C \cap D| - |A \cap B \cap C \cap D|$.

24. a. 82. **c.** 23.

25. At most 20 drivers were smoking, talking, and tuning the radio.

27. a. $[x, y, z]$, $[x, y]$. **c.** $[a, a, a, b, b, c]$, $[a, a, b]$. **e.** $[x, x, a, a, [a, a], [a, a]]$, $[x, x]$.

29. Let A and B be bags, and let m and n be the number of times x occurs in A and B, respectively. If $m \geq n$, then put $m - n$ occurrences of x in $A - B$, and if $m < n$, then do not put any occurrences of x in $A - B$.

31. a. Yes, it's the empty set.

Section 1.3

1. $\langle x, x, x \rangle$, $\langle x, x, y \rangle$, $\langle x, y, x \rangle$, $\langle y, x, x \rangle$, $\langle x, y, y \rangle$, $\langle y, x, y \rangle$, $\langle y, y, x \rangle$, $\langle y, y, y \rangle$.

2. a. $\{\langle a, a \rangle, \langle a, b \rangle, \langle b, a \rangle, \langle b, b \rangle, \langle c, a \rangle, \langle c, b \rangle\}$. **c.** $\{\langle \ \rangle\}$. **e.** $\{\langle a, a \rangle, \langle a, b \rangle, \langle a, c \rangle, \langle b, a \rangle, \langle b, b \rangle, \langle b, c \rangle, \langle c, a \rangle, \langle c, b \rangle, \langle c, c \rangle\}$.

3. a. Show that the two sets are equal by showing that each is a subset of the other. Let $\langle x, y \rangle \in (A \cup B) \times C$. Then either $x \in A$ or $x \in B$, and $y \in C$. So either $\langle x, y \rangle \in A \times C$ or $\langle x, y \rangle \in B \times C$. Thus $\langle x, y \rangle \in (A \times C) \cup (B \times C)$, and we have the containment $(A \cup B) \times C \subset (A \times C) \cup (B \times C)$. For the other containment, we'll let $\langle x, y \rangle \in (A \times C) \cup (B \times C)$. Then either $\langle x, y \rangle \in A \times C$ or $\langle x, y \rangle \in B \times C$, which says that either $x \in A$ or $x \in B$, and $y \in C$. Thus $\langle x, y \rangle \in (A \cup B) \times C$ and we have the containment $(A \times C) \cup (B \times C) \subset (A \cup B) \times C$. The two containments show that the sets are equal. **c.** We'll prove that $(A \cap B) \times C = (A \times C) \cap (B \times C)$ by showing that each side is a subset of the other. Since $A \cap B \subset A$ and $A \cap B \subset B$, it follows that $(A \cap B) \times C \subset (A \times C) \cap (B \times C)$. For the other containment, let $\langle x, y \rangle \in (A \times C) \cap (B \times C)$. Then $x \in A \cap B$ and $y \in C$, which implies that $\langle x, y \rangle \in (A \cap B) \times C$. This gives the containment $(A \times C) \cap (B \times C) \subset (A \cap B) \times C$. The two containments show that the sets are equal.

4. a. Notice that $\langle 3, 7 \rangle = \{\{3\}, \{3, 7\}\}$ and $\langle 7, 3 \rangle = \{\{7\}, \{7, 3\}\}$ and that the two sets cannot be equal. **c.** The statement $\langle x_1, x_2, x_3 \rangle = \langle y_1, y_2, y_3 \rangle$ means that $\langle \langle x_1, x_2 \rangle, x_3 \rangle = \langle \langle y_1, y_2 \rangle, y_3 \rangle$. By part (b) the latter equality is true if and only if $\langle x_1, x_2 \rangle = \langle y_1, y_2 \rangle$ and $x_3 = y_3$. One more application of part (b) yields the result that $x_i = y_i$ for each $i = 1, ..., 3$.

5. a. $\langle \{a\}, b \rangle = \{\{a\}, \{b\}\} = \{\{b\}, \{a\}\} = \langle \{b\}, a \rangle$.

7. a. $\{x, +, 3\}$. **c.** $\{0, 1, 2, 3\}$.

8. a. $R = \{\langle 2, 1, 1 \rangle, \langle 3, 1, 2 \rangle, \langle 3, 2, 1 \rangle\}$. **c.** $U = \{\langle a, 1 \rangle, \langle a, 2 \rangle, \langle b, 1 \rangle, \langle b, 2 \rangle\}$.

9. Parts $\subset \mathbb{N} \times \mathbb{N} \times \mathbb{N} \times (\mathbb{N} \times \mathbb{N} \times \mathbb{N}) \times \mathbb{N}$. Here we assume that each price is given in cents and the date is a 3-tuple of natural numbers.

11.

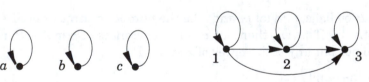

13. a.   **c.**

15. a. One answer is $a\ b\ c\ d\ e\ f$. **b.** One answer is $a\ b\ c\ e\ d\ f$.

17.

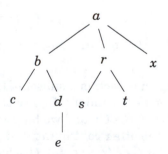

19. Here are two answers:

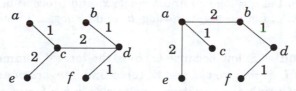

20. a $2 \cdot 5^3 - 2 \cdot 4^3 = 122$. **c** $1 \cdot 2 \cdot 4^4 = 512$.

21. $\langle a, a, a \rangle$, $\langle a, a, b \rangle$, $\langle a, b, a \rangle$, $\langle b, a, a \rangle$, $\langle a, b, b \rangle$, $\langle b, a, b \rangle$, $\langle b, b, a \rangle$, $\langle b, b, b \rangle$, $\langle \langle a \rangle \rangle$, $\langle \langle b \rangle \rangle$, $\langle a, \langle \, \rangle \rangle$, $\langle b, \langle \, \rangle \rangle$, $\langle \langle \, \rangle, a \rangle$, $\langle \langle \, \rangle, b \rangle$.

23. The graph is connected, and all vertices have even degree.

Chapter 2

Section 2.1

1. There are eight total functions of type $\{a, b, c\} \to \{1, 2\}$. For example, one function sends all elements of $\{a, b, c\}$ to 1; another function sends all elements of $\{a, b, c\}$ to 2; another sends a and b to 1 and c to 2; and so on.

2. a. O. **c.** $\{x \mid x = 4k + 3 \text{ where } k \in \mathbb{N}\}$. **e.** $\mathbb{N}$.

3. a. -5. **c.** 4.

4. a. 3. **c.** 1.

5. $(296, 872) = 8 = (-53) \cdot 296 + 18 \cdot 872$.

6. a. 3. **c.** -9.

7. a. $f(\{0, 2, 4\}) = \{0, 2, 4\}$; $f^{-1}(\{0, 2, 4\}) = \mathbb{N}_6$. **c.** $f(\{0, 5\}) = \{0, 4\}$; $f^{-1}(\{0, 5\}) = \{0, 3\}$.

8. a. $\text{floor}(x) = \text{if } x \geq 0 \text{ then } \text{trunc}(x) \text{ else if } x = \text{trunc}(x) \text{ then } x \text{ else } \text{trunc}(x - 1)$.

9. When x/y is negative, $f(x, y)$ can be different than $x \bmod y$. For example, $f(-16, 3) = -1$ and $-16 \bmod 3 = 2$.

11. a. 4. **c.** -3. **e.** $\langle \langle 4, 0 \rangle, \langle 4, 1 \rangle, \langle 4, 2 \rangle, \langle 4, 3 \rangle \rangle$.

12. $\chi_{B \cap C}(x) = \chi_B(x) \, \chi_C(x)$, and $\chi_{B-C}(x) = \chi_B(x)(\chi_B(x) - \chi_C(x))$, or more simply, $\chi_{B-C}(x) = \chi_B(x)(1 - \chi_C(x))$.

13. a. A. **c.** $\{0\}$.

14. a. If x is an integer, then $\lfloor x \rfloor = x$ and $\lfloor x + 1 \rfloor = x + 1$, which makes the desired equality true. If x is not an integer, then there is an integer n such that $n < x < n + 1$. Therefore $\lfloor x \rfloor = n$ and $\lfloor x + 1 \rfloor = n + 1$, which makes the desired equality true. **c.** If $x \in \mathbb{Z}$, then certainly $\lceil x \rceil = \lfloor x \rfloor$. If $\lceil x \rceil = \lfloor x \rfloor$, then $\lfloor x \rfloor \leq x \leq \lceil x \rceil = \lfloor x \rfloor$, which says that $x = \lceil x \rceil = \lfloor x \rfloor$. Therefore $x \in \mathbb{Z}$.

15. a. $\log_b 1 = 0$ means $b^0 = 1$, which is true. **c.** $\log_b (b^x) = x$ means $b^x = b^x$, which is true. **e.** Let $r = \log_b(x^y)$ and $s = \log_b x$, and proceed as in part (d) to show that $r = ys$. **g.** Let $r = \log_a x$, $s = \log_a b$, and $t = \log_b x$. Proceed as in (d) to show that $r = st$.

16. a. The equalities follow because (a, b) is the largest common divisor of a and b. **c.** Since $(d, a) = 1$, we can use (2.1c) to find integers m and n such that $1 = m{\cdot}d + n{\cdot}a$. Multiply the equation by b to obtain $b = b{\cdot}m{\cdot}d + b{\cdot}n{\cdot}a = b{\cdot}m{\cdot}d + a{\cdot}b{\cdot}n$. Since d divides both terms on the right side, d also divides the left side. Therefore $d \mid b$.

18. a. We'll prove both containments at once: $x \in f(E \cup F)$ iff $x = f(y)$, where $y \in E \cup F$ iff $x = f(y)$, where $y \in E$ or $y \in F$ iff $x \in f(E)$ or $x \in f(F)$ iff $x \in f(E) \cup f(F)$. **c.** If $x \in E$, then $f(x) \in f(E)$, which says that $x \in f^{-1}(f(E))$. This proves the containment.

   19. a. We'll prove both containments at once: $x \in f^{-1}(G \cup H)$ iff $f(x) \in G \cup H$ iff $f(x) \in G$ or $f(x) \in H$ iff $x \in f^{-1}(G)$ or $x \in f^{-1}(H)$ iff $x \in f^{-1}(G) \cup f^{-1}(H)$. **c.** If $x \in f(f^{-1}(G))$, then there is an element $y \in f^{-1}(G)$ such that $x = f(y)$. But the fact that $y \in f^{-1}(G)$ implies that $f(y) \in G$. Therefore $x \in G$, which proves the containment.

Section 2.2

1. a. 0, 1, 1, 2, 2, 2, 2, 3, 3, 3, 3, 3, 3, 3, 3, 4.

2. a. $2^7 \le x < 2^8$.

3. a. The ceiling returns an integer, and the floor of an integer is itself. **c.** Any x is in an interval of the form $2^n \le x < 2^{n+1}$ for some integer n. It follows that $2^n \le \text{floor}(x) < 2^{n+1}$. Taking the log of both inequalities, we obtain $n \le \log_2(x) < n + 1$ and $n \le \log_2(\text{floor}(x)) < n + 1$. Therefore $\text{floor}(\log_2(x)) = n = \text{floor}(\log_2(\text{floor}(x)))$.

4. $\text{floor}(\log_2(x)) + 1$.

5. Starting with the definition of f, we can transform the output as follows: $f(m) = \langle\langle 0, 1, ..., m\rangle, m\rangle = \langle\text{seq}(m), m\rangle = \langle\text{seq}(m), \text{id}(m)\rangle = \langle\text{seq}, \text{id}\rangle(m)$. So we can define $f = \langle\text{seq}, \text{id}\rangle$.

6. a. Let "abs" be the absolute value function. Then define $f = \text{map(abs)}$.

7. $\text{max3}(4, 9, 7) = \text{max} \circ \langle\text{max} \circ \langle\mathbf{1, 2}\rangle, \mathbf{3}\rangle(4, 9, 7)$

$$= \text{max}(\langle\text{max} \circ \langle\mathbf{1, 2}\rangle, \mathbf{3}\rangle(4, 9, 7))$$
$$= \text{max}(\text{max} \circ \langle\mathbf{1, 2}\rangle(4, 9, 7), \mathbf{3}(4, 9, 7))$$
$$= \text{max}(\text{max}(\langle\mathbf{1, 2}\rangle(4, 9, 7)), \mathbf{3}(4, 9, 7))$$
$$= \text{max}(\text{max}(\mathbf{1}(4, 9, 7), \mathbf{2}(4, 9, 7)), \mathbf{3}(4, 9, 7))$$
$$= \text{max}(\text{max}(4, 9), 7) = \text{max}(9, 7) = 9.$$

8. a. $h = + \circ \langle \text{apply} \circ \langle 1, 3 \rangle, \text{apply} \circ \langle 2, 3 \rangle \rangle$.
c. $h = + \circ \langle \text{apply} \circ \langle 1, 1 \circ 3 \rangle, \text{apply} \circ \langle 2, 2 \circ 3 \rangle \rangle$.

9. a. $\text{makeSeq} = \text{map}(+) \circ \text{dist} \circ \langle 1, \text{map}(*) \circ \text{dist} \circ \langle 2, \text{seq} \circ 3 \rangle \rangle$.

11. The function $\text{cubes}(n) = \langle 0^3, 1^3, ..., n^3 \rangle$ can be defined as cubes = map(*)∘pairs∘⟨seq, squares⟩, where squares = map(*)∘pairs∘⟨seq, seq⟩. So we can define sumCubes = insert(+)∘cubes. After substitution of these definitions, the definition for sumCubes becomes

$$\text{sumCubes} = \text{insert}(+) \circ \text{map}(*) \circ \text{pairs} \circ \langle \text{seq, squares} \rangle$$
$$= \text{insert}(+) \circ \text{map}(*) \circ \text{pairs} \circ \langle \text{seq, map}(*) \circ \text{pairs} \circ \langle \text{seq, seq} \rangle \rangle.$$

Section 2.3

1. a. $f : C \to B$, where $f(1) = x$, $f(2) = y$. **c.** $f : A \to B$, where $f(a) = x$, $f(b) = y$, $f(c) = z$.

2. a. Eight functions; no injections, six surjections, no bijections, and two with none of the properties. **c.** 27 functions; six satisfy the three properties (injective, surjective, and bijective), 21 with none of the properties.

3. The fatherOf function is not injective because some fathers have more than one child. The fatherOf function is not surjective because there are people who are not fathers.

4. a. Injective. **c.** None. **e.** Injective. **g.** Surjective. **i.** Injective. **k.** Surjective. **m.** Bijective. $f^{-1}(x) = 3x \bmod 5$.

5. a. Let $f(x) = f(y)$. Then $\frac{1}{x+1} = \frac{1}{y+1}$, which says that $x = y$, which implies that f is injective. To show that f is surjective, let $y \in (0, 1)$. Solving the equation $\frac{1}{x+1} = y$ for x yields $x = \frac{1-y}{y}$, which is a positive real number Thus $f(x) = y$, which says that f is surjective. Therefore f is a bijection.

6. a. Let $f(x) = f(y)$. Then $\frac{x}{1-x} = \frac{y}{1-y}$, which says that $x - xy = y - xy$. Cancelling $-xy$ yields $x = y$, which implies that f is injective. To show that f is surjective, let y be a positive real number. Solving the equation $\frac{x}{1-x} = y$ for x yields $x = \frac{y}{1+y}$, which is in the interval (0, 1). Thus $f(x) = y$, which says that f is surjective. Therefore f is a bijection.

7. a. Table listing: five, nine, one, eight, two, seven, three, six, four. **c.** Position 1 of the table is left blank because linear probing could not locate a position for "eight." Table listing: six, blank, one, nine, two, five, three, seven, four.

9. Assume that f and g are surjective, and let $z \in C$. Since g is surjective, there exists an element $y \in B$ such that $z = g(y)$; and since g is surjective, there exists an element $x \in A$ such that $y = f(x)$. Therefore $z = g(y) = g(f(x)) = g \circ f(x)$, so it follows that $g \circ f$ is surjective.

10. a. If $g \circ f$ is surjective, then for each element $z \in C$ there exists an element $x \in A$ such that $z = g \circ f(x)$. If we write $g \circ f(x) = g(f(x))$, it follows that $f(x)$ is an element of B such that $z = g(f(x))$. Therefore g is surjective if $g \circ f$ is surjective.

11. a. Let f be surjective, and let $b \in B$ and $c \in C$. Then there exists an element $a \in A$ such that $f(a) = \langle b, c \rangle$. But $f(a) = \langle g(a), h(a) \rangle$. Therefore $b = g(a)$ and $c = h(a)$. So g and h are surjective. Now let $A = \{1, 2, 3\}$, $B = \{4, 5\}$, and $C = \{6, 7\}$. The set $B \times C$ has four elements, and A has three elements. So there can be no surjection from A to $B \times C$.

Section 2.4

1. For each natural number n, let $A_n = \{\langle i, j \rangle \mid i + j = n\}$. Each set A_n is finite, and $\mathbb{N} \times \mathbb{N}$ is the union of the countable collection of sets A_n. Therefore $\mathbb{N} \times \mathbb{N}$ is countable by (2.7).

2. a. For each natural number i, let B_i be the set of strings in A_n with letters from the alphabet $\{a_0, a_1, ..., a_i\}$. Each B_i is finite, and A_n is the countable union of these finite sets. So A_n is countable by (2.7).

3. Suppose we draw a line in the xy-coordinate plane through the points $\langle 0, a \rangle$ and $\langle 1, b \rangle$. The line segment between these two points is described by the function $f : (0, 1) \to (a, b)$, where $f(x) = (b - a)x + a$. Since f is a bijection, it follows that $|(a, b)| = |(0, 1)|$.

5. One solution is to recall that the tangent function $\tan : (-\pi/2, \pi/2) \to \mathbb{R}$ is a bijection. So $|\mathbb{R}| = |(-\pi/2, \pi/2)|$, and from the previous exercise (letting $a = -\pi/2$ and $b = \pi/2$) we have $|(-\pi/2, \pi/2)| = |(0,1)|$. For another solution, convince yourself that the function $g : (0, 1) \to \mathbb{R}$ defined by

$$g(x) = \frac{x - 1/2}{x(x - 1)}$$

is a bijection.

7. Since A^* is countably infinite, there is a bijection between A^* and $\mathbb{N}$. So the elements in a finite subset of A^* correspond, via the bijection, to elements in a finite subset of $\mathbb{N}$, and conversely. So we have a bijection between $F(A^*)$ and $F(\mathbb{N})$. Since $F(\mathbb{N})$ is countable, so is $F(A^*)$.

9. Assume that the set of functions is countable. Then we can list all the functions as a countable sequence $f_0, f_1, f_2, ..., f_n, ...$. Each function f_n can be represented by its stream of values $\langle f_n(0), f_n(1), ... \rangle$. Therefore (2.13) implies that there is some function that is not in the sequence. This contradiction gives the result. *Note*: Another way to proceed is to actually construct a function that is not in the list, as outlined in the proof of (2.13). For example, we could define $g : \mathbb{N} \to \mathbb{N}$ as $g(n) = f_n(n) + 1$. g differs from each listed function f_n at its diagonal value $f_n(n)$. So g can't be in the list.

Chapter 3

Section 3.1

1. a. The basis case is the statement $1 \in$ Odd, and the induction case states that if $x \in$ Odd, then succ(succ(x)) $\in$ Odd. **c.** Basis: 4, $3 \in S$. Induction: If $x \in S$ then $x + 3 \in S$.

2. $4 = 3 \cup \{3\} = 2 \cup \{2\} \cup \{3\} = 1 \cup \{1\} \cup \{2\} \cup \{3\} = 0 \cup \{0\} \cup \{1\} \cup \{2\} \cup \{3\} = \varnothing \cup \{0\} \cup \{1\} \cup \{2\} \cup \{3\} = \{0, 1, 2, 3\}$.

3. a. $\langle\langle \, \rangle\rangle$. **c.** $\langle\langle \, \rangle, a\rangle$. **e.** $\langle\langle a\rangle, a, b, c\rangle$. **g.** $\langle b, a\rangle$.

4. a. cons(cons(a, $\langle \, \rangle$), cons(cons(b, $\langle \, \rangle$), $\langle \, \rangle$)).

5. a. For the base case, put $\langle \, \rangle \in$ Even[A]. For the induction case, if $L \in$ Even[A] and $a, b \in A$, then put cons(a, cons(b, L)) $\in$ Even[A].

6. Basis: $\langle \, \rangle$, $\langle a\rangle$, $\langle b\rangle \in S$. Induction: If $x \in S$ and $x \neq \langle \, \rangle$, then if head($x$) = a then put cons(b, x) $\in S$, else put cons(a, x) $\in S$.

7. Basis: 0, $1 \in B$. Induction: If $x \in B$ and head(x) = 1, then put the following four strings in B: $x \cdot 1 \cdot 0$, $x \cdot 1 \cdot 1$, $x \cdot 0 \cdot 1$, and $x \cdot 0 \cdot 0$.

9. a. For the basis case, put $A \subset$ Odd. For the induction case, if $a, b \in A$ and $x \in$ Odd, then put $a \cdot b \cdot x \in$ Odd. **c.** For the basis case, put $d.e \in$ Rat for each pair of decimal digits $d, e \in$ Dec. For the induction case, if $r \in$ Rat and $d \in$ Dec, then put $d \cdot r$, $r \cdot d \in$ Rat.

10. For the basis case, put $a, b, ab, ba \in S$. For the induction case, if $s \in S$ and tail(s) $\neq \Lambda$, then: if head(s) = a then $a::s \in S$ else $b::s \in S$.

11. a. For the basis case, put DecNum $\subset$ Exp, and for the induction case, if $x, y \in$ Exp, then put (x) $\in$ Exp and x+$y \in$ Exp. Note that the strings (x) and x+y are actually the constructions ($\cdot x \cdot$) and cat(x, $+\cdot y$).

12. To convert from tree notation to tuple notation, delete the word tree from the expression, and change all parentheses to tuple symbols. To convert from tuple notation to tree notation, just reverse the process except that the empty tuples should remain unchanged. **a.** $\langle\langle\langle \, \rangle, x, \langle \, \rangle\rangle, y, \langle\langle \, \rangle, z, \langle\langle \, \rangle, w, \langle \, \rangle\rangle\rangle\rangle$.

13. a. $B = \{\langle x, y\rangle \mid x, y \in \mathbb{N}$ and $x \geq y\}$.

14. a. First definition: Basis: $\langle\langle \, \rangle, \langle \, \rangle\rangle \in S$. Induction: If $\langle x, y\rangle \in S$ and $a \in A$, then $\langle a::x, y\rangle$, $\langle x, a::y\rangle \in S$. Second definition: Basis: $\langle\langle \, \rangle, L\rangle \in S$ for all $L \in$ Lists[A]. Induction: If $a \in A$ and $\langle K, L\rangle \in S$, then $\langle a::K, L\rangle \in S$. **c.** First definition: Basis: $\langle 0, \langle \, \rangle\rangle \in S$. Induction: If $\langle x, L\rangle \in S$ and $m \in \mathbb{N}$, then $\langle x, m::L\rangle$, $\langle$succ(x), $L\rangle \in S$. Second definition: Basis: $\langle 0, L\rangle \in S$ for all $L \in$ Lists[$\mathbb{N}$]. Induction: If $\langle n, L\rangle \in S$ and $n \in \mathbb{N}$, then $\langle$succ(n), $L\rangle \in S$.

15. Put $\langle \, \rangle \in E$ and all single-node trees in O as basis cases. For the induction case, consider the following four situations: (1) if $t \in O$ and $a \in A$, then put tree($t, a, \langle \rangle$) and tree($\langle \rangle, a, t$) in E; (2) if $t \in E$ and $a \in A$, then put tree($t, a, \langle \rangle$) and tree($\langle \rangle, a, t$) in O; (3) if $s, t \in O$ and $a \in A$, then put tree(s, a, t) in O; if $s \in O, t \in E$, and $a \in A$, then put tree(s, a, t) and tree(t, a, s) in E.

Section 3.2

1. a. $L \cdot M = \{bba, ab, a, abbbba, abbab, abba, bbba, bab, ba\}$.
c. $L^2 = \{\Lambda, abb, b, abbabb, abbb, babb, bb\}$.

2. a. $L = \{b, ba\}$. **c.** $L = \{a, b\}$.

3. a. The statement is true because $x \cdot \Lambda = \Lambda \cdot x = x$ for any string x. **c.** The first equality can be proved by showing the two set containments together as follows: $x \in L \cdot (M \cup N)$ iff $x = y \cdot z$, where $y \in L$ and $z \in M \cup N$ iff either $x = y \cdot z \in L \cdot M$ or $x = y \cdot z \in L \cdot N$ iff $x \in L \cdot M \cup L \cdot N$. The second equality can be proved the same way.

4. a. The statement is true because $A^0 = \{\Lambda\}$ for any language A. **b.** We'll prove the equality $L^* = L^* \cdot L^*$ by showing containment of sets. First we have $L^* = L^* \cdot \{\Lambda\}$ and $L^* \cdot \{\Lambda\} \subset L^* \cdot L^*$. Therefore $L^* \subset L^* \cdot L^*$. Next, if $x \in L^* \cdot L^*$, then $x = y \cdot z$, where $y, z \in L^*$. Then there are natural numbers m and n such that $y \in L^m$ and $z \in L^n$. Therefore $x = y \cdot z$ in L^{m+n}, which is a subset of L^*. Therefore we have the other containment $L^* \cdot L^* \subset L^*$. The equality $L^* = (L^*)^*$ is proved similarly: L^* is a subset of $(L^*)^*$ by definition. For the other containment, if $x \in (L^*)^*$, then there is a number n such that $x \in (L^*)^n$. So x is a concatenation of n strings, each one from L^*. So x is a concatenation of n strings, each from some power of L. Therefore $x \in L^*$. Therefore $(L^*)^* \subset L^*$.

5. a. $S \to DS, D \to 7, S \to DS, S \to DS, D \to 8, D \to 0, S \to D, D \to 1$. **c.** $S \Rightarrow DS \Rightarrow DDS \Rightarrow DDDS \Rightarrow DDDD \Rightarrow DDD1 \Rightarrow DD01 \Rightarrow D801 \Rightarrow 7801$.

6. a. Basis: $\Lambda \in L(G)$. Induction: If $w \in L(G)$, then put $aaw \in L(G)$.

7. a. $S \to bb \mid bbS$. **c.** $S \to \Lambda \mid abS$.

8. a. $O \to D1 \mid D3 \mid D5 \mid D7 \mid D9$, and $D \to \Lambda \mid D0 \mid D1 \mid D2 \mid D3 \mid D4 \mid D5 \mid D6 \mid D7 \mid D8 \mid D9$.

9. a. $S \to D \mid S + S \mid (S)$, and D denotes a decimal numeral.

10. $S \to \Lambda \mid aSa \mid bSb \mid cSc$.

11. a. $S \to aSb \mid \Lambda$.
c. $S \to T \mid C, T \to a\,Tc \mid b, C \to BA, B \to bB \mid \Lambda$, and $A \to aA \mid \Lambda$.

13. a. The set of all strings of balanced pairs of brackets.

14. $S \to \{E\}$ and $E \to E, T \mid T$ and $T \to i \mid \Lambda \mid S$.

15. a. $S \to A \mid AB, A \to Aa \mid a, B \to Bb \mid b$.

Section 3.3

1. We'll unfold the leftmost term in each expression:
fib(4) = fib(3) + fib(2) = fib(2) + fib(1) + fib(2) = fib(1) + fib(0) + fib(1) + fib(2)
= 1 + fib(0) + fib(1) + fib(2) = 1 + 0 + fib(1) + fib(2) = 1 + 0 + 1 + fib(2)
= 1 + 0 + 1 + fib(1) + fib(0) = 1 + 0 + 1 + 1 + fib(0) = 1 + 0 + 1 + 1 + 0 = 3.

3. Assume lists are not empty. Then "small" can be defined as follows:

small(L) = if tail(L) = $\langle\,\rangle$ then head(L)
else if head(L) < small(tail(L)) then head(L)
else small(tail(L)).

5. 1, 1, 2, 2, 3, 4, 4, 4, 5, 6, 7, 7, 8, 8, 8, 8, 9.

7. insert(f)($\langle a, b\rangle$) = $f(a, b)$, insert(f)(cons(a, L)) = $f(a,$insert(f)(L)).

9.   Equational form: last($x{\cdot}\Lambda$) = x, last($x{\cdot}s$) = last(s).

If-then form: last(s) = if tail(s) = Λ then s else last(tail(s)).

11. Modify (3.16) by adding another basis case to return 1 if the input is 1:

bin(x) = if x = 0 then 0
else if x = 1 then 1
else cat(bin(floor(x/2), x mod 2)).

13. a. In(T): **if** $T \neq \langle\,\rangle$ **then** In(left(T)); print(root(T)); In(right(T)) **fi.**

14. a. Equational form: leaves($\langle\,\rangle$) = 0, leaves(tree($\langle\,\rangle$, a, $\langle\,\rangle$)) = 1, leaves(tree(l, a, r)) = leaves(l) + leaves(r). If-then form: leaves(t) = if t = $\langle\,\rangle$ then 0 else if left(t) = right(t) = $\langle\,\rangle$ then 1 else leaves(left(t)) + leaves(right(t)).

15. Let rem(L) denote the list obtained from L by removing redundant copies of elements and keeping the rightmost occurrence of each element. Assume that headRight(L) returns the rightmost element of L and tailLeft(L) returns the list obtained from L by removing its rightmost element. Then we have

rem(L) = if L = $\langle\,\rangle$ then $\langle\,\rangle$
else cat(rem(removeAll(headRight(L), tailLeft(L))), headRight(L) :: $\langle\,\rangle$).

headRight(L) = if tail(L) = $\langle\,\rangle$ then head(L)
else headRight(tail(L)).

tailLeft(L) = if tail(L) = $\langle\,\rangle$ then $\langle\,\rangle$
else head(L) :: tailLeft(tail(L)).

16. a. isMember(x, L)   =   if L = $\langle\,\rangle$ then false
else if x = head(L) then true
else isMember(x, tail(L)).

c. areEqual(K, L) = if isSubset(K, L) then isSubset(L, K) else false.

e. intersect(K, L) = if $K = \langle\,\rangle$ then $\langle\,\rangle$
else if isMember(head(K), L) then
head(K) :: intersect(tail(K), L)
else
intersect(tail(K), L).

17. $f(0) = 0$, $f(1) = 1$, and $f(n + 2) = f(n + 1) + f(n) + \text{fib}(n + 1)\text{fib}(n)$.

18. Assume that the product of the empty list $\langle\,\rangle$ with any list is $\langle\,\rangle$. Then define product as follows:

product(A, B) =   if $A = \langle\,\rangle$ or $B = \langle\,\rangle$ then $\langle\,\rangle$
else concatenate the four lists
$\langle\langle\text{head}(A), \text{head}(B)\rangle\rangle$,
product($\langle\text{head}(A)\rangle$, tail($B$)),
product(tail(A), $\langle\text{head}(B)\rangle$), and
product(tail(A), tail(B)).

19. a. 1, 2.5, 2.05. **c.** 3, 2.166..., 2.0064.... **e.** 1, 5, 3.4.

20. a. For any stream s of numbers, let prod(n, s) denote the product of the first n elements of s. We can write prod(n, s) = if $n = 0$ then 1 else head(s)*prod($n - 1$, tail(s)). Therefore the product of the first n prime numbers is prod(n, sieve(ints(2))).

21. a. Square(x :: s) = $x \cdot x$ :: Square(s). **c.** map(f, a :: s) = $f(a)$ :: map(f, s).

22. $f(x) = x - 10$ for $x > 10$ and $f(x) = 1$ for $0 \leq x \leq 10$.

Chapter 4

Section 4.1

1. a. All. **c.** All. Yes. **e.** All. **g.** Transitive. **i.** Reflexive.

2. a. Symmetric. **c.** Reflexive and transitive.

3. The symmetric and transitive properties are conditional statements. Therefore they are always true for the vacuous cases when their hypotheses are false.

4. a. isGrandchildOf. **c.** isNephewOf.

5. isFatherOf ∘ isBrotherOf.

6. a. {$\langle a, a\rangle$, $\langle b, b\rangle$, $\langle c, c\rangle$, $\langle a, b\rangle$, $\langle b, c\rangle$}. **c.** {$\langle a, b\rangle$}.
e. {$\langle a, a\rangle$, $\langle b, b\rangle$, $\langle c, c\rangle$, $\langle a, b\rangle$}. **g.** {$\langle a, a\rangle$, $\langle b, b\rangle$, $\langle c, c\rangle$}.

7. a. Let R be reflexive. Then $a\,R\,a$ and $a\,R\,a$ for all a, which implies that a R^2 a for all a. Therefore R^2 is reflexive. **c.** Let R be transitive, and let a R^2 b

and $b\,R^2\,c$. Then $a\,R\,x$ and $x\,R\,b$, and $b\,R\,y$ and $y\,R\,c$ for some x and y. Since R is transitive, it follows that $a\,R\,b$ and $b\,R\,c$. Therefore $a\,R^2\,c$. Thus R^2 is transitive.

8. a. Let $R = \{\langle a, b\rangle, \langle b, a\rangle\}$. Then R is irreflexive, and $R^2 = \{\langle a, a\rangle, \langle b, b\rangle\}$, which is not irreflexive.

9. a. $\{\langle x, y\rangle \mid x < y - 1\}$.

10. a. $\mathbb{N} \times \mathbb{N}$. **c.** $\{\langle x, y\rangle \mid y \neq 0\} - \{\langle 0, 1\rangle\}$.

11. a. $\langle x, y\rangle \in R \circ (S \circ T)$ iff $\langle x, w\rangle \in R$ and $\langle w, y\rangle \in S \circ T$ for some w iff $\langle x, w\rangle \in R$ and $\langle w, z\rangle \in S$ and $\langle z, y\rangle \in T$ for some w and z iff $\langle x, z\rangle \in R \circ S$ and $\langle z, y\rangle \in T$ for some z iff $\langle x, y\rangle \in (R \circ S) \circ T$. **c.** If $\langle x, y\rangle \in R \circ (S \cap T)$, then $\langle x, w\rangle \in R$ and $\langle w, y\rangle \in S \cap T$ for some w. Thus $\langle x, y\rangle \in R \circ S$ and $\langle x, y\rangle \in R \circ T$, which implies that $\langle x, y\rangle \in R \circ S \cap R \circ T$.

13. $r(\varnothing) = \{\langle a, a\rangle \mid a \in A\}$, which is basic equality over A.

14. a. $\varnothing$. **c.** $\{\langle a, b\rangle, \langle b, a\rangle, \langle b, c\rangle, \langle c, b\rangle\}$.

15. a. $\varnothing$. **c.** $\{\langle a, b\rangle, \langle b, a\rangle, \langle a, a\rangle, \langle b, b\rangle\}$.

16. a. isAncestorOf. **c.** greater.

17. Since $t(\text{less}) = \text{less}$, it follows that $st(\text{less}) = s(\text{less}) = \{\langle m, n\rangle \mid m \neq n\}$. On the other hand, $ts(\text{less}) = t(\{\langle m, n\rangle \mid m \neq n\}) = \mathbb{N} \times \mathbb{N}$.

18. a. If R is reflexive, then it contains the set $\{\langle a, a\rangle \mid a \in A\}$. Since $s(R)$ and $t(R)$ contain R as a subset, it follows that they each contain $\{\langle a, a\rangle \mid a \in A\}$. **c.** Suppose R is transitive. Let $\langle a, b\rangle, \langle b, c\rangle \in r(R)$. If $a = b$ or $b = c$, then certainly $\langle a, c\rangle \in r(R)$. So suppose $a \neq b$ and $b \neq c$. Then $\langle a, b\rangle, \langle b, c\rangle \in R$. Since R is transitive, it follows that $\langle a, c\rangle \in R$, which of course also says that $\langle a, c\rangle \in r(R)$. Therefore $r(R)$ is transitive.

19. a. A proof by containment goes as follows: If $\langle a, b\rangle \in rt(R)$, then either $a = b$ or there is a sequence of elements $a = x_1, x_2, \ldots, x_n = b$ such that $\langle x_i, x_{i+1}\rangle \in R$ for $1 \leq i < n$. Since $R \subset r(R)$, we also have $\langle x_i, x_{i+1}\rangle \in r(R)$ for $1 \leq i < n$, which says that $\langle a, b\rangle \in tr(R)$. For the other containment, let $\langle a, b\rangle \in tr(R)$. If $a = b$, then $\langle a, b\rangle \in rt(R)$. If $a \neq b$, then there is a sequence of elements $a = x_1, x_2, \ldots, x_n = b$ such that $\langle x_i, x_{i+1}\rangle \in r(R)$ for $1 \leq i < n$. If $x_i = x_{i+1}$, then we can remove x_i from the sequence. So we can assume that $x_i \neq x_{i+1}$ for $1 \leq i < n$. Therefore $\langle x_i, x_{i+1}\rangle \in R$ for $1 \leq i < n$, which says that $\langle a, b\rangle \in t(R)$, and thus also $\langle a, b\rangle \in rt(R)$. **c.** If $\langle a, b\rangle \in st(R)$, then either $\langle a, b\rangle \in t(R)$ or $\langle b, a\rangle \in t(R)$. Without loss of generality we can assume that $\langle a, b\rangle \in t(R)$. Then there is a sequence of elements $a = x_1, x_2, \ldots, x_n = b$ such that $\langle x_i, x_{i+1}\rangle \in R$ for $1 \leq i < n$. Since $R \subset s(R)$, we also have $\langle x_i, x_{i+1}\rangle \in s(R)$ for $1 \leq i < n$, which says that $\langle a, b\rangle \in R$ (the symmetry also puts $\langle b, a\rangle \in ts(R)$).

21. a.   **b.**

	1	2	3	4
1	0	20	∞	5
2	∞	0	10	∞
3	∞	∞	0	10
4	∞	10	5	0

	1	2	3	4
1	0	15	10	5
2	∞	0	10	20
3	∞	20	0	10
4	∞	10	5	0

	1	2	3	4
1	0	4	4	0
2	0	0	0	3
3	0	4	0	0
4	0	0	0	0

22. Let "path" be the function to compute the list of edges on a shortest path from i to j. We'll use the "cat" function to concatenate two lists.

$$\text{path}(i, j) = \text{if } P_{ij} = 0 \text{ then } \langle\langle i, j\rangle\rangle \text{ else } \text{cat}(\text{path}(i, P_{ij}), \text{path}(P_{ij}, j)).$$

Section 4.2

1. a. Yes, it's an equivalence relation. The statement "$a + b$ is even" is the same as saying that a and b are both odd or both even. Using this fact, it's easy to prove the three properties.
c. Yes. The statement $ab > 0$ means that a and b have the same sign. Using this fact, it's easy to prove the three properties. Another way is to observe that there is a function "sign" such that $ab > 0$ if and only if $\text{sign}(a) = \text{sign}(b)$. Therefore the relation is a kernel relation, which we know is an equivalence relation.
e. Yes. Check the three properties.

3. R is reflexive and symmetric, but it's not transitive. For example, 2 R 6 and 6 R 12, but 2 $\not R$ 12.

4. a. Not reflexive: $\langle 0, 0\rangle \notin R$. Not symmetric: $\langle 0, 4\rangle \in R$ but $\langle 4, 0\rangle \notin R$. Not transitive: $-3 R 0$ and $0 R 4$, but $-3 \not R 4$. $tsr(R) = \mathbb{Z} \times \mathbb{Z}$.

5. $tsr(R) = trs(R) = rts(R)$ and $str(R) = srt(R) = rst(R)$.

7. $K_f = \{\langle x, y\rangle \mid x, y \in \mathbb{Z} \text{ and either } x = y \text{ or } x = -y\}$. $\mathbb{Z}/K_f = \{\{x, -x\} \mid x \in \mathbb{Z}\}$.

9. $K_f = \{\langle i, j\rangle \mid 0 \le i \le 11 \text{ and } 0 \le j \le 11, \text{ or } i = j \ge 12\}$. $\mathbb{N}/K_f$ consists of the equivalence class $\{0, 1, 2, 3, 4, 5, 6, 7, 8, 9, 10, 11\}$ together with the infinitely many singletons $\{12\}, \{13\}, \{14\}, \ldots$.

11. The function $s : A \to A/K_f$ defined by $s(a) = [a]$ is a surjection because every element in A/K_f has the form $[a]$ for some $a \in A$. The function $i : A/K_f \to B$ defined by $i([a]) = f(a)$ is an injection because if $i([a]) = i([b])$, then $f(a) = f(b)$. But this says that $[a] = [b]$, which implies that i is injective. To see that $f = i \circ s$, notice that $i \circ s(a) = i(s(a)) = i([a]) = f(a)$.

13. Here are two answers:

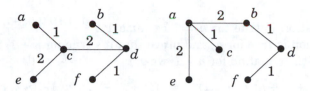

Section 4.3

1. a. False. **c.** True.

2. a. No. **c.** Yes. **e.** No.

3.   **a.**   **b.**   **c.**

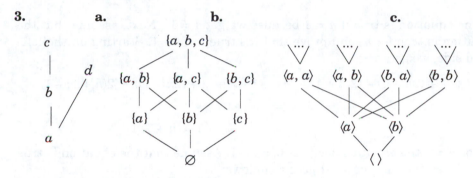

5. The glb of two elements is their greatest common divisor, and the lub is their least common multiple.

7. a. No tree has fewer than zero nodes. Therefore every descending chain of trees is finite if the order is by the number of nodes. **c.** No list has length less than zero. Therefore every descending chain of lists is finite if the order is by the length of the list.

9. Yes.

11. One possible answer is 1, 2, 4, 3, 5, 6, 7.

13. Suppose A is well-founded and S is a nonempty subset of A. If S does not have a minimal element, then there is an infinite descending chain of elements in S, which contradicts the assumption that A is well-founded. For the converse, suppose that every nonempty subset of A has a minimal element. So any descending chain of elements from A is a nonempty subset of A that must have a minimal element. Thus the descending chain must be finite. Therefore A is well-founded.

14. a. Yes. **c.** No. For example, $-2 < 1$, but $f(-2) > f(1)$. **e.** Yes. **g.** No.

Section 4.4

1. 2900.

2. a. The equation is true for $n = 1$ because $1^2 = \dfrac{1\,(1+1)\,(2\cdot 1 + 1)}{6}$. So assume that the equation is true for n, and prove that it's true for $n + 1$. Starting with the left side of the equation for $n + 1$, we get

$$1^2 + 2^2 + \dots + n^2 + (n+1)^2 \;=\; (1^2 + 2^2 + \dots + n^2) + (n+1)^2$$

$$= \;\frac{n(n+1)\,(2n+1)}{6} + (n+1)^2$$

$$= \;\frac{(n+1)\,((n+1)+1)\,(2(n+1)+1)}{6}.$$

c. The equation is true if $n = 1$ because we get $1 = 1^2$. Next, assume that the equation is true for n, and prove that it's true for $n + 1$. Starting on the left-hand side, we get

$$1 + 3 + \dots + (2n - 1) + (2(n+1) - 1) = \; (1 + 3 + \dots + (2n - 1)) + (2(n+1) - 1)$$

$$= \; n^2 + (2(n+1) - 1)$$

$$= \; n^2 + 2n + 1 = (n+1)^2.$$

3. For $n = 0$ the equation becomes $0 = 1 - 1$. Assume that the equation is true for n. Then the case for $n + 1$ goes as follows:

$$F_0 + F_1 + \dots + F_n + F_{n+1} = (F_0 + F_1 + \dots + F_n) + F_{n+1} = F_{n+2} - 1 + F_{n+1} = F_{n+3} - 1.$$

4. a. For $n = 0$ the equation becomes $2 = 3 - 1$. Assume that the equation is true for n. Then the case for $n + 1$ goes as follows:

$$L_0 + L_1 + \dots + L_n + L_{n+1} = (L_0 + L_1 + \dots + L_n) + L_{n+1} = L_{n+2} - 1 + L_{n+1} = L_{n+3} - 1.$$

5. Let $P(m, n)$ denote the equation. Induct on the variable n. For any m we have $\mathrm{sum}(m + 0) = \mathrm{sum}(m) = \mathrm{sum}(m) + \mathrm{sum}(0) + m0$. So $P(m, 0)$ is true for arbitrary m. Now assume that $P(m, n)$ is true, and prove that $P(m, n + 1)$ is true. Starting on the left-hand side we get

$$\mathrm{sum}(m + (n + 1)) \;=\; \mathrm{sum}((m + n) + 1)$$

$$= \; \mathrm{sum}(m + n) + m + n + 1$$

$$= \; \mathrm{sum}(m) + \mathrm{sum}(n) + mn + m + n + 1$$

$$= \; \mathrm{sum}(m) + \mathrm{sum}(n + 1) + m(n + 1).$$

Therefore $P(m, n + 1)$ is true. Therefore $P(m, n)$ is true for all m and n.

7. Power$(\varnothing) = \{\varnothing\}$. So a finite set with 0 elements has 2^0 subsets. Now let A be a set with $|A| = n > 0$, and assume that the statement is true for any set with fewer than n elements. We can write $A = \{x\} \cup B$, where $|B| = n - 1$. So we can write power(A) as the union of two disjoint sets: power(A) = power(B) $\cup$ {{x} $\cup$ C | C $\in$ power(B)}. Since $x \notin B$, these two sets have the same cardinality, which by induction is 2^{n-1}. In other words, we have |{{x} $\cup$ C | C $\in$ power(B)}| = |power(B)| = 2^{n-1}. Therefore |power(A)| = |power(B)| + |{{x} $\cup$ C | C $\in$ power(B)}| = 2^{n-1} + 2^{n-1} = 2^n. Therefore any finite set with n elements has 2^n subsets.

9. a. Let T be a binary tree. We know that an empty tree has no nodes. Since $g(\langle \, \rangle) = 0$, we know that the function is correct when $T = \langle \, \rangle$. For the induction part we need a well-founded ordering on binary trees. For example, let $t \prec s$ mean that t is a subtree of s. Now assume that T is a nonempty binary tree, and also assume that the function is correct for all subtrees of T. Since T is nonempty, it has the form $T = \text{tree}(L, x, R)$. We know that the number of nodes in T is equal to the number of nodes in L plus those in R plus one. The function g, when given argument T, returns $1 + g(L) + g(R)$. Since L and R are subtrees of T, it follows by assumption that $g(L)$ and $g(R)$ represent the number of nodes in L and R, respectively. Thus $g(T)$ is the number of nodes in T.

10. a. If $L = \langle x \rangle$, then forward(L) = {print(head(L)); forward(tail(L))} = {print(x); forward($\langle \, \rangle$)} = {print(x)}. We'll use the well-founded ordering based on the length of lists. Let L be a list with n elements, where $n > 1$, and assume that forward is correct for all lists with fewer than n elements. Then forward(L) = {print(head(L)); forward(tail(L))}. Since tail(L) has fewer than n elements, forward(tail(L)) correctly prints out the elements of tail(L) in the order listed. Since print(head(L)) is executed before forward(tail(L)), it follows that forward(L) is correct.

11. a. We can use well-founded induction, where $L \prec M$ if length(L) < length(M). Since an empty list is sorted and sort($\langle \, \rangle$) = $\langle \, \rangle$, it follows that the function is correct for the basis case $\langle \, \rangle$. For the induction case, assume that sort(L) is sorted for all lists L of length n, and show that sort($x :: L$) is sorted. By definition, we have sort($x :: L$) = insert(x, sort(L)). The induction assumption implies that sort(L) is sorted. Therefore insert(x, sort(L)) is sorted by the assumption in the problem. Thus sort($x :: L$)) is sorted.

13. First we must show that $f(\langle \, \rangle)$ is a binary search tree. Since $f(\langle \, \rangle) = \langle \, \rangle$ and the empty tree is a trivial binary search tree, the basis case is true. Next, we let $x :: L$ be an arbitrary list and assume that $f(L)$ is a binary search tree. Then we must show that $f(x :: L)$ is a binary search tree. Using the definition of f, we obtain $f(x :: L) = \text{insert}(x, f(L))$. Since, by assumption, $f(L)$ is a binary search tree, it follows that insert(x, $f(L)$) is also a binary search tree (remember we are assuming that the insert function is correct). Therefore $f(x :: L)$ is a binary search tree. It follows from (4.23) that $f(M)$ is a binary search tree for all lists M. QED.

15. Let $P(a, L)$ mean "removeAll(a, L) contains no occurrences of a." The definition gives removeAll($a, \langle \rangle$) = $\langle \rangle$. So $P(a, \langle \rangle)$ is true for any a. This proves the basis case. Now assume that $L \neq \langle \rangle$ and assume that $P(a, K)$ is true for all lists $K \prec L$. Show that $P(a, L)$ is true. Since there are two else clauses to the definition, we have two cases. For the first case, assume that $a = \text{head}(L)$. In this case we have removeAll(a, L) = removeAll($a, \text{tail}(L)$). The induction assumption implies that $P(a, \text{tail}(L))$ is true. Therefore $P(a, L)$ is true when $a = \text{head}(L)$. Now assume that $a \neq \text{head}(L)$. Then the definition gives

$$\text{removeAll}(a, L) = \text{head}(L) :: \text{removeAll}(a, \text{tail}(L)).$$

The induction assumption says that removeAll($a, \text{tail}(L)$) is true. Since $a \neq \text{head}(L)$, it follows that head(L) :: removeAll($a, \text{tail}(L)$) has no occurrences of a. Therefore $P(a, L)$ is true if $a \neq \text{head}(L)$. It follows from (4.23) that $P(a, L)$ is true for all elements a and all lists L. QED.

17. Let $\prec$ denote the lexicographic ordering on $\mathbb{N} \times \mathbb{N}$. Then $\mathbb{N} \times \mathbb{N}$ is a well-ordered set, hence well-founded with least element $\langle 0, 0 \rangle$. We'll use (4.23) to prove that $f(x, y)$ is defined (i.e., halts) for all $\langle x, y \rangle \in \mathbb{N} \times \mathbb{N}$. First we have $f(0, 0) = 0 + 1 = 1$. So $f(0, 0)$ is defined. Thus Step 1 of (4.23) is done. Now to Step 2. Assume that $\langle x, y \rangle \in \mathbb{N} \times \mathbb{N}$, and assume that $f(x', y')$ is defined for all $\langle x', y' \rangle$ such that $\langle x', y' \rangle \prec \langle x, y \rangle$. To finish Step 2, we must show that $f(x, y)$ is defined. The definition of $f(x, y)$ gives us three possibilities:

1. If $x = 0$, then $f(x, y) = y + 1$. Thus $f(x, y)$ is defined.

2. If $x \neq 0$ and $y = 0$, then $f(x, y) = f(x - 1, 1)$. Since $\langle x - 1, 1 \rangle \prec \langle x, y \rangle$, our assumption says that $f(x - 1, 1)$ is defined. Therefore $f(x, y)$ is defined.

3. If $x \neq 0$ and $y \neq 0$, then $f(x, y) = f(x - 1, f(x, y - 1))$. First notice that we have $\langle x, y - 1 \rangle \prec \langle x, y \rangle$. So our assumption says that $f(x, y - 1)$ is defined. Thus the pair $\langle x - 1, f(x, y - 1) \rangle$ is a valid element of $\mathbb{N} \times \mathbb{N}$. Now, since we have $\langle x - 1, f(x, y - 1) \rangle \prec \langle x, y \rangle$ our assumption again applies to say that $f(x - 1, f(x, y - 1))$ is defined. Therefore $f(x, y)$ is defined.

So Steps 1 and 2 of (4.23) have been accomplished for the statement "$f(x, y)$ is defined." Therefore $f(x, y)$ is defined for all natural numbers x and y. QED.

18. a. If $L = \langle \rangle$, then isMember(a, L) = false, which is correct. Now assume that L has length n and that isMember(a, L) is correct for all lists of length less than n. If $a = \text{head}(L)$, then isMember(a, L) = true, which is correct. So assume that $a \neq \text{head}(L)$. It follows that $a \in L$ iff $a \in \text{tail}(L)$. Since $a \neq \text{head}(L)$, it follows that isMember(a, L) = isMember($a, \text{tail}(L)$). Since tail(L) has fewer than n elements, the induction assumption says that isMember($a, \text{tail}(L)$) is correct. Therefore isMember(a, L) is correct for any list L.

19. If $x = \langle\ \rangle$, then the definition of cat implies that $\text{cat}(\langle\ \rangle, \text{cat}(y, z)) = \text{cat}(y, z)$ $= \text{cat}(\text{cat}(\langle\ \rangle, y), z)$. Now assume that the statement is true for x, and prove the statement for $a :: x$:

$$
\begin{aligned}
\text{cat}(a :: x, \text{cat}(y, z)) &= a :: \text{cat}(x, \text{cat}(y, z)) &\text{(definition)} \\
&= a :: \text{cat}(\text{cat}(x, y), z) &\text{(induction)} \\
&= \text{cat}(a :: \text{cat}(x, y), z) &\text{(definition)} \\
&= \text{cat}(\text{cat}(a :: x, y), z) &\text{(definition)}.
\end{aligned}
$$

Since the statement is true for $a :: x$ under the assumption that it is true for x, it follows by structural induction that the statement is true for all x, y, and z.

21. Let W be a well-founded set, and let S be a nonempty subset of W. We'll assume condition 2 of (4.22): Whenever an element x in W has the property that all its predecessors are elements in S, then x also is an element in S. We want to prove condition 1 of (4.22): S contains all the minimal elements of W. Suppose, by way of contradiction, that there is some minimal element $x \in W$ such that $x \notin S$. Then all predecessors of x are in S because there aren't any predecessors of x. Condition 2 of (4.22) now forces us to conclude that $x \in S$, a contradiction. Therefore condition 1 of (4.22) follows from condition 2 of (4.22).

23. a. If we can show that $f(n, 0, 1) = f(k, F_{n-k}, F_{n-k+1})$ for all $0 \le k \le n$, then for $k = 0$ we have $f(n, 0, 1) = f(0, F_n, F_{n+1}) = F_n$, by the definition of f. To prove that $f(n, 0, 1) = f(k, F_{n-k}, F_{n-k+1})$ for all $0 \le k \le n$, we'll fix n and induct on the variable k as it ranges from n down to 0. So the basis case is $k = n$. In this case we have

$$
f(n, 0, 1) = f(n, F_0, F_1) = f(k, F_{n-k}, F_{n-k+1}).
$$

For the induction case, assume that $f(n, 0, 1) = f(k, F_{n-k}, F_{n-k+1})$ for some k such that $0 < k \le n$, and prove that

$$
f(n, 0, 1) = f(k - 1, F_{n-k+1}, F_{n-k+2}).
$$

We have the following equations:

$$
\begin{aligned}
f(n, 0, 1) &= f(k, F_{n-k}, F_{n-k+1}) &\text{(induction assumption)} \\
&= f(k - 1, F_{n-k+1}, F_{n-k}, + F_{n-k+1}) &\text{(definition of } f\text{)} \\
&= f(k - 1, F_{n-k+1}, F_{n-k+2}). &\text{(definition of } F_{n-k+2}\text{)}.
\end{aligned}
$$

Therefore $f(n, 0, 1) = f(k, F_{n-k}, F_{n-k+1})$ for all $0 \le k \le n$.

Chapter 5

Section 5.1
1. a. $n^2 + 3n$.

2.

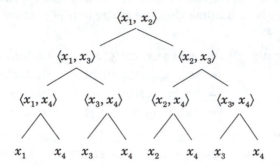

3. a. 7. **c.** 4.

5. There are 25 possibilities. Therefore a ternary pan balance algorithm must make at least three comparisons to solve the problem. One example is

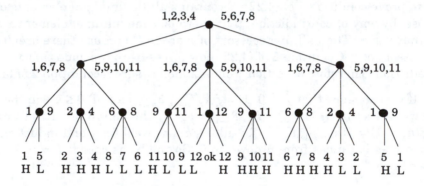

Section 5.2

1. a. 720. **c.** 30. **e.** 10.

2. a. $abc, acb, bac, bca, cab, cba$. **c.** $\{a, b, c\}$.
e. $[a, a], [a, b], [a, c], [b, b], [b, c], [c, c]$.

3. a. $P(3, 3)$. **c.** $C(3, 3)$. **e.** $C(3 + 2 - 1, 2)$.

4. The number of bag permutations of B is $\dfrac{4!}{2!2!} = 6$. They can be listed as follows: $aabb, abab, abba, bbaa, baba, baab$.

5. a. $8! = 40{,}320$. **c.** $\dfrac{6!}{2!2!1!1!} = 180$. **e.** $\dfrac{9!}{4!2!2!1!} = 3780$.

6. a. There are none. **c.** BCA, CAB.

7. $n = 7, k = 3$, and $m = 4$.

9. Either floor$(n/2) + 1$ or ceiling$(n/2) + 1$.

10. The first processor takes 0.01 second to do its job, and the second processor takes 0.005 second to do its job. Therefore 2000 operations are processed in 0.015 second, which equals 133,333 operations per second.

12. There are eight possible outcomes when three coins are tossed. **a.** 0.375. **c.** 0.875.

13. There are 36 possible outcomes. **a.** 0.1666.... **c.** 0.222....

14. a. 3.5.

16. There are $(n)(k)$ actions to schedule. Since the k actions of each process must be done in order, we can represent each process as a bag consisting of k identical elements. Assume that the bags are disjoint from each other. Then the union B of the n bags contains $(n)(k)$ elements, and each bag permuation of B is one schedule. Therefore there are as many schedules as there are bag permutations of B. That number is $(nk)!/(k!)^n$.

17. There are $\dfrac{(2n)!}{n!n!}$ total strings of length $2n$ with n zeros and n ones. Any string that is not OK must contain a zero with an equal number of zeros and ones to its left. This collection of strings has the same cardinality as the set of all strings of length $2n$ that contain $n+1$ zeros and $n-1$ ones. This latter set has cardinality $\dfrac{(2n)!}{(n+1)!(n-1)!}$. So the desired number of strings is $\dfrac{(2n)!}{n!n!} - \dfrac{(2n)!}{(n+1)!(n-1)!}$, which simplifies to $\dfrac{(2n)!}{n!(n+1)!}$.

Section 5.3

1. a. $4(n-1)$. **c.** $2^{n+2}-3$.

2. a. We can write the sum as $\sum_{i=1}^{n} i3^i$ which is an instance of formula (5.12d) with $a=3$. So the closed form is $\dfrac{3-(n+1)\,3^{n+1}+n\,3^{n+2}}{2^2}$.

3. $(1/4)n^2(n+1)^2$.

5. If D_n denotes the number of diagonals in an n-sided polygon, then we have $D_3 = 0$ and $D_n = D_{n-1} + (n-2)$. Solving the recurrence gives $D_n = \dfrac{n(n-3)}{2}$.

6. a. $a_n = \dfrac{-1}{2^{n+1}} - 2(-3)^n$. **c.** $a_n = (-1/2)(3/2)^n - n - 1$.

7. a. $a_n = 3^n + (-1)^{n+1}$. **c.** $a_n = (1/3)(2^n + (-1)^{n+1})$.

8. a. $A(x) = 4(\dfrac{1}{(1-x)^2}) - 8(\dfrac{1}{1-x}) + 4$, which yields $a_n = 4(n-1)$.

c. $A(x) = -3(\dfrac{1}{1-x}) + 4(\dfrac{1}{1-2x})$, which yields $a_n = 2^{n+2} - 3$.

9. a. Let a_n be the number of cons operations when L has length n. Then $a_0 = 0$ and $a_n = 1 + a_{n-1}$, which has solution $a_n = n$. **c.** Let a_n be the number of cons operations when L has length n. Then $a_0 = 1$, and $a_n = a_{n-1} + 5 \cdot 2^{n-1}$, which has solution $a_n = 5 \cdot 2^n - 4$.

10. a. If $n = 1$, then the equation evaluates to $2 = 2$. Assume that the equation holds for n, and show that it holds for $n + 1$. Starting with the left side for the $n + 1$ case we have

$$\frac{(1)(1)(3)(5)...(2n-3)(2n-1)}{(n+1)!} 2^{n+1}$$

$$= \frac{(1)(1)(3)(5)...(2n-3)}{n!} 2^n \frac{2n-1}{n+1} 2$$

$$= \frac{2}{n} \binom{2n-2}{n-1} \frac{2n-1}{n+1} 2 = \frac{2}{n+1} \binom{2(n+1)-2}{n+1-1}$$

which is the right side for the $n + 1$ case.

11. Let a_n denote the number of binary trees with n nodes. Then $a_0 = 1$, and $a_1 = 1$. So far, so good. Now what? Notice that a binary tree with n nodes has a root, a left subtree with i nodes, and a right subtree with $n - i - 1$ nodes, where $0 \le i \le n - 1$. Now follow the technique used in Example 8 to find the number of parenthesized expressions. For example, the number of distinct binary trees with four nodes is given by $a_4 = a_0 a_3 + a_1 a_2 + a_2 a_1 + a_3 a_0$. The general closed form is $a_n = \frac{1}{n+1} \binom{2n}{n}$.

13. Letting $F(x)$ be the generating function for F_n, we get $F(x) = \dfrac{x}{1 - x - x^2}$. The denominator factors into $1 - x - x^2 = (1 - \alpha x)(1 - \beta x)$, where

$$\alpha = \tfrac{1}{2}(1 + \sqrt{5}) \quad \text{and} \quad \beta = \tfrac{1}{2}(1 - \sqrt{5}).$$

Now use partial fractions to obtain $F(x) = \dfrac{1}{\sqrt{5}} \left(\dfrac{1}{1 - \alpha x} - \dfrac{1}{1 - \beta x} \right)$ This yields the closed formula $F_n = \dfrac{1}{\sqrt{5}} (\alpha^n - \beta^n)$.

Section 5.4

1. (Reflexive) Since $1f(n) \le f(n) \le 1f(n)$, it follows that $f(n) = \Theta(f(n))$ for every function f. (Symmetric) If $f(n) = \Theta(g(n))$, then there are nonzero constants c, d, and m such that $cg(n) \le f(n) \le dg(n)$ for all $n \ge m$. Now take the different cases for c and d. For example, we'll do the case for $c > 0$ and $d < 0$. In this case we get $-(1/d)f(n) \leq g(n) \leq (1/c)f(n)$ for all $n \geq m$. Thus $g(n) = \Theta(f(n))$. The other cases for c and d are similar. (Transitive) Assume that $f(n) = \Theta(g(n))$ and $g(n) = \Theta(h(n))$. Then we can write $cg(n) \leq f(n) \leq dg(n)$ for $n \geq m$ and $ah(n) \leq g(n) \leq bh(n)$ for $n \geq k$. Take different cases for a, b, c, and d. For example, if $c > 0$ and $d > 0$, then we have $(ca)h(n) \leq f(n) \leq (bd)h(n)$ for $n \geq \max\{m, k\}$. This says that $f(n) = \Theta(h(n))$.

2. a. The quotient $\log(kn)/\log n$ approaches 1 as n approaches infinity.

3. Let $f(n) = (n - 1)/n$. Then f is increasing, and for all $n \geq 2$ we have the inequality $(1/2)\cdot 1 \leq f(n) \leq 1\cdot 1$. Therefore $f(n) = \Theta(1)$.

5. $1 \prec \log \log n \prec \log n$.

6. a. $f(n) = \Theta(n \log n)$. Notice that $f(n) = \log(1\cdot 2\cdots n) = \log(n!)$. Now use (5.37) to approximate $n!$ and take the log of Stirlings's formula to obtain $\Theta(n \log n)$.

7. Take limits.

8. a. $5! = 120$; Stirling ≈ 118.02; diff $= 1.98$.

9. In each case, replace $n!$ by its Stirling's approximation (5.37). Then take limits.

Chapter 6

Section 6.2

1. a. $((\neg P) \wedge Q) \rightarrow (P \vee R)$. **c.** $(A \rightarrow (B \vee (((\neg C) \wedge D) \wedge E))) \rightarrow F$.

2. $(A \rightarrow B) \wedge (\neg A \rightarrow C)$.

3. a. $(P \vee Q \rightarrow \neg R) \vee \neg Q \wedge R \wedge P$.

5. $A \wedge \neg B \rightarrow \text{false} \equiv \neg(A \wedge \neg B) \vee \text{false} \equiv \neg(A \wedge \neg B) \equiv \neg A \vee \neg \neg B \equiv \neg A \vee B$ $\equiv A \rightarrow B$.

7. a. $(A \rightarrow B) \wedge (A \vee B) \equiv (\neg A \vee B) \wedge (A \vee B) \equiv (\neg A \wedge A) \vee B \equiv \text{false} \vee B \equiv B$. **c.** $A \wedge B \rightarrow C \equiv \neg(A \wedge B) \vee C \equiv (\neg A \vee \neg B) \vee C \equiv \neg A \vee (\neg B \vee C)$ $\equiv \neg A \vee (B \rightarrow C) \equiv A \rightarrow (B \rightarrow C)$. **e.** $A \rightarrow B \wedge C \equiv \neg A \vee (B \wedge C) \equiv (\neg A \vee B) \wedge (\neg A \vee C) \equiv (A \rightarrow B) \wedge (A \rightarrow C)$.

8. a. Tautology. **c.** Contingency.

9. a. One of several answers is $A = \text{false}$, $B = \text{true}$, and $C = \text{false}$. **c.** $A = \text{false}$, $B = \text{true or false}$, and $C = \text{true}$.

10. a. $(P \wedge \neg Q) \vee P$ or P. **c.** $\neg Q \vee P$. **e.** $\neg P \vee (Q \wedge R)$.

11. a. $P \wedge (\neg Q \vee P)$ or P. **c.** $\neg Q \vee P$. **e.** $(\neg P \vee Q) \wedge (\neg P \vee R)$. **g.** $(A \vee C \vee \neg E \vee F) \wedge (B \vee C \vee \neg E \vee F) \wedge (A \vee D \vee \neg E \vee F) \wedge (B \vee D \vee \neg E \vee F)$.

12. a. Full DNF: $(P \wedge Q) \vee (P \wedge \neg Q)$. Full CNF: $(P \vee \neg Q) \wedge (P \vee Q)$.

13. a. $(P \wedge Q) \vee (P \wedge \neg Q)$.
c. $(P \wedge Q) \vee (P \wedge \neg Q) \vee (\neg P \wedge Q) \vee (\neg P \wedge \neg Q)$.
e. $(P \wedge Q \wedge R) \vee (\neg P \wedge Q \wedge R) \vee (\neg P \wedge \neg Q \wedge R) \vee (\neg P \wedge Q \wedge \neg R) \vee (\neg P \wedge \neg Q \wedge \neg R)$.

14. a. $(P \vee Q) \wedge (P \vee \neg Q)$. **c.** $\neg Q \vee P$. **e.** $(P \vee Q \vee R) \wedge (P \vee Q \vee \neg R) \wedge (P \vee \neg Q \vee R) \wedge (\neg P \vee Q \vee R) \wedge (\neg P \vee \neg Q \vee R)$.

15. a. $A \vee B \equiv \neg(\neg A \wedge \neg B)$. Therefore $\{\neg, \wedge\}$ is complete because $\{\neg, \vee\}$ is complete. **c.** $\neg A \equiv A \rightarrow$ false. Therefore {false, $\rightarrow$} is a complete set because $\{\neg, \rightarrow\}$ is complete. **e.** $\neg A \equiv \text{NOR}(A, A)$, and $A \vee B \equiv \neg \text{NOR}(A, B) = \text{NOR}(\text{NOR}(A, B), \text{NOR}(A, B))$. Therefore NOR is complete because $\{\neg, \vee\}$ is a complete set of connectives.

Section 6.3

1. a. Line 2 is incorrect, since no part of W requires us to prove something of the form $A \rightarrow X$.

2. Line 6 is not correct because it uses line 3, which is in a previous subproof. Only lines 1 and 5 can be used to infer something on line 6.

3. a. Three premises: A is the premise for the proof of the conditional, whose conclusion is $B \rightarrow (C \rightarrow D)$. B is the premise for the conditional proof whose conclusion is $C \rightarrow D$. Finally, C is the premise for the proof of $C \rightarrow D$.

4. Let D mean "I am dancing," H mean "I am happy," and M mean "there is a mouse in the house." Then a proof can be written as follows:

1.	$D \rightarrow H$	P
2.	$M \vee H$	P
3.	$\neg H$	P
4.	M	2, 3, DS
5.	$\neg D$	1, 3, MT
6.	$M \wedge \neg D$	4, 5, Conj
	QED	1, 2, 3, 6, CP.

5. a.

1.	A		P
2.		B	P
3.		$A \wedge B$	1, 2, Conj
4.	$B \rightarrow A \wedge B$		2, 3, CP
	QED		1, 4, CP.

c.
1. $A \vee B \to C$ P
2. A P
3. $A \vee B$ 2, Add
4. C 1, 3, MP
    QED   1, 2, 4, CP.

e.
1. $A \vee B \to C \wedge D$ P
2. B P
3. $A \vee B$ 2, Add
4. $C \wedge D$ 1, 3, MP
5. D 4, Simp
6. $B \to D$ 2, 5, CP
    QED   1, 6, CP.

g.
1. $\neg (A \wedge B)$ P
2. $B \vee C$ P
3. $C \to D$ P
4. A P
5. $\neg A \vee \neg B$ 1, T
6. $\neg B$ 4, 5, DS
7. C 2, 6, DS
8. D 3, 7, MP
9. $A \to D$ 4, 8, CP
    QED   1, 2, 3, 9, CP.

i.
1. $A \to C$ P
2. $A \wedge B$ P
3. A 2, Simp
4. C 1, 3, MP
5. $A \wedge B \to C$ 2, 4, CP
    QED   1, 5, CP.

6.   **a.**
1. A P
2. $\neg (B \to A)$ P for IP
3. $\neg (\neg B \vee A)$ 2, T
4. $B \wedge \neg A$ 3, T
5. $\neg A$ 4, Simp
6. $A \wedge \neg A$ 1, 5, Conj
    QED   1, 2, 6, IP.

c.
1. $\neg B$ P
2. $\neg (B \rightarrow C)$ P for IP
3. $\neg (\neg B \vee C)$ 2, T
4. $B \wedge \neg C$ 3, T
5. B 4, Simp
6. $\neg B \wedge B$ 1, 5, Conj
   QED   1, 2, 6, IP.

e.
1. $A \rightarrow B$ P
2. $\neg ((A \rightarrow \neg B) \rightarrow \neg A)$ P for IP
3. $(\neg A \vee \neg B) \wedge A$ 2, T
4. A 3, Simp
5. $\neg A \vee \neg B$ 3, Simp
6. $\neg B$ 4, 5, DS
7. $\neg A$ 1, 6, MT
8. $A \wedge \neg A$ 4, 7, Conj
   QED   1, 2, 8, IP.

g.
1. $A \rightarrow B$ P
2. $B \rightarrow C$ P
3. $\neg (A \rightarrow C)$ P for IP
4. $A \wedge \neg C$ 3, T
5. A 4, Simp
6. $\neg C$ 4, Simp
7. B 1, 5, MP
8. C 2, 7, MP
9. $C \wedge \neg C$ 6, 8, Conj
   QED   1, 2, 3, 9, IP.

7. For some proofs we'll use IP in a subproof.

a.
1. A P
2. $\neg (B \rightarrow (A \wedge B))$ P for IP
3. $B \wedge \neg (A \wedge B)$ 2, T
4. B 3, Simp
5. $\neg (A \wedge B)$ 3, Simp
6. $\neg A \vee \neg B$ 5, T
7. $\neg B$ 1, 6, DS
8. $B \wedge \neg B$ 4, 7, Conj
9. false   8, T
   QED   1, 2, 9, IP.

c. 1. $A \lor B \to C$ P
2. A P
3. $\neg C$ P for IP
4. $\neg (A \lor B)$ 1, 3, MT
5. $\neg A \land \neg B$ 4, T
6. $\neg A$ 5, Simp
7. $A \land \neg A$ 2, 6, Conj
QED   1, 2, 3, 7, IP.

e. 1. $A \lor B \to C \land D$ P
2. $\neg (B \to D)$ P for IP
3. $B \land \neg D$ 2, T
4. B 3, Simp
5. $A \lor B$ 4, Add
6. $C \land D$ 1, 5, MP
7. $\neg D$ 3, Simp
8. D 6, Simp
9. $D \land \neg D$ 7, 8, Conj
QED   1, 2, 9, IP.

g. 1. $\neg (A \land B)$ P
2. $B \lor C$ P
3. $C \to D$ P
4. $\neg (A \to D)$ P for IP
5. $A \land \neg D$ 4, T
6. $\neg D$ 5, Simp
7. $\neg C$ 3, 6, MT
8. B 2, 7, DS
9. $B \to \neg A$ 1, T
10. $\neg A$ 8, 9, MP
11. A 5, Simp
12. $A \land \neg A$ 10, 11, Conj
QED   1, 2, 3, 4, 12, IP.

i. 1. $A \to (B \to C)$ P
2. B P
3. A P
4. $\neg C$ P for IP
5. $B \to C$ 1, 3, MP
6. C 2, 5, MP
7. $C \land \neg C$ 4, 6, Conj
8. $A \to C$ 3, 4, 7, IP
9. $B \to (A \to C)$ 2, 8,   CP
QED   1, 9, CP.

k. 1. $A \to C$ P
       2. A P
       3. $\neg(B \vee C)$ P for IP
       4. $\neg B \wedge \neg C$ 3, T
       5. C 1, 2, MP
       6. $\neg C$ 4, Simp
       7. $C \wedge \neg C$ 5, 6, Conj
       8. $A \wedge B \to C$ 2, 3, 7, IP
          QED   1, 8, CP.

8. **a.** 1. $(A \wedge B) \to C$ P
              2. A P
              3. B P
              4. $A \wedge B$ 2, 3, Conj
              5. C 1, 4, MP
              6. $B \to C$ 3, 5, CP
              7. $A \to (B \to C)$ 2, 6, CP
                 QED   1, 7, CP.

9. **a.** 1. $A \vee B$ P
              2. $A \to C$ P
              3. $B \to D$ P
              4. $\neg(C \vee D)$ P for IP
              5. $\neg C \wedge \neg D$ 4, T
              6. $\neg C$ 5, Simp
              7. $\neg A$ 2, 6, MT
              8. B 1, 7, DS
              9. D 3, 8, MP
             10. $\neg D$ 5, Simp
             11. $D \wedge \neg D$ 9, 10, Conj
                 QED   1, 2, 3, 4, 11, IP.

10. **a.** 1. $A \to A$ Theorem 2
              2. $(A \to A) \to (\neg A \vee A)$ Part of Axiom 5
              3. $\neg A \vee A$ 1, 2, MP
                 QED.

       c. 1. $\neg A \vee \neg \neg A$ Part (b)
                 QED.

11. a.
1.	$A \rightarrow B$	P
2.	$\neg A \vee B$	1, Axiom 5
3.	$B \rightarrow \neg \neg B$	Theorem 6
4.	$\neg A \vee B \rightarrow \neg A \vee \neg \neg B$	3, Axiom 4, MP
5.	$\neg A \vee \neg \neg B$	2, 4, MP
6.	$\neg \neg B \vee \neg A$	5, Axiom 3
7.	$\neg B \rightarrow \neg A$	6, Axiom 5
	QED	1, 7, CP.

Chapter 7

Section 7.1

1. a. $[p(0, 0) \wedge p(0, 1)] \vee [p(1, 0) \wedge p(1, 1)]$.

2. a. $\forall x \in \{0, 1\}: q(x)$. **c.** $\forall y \in \{0, 1\}: p(x, y)$.
e. Let O stand for the odd natural numbers. Then $\exists x \in O : p(x)$.

3. a. Over the natural numbers, let $e(x, y)$ mean "$x = y$," let $p(x, y)$ mean "x is a predecessor of y," and let $a = 0$. The formal wff is $\forall x (\neg e(x, a) \rightarrow \exists y\, p(y, x))$.

4. a. x is a term. Therefore $p(x)$ is a wff, and it follows that $\exists x\, p(x)$ and $\forall x\, p(x)$ are wffs. Thus $\exists x\, p(x) \rightarrow \forall x\, p(x)$ is a wff.

5. It is illegal to have an atom, $p(x)$ in this case, as an argument to a predicate.

6. a. The three occurrences of x, left to right, are free, bound, and bound. The four occurrences of y, left to right, are free, bound, bound, and free.
c. The three occurrences of x, left to right, are free, bound, and bound. Both occurrences of y are free.

7. $\forall x\, p(x, y, z) \rightarrow \exists z\, q(z)$.

8. a. One interpretation has $p(a) = $ true, in which case both $\forall x\, p(x)$ and $\exists x\, p(x)$ are true. Therefore W is true. The other interpretation has $p(a) = $ false, in which case both $\forall x\, p(x)$ and $\exists x\, p(x)$ are false. Therefore W is true.

9. a. Let the domain be the set $\{a, b\}$, and assign $p(a) = $ true and $p(b) = $ false. Finally, assign the constant $c = a$. **c** and **d.** Let $p(x, y) = $ false for all elements x and y in any domain. Then the antecedent is false for both parts (c) and (d). Therefore the both wffs are true for this interpretation. **e.** Let $D = \{a\}$, $f(a) = a$, $y = a$, and let p denote equality.

10. a. Let the domain be $\{a\}$, and let $p(a) = $ true and $c = a$. **c.** Let $D = \mathbb{N}$, let $p(x)$ mean "x is odd," and let $q(x)$ mean "x is even." Then the antecedent is true, but the consequent is false. **e.** Let $D = \mathbb{N}$, and let $p(x, y)$ mean "$y = x + 1$." Then the antecedent $\forall x\, \exists y\, p(x, y)$ is true and the consequent $\exists y\, \forall x\, p(x, y)$ is false for this interpretation.

11.a. If the domain is {a}, then either $p(a)$ = true or $p(a)$ = false. In either case, W is true.

12. a. Let {a} be the domain of the interpretation. If $p(a, a)$ = false, then W is true, since the antecedent is false. If $p(a, a)$ = true, the W is true, since the consequent is true. **c.** Let {a, b, c} be the domain. Let $p(a, a) = p(b, b) = p(c, c)$ = true and $p(a, b) = p(a, c) = p(b, c)$ = false. This assignment makes W false. Therefore W is invalid.

13. $\forall x\, p(x, x) \rightarrow$
$\forall x\, \forall y\, \forall z\, \forall w\, (p(x, y) \vee p(x, z) \vee p(x, w) \vee p(y, z) \vee p(y, w) \vee p(z, w))$.

14. a. For any domain D and any element $d \in D, p(d) \rightarrow p(d)$ is true. Therefore any interpretation is a model. **c.** If the wff is invalid, then there is some interpretation making the wff false. This says that $\forall x\, p(x)$ is true and $\exists x\, p(x)$ is false. This is a contradiction because we can't have $p(x)$ true for all x in a domain while at the same time having $p(x)$ false for some x in the domain. **e.** If the wff is not valid, then there is an interpretation with domain D for which the antecedent is true and the consequent is false. So $A(d)$ and $B(e)$ are false for some elements $d, e \in D$. Therefore $\forall x\, A(x)$ and $\forall x\, B(x)$ are false, contrary to assumption. **g** and **h.** If the antecedent is true for a domain D, then $A(d) \rightarrow B(d)$ is true for all $d \in D$. If $A(d)$ is true for all $d \in D$, then $B(d)$ is also true for all $d \in D$ by MP. Thus the consequent is true for D.

15. a. Suppose the wff is satisfiable. Then there is an interpretation that assigns c a value in its domain such that $p(c) \wedge \neg p(c)$ = true. Of course, this is impossible. Therefore the wff is unsatisfiable. **c.** Suppose the wff is satisfiable. Then there is an interpretation making $\exists x\, \forall y\, (p(x, y) \wedge \neg p(x, y))$ true. This says that there is an element d in the domain such that $\forall y\, (p(d, y) \wedge \neg p(d, y))$ is true. This says that $p(d, y) \wedge \neg p(d, y)$ is true for all y in the domain, which is impossible.

17. Assume that $A \rightarrow B$ is valid and A is also valid. Let I be an interpretation for B with domain D. Extend I to an interpretation J for A by using D to interpret all predicates, functions, free variables and constants that occur in A but not in B. So J is an interpretation for $A \rightarrow B, A$, and B. Since we are assuming that $A \rightarrow B$ and A are valid, it follows that $A \rightarrow B$ and A are true with respect to J. Therefore B is true with respect to J. But J and I are the same interpretation on B. So B is true with respect to I. Therefore I is a model for B. Since I was arbitrary, it follows that B is valid. Now we go the other direction. Assume that if A is valid, then B is valid. Let I be an interpretation for $A \rightarrow B$. Then I is also an interpretation for A and for B. Since A and B are valid, it follows that A and B are true with respect to I. Therefore $A \rightarrow B$ is true with respect to I. Therefore I is a model for $A \rightarrow B$. Since I was arbitrary, it follows that $A \rightarrow B$ is valid. QED.

Section 7.2

1. a. The left side is true for domain D iff $A(d) \wedge B(d)$ is true for all $d \in D$ iff $A(d)$ and $B(d)$ are both true for all $d \in D$ iff the right side is true for domain D. **c.** Assume that the left side is true for domain D. Then $A(d) \to B(d)$ is true for some $d \in D$. If $A(d)$ is true, then $B(d)$ is true by MP. So $\exists x\, B(x)$ is true for D. If $A(d)$ is false, then $\forall x\, A(x)$ is false. So in either case the right side is true for D. Now assume the right side is true for D. If $\forall x\, A(x)$ is true, then $\exists x\, B(x)$ is also true. This means that $A(d)$ is true for all $d \in D$ and $B(d)$ is true for some $d \in D$. Thus $A(d) \to B(d)$ is true for some $d \in D$, which says that the left side is true for D. **e.** $\exists x\, \exists y\, W(x, y)$ is true for D iff $W(d, e)$ is true for some elements $d, e \in D$ iff $\exists y\, \exists x\, W(x, y)$ is true for D.

2. All proofs use an interpretation with domain D.
a. $\forall x\, (C \wedge A(x))$ is true for D iff $C \wedge A(d)$ is true for all $d \in D$ iff C is true in D and $A(d)$ is true for all $d \in D$ iff $C \wedge \forall x\, A(x)$ is true for D.
c. $\forall x\, (C \vee A(x))$ is true for D iff $C \vee A(d)$ is true for all $d \in D$ iff C is true in D or $A(d)$ is true for all $d \in D$ iff $C \vee \forall x\, A(x)$ is true for D.
e. $\forall x\, (C \to A(x))$ is true for D iff $C \to A(d)$ is true for all $d \in D$ iff either C is false in D or $A(d)$ is true for all $d \in D$ iff $C \to \forall x\, A(x)$ is true for D.
g. $\forall x\, (A(x) \to C)$ is true for D iff $A(d) \to C$ is true for all $d \in D$ iff either $A(d)$ is false for some $d \in D$ or C is true in D iff $\exists x\, A(x) \to C$ is true for D.

3. a. $\exists x\, \forall y\, \forall z\, ((\neg p(x) \vee p(y) \vee q(z)) \wedge (\neg q(x) \vee p(y) \vee q(z)))$.
c. $\exists x\, \forall y\, \exists z\, \forall w\, (\neg p(x, y) \vee p(w, z))$.

4. a. $\exists x\, \forall y\, \forall z\, ((\neg p(x) \wedge \neg q(x)) \vee p(y) \vee q(z))$.
c. $\exists x\, \forall y\, \exists z\, \forall w\, (\neg p(x, y) \vee p(w, z))$.

5. a. Let D be the domain $\{a, b\}$. Assume that C is false, $W(a)$ is true, and $W(b)$ is false. Then $(\forall x\, W(x) \equiv C)$ is true, but $\forall x\, (W(x) \equiv C)$ is false. Therefore the statement is false.

6. a. $\forall x\, (C(x) \to R(x) \wedge F(x))$. **c.** $\forall x\, (G(x) \to S(x))$. **e.** $\exists x\, (G(x) \wedge \neg S(x))$.

Section 7.3

1. Lines 3 and 4 are errors; they apply UG to a subexpression of a larger wff.

2. a. Let $D = \mathbb{N}$, let $P(x) = $ "$x + x = x$," and $Q(x) = $ "$x + 1 = x$." Then $P(0)$ is true, and $P(1) \to Q(1)$ is also true. Therefore the antecedent of the wff is true. But $Q(x)$ is false for all x in $\mathbb{N}$. Therefore the consequent is false. Therefore the interpretation is a countermodel, and the wff is invalid.

3. a.

1.	$\forall x\, p(x)$	P
2.	$p(x)$	1, UI
3.	$\exists x\, p(x)$	2, EG
	QED	1, 3, CP.

c.
1.	$\exists x\,(p(x) \wedge q(x))$	P
2.	$p(c) \wedge q(c)$	1, EI
3.	$p(c)$	2, Simp
4.	$\exists x\,p(x)$	3, EG
5.	$q(c)$	2, Simp
6.	$\exists x\,q(x)$	5, EG
7.	$\exists x\,p(x) \wedge \exists x\,q(x)$	4, 6, Conj
	QED	1, 7, CP.

e.
1.	$\forall x\,(p(x) \rightarrow q(x))$	P
2.	$\forall x\,p(x)$	P
3.	$p(x)$	2, UI
4.	$p(x) \rightarrow q(x)$	1, UI
5.	$q(x)$	3, 4, MP
6.	$\exists x\,q(x)$	5, EG
7.	$\forall x\,p(x) \rightarrow \exists x\,q(x)$	2, 6, CP
	QED	1, 7, CP.

g.
1.	$\exists y\,\forall x\,p(x, y)$	P
2.	$\forall x\,p(x, c)$	1, EI
3.	$p(x, c)$	2, UI
4.	$\exists y\,p(x, y)$	3, EG
5.	$\forall x\,\exists y\,p(x, y)$	4, UG
	QED	1, 5, CP.

4. a.
1.	$\forall x\,p(x)$	P
2.	$\neg\,\exists x\,p(x)$	P for IP
3.	$\forall x\,\neg p(x)$	2, T
4.	$p(x)$	1, UI
5.	$\neg\,p(x)$	3, UI
6.	$p(x) \wedge \neg\,p(x)$	4, 5, Conj
7.	false	6, T
	QED	1, 2, 7, IP.

c.
1.	$\exists y\,\forall x\,p(x, y)$	P
2.	$\neg\,\forall x\,\exists y\,p(x, y)$	P for IP
3.	$\exists x\,\forall y\,\neg p(x, y)$	2, T
4.	$\forall x\,p(x, c)$	1, EI
5.	$\forall y\,\neg p(d, y)$	3, EI
6.	$p(d, c)$	4, UI
7.	$\neg\,p(d, c)$	5, UI
8.	$p(d, c) \wedge \neg\,p(d, c)$	6, 7, Conj
	QED	1, 2, 8, IP.

e.

1.	$\forall x\, p(x) \vee \forall x\, q(x)$	P
2.	$\neg\, \forall x\, (p(x) \vee q(x))$	P for IP
3.	$\exists x\, (\neg\, p(x) \wedge \neg\, q(x))$	2, T
4.	$\neg\, p(c) \wedge \neg\, q(c)$	3, EI
5.	$\neg\, p(c)$	4, Simp
6.	$\neg\, \forall x\, p(x)$	5, UI (contrapositive)
7.	$\neg\, q(c)$	4, Simp
8.	$\neg\, \forall x\, q(x)$	7, UI (contrapositive)
9.	$\neg\, \forall x\, p(x) \wedge \neg\, \forall x\, q(x)$	6, 7, Conj
10.	$\neg\, (\forall x\, p(x) \vee \forall x\, q(x))$	9, T
11.	false	1, 10, Conj, T
	QED	1, 2, 11, IP.

5. a. Let $D(x)$ mean that x is a dog, $L(x)$ mean that x likes people, $H(x)$ mean that x hates cats, and a = Rover. Then the argument can be formalized as follows:

$$\forall x\, (D(x) \to L(x) \vee H(x)) \wedge D(a) \wedge \neg\, H(a) \to \exists x\, (D(x) \wedge L(x)).$$

Proof:

1.	$\forall x\, (D(x) \to L(x) \vee H(x))$	P
2.	$D(a)$	P
3.	$\neg\, H(a)$	P
4.	$D(a) \to L(a) \vee H(a)$	1, UI
5.	$L(a) \vee H(a)$	2, 4, MP
6.	$L(a)$	3, 5, DS
7.	$D(a) \wedge L(a)$	2, 6, Conj
8.	$\exists x\, (D(x) \wedge L(x))$	7, EG
	QED	1, 2, 3, 8, CP.

c. Let $H(x)$ mean that x is a human being, $Q(x)$ mean that x is a quadruped, and $M(x)$ mean that x is a man. Then the argument is can be formalized as

$$\forall x\, (H(x) \to \neg\, Q(x)) \wedge \forall x\, (M(x) \to H(x)) \to \forall x\, (M(x) \to \neg\, Q(x)).$$

Proof:

1.	$\forall x\, (H(x) \to \neg\, Q(x))$	P
2.	$\forall x\, (M(x) \to H(x))$	P
3.	$H(x) \to \neg\, Q(x)$	1, UI
4.	$M(x) \to H(x)$	2, UI
5.	$M(x) \to \neg\, Q(x)$	3, 4, HS
6.	$\forall x\, (M(x) \to \neg\, Q(x))$	5, UG
	QED	1, 2, 6, CP.

e. Let $F(x)$ mean that x is a freshman, $S(x)$ mean that x is a sophomore, $J(x)$ mean that x is a junior, and $L(x, y)$ mean that x likes y. Then the argument can be formalized as $A \to B$, where

$A = \exists x \, (F(x) \wedge \forall y \, (S(y) \rightarrow L(x, y))) \wedge \forall x \, (F(x) \rightarrow \forall y \, (J(y) \rightarrow \neg L(x, y)))$

$B = \forall x \, (S(x) \rightarrow \neg J(x))$.

Proof:

1.	$\exists x \, (F(x) \wedge \forall y \, (S(y) \rightarrow L(x, y)))$	P
2.	$\forall x \, (F(x) \rightarrow \forall y \, (J(y) \rightarrow \neg L(x, y)))$	P
3.	$F(c) \wedge \forall y \, (S(y) \rightarrow L(c, y))$	1, EI
4.	$\forall y \, (S(y) \rightarrow L(c, y))$	3, Simp
5.	$S(x) \rightarrow L(c, x)$	4, UI
6.	$\quad S(x)$	P
7.	$\quad L(c, x)$	5, 6, MP
8.	$\quad F(c) \rightarrow \forall y \, (J(y) \rightarrow \neg L(c, y))$	2, UI
9.	$\quad F(c)$	3, Simp
10.	$\quad \forall y \, (J(y) \rightarrow \neg L(c, y))$	8, 9, MP
11.	$\quad J(x) \rightarrow \neg L(c, x)$	10, UI
12.	$\quad \neg J(x)$	7, 11, MT
13.	$S(x) \rightarrow \neg J(x)$	6, 12, CP
14.	$\forall x \, (S(x) \rightarrow \neg J(x))$	13, UG
	QED	1, 2, 14, CP.

6. First prove that the left side implies the right side, then the converse.

a.

1.	$\exists x \, \exists y \, W(x, y)$	P
2.	$\exists y \, W(c, y)$	1, EI
3.	$W(c, d)$	2, EI
4.	$\exists x \, W(x, d)$	3, EG
5.	$\exists y \, \exists x \, W(x, y)$	4, EG
	QED	1, 5, CP.

1.	$\exists y \, \exists x \, W(x, y)$	P
2.	$\exists x \, W(x, d)$	1, EI
3.	$W(c, d)$	2, EI
4.	$\exists y \, W(c, y)$	3, EG
5.	$\exists x \, \exists y \, W(x, y)$	4, EG
	QED	1, 5, CP.

c.

1.	$\exists x \, (A(x) \vee B(x))$	P
2.	$\neg \, (\exists x \, A(x) \vee \exists x \, B(x))$	P for IP
3.	$\forall x \, \neg A(x) \wedge \forall x \, \neg B(x)$	2, T
4.	$\forall x \, \neg A(x)$	3, Simp
5.	$A(c) \vee B(c)$	1, EI
6.	$\neg A(c)$	4, UI
7.	$B(c)$	5, 6, DS
8.	$\forall x \, \neg B(x)$	3, Simp
9.	$\neg B(c)$	8, UI
10.	$B(c) \wedge \neg B(c)$	7, 9, Conj
	QED	1, 2, 10, IP.

1. $\exists x\, A(x) \vee \exists x\, B(x)$ P
2. $\neg\, \exists x\, (A(x) \vee B(x))$ P for IP
3. $\forall x\, (\neg A(x) \wedge \neg B(x))$ 2, T
4. $\forall x\, \neg A(x) \wedge \forall x\, \neg B(x)$ 3, T Part b)
5. $\forall x\, \neg A(x)$ 4, Simp
6. $\neg\, \exists x\, A(x)$ 5, T
7. $\exists x\, B(x)$ 1, 6, DS
8. $\forall x\, \neg B(x)$ 4, Simp
9. $\neg\, \exists x\, B(x)$ 5, T
10. $\exists x\, B(x) \wedge \neg\, \exists x\, B(x)$ 7, 9, Conj
  QED   1, 2, 10, IP.

7.
1. $\forall x\, (\exists y\, (q(x, y) \wedge s(y)) \rightarrow \exists y\, (p(y) \wedge r(x, y)))$ P
2. $\neg\, (\neg\, \exists x\, p(x) \rightarrow \forall x\, \forall y\, (q(x, y) \rightarrow \neg\, s(y)))$ P for IP
3. $\neg\, \exists x\, p(x) \wedge \neg\, \forall x\, \forall y\, (q(x, y) \rightarrow \neg\, s(y))$ 2, T
4. $\neg\, \exists x\, p(x)$ 3, Simp
5. $\neg\, \forall x\, \forall y\, (q(x, y) \rightarrow \neg\, s(y))$ 3, Simp
6. $\exists x\, \exists y\, (q(x, y) \wedge s(y))$ 5, T
7. $\exists y\, (q(c, y) \wedge s(y))$ 6, EI
8. $\exists y\, (q(c, y) \wedge s(y)) \rightarrow \exists y\, (p(y) \wedge r(c, y))$ 1, UI
9. $\exists y\, (p(y) \wedge r(c, y))$ 7, 8, MP
10. $p(d) \wedge r(c, d)$ 9, EI
11. $p(d)$ 10, Simp
12. $\exists x\, p(x)$ 11, EG
13. false   4, 12, Conj, T
  QED   1, 2, 13, IP.

9. a. Line 2 is wrong because x is free in line 1, which is a premise. Therefore x is flagged on line 1. Thus line 1 can't be used with the UG rule to generalize x. **c.** Line 2 is wrong because $f(y)$ is not free to replace x. That is, the substitution of $f(y)$ for x yields a new bound occurrence of y. Therefore EG can't generalize to x from $f(y)$. **e.** Line 4 is wrong because c already occurs in the proof on line 3.

10. a.
1. $\exists x\, A(x)$ P
2. $A(x)$ 1, EI (wrong to use a variable)
3. $\forall x\, A(x)$ 2, UG.

c.
1. $\forall x\, (A(x) \vee B(x))$ — P
2. $\neg\, (\forall x\, A(x) \vee \forall x\, B(x))$ — P for IP
3. $\exists x\, \neg\, A(x) \wedge \exists x\, \neg\, B(x)$ — 2, T
4. $\exists x\, \neg\, A(x)$ — 3, Simp
5. $\exists x\, \neg\, B(x)$ — 4, Simp
6. $\neg\, A(c)$ — 4, EI
7. $\neg\, B(c)$ — 5, EI (wrong to use an existing constant)
8. $A(c) \vee B(c)$ — 1, UI
9. $B(c)$ — 6, 8, DS
10. false — 7, 9, Conj, T.

e.
1. $\forall x\, \exists y\, W(x, y)$ — P
2. $\exists y\, W(x, y)$ — 1, UI
3. $W(x, c)$ — 2, EI
4. $\forall x\, W(x, c)$ — 3, UG (wrong because x is subscripted)
5. $\exists y\, \forall x\, W(x, y)$ — 4, EG (may be wrong if $W(x, c)$ contains bound y).

11. a. Similar to proof of Exercise 6b.
c. Use IP in both directions. **e.** Similar to part (c). **g.** Similar to part (c).

12. a. Let $\neg\, B$ and $A \to B$ be valid wffs. Consider an arbitrary interpretation of these two wffs with domain D. Then $\neg\, B$ and $A \to B$ are true for D. Thus we can apply MT to conclude that $\neg\, A$ is true for D. Since the interpretation was arbitrary, it follows that $\neg\, A$ is valid.

13. The UI rule says that we can infer $W(t)$ from $\forall x\, W(x)$ if t is free to replace x in $W(x)$. Thus we can also infer $\neg\, W(t)$ from $\forall x\, \neg W(x)$ if t is free to replace x in $\neg W(x)$. Since $\forall x\, \neg W(x) \equiv \neg\, \exists x\, W(x)$, we can infer $\neg\, W(t)$ from $\neg\, \exists x\, W(x)$ if t is free to replace x in $\neg W(x)$. The contrapositive of inferring $\neg\, W(t)$ from $\neg\, \exists x$ $W(x)$ is to infer $\exists x\, W(x)$ from $W(t)$, which is the EG rule. The statement $W(t) = W(x)(x/t)$ is satisfied because we started with $W(x)$ and then replaced x by t. In a similar manner we can obtain the UI rule from the EG rule by using the fact that $\exists x\, \neg\, W(x) \equiv \neg\, \forall x\, W(x)$.

Chapter 8

Section 8.1

1.
1. $s = v$ — P
2. $t = w$ — P
3. $p(s, t)$ — P
4. $p(v, t)$ — 1, 3, EE
5. $p(v, w)$ — 2, 4, EE
   QED — 1, 2, 3, 5, CP.

3. We can write $Ux\, A(x)$ as either of the following two wffs:

$$\exists x\, (A(x) \wedge \forall y\, (A(y) \to x = y)),$$

$$\exists x\, A(x) \wedge \forall x\, \forall y\, (A(x) \wedge A(y) \to x = y).$$

5. a.

1.	$t = u$	P
2.	$\neg\, p(\dots t \dots)$	P
3.	$p(\dots u \dots)$	P for IP
4.	$u = t$	1, Symmetric
5.	$p(\dots t \dots)$	3, 4, EE
6.	false	2, 5, Conj, T
	QED	1, 2, 3, 6, IP.

c.

1.	$t = u$	P
2.	$p(\dots t \dots) \vee q(\dots t \dots)$	P
3.	$\neg\, (p(\dots u \dots) \vee q(\dots u \dots))$	P for IP
4.	$\neg\, p(\dots u \dots) \wedge \neg\, q(\dots u \dots)$	3, T
5.	$\neg\, p(\dots u \dots)$	4, Simp
6.	$\neg\, q(\dots u \dots)$	4, Simp
7.	$u = t$	1, Symmetric
8.	$\neg\, p(\dots t \dots)$	5, 7, EE from part (a)
9.	$\neg\, q(\dots t \dots)$	6, 7, EE from part (a)
10.	$\neg\, p(\dots t \dots) \wedge \neg\, q(\dots t \dots)$	8, 9, Conj
11.	$\neg\, (p(\dots t \dots) \vee q(\dots t \dots))$	10, T
12.	false	2, 11, Conj, T
	QED	1, 12, CP.

e.

1.	$x = y$	P
2.	$\forall z\, p(\dots x \dots)$	P
3.	$p(\dots x \dots)$	2, UI
4.	$p(\dots y \dots)$	1, 3, EE
5.	$\forall z\, p(\dots y \dots)$	4, UG
	QED	1, 2, 5, CP.

7. a. Proof of $p(x) \to \exists y\, (x = y \wedge p(y))$:

1.	$p(x)$	P
2.	$\neg\, \exists y\, (x = y \wedge p(y))$	P for IP
3.	$\forall y\, (x \neq y \vee \neg\, p(y))$	2, T
4.	$x \neq x \vee \neg\, p(x)$	3, UI
5.	$x \neq x$	1, 4, DS
6.	$x = x$	EA
7.	false	5, 6, Conj, T
	QED	1, 2, 7, IP.

Proof of $\exists y\ (x = y \wedge p(y)) \to p(x)$:

1.	$\exists y\ (x = y \wedge p(y))$	P
2.	$x = c \wedge p(c)$	1, EI
3.	$p(x)$	2, EE
	QED	1, 3, IP.

Section 8.2

1.

1. $\{odd(x + 1)\}\ y := x + 1\ \{odd(y)\}$ ⟶ AA
2. $true \wedge even(x)$ ⟶ P
3. $even(x)$ ⟶ 2, Simp
4. $odd(x + 1)$ ⟶ 3, T
5. $true \wedge even(x) \to odd(x + 1)$ ⟶ 2, 4, CP
6. $\{true \wedge even(x)\}\ y := x + 1\ \{odd(y)\}$ ⟶ 1, 5, Consequence
   QED.

2. a.

1. $\{x + b > 0\}\ y := b\ \{x + y > 0\}$ ⟶ AA
2. $\{a + b > 0\}\ x := a\ \{x + b > 0\}$ ⟶ AA
3. $a > 0 \wedge b > 0 \to a + b > 0$ ⟶ T
4. $\{a > 0 \wedge b > 0\}\ x := a\ \{x + b > 0\}$ ⟶ 2, 3, Consequence
   QED ⟶ 1, 4, Composition.

3. Use the composition rule (8.12) applied to a sequence of three statements.

a.

1. $\{temp < x\}\ y := temp\ \{y < x\ \}$ ⟶ AA
2. $\{temp < y\}\ x := y\ \{temp < x\}$ ⟶ AA
3. $\{x < y\}\ temp := x\ \{temp < y\}$ ⟶ AA
   QED ⟶ 3, 2, 1, Composition.

4. a. First, prove the correctness of the wff $\{x < 10 \wedge x \geq 5\}\ x := 4\ \{x < 5\}$:

1. $\{4 < 5\}\ x := 4\ \{x < 5\}$ ⟶ AA
2. $x < 10 \wedge x \geq 5 \to 4 < 5$ ⟶ T
3. $\{x < 10 \wedge x \geq 5\}\ x := 4\ \{x < 5\}$ ⟶ 1, 2, Consequence
   QED.

Second, prove that $x < 10 \wedge \neg\ (x \geq 5) \to x < 5$. This is a valid wff because of the equivalence $\neg\ (x \geq 5) \equiv x < 5$. Thus the original wff is correct, by the if-then rule.

c. First, prove $\{true \wedge x < y\}\ x := y\ \{x \geq y\}$:

1. $\{y \geq y\}\ x := y\ \{x \geq y\}$ ⟶ AA
2. $true \wedge x < y \to y \geq y$ ⟶ T
3. $\{true \wedge x < y\ \}\ x := y\ \{x \geq y\}$ ⟶ 1, 2, Consequence
   QED.

Second, prove that true $\wedge \neg (x < y) \to x \geq y$. This is a valid wff because of the equivalence true $\wedge \neg (x < y) \equiv \neg (x < y) \equiv (x \geq y)$. Thus the original wff is correct, by the if-then rule.

5. a. Use the if-then-else rule. Thus we must prove the two statements

$$\{\text{true} \wedge x < y\} \ \max := y \ \{\max \geq x \wedge \max \geq y\}$$
$$\{\text{true} \wedge x \geq y\} \ \max := x \ \{\max \geq x \wedge \max \geq y\}.$$

For example, the first statement can be proved as follows:

1. $\{y \geq x \wedge y \geq y\} \ \max := y \ \{\max \geq x \wedge \max \geq y\}$ ⠀⠀AA
2. ⠀⠀⠀⠀⠀⠀true $\wedge x < y$ ⠀⠀⠀⠀⠀⠀⠀⠀⠀⠀⠀⠀⠀⠀P
3. ⠀⠀⠀⠀⠀⠀⠀⠀$x < y$ ⠀⠀⠀⠀⠀⠀⠀⠀⠀⠀⠀⠀⠀⠀2, Simp
4. ⠀⠀⠀⠀⠀⠀⠀⠀$x \leq y$ ⠀⠀⠀⠀⠀⠀⠀⠀⠀⠀⠀⠀⠀⠀3, Add
5. ⠀⠀⠀⠀⠀⠀⠀⠀$y \geq y$ ⠀⠀⠀⠀⠀⠀⠀⠀⠀⠀⠀⠀⠀⠀T
6. ⠀⠀⠀⠀⠀⠀$y \geq x \wedge y \geq y$ ⠀⠀⠀⠀⠀⠀⠀⠀⠀4, 5, Conj
7. true $\wedge x < y \to y \geq x \wedge y \geq y$ ⠀⠀⠀⠀2, 6, CP
8. $\{\text{true} \wedge x < y\} \ \max := y \ \{\max \geq x \wedge \max \geq y\}$ ⠀1, 7, Consequence
⠀⠀QED.

6. a. The wff is incorrect if $x = 1$.

7. Since the wff fits the form of the while rule, we need to prove the following statement:

$$\{x \geq y \wedge \text{even}(x - y) \wedge x \neq y\} \ x := x - 1; y := y + 1 \ \{x \geq y \wedge \text{even}(x - y)\}.$$

Proof:
1. $\{x \geq y + 1 \wedge \text{even}(x - y - 1)\} \ y := y + 1 \ \{x \geq y \wedge \text{even}(x - y)\}$ ⠀AA
2. $\{x - 1 \geq y + 1 \wedge \text{even}(x - 1 - y - 1)\} \ x := x - 1 \{x \geq y + 1 \wedge \text{even}(x - y - 1)\}$ AA
3. $x \geq y \wedge \text{even}(x - y) \wedge x \neq y$ ⠀⠀⠀⠀⠀⠀⠀⠀P
4. ⠀⠀⠀$x \geq y + 2$ ⠀⠀⠀⠀⠀⠀⠀⠀⠀⠀⠀⠀⠀3, T
5. ⠀⠀⠀$x - 1 \geq y + 1$ ⠀⠀⠀⠀⠀⠀⠀⠀⠀⠀⠀4, T
6. ⠀⠀⠀$\text{even}(x - 1 - y - 1)$ ⠀⠀⠀⠀⠀⠀⠀⠀⠀3, T
7. $x \geq y \wedge \text{even}(x - y) \wedge x \neq y \to x - 1 \geq y + 1 \wedge \text{even}(x - 1 - y - 1)$ ⠀⠀3, 6, CP
⠀⠀QED ⠀⠀⠀⠀⠀⠀⠀⠀⠀⠀⠀⠀⠀1, 2, 7, Consequence, Composition.

Now the result follows from the while rule.

8. a. The postcondition $i = \text{floor}(x)$ is equivalent to $i \leq x \wedge x < i + 1$. This statement has the form $Q \wedge \neg C$, where C is the condition of the while loop and Q is the suggested loop invariant. To show that the while loop is correct with respect to Q, show that $\{Q \wedge C\} \ i := i + 1 \ \{Q\}$ is correct. Once this is done, show that $\{x \geq 0\} \ i := 0 \ \{Q\}$ is correct. **c.** The given wff fits the form of the if-then-else rule. Therefore we need to prove the following two wffs:

$$\{\text{true} \wedge x \geq 0\} \ S_1 \ \{i = \text{floor}(x)\} \quad \text{and} \quad \{\text{true} \wedge x < 0\} \ S_2 \ \{i = \text{floor}(x)\}.$$

These two wffs are equivalent to the two wffs of parts (a) and (b). Therefore the given wff is correct.

9. Let Q be the suggested loop invariant. The postcondition is equivalent to $Q \wedge \neg C$, where C is the while loop condition. Therefore the program can be proven correct by proving the validity of the following two wffs:

$$\{Q \wedge C\}\ i := i + 1;\ s := s + i\ \{Q\} \quad \text{and} \quad \{n \geq 0\}\ i := 0; s := 0;\ \{Q\}.$$

11. Letting Q denote the loop invariant, the while loop can be proved correct with respect to Q by proving the following wff:

$$\{Q \wedge x \neq y\}\ \textbf{if}\ x > y\ \textbf{then}\ x := x - y\ \textbf{else}\ y := y - x\ \{Q\}.$$

The parts of the program before and after the while loop can be proved correct by proving the following two wffs:

$$\{a > 0 \wedge b > 0\}\ x := a; y := b\ \{Q\},$$

$$\{Q \wedge \neg (x \neq y)\}\ great := x\ \{(a, b) = great\}.$$

13. a. $\{(\textbf{if}\ j = i - 1\ \textbf{then}\ 24\ \textbf{else}\ a[j]) = 24\}$.
c. We obtain the precondition

$$\{(\textbf{if}\ i = j - 1\ \textbf{then}\ 12\ \textbf{else}\ (\textbf{if}\ i = i + 1\ \textbf{then}\ 25\ \textbf{else}\ a[i])) = 12$$
$$\wedge\ (\textbf{if}\ j = j - 1\ \textbf{then}\ 12\ \textbf{else}\ (\textbf{if}\ j = i + 1\ \textbf{then}\ 25\ \textbf{else}\ a[j])) = 25)\}.$$

Since it is impossible to have $i = i + 1$ and $j = j - 1$, the precondition can be simplified to

$$\{(\textbf{if}\ i = j - 1\ \textbf{then}\ 12\ \textbf{else}\ a[i]) = 12 \wedge (\textbf{if}\ j = i + 1\ \textbf{then}\ 25\ \textbf{else}\ a[j]) = 25)\}.$$

14. a. 1. $\{(\textbf{if}\ j = i - 1\ \textbf{then}\ 24\ \textbf{else}\ \text{a}[j]) = 24\}\ \text{a}[i - 1] := 24\ \{a[j] = 24\}$ AAA
    2. $i = j + 1 \wedge a[j] = 39$ P
    3. $i = j + 1$ 2, Simp
    4. $24 = 24$ T
    5. $i = j + 1 \rightarrow 24 = 24$ T (true conclusion)
    6. $i \neq j + 1 \rightarrow a[j] = 24$ 3, T (false premise)
    7. $(\textbf{if}\ j = i - 1\ \textbf{then}\ 24\ \textbf{else}\ \text{a}[j]) = 24$ 5, 6, Conj
    8. $i = j + 1 \wedge a[j] = 39 \rightarrow (\textbf{if}\ j = i - 1\ \textbf{then}\ 24\ \textbf{else}\ \text{a}[j]) = 24$ 2, 7, CP
       QED   1, 8, Consequence.

c. 1. $\{(\textbf{if}\ i = j - 1\ \textbf{then}\ 12\ \textbf{else}\ a[i]) = 12 \wedge (\textbf{if}\ j = j - 1\ \textbf{then}\ 12\ \textbf{else}\ a[j]) = 25\}$
      $a[j - 1] := 12$
      $\{a[i] = 12 \wedge a[j] = 25\}$ AAA
    2. $\{(\textbf{if}\ i = j - 1\ \textbf{then}\ 12\ \textbf{else}\ a[i]) = 12 \wedge a[j] = 25\}$
      $a[j - 1] := 12$
      $\{a[i] = 12 \wedge a[j] = 25\}$ 1, T

3. $\{(\text{if } i = j - 1 \text{ then } 12 \text{ else } (\text{if } i = i + 1 \text{ then } 25 \text{ else } a[i])) = 12$
    $\wedge (\text{if } j = i + 1 \text{ then } 25 \text{ else } a[j]) = 25\}$
    $a[i + 1] := 25$
    $\{(\text{if } i = j - 1 \text{ then } 12 \text{ else } a[i]) = 12 \wedge a[j]) = 25\}$ AAA

4. $\{(\text{if } i = j - 1 \text{ then } 12 \text{ else } a[i]) = 12 \wedge (\text{if } j = i + 1 \text{ then } 25 \text{ else } a[j]) = 25\}$
    $a[i + 1] := 25$
    $\{(\text{if } i = j - 1 \text{ then } 12 \text{ else } a[i]) = 12 \wedge a[j]) = 25\}$ 3, T

5. $i = j - 1 \wedge a[i] = 25 \wedge a[j] = 12$ P

6. $i = j - 1$ 2, Simp

7. $(\text{if } i = j - 1 \text{ then } 12 \text{ else } a[i]) = 12$
         $\wedge (\text{if } j = i + 1 \text{ then } 25 \text{ else } a[j]) = 25$ 6, T

8. $i = j - 1 \wedge a[i] = 25 \wedge a[j] = 12$
    $\rightarrow \quad (\text{if } i = j - 1 \text{ then } 12 \text{ else } a[i]) = 12$
        $\wedge (\text{if } j = i + 1 \text{ then } 25 \text{ else } a[j]) = 25$ 5, 7, CP
    QED   2, 4, Consequence, Composition.

15. a. After applying AAA to the postcondition and assignment, we obtain the condition even$(a[i] + 1)$. It is clear that the precondition even$(a[i])$ does not imply even$(a[i] + 1)$. **c.** After applying AAA twice to the postcondition and two assignments, we obtain the condition

$$\forall j \, (1 \leq j \leq 5 \rightarrow (\text{if } j = 3 \text{ then } 355 \text{ else } a[j]) = 23).$$

This wff is the conjunction of five propositions, one for each j, where $1 \leq j \leq 5$. For $j = 3$ we obtain the proposition

$$(1 \leq 3 \leq 5 \rightarrow (\text{if } 3 = 3 \text{ then } 355 \text{ else } a[j]) = 23),$$

which is equivalent to the false statement $(1 \leq 3 \leq 5 \rightarrow 355 = 23)$. Therefore the given precondition cannot imply the obtained condition.

16. a. Let the well-founded set be $\mathbb{N}$, and define $f : \text{States} \rightarrow \mathbb{N}$ by $f(i, x) = x - i$. If $s = \langle i, x \rangle$, then after the execution of the loop body the state will be $t = \langle i + 1, x \rangle$. Thus $f(s) = x - i$ and $f(t) = x - i - 1$. With these interpretations, (8.19) can be written as follows:

$$i \leq x \wedge i < x \wedge x - i \in \mathbb{N} \rightarrow x - i - 1 \in \mathbb{N} \wedge x - i > x - i - 1.$$

Proof: 1. $i \leq x \wedge i < x \wedge x - i \in \mathbb{N}$ P

      2. $i < x$ 1, Simp

      3. $x - i > 0$ 2, T

      4. $x - i - 1 \geq 0$ 3, T

      5. $x - i > x - i - 1$ T

      6. $x - i - 1 \in \mathbb{N} \wedge x - i > x - i - 1$ 4, 5, Conj
         QED   1, 6, CP.

17. Let P be the set of positive integers, and define $f :$ States $\rightarrow P$ by $f(x, y) =$ $x + y$. If $s = \langle x, y \rangle$, then the state after the execution of the loop body will depend on whether $x < y$. If $x < y$, then $t = \langle x, y - x \rangle$, which gives $f(t) = x$. Otherwise, if $x > y$, then $t = \langle x - y, y \rangle$, which gives $f(t) = y$. So in either case, $f(t) < f(s)$ and $f(t) \in P$. This amounts to an informal proof of statement (8.19): true $\wedge \; x \neq y \wedge f(s) \in \mathbb{N} \rightarrow f(t) \in \mathbb{N} \wedge f(s) > f(t)$.

Section 8.3

1. a. Second. **c.** Fifth. **e.** Third. **g.** Third. **i.** Fourth.

2. a. $\exists A \; \exists B \; \forall x \; ((A(x) \rightarrow \neg B(x)) \wedge (B(x) \rightarrow \neg A(x))$ or $\exists A \; \exists B \; \forall x \neg (A(x) \wedge B(x))$.

3. a. Let S be state and C be city. Then we can write $\forall S \; \exists C \; S(C) \wedge (C =$ Springfield$)$. The wff is second order. **c.** Let $H, R, S, B,$ and A stand for house, room, shelf, book, and author, respectively. Then we can write the statement as $\exists H \; \exists R \; \exists S \; \exists B \; (H(R) \wedge R(S) \wedge S(B) \wedge A(B,$ Thoreau$))$. The wff is fourth order. **e.** The statement can be expressed as follows:
$\exists S \; \exists A \; \exists B \; (\forall x \; (A(x) \vee B(x) \rightarrow S(x)) \wedge \forall x \; (S(x) \rightarrow A(x) \vee B(x)) \wedge \forall x \neg (A(x) \wedge B(x)))$. The wff is second order.

5. Think of $S(x)$ as $x \in S$. **a.** For any domain D the antecedent is false because S can be the empty set. Thus the wff is true for all domains. **c.** For any domain D the consequent is true because S can be chosen as D. Thus the wff is true for all domains.

6. a. Assume that the statement is false. Then there is some line L containing every point. Now Axiom 3 says that there are three distinct points not on the same line. This is a contradiction. Thus the statement is true. **c.** Let w be a point. By Axiom 3 there is another point x, $x \neq w$. By Axiom 1 there is a line L on x and w. By part (a) there is a point z not on L. By Axiom 1 there is a line M on w and z. Since z is on M and z is not on L, it follows that $L \neq M$. QED.

7. Here are some sample formalizations.

a. $\forall L \; \exists x \neg L(x)$.

Proof:
1.	$\neg \forall L \; \exists x \neg L(x)$	P for IP
2.	$\exists L \forall x \; L(x)$	1, T
3.	$\forall x \; l(x)$	2, EI
4.	Axiom 3	
5.	$l(a) \wedge l(b) \rightarrow \neg \, l(c)$	4, EI, EI, EI, Simp, UI
6.	$l(a) \wedge l(b)$	3, UI, UI, Conj
7.	$\neg \, l(c)$	5, 6, MP
8.	$l(c)$	3, UI
9.	false	7, 8, Conj, T
	QED	1, 9, IP.

c. $\forall x \, \exists L \, \exists M \, (L(x) \wedge M(x) \wedge \exists y \, (\neg L(y) \wedge M(y)))$. We give two proofs. The first uses part (a), and the second does not.

Proof (using part (a)):

1.	$\neg (\forall x \, \exists L \, \exists M \, (L(x) \wedge M(x) \wedge \exists y \, (\neg L(y) \wedge M(y))))$	*P* for IP
2.	$\exists x \, \forall L \, \forall M \, (\neg (L(x) \wedge M(x)) \vee \forall y \, (L(y) \vee \neg M(y)))$	1, *T*
3.	$\forall L \, \forall M \, (\neg (L(a) \wedge M(a)) \vee \forall y \, (L(y) \vee \neg M(y)))$	2, EI
4.	Axiom 3	
5.	$b \neq c \wedge b \neq d \wedge c \neq d \wedge \forall L \, (L(b) \wedge L(c) \rightarrow \neg L(d))$	4, EI, EI, EI
6.	$a \neq b$	*T*
7.	$a \neq b \rightarrow \exists L \, (L(a) \wedge L(b))$	Axiom 1, UI, UI
8.	$\exists L \, (L(a) \wedge L(b))$	6, 7, MP
9.	$l(a) \wedge l(b)$	8, EI
10.	$\forall L \, \exists x \, \neg L(x)$	Part (a)
11.	$\exists x \, \neg l(x)$	10, UI
12.	$\neg l(e)$	11, EI
13.	$a \neq e$	*T*
14.	$a \neq e \rightarrow \exists L \, (L(a) \wedge L(e))$	Axiom 1, UI, UI
15.	$\exists L \, (L(a) \wedge L(e))$	13, 14, MP
16.	$m(a) \wedge m(e)$	15, EI
17.	$l(a) \wedge m(a)$	9, Simp, 16, Simp, Conj
18.	$\neg (l(a) \wedge m(a)) \vee \forall y \, (l(y) \vee \neg m(y))$	3, UI, UI
19.	$\forall y \, (l(y) \vee \neg m(y))$	17, 18, DS
20.	$l(e) \vee \neg m(e)$	19, UI
21.	$\neg m(e)$	12, 20, DS
22.	false	16, Simp, 21, Conj *T*
	QED	1, 22, CP.

Proof (without using part (a)):

1.	$\neg (\forall x \, \exists L \, \exists M \, (L(x) \wedge M(x) \wedge \exists y \, (\neg L(y) \wedge M(y))))$	*P* for IP
2.	$\exists x \, \forall L \, \forall M \, (\neg (L(x) \wedge M(x)) \vee \forall y \, (L(y) \vee \neg M(y)))$	1, *T*
3.	$\forall L \, \forall M \, (\neg (L(a) \wedge M(a)) \vee \forall y \, (L(y) \vee \neg M(y)))$	2, EI, UI, UI
4.	Axiom 3	
5.	$b \neq c \wedge b \neq d \wedge c \neq d \wedge \forall L \, (L(b) \wedge L(c) \rightarrow \neg L(d))$	4, EI, EI, EI
6.	$a \neq b$	*T*
7.	$a \neq b \rightarrow \exists L \, (L(a) \wedge L(b))$	axiom 1, UI, UI
8.	$\exists L \, (L(a) \wedge L(b))$	6, 7, MP
9.	$l(a) \wedge l(b)$	8, EI
10.	$a \neq c$	*T*
11.	$a \neq c \rightarrow \exists L \, (L(a) \wedge L(c))$	axiom 1, UI, UI
12.	$\exists L \, (L(a) \wedge L(c))$	10, 11, MP
13.	$m(a) \wedge m(c)$	12, EI
14.	$a \neq d$	*T*
15.	$a \neq d \rightarrow \exists L \, (L(a) \wedge L(d))$	axiom 1, UI, UI

16. $\exists L\ (L(a) \wedge L(d))$	15, 16, MP
17. $n(a) \wedge n(d)$	16, EI
18. $l(a) \wedge m(a)$	9, Simp, 13, Simp, Conj
19. $\neg\ (l(a) \wedge m(a)) \vee \forall y\ (l(y) \vee \neg\ m(y))$	3, UI, UI
20. $\forall y\ (l(y) \vee \neg m(y))$	18, 19, DS
21. $l(c) \vee \neg\ m(c))$	20, UI
22. $l(c)$	13, Simp, 21, DS
23. $l(a) \wedge n(a)$	9, Simp, 13, Simp, Conj
24. $\neg\ (l(a) \wedge n(a)) \vee \forall y\ (l(y) \vee \neg\ n(y))$	3, UI, UI
25. $\forall y\ (l(y) \vee \neg n(y))$	23, 24, DS
26. $l(d) \vee \neg\ n(d)$	25, UI
27. $l(d)$	17, Simp, 26, DS
28. $l(b) \wedge l(c)$	9, Simp, 22, Conj
29. $\forall L\ (L(b) \wedge L(c) \rightarrow \neg\ L(d))$	5, Simp
30. $l(b) \wedge l(c) \rightarrow \neg\ l(d)$	29, UI
31. $\neg\ l(d)$	28, 30, MP
32. false	27, 31, Conj, T
QED	1, 32, IP.

Chapter 9

Section 9.1

1. a. isChildOf$(x, y) \leftarrow$ isParentOf(y, x).
c. isGreatGrandParentOf$(x, y) \leftarrow$ isParentOf(x, w), isParentOf(w, z), isParentOf(z, y).

2. a. The following definition will work if $x \neq y$:

$$\text{isSiblingOf}(x, y) \leftarrow \text{isParentOf}(z, x), \text{isParentOf}(z, y).$$

c. Let s denote isSecondCousinOf. Two possible definitions are

$s(x, y) \leftarrow$ isParentOf(z, x), isParentOf(w, y), isCousinOf(z, w)

$s(x, y) \leftarrow$ isGreatGrandParentOf(z, x), isGreatGrandParentOf(z, y).

3. The $g(v, w)$ match forces the two equalities, $v = x$ and $w = y$. Try to match $p(x, z)$ first, starting at the top of the list. If it matches some fact, then mark the position of that fact. Suppose a yes answer is eventually found. Now continue the process as before, by trying to match $g(v, w)$, but when we want to match $p(x, z)$, we don't start at the top of the list. Instead, we start at the statement below the previously marked statement.

Section 9.2

1. a. $(A \vee C \vee D) \wedge (B \vee C \vee D)$.
c. $\forall x\ (\neg\ p(x, c) \vee q(x))$.
e. $\forall x\ \forall y\ (p(x, y) \vee q(x, y, f(x, y)))$.

2. $p \vee \neg p$ and $p \vee \neg p \vee q \vee q$.

3. a.
1.	$A \vee B$	P
2.	$\neg A$	P
3.	$\neg B \vee C$	P
4.	$\neg C$	P
5.	B	$1, 2, R$
6.	$\neg B$	$3, 4, R$
7.	$\square$	$5, 6, R.$

c.
1.	$A \vee B$	P
2.	$A \vee \neg C$	P
3.	$\neg A \vee C$	P
4.	$\neg A \vee \neg B$	P
5.	$C \vee \neg B$	P
6.	$\neg C \vee B$	P
7.	$B \vee C$	$1, 3, R$
8.	$B \vee B$	$6, 7, R$
9.	$\neg A$	$4, 8, R$
10.	$\neg C$	$2, 9, R$
11.	$\neg B$	$5, 10, R$
12.	A	$1, 11, R$
13.	$\square$	$9, 12, R.$

4. a. $\{y/x\}$. **c.** $\{y/a\}$. **e.** $\{x/f(a), y/f(b), z/b\}$.

5. a. $\{x/f(a, b), v/f(y, a), z/y\}$ or $\{x/f(a, b), v/f(z, a), y/z\}$. **c.** $\{x/g(a), z/g(b), y/b\}$.

6. Make sure the clauses to be resolved have distinct sets of variables. The answers are $p(x) \vee \neg p(f(a))$ and $p(x) \vee \neg p(f(a)) \vee q(x) \vee q(f(a))$.

7. a.
1.	$p(x)$	P
2.	$q(y, a) \vee \neg p(a)$	P
3.	$\neg q(a, a)$	P
4.	$\neg p(a)$	$2, 3, R, \{y/a\}$
5.	$\square$	$1, 4, R, \{x/a\}$
	QED.	

c.
1.	$p(a) \vee p(x)$	P
2.	$\neg p(a) \vee \neg p(y)$	P
3.	$\square$	$1, 2, R, \{x/a, y/a\}$
	QED.	

e. Number the clauses 1, 2, and 3. Resolve 2 with 3 by unifying all four of the p atoms to obtain the clause $\neg q(a) \vee \neg q(a)$. Resolve this clause with 1 to obtain the empty clause.

8. a. After negating the statement and putting the result in clausal form, we obtain the following proof:

1.	$A \lor B$	P
2.	$\neg A$	P
3.	$\neg B$	P
4.	B	$1, 2, R$
5.	$\square$	$3, 4, R$, QED.

c. After negating the statement and putting the result in clausal form, we obtain the following proof:

1.	$p \lor q$	P
2.	$\neg q \lor r$	P
3.	$\neg r \lor s$	P
4.	$\neg p$	P
5.	$\neg s$	P
6.	$\neg r$	$3, 5, R$
7.	$\neg q$	$2, 6, R$
8.	q	$1, 4, R$
9.	$\square$	$7, 8, R$, QED.

9. a. After negating the statement and putting the result in clausal form, we obtain the following proof:

1.	$p(x)$	P
2.	$\neg p(y)$	P
3.	$\square$	$1, 2, R, \{x/y\}$ QED.

c. After negating the statement and putting the result in clausal form, we obtain the following proof:

1.	$p(x, a)$	P
2.	$\neg p(b, y)$	P
3.	$\square$	$1, 2, R, \{x/b, y/a\}$ QED.

e. After negating the statement and putting the result in clausal form, we obtain the following proof:

1.	$p(x) \lor q(y)$	P
2.	$\neg p(a)$	P
3.	$\neg q(a)$	P
4.	$q(y)$	$1, 2, R, \{x/a\}$
5.	$\square$	$3, 4, R, \{y/a\}$ QED.

10. a. In first-order predicate calculus the argument can be written as the following wff:

$$\forall x\, (C(x) \to P(x)) \land \exists x\, (C(x) \land L(x)) \to \exists x\, (P(x) \land L(x)),$$

where $C(x)$ means that x is a computer science major, $P(x)$ means that x is a person, and $L(x)$ means that x is a logical thinker. After negating the wff and transforming the result into clausal form, we obtain the proof:

1. $\neg C(x) \vee P(x)$ P
2. $C(a)$ P
3. $L(a)$ P
4. $\neg P(z) \vee \neg L(z)$ P
5. $\neg P(a)$ $3, 4, R, \{z/a\}$
6. $\neg C(a)$ $1, 5, R, \{x/a\}$
7. $\square$ $2, 6, R, \{\}$ QED.

11. a. Let $D(x)$ mean that x is a dog, $L(x)$ mean that x likes people, $H(x)$ mean that x hates cats, and a = Rover. Then the argument can be formalized as follows:

$$\forall x\,(D(x) \to L(x) \vee H(x)) \wedge D(a) \wedge \neg H(a) \to \exists x\,(D(x) \wedge L(x)).$$

After negating the wff and transforming the result into clausal form, we obtain the proof:

1. $\neg D(x) \vee L(x) \vee H(x)$ P
2. $D(a)$ P
3. $\neg H(a)$ P
4. $\neg D(y) \vee \neg L(y)$ P
5. $L(a) \vee H(a)$ $1, 2, R, \{x/a\}$
6. $L(a)$ $3, 5, R, \{\}$
7. $\neg D(a)$ $4, 6, R, \{y/a\}$
8. $\square$ $2, 7, R, \{\}$ QED.

c. Let $H(x)$ mean that x is a human being, $Q(x)$ mean that x is a quadruped, and $M(x)$ mean that x is a man. Then the argument can be formalized as

$$\forall x\,(H(x) \to \neg Q(x)) \wedge \forall x\,(M(x) \to H(x)) \to \forall x\,(M(x) \to \neg Q(x)).$$

After negating the wff and transforming the result into clausal form, we obtain the proof:

1. $\neg H(x) \vee \neg Q(x)$ P
2. $\neg M(y) \vee H(y)$ P
3. $M(a)$ P
4. $Q(a)$ P
5. $H(a)$ $2, 3, R, \{y/a\}$
6. $\neg O(a)$ $1, 5, R, \{x/a\}$
7. $\square$ $4, 6, R, \{\}$ QED.

e. Let $F(x)$ mean that x is a freshman, $S(x)$ mean that x is a sophomore, $J(x)$ mean that x is a junior, and $L(x, y)$ mean that x likes y. Then the argument can be formalized as $A \rightarrow B$, where

$$A = \exists x \, (F(x) \wedge \forall y \, (S(y) \rightarrow L(x, y))) \wedge \forall x \, (F(x) \rightarrow \forall y \, (J(y) \rightarrow \neg L(x, y)))$$

and $B = \forall x \, (S(x) \rightarrow \neg J(x))$. After negating the wff and transforming the result into clausal form, we obtain the proof:

1.	$F(a)$	P
2.	$\neg S(x) \vee L(a, x)$	P
3.	$\neg F(y) \vee \neg J(z) \vee \neg L(y, z)$	P
4.	$S(b)$	P
5.	$J(b)$	P
6.	$\neg J(z) \vee \neg L(a, z)$	$1, 3, R, \{y/a\}$
7.	$\neg L(a, b)$	$5, 6, R, \{z/b\}$
8.	$\neg S(b)$	$2, 7, R, \{x/b\}$
9.	$\square$	$4, 8, R, \{\,\}$ QED.

12. a. We need to show that $x(\theta\sigma) = (x\theta)\sigma$ for each variable x in E. First, suppose $x/t \in \theta$ for some term t. If $x = t\sigma$, then $x(\theta\sigma) = x$ because the binding $x/t\sigma$ has been removed from $\theta\sigma$. But since $x/t \in \theta$, it follows that $x\theta = t$. Now apply σ to both sides to obtain $(x\theta)\sigma = t\sigma = x$. Therefore $x(\theta\sigma) = x = (x\theta)\sigma$. If $x \neq t\sigma$, then $x(\theta\sigma) = t\sigma = (x\theta)\sigma$. Second, suppose that $x/t \in \sigma$ and x does not occur as a numerator of θ. Then $x(\theta\sigma) = t = x\sigma = (x\theta)\sigma$. Lastly, if x does not occur as a numerator of either σ or θ, then the substitutions have no effect on x. Thus $x(\theta\sigma)$ $= x = (x\theta)\sigma$. **c.** If $x/t \in \theta$, then $x/t = x/t\varepsilon$, so it follows from the definition of composition that $\theta = \theta\varepsilon$. For any variable x we have $x(\varepsilon\theta) = (x\varepsilon)\theta = x\theta$. Therefore $\theta\varepsilon$ $= \varepsilon\theta = \theta$. **e.** The proof follows:

$$(A \cup B)\theta = \{E\theta \mid E \in A \cup B\} = \{E\theta \mid E \in A\} \cup \{E\theta \mid E \in B\} = A\theta \cup B\theta.$$

Section 9.3

1. a.
	1.	$p(a, b) \leftarrow$	P
	2.	$p(a, c) \leftarrow$	P
	3.	$p(b, d) \leftarrow$	P
	4.	$p(c, e) \leftarrow$	P
	5.	$g(x, y) \leftarrow p(x, z), p(z, y)$	P
	6.	$\leftarrow g(a, w)$	P initial goal
	7.	$\leftarrow p(a, z), p(z, y)$	$5, 6, R, \theta_1 = \{x/a, w/y\}.$
	8.	$\leftarrow p(b, y)$	$1, 7, R, \theta_2 = \{z/b\}$
	9.	$\square$	$3, 8, R, \theta_3 = \{y/d\}$ QED.

b.

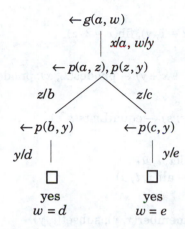

2. a.

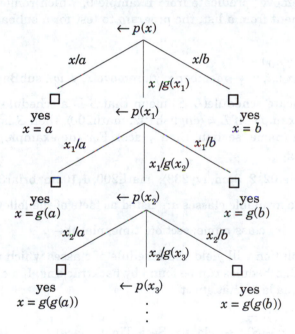

c. $\{g^n(a) \mid n \in \mathbb{N}\}$.

3. a. The program returns the answer yes.

4. a. The symmetric closure s can be defined by the two clause program:

$$s(x, y) \leftarrow r(x, y)$$
$$s(x, y) \leftarrow r(y, x).$$

5. a. $\text{fib}(0, 1) \leftarrow$
$\text{fib}(1, 1) \leftarrow$
$\text{fib}(x, y + z) \leftarrow \text{fib}(x - 1, y), \text{fib}(x - 2, z).$

c. $\text{pnodes}(\langle\,\rangle, 0) \leftarrow$
$\text{pnodes}(\langle L, a, R \rangle, 1 + x + y) \leftarrow \text{pnodes}(L, x), \text{pnodes}(R, y).$

6. a. $\text{equalLists}(\langle\,\rangle, \langle\,\rangle) \leftarrow$
$\text{equalLists}(x :: t, x :: s) \leftarrow \text{equalLists}(t, s).$

c. $\text{all}(x, \langle\,\rangle, \langle\,\rangle) \leftarrow$
$\text{all}(x, x :: t, u) \leftarrow \text{all}(x, t, u)$
$\text{all}(x, y :: t, y :: u) \leftarrow \text{all}(x, t, u).$

e. $\text{subset}(\langle\,\rangle, y) \leftarrow$
$\text{subset}(x :: t, y) \leftarrow \text{member}(x, y), \text{subset}(t, y).$

g. Using the "remove" predicate from Example 9, which removes one occurrence of an element from a list, the program to test for a subbag can be written as follows:

$\text{subBag}(\langle\,\rangle, y) \leftarrow$
$\text{subBag}(x :: t, y) \leftarrow \text{member}(x, y), \text{remove}(x, y, w), \text{subBag}(t, w).$

7. Let the predicate schedule(L, S) mean that S is a schedule for the list of classes L. For example, if $L = \langle \text{english102}, \text{math200} \rangle$, then S is a list of 4-tuples of the form $\langle \text{name}, \text{section}, \text{time}, \text{place} \rangle$. For the example, S might look like the following:

$$\langle\langle \text{english102}, 2, \text{3pm}, \text{ivy238} \rangle, \langle \text{math200}, 1, \text{10am}, \text{briar315} \rangle\rangle.$$

Assume that the available classes are listed as facts of the following form:

$$\text{class(name, section, time, place)} \leftarrow .$$

The following solution will yield one schedule of classes which might contain time conflicts. All schedules can be found by backtracking. If a class cannot be found, a note is made to that effect.

$\text{schedule}(\langle\,\rangle, \langle\,\rangle) \leftarrow$
$\text{schedule}(x :: y, S) \quad \leftarrow \text{class}(x, \text{Sect}, \text{Time}, \text{Place}),$
$\qquad\qquad\qquad\qquad \text{schedule}(y, T),$
$\qquad\qquad\qquad\qquad \text{cons}(\langle x, \text{Sect}, \text{Time}, \text{Place} \rangle, T, S)$
$\text{schedule}(x :: y, \langle \text{unfillable} \rangle) \leftarrow .$

8. Let letters(A, L) mean that L is the list of propositional letters that occur in the wff A. Let replace(p, true, A, B) mean $B = A(p/\text{true})$. Then we can start the process for a wff A with the goal $\leftarrow \text{tautology}(A, \text{Answer})$, where A is a tautology if Answer = true. The initial definitions might go like the following, where capital letters denote variables:

> tautology(A, Answer) ← letters(A, L), evaluate(A, L, Answer)

> evaluate(A, $\langle\ \rangle$, Y) ← value(A, Y)
> evaluate(A, $H :: T$, Answer)   ←   replace(H, true, A, B),
>   replace(H, false, A, C),
>   value($B \wedge C$, Answer).

When "value" is called, $B \wedge C$ is a proposition containing only true and false terms. The definition for the "replace" predicate might include some clauses like the following:

> replace(X, true, X, true) ←
> replace(X, true, $\neg X$, false) ←
> replace(X, true, $\neg A$, $\neg B$) ← replace(X, true, A, B)
> replace(X, true, $A \wedge X$, B) ← replace(X, true, A, B)
> replace(X, true, $X \wedge A$, B) ← replace(X, true, A, B)
> replace(X, true, $A \wedge B$, $C \wedge D$)   ←   replace(X, true, A, C),
>   replace(X, true, B, D).

Continue by writing the clauses for the false case and for the other operators $\vee$ and $\rightarrow$. The first few clauses for the "value" predicate might include some clauses like the following:

> value(true, true) ←
> value(false, false) ←
> value($\neg$ true, false) ←
> value($\neg$ false, true) ←
> value($\neg X$, Y) ← value(X, A), value ($\neg A$, Y)
> value(false $\wedge X$, false) ←
> value($X \wedge$ false, false) ←
> value(true $\wedge X$, Y) ← value(X, Y)
> value($X \wedge$ true, Y) ← value(X, Y)
> value($X \wedge Y$, Z) ← value(X, U), value(Y, V), value($U \wedge V$, Z).

Continue by writing the clauses to find the value of expressions containing the operators $\vee$ and $\rightarrow$. The predicate to construct the list of propositional letters in a wff might start off something like the following:

> letters(X, $\langle X \rangle$) ← atom(X).
> letters($X \wedge Y$, Z) ← letters(X, U), letters(Y, V), cat(U, V, Z).

Continue by writing the clauses for the other operations.

Chapter 10

Section 10.1

1. The zero is m because $\min(x, m) = \min(m, x) = m$ for all $x \in A$. The identity is n because $\min(x, n) = \min(n, x) = x$ for all $x \in A$. If $x, y \in A$ and $\min(x, y) = n$, then x and y are inverses of each other. Since n is the largest element of A, it follows that n is the only element with an inverse.

2. a. No; no; no. **c.** True; false; false is its own inverse.

3. $S = \{a, f(a), f^2(a), f^3(a), f^4(a)\}$.

4. a. An element z is a zero if both row z and column z contain only the element z. **c.** If x is an identity, then an element y has a right and left inverse w if x occurs in row y column w and also in row w column y of the table.

5. a.

$\circ$	a	b	c	d
a	a	b	c	d
b	b	c	d	d
c	c	d	b	b
d	d	a	b	c

Notice that $d \circ b = a$, but $b \circ d \neq a$. So b and d have one-sided inverses but not inverses (two-sided).

c.

$\circ$	a	b	c	d
a	a	a	a	a
b	a	c	d	b
c	a	d	a	b
d	a	a	b	c

Notice that $(b \circ b) \circ c = c \circ c = a$ and $b \circ (b \circ c) = b \circ d = b$. Therefore $\circ$ is not associative.

e.

$\circ$	a	b	c	d
a	a	b	c	d
b	b	a	a	a
c	c	a	a	a
d	d	a	a	a

Notice that $(b \circ b) \circ c = a \circ c = c$ and $b \circ (b \circ c) = b \circ a = b$. Therefore $\circ$ is not associative.

6. a.

$\circ$	a	b
a	a	b
b	b	a

c.

$\circ$	a	b
a	b	b
b	b	b

7. Suppose the elements of table T are numbers 1, ..., n. Check the equation $T(i, T(j, k)) = T(T(i, j), k)$ for all values of $i, j,$ and k between 1 and n.

8. a. $y = y \circ e = y \circ (x \circ x^{-1}) = (y \circ x) \circ x^{-1} = (z \circ x) \circ x^{-1} = z \circ (x \circ x^{-1}) = z \circ e = z.$

9. The tables for $+_5$ and $\cdot_5$ are given prior to Example 6. Associativity and commutativity follow from the properties of regular addition and multiplication of natural numbers. The tables show that 0 is the identity for $+_5$, and 1 is the identity for $\cdot_5$, and each element has an inverse with respect to $+_5$, and each nonzero element has an inverse with respect to $\cdot_5$. For example, $2 \cdot_5 3 = 3 \cdot_5 2 = 1$, which shows that 2 and 3 are inverses with respect to $\cdot_5$.

Section 10.2

1. No. Notice that $A \cdot B \neq B \cdot A$ because $A - B \neq B - A$. Similarly, 0 is not an identity for +, and 1 is not an identity for $\cdot$.

2. a. $x + x y = x(1 + y) = x 1 = x.$ **c.** $x + \bar{x}y = (x + \bar{x})(x + y) = 1(x + y) = x + y.$

3. $\bar{e} = \overline{\overline{yz} + y\overline{z}} = \overline{\overline{yz}} \, \overline{y\overline{z}} = (y + \bar{z})(\bar{y} + z) = \bar{y}\,\bar{z} + y\,z.$

4. a. $\bar{x} + \bar{y} + xyz = \overline{xy} + (xy)z = \overline{xy} + z = \bar{x} + \bar{y} + z.$

5. a. $x + y.$ **c.** $\bar{y}(x + z).$ **e.** $x\,y + z.$

6. a.

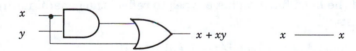

c.

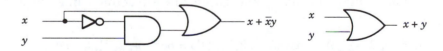

7. a. $x \, 0.$ **c.** $(x + y)(x + z).$ **e.** $y (\bar{x} + z).$

9. Show that $\bar{a} + \bar{b}$ acts like the complement of ab. In other words, show that

$$(ab) + (\bar{a} + \bar{b}) = 1 \text{ and } (ab)(\bar{a} + \bar{b})) = 0.$$

The result then follows from (10.8). For the first equation we have

$$(ab) + (\bar{a} + \bar{b}) = (a + \bar{a} + \bar{b})(b + \bar{a} + \bar{b}) = (1 + \bar{b})(1 + \bar{a}) = (1)(1) = 1.$$

For the second equation we have

$$(ab)(\bar{a} + \bar{b}) = ab\bar{a} + ab\bar{b} = 0 + 0 = 0.$$

11.a. Since $x = xx$, we have $x \preceq x$. So $\preceq$ is reflexive. If $x \preceq y$ and $y \preceq x$, then $x = xy$ and $y = yx$. Therefore $x = xy = yx = y$. Thus $\preceq$ is antisymmetric. If $x \preceq y$ and $y \preceq z$, then $x = xy$ and $y = yz$. Therefore $x = xy = x(yz) = (xy)z = xz$. So $x \preceq z$. Thus $\preceq$ is transitive.

b. Since $a \preceq b$ means $a = ab$, it will suffice to show that $a = ab$ iff $b = a + b$. If $a = ab$, then we have $a + b = ab + b = ab + 1b = (a + 1)b = 1b = b$. So if $a = ab$, then $b = a + b$. Now assume that $b = a + b$. Then we have $ab = a(a + b) = aa + ab = a + ab = a1 + ab = a(1 + b) = a1 = a$. So if $b = a + b$, then $a = ab$.

13. Since p occurs more than once in the factorization of n, it follows that n/p still contains at least one factor of p. For example, if $n = p^2q$, then $n/p = pq$. So lcm$(p, n/p) = n/p$, which is not equal to n (the unit of the algebra). Similarly, gcd$(p, n/p) = p$, which is not 1 (the zero of the algebra). So properties of part 3 of the definition of a Boolean algebra fail to hold.

Section 10.3

1. monus$(x, 0) = x$, monus$(0, y) = 0$, monus$(s(x), s(y)) = $ monus(x, y), where $s(x)$ denotes the successor of x.

3. reverse$(L) = $ if isEmptyL(L) then L
                    else cat(reverse(tail(L)), $\langle$head$(L)\rangle$).

5. Let GenLists[A] denote the set of general lists over A. The operations for general lists are similar to those for lists. The main difference is that the cons function and the head function have types to reflect the general nature of elements in a list:

> cons: $A \cup$ GenLists[A] $\rightarrow$ GenLists[A],
> head: GenLists[A] $\rightarrow$ GenLists[A] $\cup A$.

The functions length, member, and so on are all defined like those for Lists[A].

7. post$(\langle 4, 5, -, 2, + \rangle, \langle\,\rangle) = $ post$(\langle 5, -, 2, + \rangle, \langle 4 \rangle) = $ post$(\langle -, 2, + \rangle, \langle 5, 4 \rangle)$
$= $ post$(\langle 2, + \rangle,$ eval$(-, \langle 5, 4 \rangle)) = $ post$(\langle 2, + \rangle, \langle -1 \rangle) = $ post$(\langle + \rangle, \langle 2, -1 \rangle)$
$= $ post$(\langle\,\rangle,$ eval$(+, \langle 2, -1 \rangle)) = $ post$(\langle\,\rangle, \langle 1 \rangle) = 1$.

9. In equational form we have preorder(emptyTree) = emptyQ, and preorder(tree(L, x, R)) = apQ(addQ$(x,$ emptyQ), apQ(preorder(L), preorder(R))).

11. Remove (all occurrences of an element from a stack).

13. Let D be the set of deques over the set A. Then the carriers should be A, D, and Boolean. The operators can be defined as follows:

$$\text{emptyD} \in D,$$
$$\text{isEmptyD: } D \to \text{Boolean},$$
$$\text{addLeft: } A \times D \to D,$$
$$\text{addRight: } D \times A \to D,$$
$$\text{left: } D \to A,$$
$$\text{right: } D \to A,$$
$$\text{deleLeft: } D \to D,$$
$$\text{deleRight: } D \to D.$$

With axioms:

$$\text{isEmptyD(emptyD)} = \text{true},$$
$$\text{isEmptyD(addLeft}(a, d)) = \text{isEmptyD(addRight}(d, a)) = \text{false},$$
$$\text{left(addLeft}(a, d)) = \text{right(addRight}(d, a)) = a,$$
$$\text{left(addRight}(d, a)) = \text{ if isEmptyD}(d) \text{ then } a$$
$$\text{else left}(d),$$
$$\text{right(addLeft}(a, d)) = \text{ if isEmptyD}(d) \text{ then } a$$
$$\text{else right}(d),$$
$$\text{deleLeft(addLeft}(a, d)) = \text{deleRight(addRight}(d, a)) = d,$$
$$\text{deleLeft(addRight}(d, a)) = \text{ if isEmptyD}(d) \text{ then emptyD}$$
$$\text{else addRight(deleLeft}(d), a),$$
$$\text{deleRight(addLeft}(a, d)) = \text{ if isEmptyD}(d) \text{ then emptyD}$$
$$\text{else addLeft}(a, \text{deleRight}(d)).$$

15. Let

$$Q[A] = D[A],$$
$$\text{emptyQ} = \text{emptyD},$$
$$\text{isEmptyQ} = \text{isEmptyD},$$
$$\text{frontQ} = \text{left},$$
$$\text{deleQ} = \text{deleLeft},$$
$$\text{addQ}(a, q) = \text{addRight}(q, a).$$

Then the axioms are proved as follows:

$$\text{isEmptyQ(emptyQ)} \quad = \quad \text{isEmptyD(emptyD)} = \text{true}.$$
$$\text{isEmptyQ(addQ}(a, q)) \quad = \quad \text{isEmptyD(addRight}(q, a)) = \text{false}.$$
$$\text{frontQ(addQ}(a, q)) \quad = \quad \text{left(addRight}(q, a))$$
$$= \quad \text{if isEmptyD}(q) \text{ then } a \text{ else left}(q)$$
$$= \quad \text{if isEmptyQ}(q) \text{ then } a \text{ else frontQ}(q).$$
$$\text{delQ(addQ}(a, q)) \quad = \quad \text{deleLeft(addRight}(q, a))$$
$$= \quad \text{if isEmptyD}(q) \text{ then emptyD}$$
$$\text{else addRight(deleLeft}(q), a)$$
$$= \quad \text{if isEmptyQ}(q) \text{ then emptyQ}$$
$$\text{else addQ}(a, \text{deleQ}(q)).$$

17. a. Let $P(x)$ denote the statement "plus$(x, s(y)) = s($plus$(x, y))$ for all $y \in \mathbb{N}$."
Certainly $P(0)$ is true because plus$(0, s(y)) = s(y) = s($plus$(0, y))$. So assume
that $P(x)$ is true, and prove that $P(s(x))$ is true. We can evaluate each expression in the statement of $P(s(x))$ as follows:

$$
\begin{aligned}
\text{plus}(s(x), s(y)) &= s(\text{plus}(p(s(x)), s(y))) & \text{(by definition of plus)} \\
&= s(\text{plus}(x, s(y))) & \text{(since } p(s(x)) = x) \\
&= s(s(\text{plus}(x, y))) & \text{(by induction)},
\end{aligned}
$$

and

$$
\begin{aligned}
s(\text{plus}(s(x), y)) &= s(s(\text{plus}(p(s(x)), y))) & \text{(by definition of plus)} \\
&= s(s(\text{plus}(x, y))) & \text{(since } p(s(x)) = x).
\end{aligned}
$$

Both expressions are equal. So $P(s(x))$ is true. QED.

b. Let $P(x)$ denote the statement "plus$(x, y) = add(x, y)$ for all $y \in \mathbb{N}$."
Certainly $P(0)$ is true because plus$(0, y) = y = add(0, y)$ from the two definitions. So assume that $P(x)$ is true, and prove that $P(s(x))$ is true. Starting with
the expression plus$(s(x), y)$ in $P(s(x))$, we obtain the other expression as follows:

$$
\begin{aligned}
\text{plus}(s(x), y) &= s(\text{plus}(p(s(x)), y)) & \text{(definition of plus)} \\
&= s(\text{plus}(x, y)) & \text{(since } p(s(\text{x})) = x) \\
&= \text{plus}(x, s(y)) & \text{(from part (a))} \\
&= \text{add}(x, s(y)) & \text{(induction)} \\
&= \text{add}(p(s(x)), s(y)) & \text{(since } p(s(\text{x})) = x) \\
&= \text{add}(s(x), y) & \text{(definition of add)}.
\end{aligned}
$$

So $P(s(x))$ is true. QED.

18. Induction will be with respect to the length of y. We'll use the notation $y{:}a$
for addQ(a, y). For the basis case we have the following equations, where $y = $emptyQ:

$$
\begin{aligned}
\text{apQ}(x, \text{emptyQ}{:}a) &= \text{apQ}(x{:}\text{front}(\text{emptyQ}{:}a), \text{delQ}(\text{emptyQ}{:}a)) & \text{(def of apQ)} \\
&= \text{apQ}(x{:}a, \text{emptyQ}) & \text{(simplify)} \\
&= x{:}a & \text{(simplify)} \\
&= \text{apQ}(x{:}a, \text{emptyQ})) & \text{(def of apQ)}.
\end{aligned}
$$

For the induction case, assume that the equation is true for all queues y
having length n, and show that the equation true for the queue $y{:}b$, having
length $n + 1$. Starting with the left side of the equation, we have

$$
\begin{aligned}
\text{apQ}(x, y{:}b{:}a) &= \text{apQ}(x{:}\text{front}(y{:}b{:}a), \text{delQ}(y{:}b{:}a)) & \text{(def of apQ)} \\
&= \text{apQ}(x{:}\text{front}(y{:}b), \text{delQ}(y{:}b{:}a)) & (\text{front}(y{:}b{:}a) = \text{front}(y{:}b)) \\
&= \text{apQ}(x{:}\text{front}(y{:}b), \text{delQ}(y{:}b){:}a) & (\text{delQ}(y{:}b{:}a) = \text{delQ}(y{:}b){:}a) \\
&= \text{apQ}(x{:}\text{front}(y{:}b), \text{delQ}(y{:}b)){:}a & \text{(induction)} \\
&= \text{apQ}(x, y{:}b){:}a & \text{(def of apQ)}.
\end{aligned}
$$

Section 10.4

1. a. $\{\langle 1, a, \#, M \rangle, \langle 2, a, *, M \rangle, \langle 3, a, \%, M \rangle\}$. **c.** $\{\langle 1, a, \#, M, x \rangle, \langle 1, a, \#, M, z \rangle\}$.
e. $\{\langle a, M \rangle, \langle b, M \rangle\}$.

2. a. $t \in \text{sel}_{A=a}(\text{sel}_{B=b}(R))$ iff $t \in \text{sel}_{B=b}(R)$ and $t(A) = a$ iff $t \in R$ and $t(B) = b$ and $t(A) = a$ iff $t \in \text{sel}_{A=a}(R)$ and $t(B) = b$ iff $t \in \text{sel}_{B=b}(\text{sel}_{A=a}(R))$. **c.** Follows directly from the definition: $t \in R \bowtie S$ iff there exist $r \in R$ and $s \in S$ such that $t(a) = r(a)$ for all $a \in I$ and $t(b) = s(b)$ for all $b \in J$ iff $t \in S \bowtie R$.

3. $s \in \text{proj}_X(\text{sel}_{A=a}(R))$ iff there exists $t \in \text{sel}_{A=a}(R)$ such that $s(A) = t(A)$ for all $A \in X$ iff there exists $t \in R$ such that $s(A) = t(A) = a$ for all $A \in X$ iff $s \in \text{proj}_X(R)$ and $s(A) = a$ iff $s \in \text{sel}_{A=a}(\text{proj}_X(R))$.

4. a. Single node tree with root a. **c.** Let the tree for $t_1 + t_2$ have root $+$ and two subtrees t_1 and t_2.

5. seqPairs = eq0 $\rightarrow$ ~$\langle\langle 0, 0 \rangle\rangle$; apndr @ [seqPairs @ sub1, [id, id]].

7. dotProduct = !+ @ &* @ pairs.

8. a. For any pair of numbers $\langle m, n \rangle$, all three expressions compute the value of the expression $m + n$.

9. If c returns 0, then $* @ [a, g @ [b, c]] = * @ [a, b] = g @ [*@ [a, b], c]$, which proves the basis case. Now assume that c returns a positive number and (10.12) holds for sub1 @ c. We'll prove that (10.12) holds for c as follows, starting with the left side:

$$
\begin{aligned}
* @ [a, g @ [b, c]] \quad &= * @ [a, (\text{eq0} @ 2 \rightarrow 1; g @ [*, \text{sub1} @ 2]) @ [b, c]] \quad \text{(def of } g) \\
&= * @ [a, g @ [* @ [b, c], \text{sub1} @c]] \quad \text{(eq0} @ c = \text{false)} \\
&= g @ [*@[a, * @ [b, c]], \text{sub1} @c] \quad \text{(induction)}.
\end{aligned}
$$

Now look at the right side:

$$
\begin{aligned}
g @ [* @ [a, b], c] \quad &= g @ [*, \text{sub1} @ 2] @ [* @ [a, b], c] \quad \text{(def of } g) \\
&= g @ [*@[* @ [a, b], c], \text{sub1} @c].
\end{aligned}
$$

It follows that the two sides are equal because multiplication is associative:

$$*@[a, * @ [b, c]] = *@[* @ [a, b], c].$$

Section 10.5

1. a. 99.

2. For example, we could choose $m = 20$ and $n = 21$. With this choice, we could choose $r = 398 \mod 20 = 18$ and $s = 398 \mod 21 = 20$.

3. a. [3]. **c.** [1].

4. a. Yes. **c.** No. $4 +_9 6 = 1 \notin \{0, 2, 4, 6, 8\}$.

5. a. $\{0, 6\}$. **c.** $\mathbb{N}_{12}$.

7. $f(\Lambda) = \text{length}(\Lambda) = 0$ and if x and y are arbitrary strings in A^*, then

$$f(\text{cat}(x, y)) = \text{length}(\text{cat}(x, y)) = \text{length}(x) + \text{length}(y) = f(x) + f(y).$$

9. Define the function $f : \{a, b, c\} \to \mathbb{N}_3$ by $f(a) = 1$, $f(b) = 0$, and $f(c) = 2$. The table for $\circ$ is symmetric about the main diagonal, which says $\circ$ is commutative. Now we need to show that $f(x \circ y) = f(x) +_3 f(y)$ for all $x, y \in \{a, b, c\}$. There are nine equations to check. For example,

$$f(a \circ c) = f(b) = 0 = 1 +_3 2 = f(a) +_3 f(c).$$

10. a. $\{(ab)^n b \mid n \in \mathbb{N}\}$. **c.** $\varnothing$. **e.** $\{ba^n \mid n \in \mathbb{N}\}$.

Bibliography

In addition to the books and papers specifically referenced in this book, we've also included some general references.

Andrews, P. B., *An Introduction to Mathematical Logic and Type Theory: To Truth Through Proof*. Academic Press, New York, 1986.

Appel, K., and W. Haken, Every planar map is four colorable. *Bulletin of the American Mathematical Society 82* (1976), 711–712.

Appel, K., and W. Haken, The solution of the four-color-map problem. *Scientific American 237* (1977), 108–121.

Apt, K. R., Ten years of Hoare's logic: A survey—Part 1. *ACM Transactions on Programming Languages and Systems 3* (1981), 431–483.

Backus, J., Can programming be liberated from the von Neumann style? A functional style and its algebra of programs. *Communications of the ACM 21* (1978), 613–641.

Brassard, G., and P. Bratley, *Algorithmics: Theory and Practice*. Prentice-Hall, Englewood Cliffs, N. J., 1988.

Chang, C., and R. C. Lee, *Symbolic Logic and Mechanical Theorem Proving*. Academic Press, New York, 1973.

Cichelli, R. J., Minimal perfect hash functions made simple. *Communications of the ACM 23* (1980), 17–19.

Coppersmith, D., and S. Winograd, Matrix multiplication via arithmetic progressions. *Proceedings of 19th Annual ACM Symposium on the Theory of Computing* (1987), 1–6.

Delong, H., *A Profile of Mathematical Logic*. Addison-Wesley, Reading, Mass., 1970.

Floyd, R. W., Algorithm 97: Shortest path. *Communications of the ACM 5* (1962), 345.

Floyd, R. W., Assigning meanings to programs. *Proceedings AMS Symposium Applied Mathematics*, *19*, AMS, Providence R.I., 1967, pp. 19–31.

Galler, B. A., and M. J. Fischer, An improved equivalence algorithm. *Communications of the ACM 7* (1964), 301–303.

Gentzen, G., Untersuchungen uber das logische Schliessen. *Mathematische Zeitschrift 39* (1935), 176–210, 405–431; English translation: Investigation into logical deduction, *The Collected Papers of Gerhard Gentzen*, ed. M. E. Szabo. North-Holland, Amsterdam, 1969, pp. 68–131.

Gödel, K., Die Vollständigkeit der Axiome des logischen Funktionenkalküls. *Monatshefte für Mathematic und Physik 37* (1930), 349–360.

Gödel, K., Über formal unentscheidbare Sätze der Principia Mathematica und verwandter Systeme I. *Monatshefte für Mathematic und Physik 3 8* (1931), 173–198.

Graham, R. L., D. E. Knuth, and O. Patashnik, *Concrete Mathematics*. Addison-Wesley, Reading, Mass., 1989.

Halmos, P. R., *Naive Set Theory*. Van Nostrand, New York, 1960.

Hamilton, A. G., *Logic for Mathematicians*. Cambridge University Press, New York, 1978.

Hein, J. L., A declarative laboratory approach for discrete structures, logic, and computability. *ACM SIGCSE Bulletin 25*, 3 (1993), 19–25.

Hennessy, M., *The Algebraic Theory of Processes*. The MIT Press, Cambridge, Mass., 1988.

Hilbert, D., and W. Ackermann, *Principles of Mathematical Logic*. (1938). Translated by Lewis M. Hammond, George G. Leckie, and F. Steinhardt. Edited by Robert E. Luce. Chelsea, New York, 1950.

Hoare, C.A.R., An axiomatic basis for computer programming. *Communications of the ACM 12* (1969), 576–583.

Knuth, D. E., *The Art of Computer Programming*. Volume 1: *Fundamental Algorithms*. Addison-Wesley, Reading, Mass., 1968; second edition, 1973.

Knuth, D. E., Two notes on notation. *The American Mathematical Monthly 99* (1992), 403–422.

Kruskal, J. B., Jr., On the shortest spanning subtree of a graph and the traveling salesman problem. *Proceedings of the American Mathematical Society 7* (1956), 48–50.

Liu, C. L., *Introduction to Combinatorial Mathematics*. McGraw-Hill, New York, 1968.

Lukasiewicz, J., *Elementary Logiki Matematycznej*. PWN (Polish Scientific Publishers), 1929; translated as *Elements of Mathematical Logic*, Pergamon, Elmsford, N. Y., 1963.

Mallows, C. L., Conway's challenge sequence. *The American Mathematical Monthly 98* (1991), 5–20.

Mendelson, E., *Introduction to Mathematical Logic*. Van Nostrand, New York, 1964.

Nagel, E., and J. R. Newman, *Gödel's Proof*. New York University Press, New York, 1958.

Pan, V., Strassen's algorithm is not optimal. *Proceedings of 19th Annual IEEE Symposium on the Foundations of Computer Science* (1978), 166–176.

Paterson, M. S.; and M. N. Wegman, Linear Unification. *Journal of Computer and Systems Sciences 16* (1978), 158–167.

Paulson, L. C., *Logic and Computation*. Cambridge University Press, New York, 1987.

Prim, R. C., Shortest connection networks and some generalizations. *Bell System Technical Journal 36* (1957), 1389–1401.

Robinson, J. A., A machine-oriented logic based on the resolution principle. *Journal of the ACM 12* (1965), 23–41.

Schoenfield, J. R., *Mathematical Logic*. Addison-Wesley, Reading, Mass., 1967.

Skolem, T., Uber de mathematische logik. *Norsk Matematisk Tidsskrift 10* (1928), 125–142. Translated in ed. Jean van Heijenoort. *From Frege to Godel: A Source Book in Mathematical Logic 1879–1931*, Harvard University Press, Cambridge, Mass., 1967, pp. 508–524.

Snyder, W., and J. Gallier, Higher-order unification revisited: Complete sets of transformations. *Journal of Symbolic Computation 8* (1989), 101–140.

Stanat, D. F., and D. F. McAllister, *Discrete Mathematics in Computer Science*. Prentice-Hall, Englewood Cliffs, N. J., 1977.

Strassen, V., Gaussian elimination is not optimal. *Numerische Mathematik 13* (1969), 354–356.

Suppes, P., *Introduction to Logic*. Van Nostrand, New York, 1957.

Warren, D. S., Memoing for logic programs. *Communications of the ACM 35* (1992), 93–111.

Warshall, S., A theorem on Boolean matrices, *Journal of the ACM 9* (1962), 11–12.

Whitehead, A. N., and B. Russell, *Principia Mathematica*. Cambridge University Press, New York, 1910.

Wos, L., R. Overbeek, E. Lusk, and J. Boyle, *Automated Reasoning: Introduction and Applications*. Prentice-Hall, Englewood Cliffs, N. J., 1984.

Greek Alphabet

A	α	alpha
B	β	beta
Γ	γ	gamma
Δ	δ	delta
E	ε	epsilon
Z	ζ	zeta
H	η	eta
Θ	θ	theta
I	ι	iota
K	κ	kappa
Λ	λ	lambda
M	μ	mu
N	ν	nu
Ξ	ξ	xi
O	o	omicron
Π	π	pi
P	ρ	rho
Σ	σ	sigma
T	τ	tau
Y	υ	upsilon
Φ	ϕ	phi
X	χ	chi
Ψ	ψ	psi
Ω	ω	omega

Symbol Glossary

Each symbol or expression is listed with a short definition and the page number where it first occurs. The list is ordered by page number.

$m \mid n$	m divides n with no remainder	5
$m \nmid n$	m does not divide n	5
$x \in S$	x is an element of S	10
$x \notin S$	x is not an element of S	11
...	ellipsis	11
$\varnothing$	the empty set	11
$\mathbb{N}$	natural numbers	12
$\mathbb{Z}$	integers	12
$\mathbb{Q}$	rational numbers	12
$\mathbb{R}$	real numbers	12
$\{x \mid P\}$	set of all x satisfying property P	12
$A \subset B$	A is a subset of B	13
$A \not\subset B$	A is not a subset of B	14
$A \cup B$	A union B	16
$\cup A_i$	union of the sets A_i	17
$A \cap B$	A intersection B	18
$\cap A_i$	intersection of the sets A_i	19
$A - B$	difference: elements in A but not B	20
$A \oplus B$	symmetric difference: $(A - B) \cup (B - A)$	21
A'	complement of A	21
$\lvert S \rvert$	cardinality of S	22
$[a, b, b, a]$	bag, or multiset, of four elements	25
$\langle x, y, x \rangle$	tuple of three elements	31
$\langle \, \rangle$	empty tuple	32
$A \times B$	product of A and B: $\{\langle a, b \rangle \mid a \in A \text{ and } b \in B\}$	33
$\langle x, y, x \rangle$	list of three elements	35

$\langle\,\rangle$	empty list	35
Lists[A]	lists over A	36
GenLists[A]	generalized lists over A	36
Λ	empty string	37
$|s|$	length of string s	37
A^*	strings over alphabet A	38
$x \, R \, y$	x related by R to y	42
$f : A \to B$	function type: f has domain A and codomain B	62
$f(C)$	image of C under f	64
$f^{-1}(D)$	pre-image of D under f	64
$\lfloor x \rfloor$	floor of x: largest integer $\leq x$	67
$\lceil x \rceil$	ceiling of x: smallest integer $\geq x$	67
(a, b)	greatest common divisor of a and b	67
$a \bmod b$	remainder upon division of a by b	71
$\mathbb{N}_n$	the set $\{0, 1, ..., n-1\}$	72
χ_B	characteristic function for subset B	73
$f \circ g$	composition of functions f and g	79
f^{-1}	inverse of bijective function f	94
cons(x, t)	list with head x and tail t	116
$x :: t$	list with head x and tail t	117
$a \cdot s$	string with head a and tail s	120
tree(L, x, R)	binary tree with root x and subtrees L and R	123
BinTrees[A]	set of binary trees over A	123
$L \cdot M$	product of languages L and M	131
L^0	the language $\{\Lambda\}$	132
L^n	product of language L with itself n times	132
L^*	closure of language L	133
L^+	positive closure of language L	133
$A \to \alpha$	grammar production	135
$A \to \alpha \mid \beta$	grammar productions $A \to \alpha$ and $A \to \beta$	137
$A \Rightarrow \alpha$	A derives α in one step	138
$A \Rightarrow^+ \alpha$	A derives α in one or more steps	138
$A \Rightarrow^* \alpha$	A derives α in zero or more steps	138
$L(G)$	language of grammar G	139
$\Sigma\, a_i$	sum of the numbers a_i	154
$\Pi\, a_i$	product of the numbers a_i	154
$n!$	n factorial: $n \cdot (n-1) \cdots 1$	155
$R \circ S$	composition of binary relations R and S	179
$r(R)$	reflexive closure of R	183
$s(R)$	symmetric closure of R	183
$t(R)$	transitive closure of R	183

R^c	converse of relation R	184
R^+	transitive closure of R	187
R^*	reflexive transitive closure of R	187
$[x]$	equivalence class of things equivalent to x	199
S/R	partition of S by the equivalence relation R	199
$\mathrm{tsr}(R)$	smallest equivalence relation containing R	204
K_f	kernel relation defined by f	207
$\langle A, \preceq \rangle$	reflexive partially ordered set	216
$\langle A, \prec \rangle$	irreflexive partially ordered set	216
$x \prec y$	x is less than y or x is a predecessor of y	217
$x \preceq y$	$x \prec y$ or $x = y$	217
W_A	worst case function for algorithm A	254
$P(n, r)$	number of permutations of n things taken r at a time	262
$C(n, r)$	number of combinations of n things taken r at a time	266
$\binom{n}{r}$	binomial coefficient symbol	266
$\Theta(f)$	big theta: same growth rate as f	299
$o(f)$	little oh: lower growth rate than f	303
$O(f)$	big oh: same as or lower growth rate than f	304
$\Omega(f)$	big omega: same as or higher growth rate than f	305
$\neg P$	logical negation of P	313
$P \wedge Q$	logical conjunction of P and Q	313
$P \vee Q$	logical disjunction of P and Q	313
$P \to Q$	logical conditional: P implies Q	313
$P \equiv Q$	logical equivalence of P and Q	317
$\therefore$	therefore	333
$\vdash W$	turnstile to denote W is a theorem	352
$\exists x$	existential quantifier: there is an x	357
$\forall x$	universal quantifier: for all x	357
$W(x/t)$	wff obtained from W by replacing free x's by t	363
x/t	binding of the variable x to the term t	363
$W(x)$	W contains a free variable x	363
c_x	x is a subscripted variable	394
$\{P\}\, S\, \{Q\}$	S is correct for the precondition P and the postcondition Q	420
$\square$	empty clause: a contradiction	454
$\{x/t, y/s\}$	substitution containing two bindings	462
ε	empty substitution	462
$E\theta$	instance of E: substitution θ applied to E	462

$\theta\sigma$	composition of substitutions θ and σ	463
$C\theta - N$	remove all occurrences of N from clause $C\theta$	469
$R(S)$	resolution of clauses in the set S	471
$\leftarrow A$	logic program goal: is A true?	480
$C \leftarrow$	logic program fact: C is true	481
$C \leftarrow A, B$	logic program conditional: C if A and B	481
$\langle A; s, a \rangle$	algebra with carrier A and operations s and a	511
$\bar{x}$	complement of Boolean algebra variable x	523

x, y →⟩— $x \cdot y$ AND gate   528

x, y →⟩— $x + y$ OR gate   528

x —▷o— $\bar{x}$ NOT gate   528

$\mathrm{sel}_{A=a}(R)$	select all tuples of R whose Ath element is a	552
$\mathrm{proj}_X(R)$	projection of R to tuples indexed by the set X	552
$R \bowtie S$	join of relations R and S	553
$x \equiv y \pmod{n}$	$x \bmod n = y \bmod n$	564
$x \equiv_n y$	$x \bmod n = y \bmod n$	564

Index